Printing Estimating

Principles and Practices

Dedication

To Joanne and Lauren for their continuing encouragement, support, and love.

Printing Estimating

Principles and Practices

Third Edition

Philip Kent Ruggles

California Polytechnic State Universitry

DELMAR PUBLISHERS INC.®

NOTICE TO THE READER

Publisher does not warrant or guarantee any of the products described herein or perform any independent analysis in connection with any of the product information contained herein. Publisher does not assume, and expressly disclaims, any obligation to obtain and include information other than that provided to it by the manufacturer.

The reader is expressly warned to consider and adopt all safety precautions that might be indicated by the activities described herein and to avoid all potential hazards. By following the instructions contained herein, the reader willingly assumes all risks in connection with such instructions.

The publisher makes no representations or warranties of any kind, including but not limited to, the warranties of fitness for particular purpose or merchantability, nor are any such representations implied with respect to the material set forth herein, and the publisher takes no responsibility with respect to such material. The publisher shall not be liable for any special, consequential or exemplary damages resulting, in whole or in part, from the readers' use of, or reliance upon, this material.

Cover design and paper illustration by
Mary Beth Vought
Cover Photo by Nancy Gwarek

For information address Delmar Publishers Inc.
Two Computer Drive West, Box 15-015
Albany, New York 12212

Delmar Staff
Senior Administrative Editor: Michael McDermott
Project Editor: Carol Micheli
Art Coordinator: Mike Nelson
Design Coordinator: Susan C. Mathews

Printed in the United States of America
Published simultaneously in Canada
by Nelson Canada
a division of The Thomson Corporation

10 9 8 7 6 5 4 3

Library of Congress Cataloging-in-Publication Data

Ruggles, Philip Kent.
Printing estimating: principles and practices/Philip Kent Ruggles.—3rd ed.
p. cm
ISBN 0-8273-3805-8
1. Printing, Practical—Estimates. I. Title
Z245.R84 1990
686.2'068—dc20 90-33108
CIP

Contents

Preface

As with both the first and second editions of this book, the drafting of the third edition of this work has truly been a labor of love. It has also been a provocative and challenging task.

If the previous two editions of *Printing Estimating Principles and Practices* are predictors of where this text will ultimately be used, then this text will address two primary needs. First, it will serve as a textbook for schools, colleges, universities, or any program of study which addresses the subject of estimating, costing, pricing and profitability for the printing industry. The text is organized to provide up-to-date information for graphic communication students and teachers, and anyone wishing to expand their knowledge of this subject. Second, it will become a reference book for the industry as a whole, including professional estimators, production planners, customer service and sales representatives, printing managers at all levels, and computer vendors serving the industry. In this regard, the text details the manufacturing components of current imaging and printing technologies so that they can be cost estimated accurately, leading to improved cost controls, financially realistic production techniques and enhanced company profitability throughout the industry.

This text is an almost completely new book compared to the previous edition. Perhaps the most significant, single difference is the accelerated and pervasive influence of electronic systems on the industry, particularly related in the prepress production area. The "Macintosh Revolution," which began in January 1984 with the introduction of the Macintosh during Super Bowl XVIII ads, has greatly influenced the prepress segment of our business, not only through the unique architecture of Mac hardware but with the advent of "mouse" technology, pull-down menus and simplified, user-friendly operating procedures. A significant evolution of this technology has been desktop publishing, which, in just a few years, has had major impact on the production of typesetting, mechanical art and other prepress functions. Macintosh concepts have also provided the impetus for development and implementation of highly sophisticated electronic prepress platforms including Color Electronic Prepress Systems (or CEPS) and intermediate electronic solutions such as the Gerber AutoPrep 5000.

This book reflects this author's discussions with literally hundreds of industry professionals including estimators, production planners, customer service representatives, sales representatives, production employees, company owners and printing industry computer vendors over the past five years. Further, it reflects numerous conversations with a widely diverse group of industry experts—many of whom are also personal friends—related to the rapidly-changing operating environment, production conditions, technology and business and economic conditions which affect our industry. Also included is valuable input from many of this author's teaching colleagues at Cal Poly State University and other academic professionals from universities across the U.S., as well as comments from hundreds of students who have diligently pursued estimating careers having been exposed to the subject while studying at Cal Poly State University. To all of the above, too numerous to mention by name, the author owes a heartfelt thanks.

This author sincerely appreciates the information provided by the major national associations serving the printing and allied industries including the National Association of Printers and Lithographers, Printing Industries of America and the National Association of Quick Printers. Specific thanks to Chuck Alessandrini and Rhona Bronson at NAPL, Ray Roper and Bill Teare at PIA and Brian Pugliese at NAQP.

Thanks also to PIA of Southern California's Executive Director Bob Lindgren, and Frank Iannuzzi and Tom Stodola in their Management Services Department for permission to use material from their excellent *1988-89 Southern California Blue Book of Production Standards and Costs.* Other association executives who provided input to this third edition included Nolan Moore and Joe Polanco of Printing Industries of Texas.

Certain industry professionals (listed alphabetically) deserve thanks for their direct input and information for this text including Dave Allen and Donna Peralta of Unisource Paper, Virgil Busto of *Instant and Small Commercial Printer,* John Doyle of Walker Graphics/Techtron, Bob Gans of Gans Ink and Supply Co., Mark Geannette and Leonard Bacharach of Crosfield Electronics, Inc., Bob Hall of *Quick Printing Magazine*, Jake Swenson of Moore Business Forms and Roger Ynostroza of *Graphic Arts Monthly.*

Thanks also to Professors Bill Bidez, George Elliott, and Malcolm Keif for their diligent review and input of the third edition drafts from an academic perspective, and to professional estimators Karen Habendank-Matson and Tom Martinelli for their industry view. Thanks as well to Rod Piechowski for his review of Chapter 14 which address the small commercial printer and quick printer.

This author would like to extend sincere appreciation to Delmar Publishers' Project Editor Carol Micheli for her diligent work on this text, and to Senior Administrative Editor Mike McDermott. Thanks also to Delmar's President, Greg Spatz, for ensuring that this book has become a reality during 1990.

Finally, this third edition is lovingly dedicated to my father, the late Arthur Ruggles, who introduced me to the world of printing as a youth and provided me with the opportunity to love this industry as much as he did. For more than 35 years he owned and operated two small, interconnected, commercial establishments. As a boy and young man, I watched and worked in his printing plant, which we fondly knew as the"shop." Those early years allowed me to experience the industry from many perspectives, particularly from both business and printing technology vantage points. My father taught me basic paper mathematics at age fourteen and introduced me to the fundamentals of printing estimating about age sixteen. Throughout his almost four decades in the printing business, he continuously exhibited thoughtfulness, patience, and a consuming passion for the art and science of printing. It goes without saving that without the encouragement, enthusiasm and contagious interest my father demonstrated over the years, it is unlikely that this book—or any previous edition of this work—would ever have been written.

PKR
San Luis Obispo, CA
August, 1990

The Scope and Function of Printing Estimating

1.1 Introduction to Cost Estimating for Printing

Printing estimating is a cost-finding procedure for printing management. Estimating typically requires two sequential steps:

The determination of a plan of production to make the product,
The assignment of costs using the defined production plan.

In the first step, production planning, the estimator (or job planner in a large facility) must break the product to be manufactured into component segments, as will be necessary when the product is produced. Once this task is completed, the estimator must identify those production segments that can be best completed in his or her own printing plant and determine the segments that can be made more efficiently outside the production facility.

The second step in cost estimating involves the assignment of standard production times and budgeted hour cost factors to determine the costs that will be incurred in the manufacture of the product. The plan of production is closely followed. The estimator assigns the standard production values, in hours or decimal equivalents of hours, based on preestablished, accurate data representative of plant output. After the standards have been accurately applied—for example, the time needed to expose and process a lithographic plate—the assignment of cost values using a budgeted hour cost figure is then made. Thus, the cost to make a lithographic plate = estimated time (hours) for exposure and development × budgeted hour cost rate. The costs of materials, such as the lithographic plate itself, are determined separately and added to the platemaking cost. The total sum represents the cost to manufacture a lithographic plate for a customer. Later, top management will add profit to the total, and the selling price for platemaking then can be quoted to the customer. The formula used for cost estimating is:

(standard production time × budgeted hour cost rate) + material costs
+ buyout costs + profit = selling price of job

Some large printing manufacturing operations, with complex production equipment and specialized product lines, keep these two estimating duties separate. They have job planners, or production planners, who initially determine the best production sequence for the work. Once this task is completed, the planned job sequence is given to an estimator for application of time and cost data. In some plants, the job planner/production planner may also follow up work-in-progress as an affiliated job duty. The estimator may take on other responsibilities also such as review and verification of production standards and cost values. For the most part, this split in estimating duties is desirable because it allows for greater specialization in each of the functions, which translates into greater accuracy in estimates.

Many smaller printing plants and quick printers—due to their immediate customer requirements, fast turn-around of printing, and standardized products—don't cost estimate on a job-by-job basis, as described. For them price lists, counter pricing books, other counter reference materials and computer-assisted counter systems are available. It is important to note that cost estimating still has much relevance to these smaller firms, since the process may be the only way they can ensure that their price list prices are accurate as to the recovery of labor and material costs.

As a point of clarification, in this book the term estimator describes the person who completes both job planning and cost assignment functions, unless otherwise specified. Estimating is the process by which these planning and cost values are assigned by the estimator.

If estimators are unfamiliar with a particular manufacturing segment or operation, they may request help from colleagues or some other group or individual more knowledgeable in that particular area. As estimators become more experienced, or as the job components become more familiar or less detailed, they may personally determine the entire sequence of manufacturing steps necessary to produce a specified product.

1.2 The Interrelationship of Cost Estimating and Other Plant Duties

The estimator's assigned duties may vary depending upon the size of the printing plant, the specialization of product and process, the size and quality of the estimating staff, the type of estimating procedures used by the company, the speed by which estimates are completed, and the computer equipment available to aid the estimating staff both to increase the number of estimates and to reduce estimating time. Company management typically determines the estimator's job mix.

Smaller printing plants and quick printers may have no estimator, yet may have one individual assigned to handle customer pricing, material purchasing, buyouts from other printing service organizations, price list analysis, etc. Larger commercial printing plants and high volume printers typically use full-time estimators to handle estimating plus other duties; some very large printing firms use estimating specialists—that is estimators who concentrate on a particular product or process segment. Usually, the larger the estimating staff, the more specialized the task of each individual. If various estimating and pricing systems are intermixed, the job duties of the estimator may vary accordingly. The following sections summarize the segments in a printing company with which the typical printing estimator must function.

Production Scheduling and Control

In smaller printing companies and quick printers, it is common for the estimator or person assigned with the pricing duties to schedule work into production on a basis that factors in the immediate results desired by the customer. As the company grows or volume increases, scheduling of jobs may necessitate the development of a production control system and more complex interfacing of production and delivery.

It is common for estimators to perform certain duties related to the scheduling of jobs into plant production or work as a part of production control. With the labor force at a premium, this assignment appears to be a reasonable extension of the estimators' duties since they are directly concerned with such information when estimating. In larger commercial printing firms and high volume plants, as the diversity of products and processes increases, estimators may become more specialized. In such circumstances, they may have some indirect responsibility with production control or may schedule certain jobs, but usually are busy enough with estimating duties alone.

In order to ensure better communications, many larger printing operations use production coordinators or customer service coordinators who work with sales representatives, estimators, and production control and customers to ensure the smooth flow of jobs through the plant. The primary focus of such coordinators is to minimize errors on jobs, maximize production in the plant and service customer needs fully. Of course, such personnel are an additional expense to the printing firm.

Purchasing

Because estimators are required to determine the cost of materials necessary to complete printing orders, some plants require that estimators handle purchasing duties in addition to their estimating responsibilities.

Quick printers and smaller commercial printers typically have purchasing duties completed by the manager of the plant, who may also do some estimating. As the company volume increases, estimators seem ideal to take over such purchasing duties, since they must calculate material costs as a portion of the estimate anyway, so they will plan and order what is needed as jobs require. For this reason, estimators must remain current with respect to various discounts and bargains offered by suppliers.

In fact, estimators may be the only plant representatives dealing with such suppliers continually. Large companies, who usually benefit from combined purchasing power in many supply categories, may have a purchasing agent or department in the plant. Purchasing must then be a coordinated effort with the estimators, who determine material costs, so that inventories or supplies do not become burdensome.

Sales

The relationship between estimating and sales is direct—the sales representative visits the customer, determines the job specifications for the customer's order, then requests the estimator to prepare the estimate. Both sales and estimating are different yet vital aspects of a printing business, and there should be a recognized compatibility between the two. The most important duty that sales representatives can perform is to always obtain exact and complete job specifications from the customer. Without specifications, estimating is like putting a puzzle together with some of the pieces missing. Estimators, likewise, can aid the sales representatives by completing estimates quickly so that the sales representatives can provide quick responses to customers. Experienced estimators and sales representatives understand the mutual coordination necessary.

Some managers believe that the duties of the estimators and sales representatives should be integrated—that estimators should handle selling duties while the sales representatives should estimate. In some plants, estimators may sell reorder jobs and walk-in business. In other plants, sales representatives may carry estimating data with them so that they may estimate in the field. The specific mixture of estimating and sales is best determined within each plant individually. Considerable problems may arise if the job duties of each are not clear or if management appears more supportive of either sales or estimating.

Printing brokers represent one way for printers to increase sales volume without additional sales personnel or without increasing sales costs. In sum, a printing broker is a printing sales representative working independently of any printing company. After selling a print job, the broker locates a printer to do it, and when it is completed, he or she delivers it and bills the customer. The printer then bills the broker, from whom the order was placed. A printing broker's success is driven by his or her skills at providing superior customer service and representing competition to the conventional printing sales representative. The printing brokerage business has grown in the past ten years, particularly for the smaller commercial printer and quick printer. For those companies with estimators, working with brokers—just as working with their own sales representatives—has become more common and generally more accepted.

Customer Service

A growing trend among progressive printers, particularly larger printers and those which are very customer-focused, is the addition of customer service representatives (also called "CSRs"), as a vital link between customer and company.

While duties may vary from company to company CSRs generally serve as a liaison between customers, sales representatives, estimating and production. Individuals serving in CSR positions work to smooth out and coordinate production elements of customer's jobs, deal with customer problems and complaints, help in such areas as press cheeks and direct customer contact with the production of the job.

In terms of estimating, some companies have the customer service representatives actually do estimating as part of their job duties, thus requiring that the CSR be very familiar with production sequences, job specification completion, computer-assisted estimating systems used in the company and proposal and quotation procedures used by the company. In fact, printers who require their CSRs to estimate generally ask the sales representative to work closely with the customer in a sales capacity only, and once the job is fairly well established and the customer is sold, the sales rep "hands off" the customer to a CSR. This allows the sales representative to maintain a larger group of clients and to pursue new clients, while ensuring the important company connection via the CSR.

Accounting and Finance

Because of the mathematical interrelationship of estimating to the variety of plant production and operating functions, estimators may be asked to aid in the accounting and financial area. There is a diversified array of accounting duties necessary in the operation of any plant, and as a general rule an accountant will be engaged to handle these duties. However, the estimators' knowledge and background in cost accounting is beneficial, perhaps even necessary, for accurate development of good costing systems with which they must work. In some plants, financial matters—such as cash flow, banking practices, equipment leasing, and capital borrowing—and similar decision-making concerns may be part of the estimators' duties.

Management

In most plants, the estimating department is staff, or advisory, to management as distinguished from line, or production-related. In this advisory capacity, estimators may sometimes be asked to take part in certain decision-making efforts. As with the finance area, such decisions usually relate to matters dealing with the mathematics of the business operation, such as production data and costing information. For example, if top management is considering the purchase of a new lithographic press, they may ask the estimator to compare outputs of the different presses considered for purchase and/or to compare them with existing equipment.

Note that the duties of estimating provide an ideal training ground for other management positions in a printing business. Many printing plants, especially those of medium size and larger, may well begin new employees in the estimating department. Here, the new employees will immediately be in touch with a diversified array of plant data and begin to familiarize themselves with many important segments of printing manufacturing. All this information is necessary to become a top management candidate.

1.3 What the Estimator Needs to Know

Successful printing estimators must be knowledgeable in many areas. A discussion of these areas follows.

Production Sequences

In the modern printing plant, there are dozens of available procedures and techniques to produce the printed image. Estimators should have a thorough, working knowledge of all such procedures applicable to their plant. They should remain abreast of new technology that could continue to change production phases. They should, if possible, have hands-on experience in production prior to estimating. Estimators should be aware of plant production bottlenecks where production is limited or slowed.

Printing Equipment

Estimators should understand all equipment parameters—speeds of output, adjustments made during setup and operation, sizes and types of products produced, and maintenance information, to name a few. If possible, they should have practical experience with all major plant equipment.

Operating Personnel and Personnel Politics

All company personnel policies should be understood when estimating—coffee breaks, vacation and holiday times, overtime and other pay adjustments, and appropriate union contract parameters (when applicable). When possible, estimators should know personally the operating personnel. Furthermore, estimators should always remember that production data are derived as a combination of employees, machines, and materials.

Derivation of Production Data

Estimators should thoroughly understand the standard production data used for estimating and their derivation. As a general rule, such information comes from plant historical records. It may also be available in published form or from projections of new equipment performance. Knowing the derivation of such production data, estimators must then understand completely the data's application and practical use when preparing an estimate (see Section 1.5).

Cost Information

Cost information is used in conjunction with production data to determine the cost to manufacture the product. Estimators should understand the derivation of budgeted hour cost rates. They should know both fixed and variable contributions to the costing system (see Section 1.5).

Material Costs and Buyouts

Material costs are costs of all appropriate materials needed to complete the customer's order. Materials usually include paper, ink, film, and plates. *Buyouts* are products, goods, or services purchased from a trade house—that is, a company that does only specific types of work for the printing industry. Typically trade houses are categorized into typesetting, binding, color separation, and specialty work segments. Thus, if typesetting is to be completed outside the plant, it would be ordered from the appropriate trade house and be considered a buyout in the estimate. The estimator should understand material purchasing and inventory requirements and all company policies related to buyouts. With increased specialization in the printing and publishing industry, buyouts of many segments of any given printing job are common.

Computers

Due to the growing impact of computer assisted estimating (CAE), estimators should understand how to work with computers and understand application of computers to the estimating function. Estimators are not required to become programmers or computer technologists, but they should learn to use yet another tool that will have great impact on their jobs in the immediate future.

Mathematics

Estimators must be comfortable with mathematics. They should have a good understanding of addition, subtraction, multiplication, and division. Estimators should also understand elementary algebra and should have a practical, working knowledge of electronic calculators and other similar devices. They should develop a commonsense relationship with number values so as to minimize decimal errors and other typical mistakes. As estimators become more experienced, they should work to develop a yardstick ability as they estimate that allows them to compare mentally all or part of an estimate-in-process to see if it is in line with previous estimates of similar type.

Customers

While the estimator's contact with customers and print buyers varies from very little to extensive, some estimators work with customers on a frequent basis. Usually the estimator initiates this process by calling the customer or transmitting a FAX to the customer. Usually the request is to get unknown information for a pending estimate or to confirm a particular job specification which was unclear, incorrect or requires modification. The primary reason for such direct contact is to speed up the estimating process. Some estimators are very good at dealing with customers and buyers, to the point that they prefer to take job specifications directly from the customer or buyer instead of having such job specs go through the sales representatives. Other estimators prefer not to deal at all with customers and insist that the sales representative provide full job specifications before estimating can be completed. If a printing company has a customer service department, direct customer contact by the estimator is usually minimized and all required information is obtained through the customer service representative.

1.4 General Procedure for Selling, Estimating, and Quoting Printed Work

Request for Estimate

The estimating chain begins with the printing sales representative who contacts customers to determine their printing needs. In consultation with the customer, the sales representative completes a *job specification form* that details all factors necessary to completely describe the desired product. A sample of the work is helpful also. The sales representative then returns to the plant and transfers this information, with clarification when needed, to a *request for estimate form* that accurately details and clarifies the job specifications. The request for estimate is then forwarded to the estimator. Many plants combine the job specification form and the request for estimate, thus saving the sales representative's time in repeating such information and allowing the sales representative more time in the actual sales contact.

Cost Estimate

After the estimator receives the request for estimate form, he or she thoroughly reviews the desired product and then determines the *manufacturing segments*—that is, the appropriate manufacturing steps to produce it completely. These segments are normally recorded on an *estimate blank* (or form). Once the manufacturing segments are identified, the estimator assigns production times and appropriate budgeted hour cost values. When the cost values are summed, they represent the manufacturing cost for the job. Such important factors as the required number of copies (quantity) and the general quality level desired must be taken into consideration since they will have a bearing on production time and cost. After all estimating is complete, the estimate should be double-checked for errors that might have occurred during the procedure.

Job Price

Once the cost estimate has been verified as accurate, it is forwarded to top management to determine the price the customer will pay (selling price). Pricing is the addition of an established dollar amount that will provide a profit for the company. The top management representative may be a sales manager, vice president for sales, or perhaps the company president. Sometimes management will ask the estimator to price the job on the basis of a profit markup percentage, a fixed dollar amount, or on some other basis determined by management to provide a suitable profit. The estimator then calculates the total cost to manufacture the product and adds on the specified amount for profit.

Job Proposal and Quotation

After the price assignment has been made, a *proposal* is typewritten and mailed to the customer, delivered by the sales representative or sent by FAX. A proposal is an offer made to a customer for production of printing goods and services that specifically states all job requirements, specifications, and prices for such goods and services. This proposal may be accepted, rejected, or modified by the customer. Once the customer and management have reached an agreement regarding all aspects of the job, a quotation will be issued to the customer. A *quotation* is a binding agreement with specific prices and quantities indicated and is a legal contract between both parties. Many plants prefer to combine the functions of the proposal and quotation, using them as one document.

Many commercial printing operations—that is, those companies doing general printing of all types including advertising and promotional items—depart from these described procedures depending upon their size, type of customer, product line, and general business philosophy. Chapter 4 investigates this area in greater depth.

1.5 Standard Production Time and Budgeted Hour Cost Rates

Accurate cost estimates are a product of standard production time and budgeted cost rates. It is important that the reader understand these terms. Chapter 3 covers both of these subjects in greater detail.

Standard Production Time

Every printing plant—quick printer, commercial printer, or high volume printer—has a certain amount of equipment necessary to produce some portion of the final product. Some of this equipment may use sophisticated electronics and operate at high speeds, while other equipment may be older and slower in output. In addition, skilled employees are needed to operate such equipment, even though their abilities can vary tremendously depending on their individual training and experience. Standard production time is an average of the output per unit of time for each working area in the plant. Such standards may be expressed in hours per 1,000 impressions, number of sheets per hour, or any similar output as a function of time. Standard production time is used to determine the amount of time a particular job will take.

Consider the operation of a small vertical process camera. The cameraperson who is skilled at this job can expose and process a 10 x 12 inch film negative of typematter at a rate of one film image each six clock minutes. The standard production time may be restated in decimal hours as 0.10 hour per film negative. In this time period, the cameraperson must place the copy in the camera copyboard, adjust for enlargement or reduction, set the timer, position the film in the vacuum of the camera, expose, develop, and fix, then wash, squeegee, and hang the film sheet to dry. Should any additional operation or procedure enter into the work sequence as described, the standard time must be reevaluated. For example, the use of automatic film processing equipment will save the cameraperson processing time, allowing for increased production per unit of time and making it necessary to compute a new standard production time. Because many such changes occur continually throughout the printing plant, standard production times must be constantly monitored and updated.

It is important to point out that such standard rates are averages of output per unit of time. Machinery is reasonably consistent, but most production workers exhibit differing levels of output on a day-to-day basis. When operators and machines work in combination, no standard is completely precise. Such averaging allows for one set of standards to reasonably cover the operator-machine combination under normal operating conditions. Significant variations require the establishment of new standards.

Budgeted Hour Cost Rates

Every printing plant can be segregated into a number of production centers, such as a printing press or process camera, wherein costs are regularly incurred. Some of these costs remain constant regardless of output in the center. These costs are termed *fixed costs* and include items such as insurance, taxes, and rent. Other costs, such as the cost of labor, are termed *variable costs* and are generated only when production actually occurs in the center. Very simply, the *budgeted hour cost rate* (BHR) is the cost assigned for all fixed and variable costs incurred in the operation of a specific manufacturing center on an hourly time basis. Material costs that can be specifically identified, such as paper, film, plates, and ink, are not included in budgeted hour cost techniques. They must be determined and charge to the customer separately.

Consider the process camera example discussed with respect to standard production times. Let us assume that the cameraperson must produce 10 camera negatives for a customer. At the decimal hour standard production rate of 0.10 hour per negative, it will take 1 hour to complete all work on the 10 images. If the operating cost—that is, the budgeted hour rate for the small vertical camera—has been determined to be $30 per hour, then this customer would owe $30 for the use of the camera, cameraperson, and all associated facilities. Add to this a material cost of $8 for film and processing chemistry, and the customer would owe $38 for camera work and all materials. To complete this order, additional manufacturing steps such as image assembly, platemaking, presswork, and bindery must follow. Each of these would also be evaluated at a production standard multiplied by the determined budgeted hour cost rate. Summed together, with material costs included, this total cost would represent the internal manufacturing expense for each segment of the customer's order.

Both standard production times and budgeted hour rates are important concepts. Each concept will be discussed in greater depth in Chapter 3.

1.6 Additional Methods Used to Estimate Printing

In addition to, or in place of, the described cost-estimating procedure, the following eight procedures are used to estimate and price printing. One major difference between these techniques and cost estimating is that each is based on the selling price of the job. Selling price is defined as the total dollar amount the customer will pay when the job is completed. It is represented by the sum of all manufacturing, material, and overhead costs, plus profit.

When estimating using selling price as a base (*price estimating*), the profit markup step is removed from the estimating procedure. In addition, detailing of specific production segments is not completed. These two major changes allow price estimating to be a much faster procedure when compared to cost estimating. Using this method translates into more estimates per estimator per day and the ability of the company to quote on more printing orders. It should be noted that major disadvantages include the fact that price-estimating procedures have a built-in profit margin, thus making it difficult for management to modify the profit on a particular job. In addition, no production times are assigned during estimating, so when (and if) the job begins production, scheduling it into plant workflow is not easily completed.

Figure 1.1 summarizes the general characteristics of each method with respect to accuracy, ease of system use, system complexity, ease of modification, computer adaptability, training time and shop performance data. This is covered in more detail in Chapter 14, but also general estimating methods more oriented to the quick printer or smaller commercial printer are covered.

Tailored Price Lists

Estimating from price lists is very common in all areas of the printing industry—quick printer, commercial printer, and high volume printer—so long as the product line of the price list item is standardized. For example, many printers offer standard sizes and colors of paper, standard colors of ink, defined quantities and routine production making price lists convenient, reliable and fast. Depending on competition and other factors, price lists may be printed and distributed to customers, so that they can use the price list to determine the price of the work they are contemplating, prior to contact with the printer. When price lists are to be passed out to customers, they should be dated with the statement that "Prices may change without notice," thus protecting the printer from problems regarding price changes and out-of-date lists.

Initially, price lists should be developed using cost-estimating techniques. Tailoring the price to the particular printing company is very important, since it is likely that costs of production, labor and materials may vary and full cost recovery plus a profit is desired. Revision of the price list should also be completed using cost-estimating, especially if new equipment or production changes have been made. A cost-of-living revision may be done as a percentage markup of existing prices; however, this percentage should be determined accurately to reflect such changes in cost.

There is no question that estimating from a price list is fast and convenient for customer and company alike. These advantages are significant, but price list estimating does not provide any information about production times, which is a disadvantage, perhaps, for large or complex jobs. Also, price lists are very accurate for work that is locked into the established product lines of the company (and therefore uses standard production techniques), but should not be used for jobs that deviate to any extent. Should a customer desire a product not specifically covered in the price list, it should be cost estimated for accuracy.

Large printing companies that offer standardized products, such as labels or business forms, normally develop a price list book that covers literally all products they produce. This book is carried by each sales representative, who has been trained in its use. When a customer desires a price for a certain quantity and type of label, for example, the sales representative is able to determine a firm price within a short time. Quick price estimating is a tremen-

	System accuracy	Easy to use/ speed of use	Complexity of basic system	Ease of modification	Computer-adaptability	Time to learn/ training time	Provides shop performance data
Cost estimating	excellent	manual: slow computer: fast	fairly high	good	excellent	may be extensive	yes (when job costing included)
Tailored price lists	excellent (when based on cost estimating)	moderate to fast	varies with system design	varies with system design	good	moderate to low	not extensive
Standard price books *(Franklin Catalog)*	good to fair	moderate to fast	fairly low	done by vendor	fair	moderate to low	not extensive
Past work basis	good	good	varies with system	good	good (but based on system design)	moderate to low	not extensive
Using the competition's price	poor	varies with system	varies with system	can't easily modify	varies with system	varies with system	not extensive
Ratio systems (paper ratio or other index)	good to fair (depends on index)	moderate to fast	moderate	good	varies with system	varies with system	not extensive
Chargeback	good	data tracked with job	low	varies with system design	varies with system	varies with system	yes
Customer's ability and willingness to pay	poor	moderate to fast	low	easily modified	none	hard to determine	none
Intuitive method	poor	fast	low	low	none	hard to determine	none

Figure 1.1

Comparison chart for various methods for estimating and pricing commercial printing.

dous aid in selling standardized products. Of course, such estimating—completed in the customer's office under less than ideal conditions—can be embarrassing if an error is made or a price is misquoted by the sales representative.

Price lists are an extremely valuable estimating and pricing tool when they are accurately developed and updated, and used for the product lines for which they were designed. As specialization increases in the printing industry, estimating/pricing jobs from tailored price lists will become more frequent, including the development of computer programs to do this task.

Pricing by Standard Price Books or Standard Catalogs

Two standard catalogs oriented to the quick printer or smaller commercial printer serve the printing industry: the *Franklin Catalog* and the *Counter Price Book*.

The *Franklin Catalog* is widely distributed throughout the U.S. and has been a major printing pricing tool since the 1930s. Using *Franklin*, almost any simple type of printing can be estimated quickly and easily. Estimating with *Franklin* is completed based on selling price dollars, which may then be quoted to the customer directly from the catalog, or adjusted to meet competitive needs.

The most popular of the *Franklin Catalogs* is the Offset Catalog, which is leased from Franklin for an annual fee of approximately $75, and includes updated sections mailed to the user four times yearly. The sections, which are categorized by type of work or type of paper used, and cross-referenced by size of product and level of quality, are held in a three-ring binder. There are also sections containing instructions for using *Franklin*, hour rates and various operational times.

The procedure for estimating is reasonably simple. First the estimator identifies the proper catalog section, then the specific page in the section and finally the specific table on the page, based on the finish size of the product and the level of quality expected. Table prices are based on the grade of paper used, typically given as the cost per 1,000 sheets of a specific size and kind of paper and the quantity desired by the customer. Using the proper grade and quantity figures, the estimator cross-references on the table to locate the selling price for the job. Extra operations—for example, additional colors of ink, printing on the reverse side of the sheet, or numbering—are then added as they apply. The selling price from the table and any additions then become the full selling price for the desired product at the specified quantity level.

Specific procedures for using *Franklin* are provided in the instructions portion of the book. These instructions include a step-by-step procedure for estimating with Franklin, as well as details concerning specific types of products and processes that apply.

Compared to cost-estimating procedures, catalog pricing with *Franklin* is fast and convenient. It also provides a uniform pricing base from which to sell printing. Franklin is most popular with smaller printers who provide a fairly wide array of more typical printed products, and who do not product-specialize to any great extent. These companies do not have the time or need to completely cost estimate each job. They have found that *Franklin* provides speed and convenience through a generally uncomplicated process, and since they are making money they feel that the system is accurate and reliable enough to meet their needs.

One of the ways *Franklin* has streamlined the use of their catalog has been through the conversion of the manual "look-up" tables to floppy computer disks. The original version—called Comp-Est—was released in 1983. The floppy disk program, purchased from Franklin Estimating Systems, is available to run with MS-DOS, CP/M or M-BASIC operating systems. The cost is approximately $1,000 per year for the first year (including user's manual), and a $250 per year renewal fee for providing continuing services. Smaller commercial printers and quick printers are the major users of Franklin's Comp-Est system.

Bill Friday's *Counter Pricing Book* can be found in quick printing firms throughout the nation as the major pricing tool. Offered through Prudential Publishing Company, the book is available on a lease basis, allowing for updates throughout the year. The primary purpose of the book is to provide quick, consistent pricing for customers who order fast-turnaround products, largely available through a quick printer or smaller commercial printer.

Prices in the book are based on jobs produced on normal quick printing equipment, such as an Itek-type camera/platemaker and an A.B. Dick press. Price codes are used—"A" for average prices based on national averages and pricing formulas, "L" for 10% below average, and "H" for 10% above average. Every price page is also date coded. High and low sets of prices allow the user to move 10% up or down. Printed items covered in the book include a wide array of papers and products: bond letterheads, carbonless forms, cover stocks, envelopes, index bristol, offset book for various product sizes, padding, scoring, and typesetting.

The prices in the *Counter Price Book* represent the publisher's best judgment as to fair selling prices. No specific amount of profit is indicated or guaranteed; the efficiency of the printer's operation will affect the cost of production and determine the amount of profit to be obtained, or loss to be absorbed.

Standard catalogs have certain generally recognized drawbacks. First, there is no identified production plan within the system, which means that no production times for scheduling the work in the plant are available. Second, no modifications for price exist for differences in production from one plant to the next or even within the same plant. To illustrate, a letterhead might run 1-up (one image on each sheet of paper) in Plant A and 4-up (four identical images on one sheet of paper, printed simultaneously) in Plant B, but each company using Franklin would quote an identical price. The reality of this example is the production costs may be significantly less for Plant B than Plant A since it would take less time for Plant B to complete the press run to produce the product. Passing on the savings of reduced production costs to the customer increases sales and is a good business practice.

Also, prices used in standard catalogs are the same for all parts of the United States. Printing managers recognize that geographic location has a direct bearing on their costs and eventually on the prices they charge for their products. Since standard catalog prices do not vary geographically, each plant/user must adjust prices relative to their own geographic region.

Pricing on the Basis of Past Work

A common technique for estimating in the printing industry, especially for a mix of products produced in the same plant, is pricing on the basis of past work. Essentially, the estimator and sales representative establish and maintain a file of past jobs that have been profitably produced and for which production is geared. The salesperson uses this sample portfolio during the sales contact with the customer. Frequently, the client will select a product from the sample, with a request for a color change or perhaps a slightly different page size. Back at the plant, the estimator will pull the original estimate for this job, review it, and make necessary changes to reflect the modifications desired by the customer. The sales representative will then return to the customer with a proposal stating exact prices for various quantity levels. If changes are limited, the procedure is fast and reasonably accurate. The technique leads toward product specialization that may be most profitable for the company in the long term.

In some cases, the procedure is modified somewhat: The estimator maintains a card file system of past job quotations. The sales representative then sells work without a sample portfolio and brings the work in to be estimated. At this point, the estimator refers to the card file for a similar job done perhaps three months ago and quotes that former price or one close to it. The filing system used must be indexed and cross-referenced to be a timesaver for the estimator.

It is important to note that cost estimating should be used to periodically verify the costs upon which the system is based, which may lead to price changes. Also, as new products are added to the portfolio, cost estimating should be used to establish the initial prices.

Pricing by Competition

Using the prices charged by competing printers as a basis for estimating is fairly common in the printing industry. Such information is available from many sources—from published price lists, from customers who are checking around for the best price on their job, from supply salespeople, from former employees of that business, or perhaps from table talk at a local association meeting. Once the information is obtained, it may be used to adjust or establish a price for a job under consideration.

Two major problems exist when pricing by competition. First, it is exceptionally difficult to verify if the obtained prices are accurate for the described product and quantity. How can the accuracy be checked? Second, there may be little, if any, resemblance between the companies comparing prices. Each company may have entirely different types of equipment, different production techniques, and different types of personnel with varying levels of experience. There may also be tremendously different accounting, costing, and estimating procedures. If the competitors have very little in common, then costs and prices will tend to fluctuate significantly.

The advantage of pricing by competition is that it is a simple procedure once the information concerning the competitor's prices has been obtained. It is possible to undercut the competition's price and subsequently increase the volume of work by winning those jobs that normally would go elsewhere. Of course, while production may be up, costs to cover such production may not be recovered because they were cut to beat the competition. The net effect, should this happen, is bare bones survival or future dissolution of the company through bankruptcy. Consider this situation: If each printing plant in the community priced work by comparison with the competition, a price-cutting cycle could easily begin. Those plants that refused to cut prices would lose business to the firms that were offering the best deal. Those plants that cut prices would be doing a tremendous volume of work, perhaps much of it at cost or below. The net effect of price cutting is a vicious cycle that can hurt the entire printing community.

It is safe to state that most estimators and sales representatives, at least once in a while, attempt to meet or beat the competition's price on a job they consider important. In fact, it is reasonable to assume that some printing plants price their work as much as possible on competitors' bids and may even make money if the competitors know their costs and pricing structure well. Consistently making a profit when pricing by competition, however, is more a product of luck than of skill.

Pricing Using a Ratio System

This estimating technique involves establishing a ratio of paper or some other job component to the selling price of the job. The most common ratio component used is the cost of paper, since it is the material found consistently in most printing orders. To use the ratio system with paper as the component item, the cost of paper for an order is determined; then the estimator multiplies this cost by a specific ratio figure—say, four—thus calculating the selling price for the printing order as a function of paper cost.

This technique does not reflect the actual costs incurred during manufacture of the product, but is accepted because it is very easy to execute, and, in the view of some plant owners, it is accurate enough for their needs. The key to its accuracy is to carefully identify the component item common to the products produced by the company. The next step is to develop a ratio figure that most accurately provides for full job-cost recovery and profit. It is possible to increase the accuracy of the system by setting up the ratio based on past jobs done by the company which have been profitable. To ensure the system's accuracy, all jobs estimated using the ratio system should be as similar as possible in terms of manufacturing procedures to the jobs on which the index is based—that is similar in stripping, typesetting, art production, etc.

Pricing Using a Ratio System

This estimating technique involves establishing a ratio of paper or some other job component to the selling price of the job. The most common ratio component used is the cost of paper, since it is the material found consistently in most printing orders. For example, to use the ratio system with paper as the component item, the cost of paper for an order is determined, then the estimator multiplies this cost by a specific ratio figure—for example, four—thus calculating the selling price for the printing order as a function of paper cost.

This technique does not reflect the actual costs incurred during manufacture of the product, but is accepted because it is very easy to execute and, in the view of some plant owners, is accurate enough for their needs. The key is to carefully identify the component item common to the products produced by the company, and then develop a ratio figure that most accurately provides for full job-cost recovery and profit. It is possible to increase the accuracy of the system by setting up the ratio based on past jobs done by the company which have been profitable. To ensure the system's accuracy, all jobs estimated using the ratio system should be as similar as possible in terms of manufacturing procedures to the jobs on which the index is based—that is similar in stripping, typesetting, art production, etc.

Pricing by the Chargeback System

Some commercial printing plants, and many in-plant operations, use the chargeback system to price the value of their printing. The chargeback procedure uses actual production and material cost data accumulated as the job is completed, as opposed to estimating what is expected to occur before the job ever begins production. Actual time and material costs are tracked during production and summed at the completion of the job. In the commercial printing segment, a reasonable profit is added to the actual costs to arrive at a final selling price for the customer. In-plant shops, doing captive printing for a parent company, may or may not add extra dollars for profit.

An advantage of chargeback is that the customer pays only for the actual time worked and materials consumed for a job, eliminating any differences between the estimated cost and actual cost of the job. Essentially, chargeback procedures eliminate or reduce the estimator's involvement in such jobs since prices are based on actual and not estimated times. The estimator's time can be spent on other jobs, which is another advantage.

One disadvantage of chargeback is that since no estimate is prepared, no proposal or quotation is offered. Thus, customers may be surprised at the final price of a job and consider it to be excessive. Without a proposal or quotation stating specific dollar amounts, the question of whether a contractual agreement exists between the printer and the customer is possible.

Generally, commercial printers use the chargeback system only with customers who are established and wish to expedite production on their jobs, especially on repeat printing orders or routine jobs. In such cases, an established climate of trust exists between printer and customer. In-plant printing operations, such as a printing facility contained within an insurance company doing work only for that parent operation, typically finds the chargeback system desirable since the customers—departments within the parent company—use the procedure as a budgetary matter. Thus, when printing is done for a department, the cost is tracked during production and then charged back against that department's budget or operating fund. The printing department's income is then generated by the volume of chargeback work done for the various departments it supports within the parent organization.

Pricing by the Customer's Ability and Willingness to Pay

This pricing procedure is extremely subjective. Customers are evaluated by the sales representative or company management with respect to willingness and ability to pay for a job. When possible, some type of index, such as a credit rating or other evidence of financial dependability, is used as a base for arbitrary establishment of selling price. Once this price is communicated to the customers, they may accept, reject, or attempt to negotiate the figure downward. Because the customer, not the costs incurred in producing the product, is the focal point of the pricing system, it is not unusual for the quoted price to be higher than if normal estimating procedures were used.

Plants that price in this manner may not want to take the time to estimate the job, or they may have no defined procedure by which estimates can be accurately completed. Pricing exclusively by the customer's ability or willingness to pay for the work appeals largely to the entrepreneurial printer and may border on guesswork, depending on the method by which the factors of "willingness to pay" and "ability to pay" are determined. However, there is merit to the concept in that the printer can subjectively price on what the "market will bear"—based on his own feelings and observations—and can therefore earn substantial profits if such observations and feelings about customers are correct.

Pricing by Intuition or Guesswork

Some printing managers believe that definitive cost estimating is a waste of time. They base this reasoning on the fact that estimating requires standard times and budgeted costs, both of which are actually averages; that many mistakes occur during the estimating process; and that too many changes occur during production that simply cannot be determined during estimating. So instead of estimating at all, these managers use intuition to determine a reasonable price for each job. In some cases, they may review the production requirements that are needed to do the job and include that in their guess. In other cases, they may quote a price based on quantity or type of product with no other input at all.

It is inconceivable to think that guessing at a price for a printed product is a reasonable pricing procedure. Nonetheless, it is a technique practiced today by those who believe that intuition is better than averaging when pricing printing or by those faced with an excessively large number of estimates to complete in a limited amount of time.

1.7 The Estimator's Working Environment

The following addresses the working environment for the typical commercial printing estimator. For quick printers and commercial printers doing fast-turnaround work, the normal pricing process is usually completed at the counter—manually, with an electronic calculator, or using a countertop computer system—or at the time the order is tallied at the cash register. In instances where the quick print customer requests a competitive price prior to placing the order, such pricing may be completed by the counter employee and given to the owner/operator or store manager for approval. When such orders are complex or of fairly high quantities, price differences between quick printing production and commercial "metal plate" production may be a customer consideration. Further information on counter pricing will be provided in Chapter 14 which deals with estimating and pricing for the quick printer.

Office Location and Equipment for the Commercial Printing Estimator

The physical location of the commercial printing estimator's office is important because it services sales representatives, customer service representatives, management and customers. It must be easily accessible, yet not in the center of traffic or major plant activities. It should be quiet, so location immediately adjacent to production lines may be undesirable. When possible, outside windows should be used to provide natural lighting. Artificial white lighting of normal intensity is also needed. A complete heating/air conditioning system should provide a comfortable working environment.

The size of the estimator's office is important The office should be large enough for conferences with sales representatives, customers, other estimators, and management. Thus, a large table with chairs should be available. In some plants, each estimator will have a conference room; in other plants, the estimator's offices may be located adjacent to a single conference room shared by others. Because layout and inspection of artwork and printed press sheets are completed in the estimator's office, there must be ample space to spread out large sheets.

Each office should be furnished with a large desk and an additional large table for greater working space. Ample filing space in closed cabinets for job tickets and other documents should be easily accessible. A bookcase should be available for storage of reference material. Some estimators also prefer a filing card system by which previous jobs can be recorded for future reference. A telephone is a very important feature and should be located where is accessible to visitors as well as convenient to the estimator.

If financially practical, the office should be carpeted. Carpeting soundproofs the room, provides insulation, and adds decorative flavor. Stylized furniture and framed artwork are additional options. The more appealing the estimator's surroundings, the more enjoyable the working situation.

Tools

The estimator must have certain tools to insure accurate estimates. Normal office supplies such as pencils, pens, erasers, and graph paper are needed, and convenient access to a copying machine is most helpful. A typewriter should be provided for typing proposals and quotations to be sent to customers.

Each estimator should have access to two types of calculators: a printing calculator and a pocket type. The printing calculator provides a paper tape record for all input figures and operations that can be rechecked at a future time. Printing calculators are desk models that operate on normal electrical current. The portable pocket calculator, which may be solar or battery powered, is hand held and used for calculations that require no future record. Many models have a built-in memory that allows for storage of intermediate figures during calculations. A pocket calculator with only one memory can speed up estimating significantly.

Many estimators use computer-assisted estimating systems as an aid in completing estimates, proposals and quotations, as well as for order entry, production management and job costing. The estimator's office should provide a comfortable work area for the computer system, including ergonomic considerations that relate to working with CAE system for long, unbroken time periods. Chapter 5 provides an in-depth discussion of the computer in the printing company, including CAE procedures.

Other tools are also needed for estimating. A magnifying glass (or loop) for the inspection of samples and press sheets, a paper thickness gauge, a proportion scale for fitting artwork, rulers and tape measure for accurate measurements, and gauges for determining screen rulings and angles of screens are all standard tools for estimating. In addition, reference books containing production standards, costing information, and process considerations are very helpful. Providing the estimator with an efficient, pleasing office and the proper tools and resources necessary for executing estimates is very important if accuracy in estimating is desired.

1.8 Selecting an Estimator or Estimating Trainee

Preparing a Job Description

The initial step in selecting a printing estimator or trainee is the development of a job description that specifically states the duties that the particular individual will have in the organization. For example, an estimator's duties in a small printing plant (which may have 1 estimator for perhaps 15 production and management personnel) may include office management duties and over-the-counter sales. In a large plant, estimators may have far less breadth of duties, but may be required to specifically concentrate their estimating skills in production planning or, perhaps, in estimating only a certain product line or process. Regardless of the scope, it is important to determine accurately the range of duties that the estimator will complete. Hiring is then a matter of fitting each applicant into as many categories as possible.

For even greater clarification, it may be beneficial to write an expanded job description that details even the most infrequent duties the estimator will perform. The more specifically the job is identified, the easier it will be to determine who should be hired.

Advertising

Once the job description is complete, it should be used in advertising and external communication with potential job applicants. Depending upon available funds and the importance assigned to this hiring by top management, advertising may be done either in a limited geographic region or on a national basis. A local newspaper may be one advertising source, but it may not provide the best coverage, especially when attempting to reach applicants who completely meet the job description. In order to reach experienced applicants, advertising is often done in trade journals, such as *Graphic Arts Monthly, Printing Impressions,* and *American Printer*, as well as numerous regional graphic arts publications. A general rule is the more applicants, the better the selection, and the better the selection, the few the training costs incurred once an individual is hired.

Hiring Within or Outside the Plant

A very good source for an estimating trainee is from within the printing plant—for example, an employee currently doing another job who has expressed an interest in such a position. A working supervisor or management trainee could be possible sources. Perhaps the major advantage of hiring from within is the fact that the individual under consideration is known to management. His or her general on-the-job performance, attitudes, personality, and printing background have been observed on a day-to-day basis. One of the most significant disadvantages when considering current employees relates to the impact of a negative decision. If not accepted for the estimating position, the employee may be discouraged and disappointed, ultimately becoming an unhappy and unproductive worker over time.

When there are no internal applicants or when a wider selection of applicants is desired, recruiting outside the plant is the typical procedure. Such external sources might include a college or university graphic communications program and attraction of trainees or experienced estimators from the industry. Recruiting is generally done using classified advertisements in newspapers and trade magazines.

Hiring an individual who is unknown to company management requires careful assessment of the applicant's background, knowledge of printing, personality, and potential for success as an estimator or estimating trainee. Past job history is very important. In most cases, the applicants under final consideration should be invited to the plant, which necessitates travel and boarding costs. Testing of the final applicants may also be desired, which is an additional expense.

Weighted Rating System

The interviewing and hiring of the right person for an estimating position is an art in itself. To aid in this endeavor, the weighted rating system in Figure 1.2 is provided. The criteria in the left column cover the major areas of importance for estimating: knowledge of printing, mathematical ability, technical printing competence, personality factors, logic ability, educational background, and computer knowledge. Note that the importance of each has been weighted with respect to its relative importance in estimating. Of course, should the expanded job description indicate that other factors are of greater importance or the weight of one factor far outweighs others, the chart should be so modified. These decisions are best made by top management prior to interviewing and testing.

When using the weighted system, which is an attempt to objectively categorize and define subjective values, the interviewers must attempt to accurately determine the levels of competence for each candidate. Point values for each category must be fairly assigned and then totaled for final assessment and deci-

	Weight Factor	Excellent (8)	Good (6)	Fair (4)	Poor* (2)	Totals (factor × pts.)
Knowledge of printing: Includes applicant's understanding of processes and products, and knowledge of workflow for printing manufacturing.	3.0					
Mathematical ability: Assessment of the applicant's knowledge, understanding and use of mathematics. May be related to high school or collegiate test scores.	2.5					
Technical printing competence: The applicant's awareness and understanding of technical factors involved in the production of printed materials; includes knowledge of materials used, costs and job components.	2.5					
Personality factors: An assessment of the individual's basic personality profile—maturity, ability to work independently, level of frustration in solving problems, general character and disposition, etc.	2.0					
Logic ability: Includes assessment of the applicant's ability to think and speak in a logical manner; effective and clear reasoning capacity; may relate to mathematical ability closely	2.0					
Educational background: Relates to the applicant's ability to deal effectively with theoretical as well as practical problems; grades earned may be a good indicator of future success	1.0					
Computer knowledge: Relates to the individual's understanding of basic computer applications and the functions and operations of a computer system; not necessarily related to programming skills	0.5					
Column totals		____	____	____	____	________

Total point evaluation:

Excellent selection	95-108 points
Candidate has possibilities	68-94 points
Questionable—could go either way	41-67 points
Don't hire as estimator/planner	Under 40 points

*If no reasonable rating can be made, add a zero.

Figure 1.2

Sample of weighted rating scale for evaluating printing estimator candidates.

sion on hiring. At the bottom of the chart is a total point evaluation to aid management in assessing the choice that must be made.

Interview and Testing

It is vital that the interviewer rate every candidate consistently and accurately. Inputs may not be limited to personal interviews. Discussions with former teachers and employers as well as educational transcripts (which also contain general aptitude test results), will provide additional source material. All inputs must be as unbiased and reasonable as possible. Should the interviewer be unable to rate the candidate confidently in three or more categories (of which two of those three should be among the last for categories listed in Figure 1.2), further information from additional sources may be necessary.

The subject of testing deserves discussion. While there are few tests designed specifically for the selection of an estimator, there are tests available for purchase in most of the category areas. The Graphic Arts Technical Foundation (Pittsburgh, PA) and Educational Testing Service (Princeton, NJ) are two sources for such tests. Of course, most employment services and professional psychologists provide testing in their respective disciplines. If tests are administered by the interviewer, they should be reviewed and evaluated prior to actual use with applicants. Sometimes the best way to do this is for the interviewer or a top management candidate to take the test personally. It must be noted that test results, for obvious reasons, cannot provide a total basis from which to make any hiring decision. They reflect only what the test comprises and may have built-in biases themselves.

If the weighted rating procedure, as illustrated in Figure 1.2, is used as a test, remember that the weight factor can be modified to better reflect the type of individual to be hired, the job description, and other contingencies. Total point evaluation may also be reassessed if desired.

1.9 Developing an Estimating Training Program

One of the most difficult problems with printing estimating is the continual turnover of estimating staff. This is to be expected for a number of reasons. First, estimating does serve as an ideal entry-level training ground for operating management, as well as a good place for the customer service or future sales representative to "learn the ropes" about how the printing company functions. Second, estimating requires a continual application of numbers, much as an accountant works with figures, making it an undesirable job for some types of people in the long run. Third, estimating requires a constant attention to detail which, in turn, make it a somewhat stressful job over the long-term. Fourth, the salaries paid to estimators may not always reflect the responsibility, knowledge, and stress required of the job; this sometimes leads to estimators moving into customer service, sales, or production management positions.

Figure 1.3 provides a flowchart with step-by-step points for training a printing estimator. While it is general in scope, the framework of how such training should be completed is outlined. The following explain the steps in further detail.

Step 1 Once it has been determined that an estimator trainee is needed, a person with the proper aptitude and interest level should be carefully selected. In many cases, selecting someone from inside the plant—a person with known skills, personality, and work record—proves less problematic in the long term.

Step 2 Basic estimating theory is best taught away from the plant, by skilled industry professionals. Seminars, workshops, and short courses offer one form of this basic instruction. Many PIA regional offices offer ten to twenty week basic courses in estimating, usually at night or on weekends. If courses are not available, instruction must be provided by the estimating supervisor, who is responsible for training.

Step 3 Trainees must immediately learn the rudiments of production standard development—known as job costing—and how to develop budgeted hour costs. If possible, the trainee should learn to build BHRs on a computer spreadsheet, tailored to the printing firm. To learn job costing and production standards, the trainee might work for a brief time in the production management area, scheduling jobs and dealing with an overview of company production.

Step 4 The trainee divides his workday. As an example, he could spend mornings in prepress, press and bindery for two weeks each, and afternoons estimating. If one area was to be selected as "most important" for the trainee, it would be image assembly, where most author's alterations and house errors occur.

Step 5 After completing the six week "back shop" working commitment—two weeks of half-day training in prepress, press, and bindery—the trainee will begin full-time estimating duties. Manual estimating procedures should continue to be used, with movement to computer-assisted estimating as the supervisor and trainee determine. While manual estimating training is slow and laborious, the shift to computers will be much easier if manual estimating procedures are emphasized initially.

The speed with which the training progresses will be conditional on many factors, including the trainee's interest and ability in learning the procedures, previous experience and knowledge of printing, desire to be an estimator, and his or her focus on completing the training. If at any point the trainee begins to lose interest or focus, the training leader should pursue reasons as to this.

Step 6 Once manual estimating is mastered, the trainee should be introduced to the computer-assisted estimating system used by the company. It is best to begin CAE training with an introduction to the system, which may be

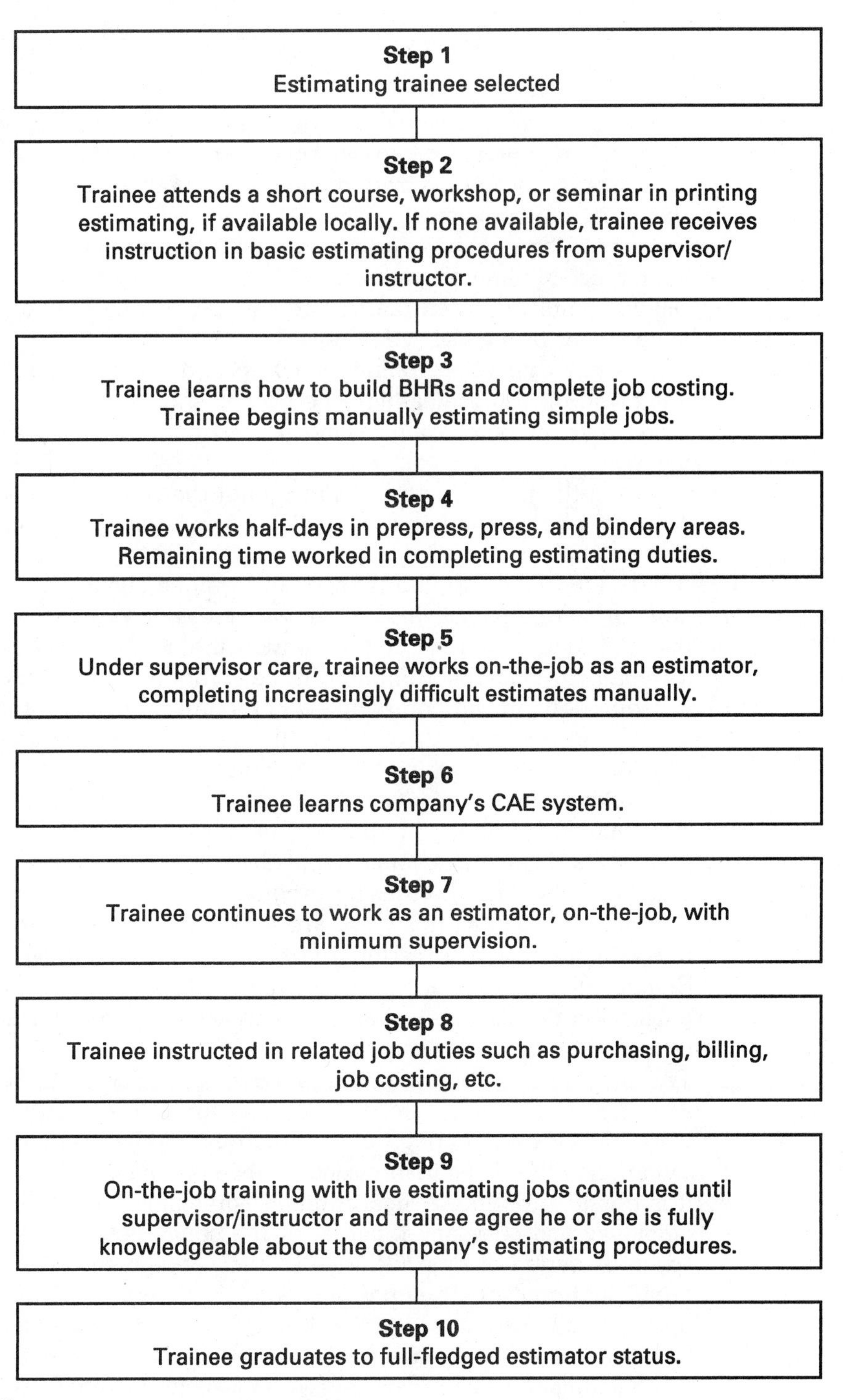

Figure 1.3

Flowchart for training an estimator.

most effectively taught by the vendor of the computer system. Budgeted hour costs and production standard databases should be clearly understood by this time, and the process of moving from manual to computer-assisted estimating should be fairly easy to complete.

Step 7 Training continues on-the-job, with the trainee working on all kinds of estimates. Minimum supervision should be required at this point. Training will be completed when the trainee can handle any type of estimate, whether by computer or manual procedures.

Step 8 Once estimating training is completed—or perhaps during the latter phases of on-the-job estimating—the trainee should be instructed in related job duties such as purchasing, billing, job costing, inventory control, or report generation.

Step 9 On-the-job training continues until the trainee and supervisor agree that the trainee is fully knowledgeable and capable in performing estimating duties for the company

Step 10 Once the training is over, the trainee should "graduate" to master estimator status. A pay raise or salary increase does wonders to congratulate the full-fledged estimator and will ensure the newly trained estimator's interest in remaining with the company. In celebration of this event—along with other promotions in the company—a company-sponsored luncheon or dinner might be appropriate. A framed certificate also might be given to the trained estimator.

Some additional training points:

1. The estimating training program should have focus and follow a clear plan. This document should detail the major points and provide a time breakdown for the training program.
2. Use of texts such as this book are important resources for training the estimator. *The Printing Estimating Workbook* by Philip K. Ruggles (P.O. Box 46, San Luis Obispo, CA 93406)—a collateral publication to this book—is an excellent additional tool for the novice estimator or estimating trainee.
3. Training should be a continuous process, supported by the company, at company expense, and on company time. If estimators have to work "double-duty" during training, there should be overtime compensation and other benefits provided by the company.
4. It is possible to write a legal training agreement which will lock the employee and company together at the conclusion of training for a specified period of time. So long as such an agreement is not overbroad in terms of its conditions, and is entered into clearly and openly by the estimator trainee, it should be generally acceptable. A lawyer will likely be needed to aid in drafting such a document.
5. If the company has a junior college, university, or other academic facility close by, there may be faculty who would serve as special resources for estimating training. Also consider any type of estimating training program offered through a national association such as Na-

tional Association of Printers and Lithographers or Printing Industries of America. Such programs are usually jam-packed with information and can be of great educational value.

6. If your computer vendor is willing to provide training for entry-level trainees—even for a fee—serious consideration should be to given to using it. While vendor training is specific to their own computer equipment, it may be much faster, more convenient, and more direct than learning the computer system through trial-and-error

1.10 Some Rules for the Working Printing Estimator

There are major areas of difficulty that the printing estimator typically encounters. The following guidelines, arranged in no particular order, represent ways to avoid such difficulties:

1. Before estimating any job, make certain it is clear who will supply artwork for the job and establish generally the quality level of that artwork so that your plant production will not be adversely affected.
2. Determine your customer's needs exactly.
3. Use up-to-date production standards, budgeted hour cost rates, and material costs.
4. Always include a profit on each job.
5. Try to have all estimates double-checked before the proposal/quotation is issued.
6. Use a good filing system and keep readable copies of all quotations/ proposals, estimating worksheets, and so forth.
7. Make minimum assumptions when estimating by requiring complete job specifications from your sales representative.
8. Estimate the job using the most typical or probable production sequence.
9. Do not make arbitrary exceptions to established rules and plant policies.
10. When estimating, try to work slowly and carefully. Do not estimate in the customer's presence or with the customer standing over your shoulder.
11. Do not allow yourself to be pressured by your sales representatives or company management.
12. Carefully establish delivery dates and be certain you keep them.
13. If buyouts from trade suppliers are a part of the estimate, do not guess at these, but get firm estimates from the suppliers involved.
14. Try not to use cost plus straight percentage to determine the price of any job. Use a percentage markup system or other method that relates price of the job to market factors for that particular customer.

Marketing, Pricing and Profit for the Quick Printer, Commercial Printer and High Volume Printer

2.1 Introduction

Since the mid-1970s there have been three evolving conditions or factors that have had an increasingly significant effect on printing and allied industries. The first is the recognition of marketing and sales as essential keys to printing company success, which has lead to the evolution of various marketing tools and sales techniques. The focus has been on specialization of product and process and enhanced customer service. The second is the customer's acceptance of the quick printer, making this segment of the industry a major and dynamic force. This has, in turn, changed the competitive relationship between commercial printing and quick printing. The third is the now-accelerating evolution of electronic imaging equipment, controls, computers and other electronic gear, effectively converting the industry from a "craftsmanship" focus to an industry of "technology" dominance.

This chapter investigates the interrelationship between quick printing, commercial printing, and high volume printing. Sections deal with the characteristics of these groups, general marketing and sales considerations for each, and the pricing and profitability issues affecting each group and the industry as a whole. Chapter 14 will do an in-depth expansion on the pricing and estimating procedures for the quick or smaller commercial printer. Chapter 15 will detail the techniques used by high volume printers—web plants—to estimate and price their work.

The job of the estimator—those employees in the industry with the responsibility to plan production and "crunch the numbers" with respect to cost, price, and profitability—relates (directly or indirectly) to sales, marketing, pricing and profits earned by the printing company. He or she must deal closely with the printing sales representative and customer service representative because of their mutual need for complete job specifications on which the estimate will be based. The type of product specialties offered by the printer, based on the market niches the printer has elected to pursue, and the resultant costs generated by these specialized products and markets become important to the estimator, who is generally responsible for ensuring a fairly consistent cost-price strategy for the company. The estimator may also aid management in determining the cost effectiveness of specialized products under consideration as new markets by the company. Pricing, typically a top management discipline, is sometimes done by the estimator based on prior decisions by the company president, the sales manager or another top company official.

The estimator often uses information from industry trade associations including Printing Industries of America (PIA) and its regional affiliate offices, the National Association of Printers and Lithographers (NAPL), the National Association of Quick Printers (NAQP), and Graphic Arts Technical Foundation (GATF); each will be briefly summarized at the end of this chapter. These important industry-supported groups provide vital services to printing managers everywhere, including industrial and human relations, credit referrals, costing and production data, individual consultation on business matters, technical assistance, and many other valuable aids.

2.2 General Characteristics of the Printing Industry

Over the past 30 years printing plants across the United States have varied with respect to equipment availability, production output, management philosophy, skill level of personnel, labor and factory costs and in dozens of additional ways. However, there are seven generally recognized characteristics common to all firms engaging in the sales and production of printed goods across the United States.

The Printing Industry is Dominated by Smaller Businesses

While figures regarding the size of printing industry vary it is generally acknowledged that there were approximately 40,000 independent printing firms engaging in commercial and advertising printing across the United States in 1988, with approximately 50,000 operating plants. In addition, there were an estimated 25,000 in-plant printing firms who serve the graphic needs of a parent company (such as an insurance company or bank) and another 25,000 quick printing firms in convenient locations throughout the country. In sum, there are an estimated 100,000 firms engaged in some form of graphic reproduction.

Studies have shown that the majority of all these firms—commercial printer, quick printer, and in-plant printer—are classified as small and employ 20 workers or less. Over the years, the U.S. Department of Commerce statistics have indicated that printing leads all manufacturers in terms of the highest number of small business establishments. While differing estimates of company size exist, the average quick printer has about six employees, and the average in-plant printer has slightly over eight workers. Viewed from another perspective: The 1989 *Graphic Arts Monthly* "GAM 101: Official Ranking of the Top 101 Printing Firms in the U.S." shows the largest U.S. printer (R. R. Donnelley & Sons) with 24,000 employees, while the 101st firm has 175 workers. Since there are about 100,000 firms engaged in printing in the U.S., and the bottom point of the top 100 ends at less than 200 employees, it becomes clear that our industry is dominated by smaller firms.

The Printing Industry has Generally Low Entry Barriers

While this characteristic is not as true today as in the 1960s and 1970s, it is still possible to open and successfully operate a printing establishment on a modest amount of capital. Examples might include the person who buys a used duplicator press and, in his or her basement or garage, prints jobs during weekends. Or consider the young entrepreneur who, with little operating cash, begins to print T-shirts by screen printing techniques. Given such low barriers, hundreds of small printing businesses open yearly only to replace other such printing firms that were not successful and forced to close. Such business turnover is constant in the printing industry.

Printing is a Materials-, Labor- and Capital-Intensive Business

Literally all printing establishments operate in a material-intensive, labor-intensive and capital-intensive environment. Statistics from the 1988 PIA Ratio Studies demonstrate that approximately 35% of every sales dollar goes to material and outside purchases, with another 25% going to labor and factory costs. Thus, when summed together, approximately 60% of each sales dollar goes toward "hard" costs which do not contribute to profit and represent a "pass through" relationship, from customer to printer to supplier or employee. For example when the customer pays his $100 printing bill for an average job, about $35 will be paid directly to the paper company or other suppliers, and another $25 for out-of-pocket costs to the employees who produced the job. Since the payouts are pass-throughs, there is no contribution of these dollar amounts to profit. This concept—called value added—demonstrates that printing may not be extremely profitable unless a typical printer finds ways to improve the value of the product to the customer beyond the high material and buyout costs.

There is no question that the industry has become more capital-intensive in the past ten years as well. For example, in the 1970s the emphasis on "major" purchases for the typical printer was in the pressroom, and presses thus represented the focus of where we worked hardest to schedule production efficiently. However, in the 1980s, with increased interest in electronic prepress systems and sophisticated bindery operations, all areas of a printing company are now considered capital-intensive to most knowledgeable industry consultants and observers.

Finally, there is the issue of printing as a labor-intensive industry. While it is true that numerous labor-saving devices—particularly electronic equipment—are now commonly available to a printer of any size, instead of reducing the number of people needed to produce a typical printing job, the number of people has remained largely the same. Furthermore, these employees are no longer recognized as just "skilled craftsmen," but recognized more as "technicians" dealing with complex, highly sophisticated equipment.

Printing Is Both Manufacturing and Service. The actual production of the customer's product in the plant is manufacturing. However, the sales function and marketing aspects are essentially services to the customer. Such services might include ideas for a new brochure, design of a corporate logo, photographic work, or a new business form design—with arrangements for these services made and coordinated by the printing salesperson. Additional services—such as gluing in an insert, special wrapping techniques, or unusual perforation or scoring procedures—may be provided during product manufacture using special equipment. To thousands of purchasers of printed goods, such services are considered indispensable.

There Are Thousands of Demand Categories for Printed Products. Even to the most casual observer, this characteristic should not be surprising. All modern business establishments require printing—packages in the supermarket or drugstore, forms for the operation of an insurance agency, textbooks for education, and so on—

the list is almost infinite. In addition, many of the products are tailor-made specifically for a customer's own purposes and requirements.

Intense Competition Between Companies is Common

Owing to the five previous factors, the competition is unusually intense between printing companies. There are thousands of small printing businesses, each attempting to serve many demand categories, very few with exactly the same equipment array, all working in a materials-, labor-, and capital-intensive environment, with many unaware of their true costs or the pricing procedures, driven by numerous subjective issues. Furthermore, there is a constant turnover of new, smaller printing establishments that are willing to cut prices (perhaps below costs, if they know them) simply to survive and larger, high volume printers, with presses to fill and required volume needed simply to break even.

This intense competition is increased even more by the following additional factors which have become more clear during the 1980s:

Entrepreneurial vs Investment Community Philosophies. There are two uniquely different styles, or philosophies, under which all printing companies operate. The first—the entrepreneurial philosophy—is used mostly by small and medium-size firms. This style provides that the owner/operator has little interest in profit maximization, thereby justifying the owner to take any type of printing job, do it,then charge whatever he can get for the work, regardless of actual production cost, level of difficulty, value added, services provided, etc. The entrepreneurial philosophy is typically the type of strategy used by smaller companies to "make ends meet," and is generally measured at the end of a weekly (or other) period by determining how much is left in the company checking account after all the bills are paid. The amount left over is then available to the owners to meet their own, personal obligations.

On the other end of the scale are those printers who carefully operate their businesses in such a manner that they analyze all costs, have an identified marketing plan, coordinate production, estimating and sales, use computers to track job costs, and make use of dozens of other tools and procedures to ensure their success. They have been termed "investment community" because they have instituted such controlled procedures and techniques in order to grow; since getting bigger requires that the company obtain funds, the printing firm must be prepared to demonstrate solvency to the bankers and lending institutions.

Understanding the differences between the entrepreneurial printer and the investment community printer is critical to understanding the competitive pressures at work in the marketplace. For example, smaller commercial printers and quick printers have thus become much more oriented to competing with each other, and have blended together to represent to the customer the source for shorter run, fast-turnaround printing in the 1980s. In fact, both the quick printer and the small commercial printer attempt to track the same type of customer, requiring them to compete "head to head" causing undesirable price cutting in some cases.

The investment community printer, on the other hand, is not without his own competitive pressures from other large printers and publishers, in-plant printers who have a portion of their overhead costs covered by a parent company, foreign printers who began to move into the American printing scene with vigor in the mid-1980s (particularly on the west coast), and from mergers of two smaller printers to one larger one. In sum, competitive pressures exist for printing companies of all sizes and kinds, and with either operating philosophy.

Subtle Market Delineations are Closing the Gap Between Quick Printer, Commercial Printer and High Volume Printer. The 1980s has become a veritable open market situation for the printing industry because of the blending of quick printing, commercial printing and higher volume printers. Today the quick printer—producer of short-run, rush work—is directing an upscale move to produce fast-turnaround multicolor work by the early 1990s. Many commercial printers—largely using sheetfed press equipment—have abandoned the quick printer's "low-end" market, and have moved toward medium- and longer-run, process color plus coatings, investing in significant equipment and personnel to ensure their own market niches. The higher volume printer—the web printer—has begun to move toward specific products and quantity levels which fit the equipment currently on line, that can't be easily produced by the sheetfed commercial printer nor by gravure and flexographic processes, which are also competition.

As we enter the 1990s the commercial printer still competes with the web printer on some types of work, and the quick printer competes with the commercial printer as well. Yet in the longer term, by the year 2000 it is anticipated that the commercial printer may be "squeezed" by the quick printer on the low end and high volume printer on the high end, surviving through providing excellence throughout the business, reacting dynamically to customer needs and wishes and carefully selecting markets and niches where exceptional customer strength prevails.

The Level of Technology Varies Between Type of Printing Operation. Another view of the way the printing company of the 1990s operates is reflected in the relative level of technology incorporated in how printing companies produce and manage their companies. It is generally true that the smaller printer, due to his relative size and entrepreneurial nature, exhibits "low tech" dynamics. This smaller printer simply can't afford and may not wish to enter the technology race, largely oriented toward computers and electronic imaging systems. This view is similar for the quick printer, since a fairly sizable portion of the printing work they will do in the 1990s will be electronic via copying or laser-driven units. However, the quick printer will probably purchase very specific electronic equipment to serve very specific customer demands, making this equipment limited to a relative"low-tech" orientation other than repairs and maintenance.

The commercial printer of the 1990s will likely exhibit a "medium- or high-tech" orientation, purchasing whatever equipment is necessary to serve the markets and customers they have chosen. For example, it is conceivable that some commercial printers will continue to complete image assembly using manual, table-stripping, while others may use electronic mask cutting devices or perhaps a Gerber AutoPrep stripping system. These same commercial printers, to be competitive, will likely be forced to provide five- to ten-color, two-sided work, meaning their press equipment will be highly electronic and their operators highly skilled technicians.

The higher volume web plant will probably be an extremely "high-tech" business, with major capital investment driving the necessity to produce vast amounts of printing in a fairly fast turnaround mode. It is conceivable that the higher volume printer will find the need to do much of their work on a contract basis with customers, ensuring longer term printing volumes and fairly stable prices.

There are Defined Price Ceilings on Many Printed Goods

While there is no question that printers serving any given market niche could raise prices if they desired to, such price increases may jeopardize the profit picture of the business. For example, if the management of Plant A decided to raise prices even a slight amount, the competition at Plant B would use this price umbrella and underbid Plant A. This action would result in a loss of volume for Plant A, which translates into a loss in dollar profits. If the volume reduction is greater than the slight increase in price, the future survival of Plant A is not good. Such artificially defined price ceilings—particularly in demand categories where competition is intense—have the effect of keeping profits reasonably low, since price is a major competitor to profitability of the printing company.

Section 2.5 of this chapter deals with pricing and profitability in depth.

2.3 Dynamic Factors Affecting the Printer of the 1990s

While the market delineation between quick printer, commercial printer, and high volume printer is not completely clear, there are a number of factors affecting all or specific segments of the printing industry. It is these factors that may have significant bearing on the direction, size, and profitability for the industry into the year 2000.

The Overcapacity Issue

There is disagreement among printing managers regarding whether there is excess production capacity in the industry or too little volume. Generally the discussion centers around the capacity in the pressroom, particularly for higher volume printers who are largely web companies.

The advantage to any printer with the fastest or biggest press is that such speed or size capacity allows the company to more efficiently produce the product, thus reducing costs and thereby reducing price to the client. However such excess capacity also has a tendency to reduce price below cost in situations where a printer simply has purchased or owns too much press capacity and must "fill presses" to recover as much fixed cost as possible. Whether there is too much capacity or too little volume of work remains as a major issue for the 1990s printer, particularly affecting higher volume companies. As smaller printers grow, or as their markets become saturated—such as with the market-responsive quick printer—the issue of excess capacity even at lower volume levels may become a major problem.

Foreign Competition

There is no question that printers in other countries—particularly in Europe and Asia—have recognized the huge printing market potential which exists in America. The major advantage a print buyer may experience when dealing with an out-of-country printer is a substantially reduced price on the work performed; the major disadvantages include longer turnaround time and loss of control of the product, since it will be produced thousands of miles from the buyer's location.

The relative value and strength of the dollar has also had an impact on increasing foreign competition, as have legislative issues removing trade barriers, such as the cancellation of the Manufacturing Clause of the Copyright Act in 1986.

Changes in Web Production

As will be noted in Chapter 15 and later in this chapter, web printing has become a very popular process specialization. While there are many issues favoring web production over sheetfed printing, three points deserve mention. First, there have been major electronic improvements in web makeready systems and press controls during printing, thereby reducing both fixed costs and variable costs during production. This means less customer dollars to "get to press" and makes web lithography an extremely competitive process with longer-run sheetfed jobs. Second, there has been a movement toward full width webs and away from narrow width webs. This means that with one press revolution, the full width web can produce twice as much product as the half width web is capable of producing. Third, web press speeds have moved into the 2,000 feet per minute range with newer web presses, providing delivery of the printed product at an approximately 50% faster rate than with older web presses.

Accelerated Movement Toward Electronic Prepress Systems

While all areas of the printing industry have experienced the impact of electronic technology since 1980, the prepress segment has experienced the most change. Prepress includes all production areas prior to press and incorporates the production of artwork, camera production of film and other photographic materials, image assembly, proofing, and platemaking. For a plant producing process color using manual production techniques, the prepress area has been very labor-intensive and materials-intensive and less capital-intensive. Obtaining quality, skilled employees has been a major problem during the 1980s.

Electronic prepress systems have evolved through simultaneous development on three technological fronts: conversion from photographic typesetting to digitized typesetting systems (including desktop publishing systems) improvement of photographic scanning equipment from analog machines to fully digital devices, and the development of image manipulation systems, commonly known as Color Electronic Prepress Systems (CEPS) such as Scitex, Crosfield and Hell. The reality of such systems is that they have been very expensive to purchase and require extremely knowledgeable technicians to operate. For that reason CEPS have been utilized by trade suppliers specializing in providing completed prepress materials to the industry or by larger, "high-tech" printers who could afford the investment from a productivity standpoint.

By the year 2000 it is anticipated that the cost of such electronic systems will fall making these units more acceptable to the "medium-tech" printer. While opinion varies as to the ultimate impact this will have on all areas of the printing industry, there is no question that printers will have no choice but to consider moving in the direction of such technological advances simply to remain competitive.

Mergers and Acquisitions

The merger/acquisition process accelerated in the 1980s and is expected to continue into the 1990s for a number of reasons. First, foreign companies have recognized the advantage of buying an American printing business and currency rates have shifted so that it is economically advantageous for them to do so. Second, some printing firms, in trouble because of a poor cash position or for other financial reasons, are ripe for being purchased, with their counterpart cash-rich printing companies looking for something to buy. Third, more than ever before printing has been recognized as a way to make money, thereby attracting investors. Fourth, there has been an acceleration in leveraged buyouts, affecting printing as much as other businesses.

It is possible that mergers and acquisitions could reduce the total number of printing firms in the U.S. substantially in the next twenty or so years. There is no doubt that the result of mergers and acquisitions will have a significant, long-term bearing on the overall printing and publishing industry, perhaps reshaping certain markets and working to change the composition of the printing business as we know it in 1990.

Adoption of Statistical Process Control and Just-In-Time Management Systems

During the late 1980s, a growing number of progressive printing companies throughout the U.S. began to adopt two important concepts as foundations upon which they would operate in the 1990s: Statistical Process Control (SPC) and Just-In-Time (JIT) management.

SPC was first practiced in Japan and is based on W. Edwards Deming's quality control procedures, including his "quality circle" concept. The purpose of SPC is to improve the quality of any product by first establishing and maintaining a set level of quality conditions, then providing shop-floor employees with "tools" to ensure that the quality of whatever product they produce is maintained. Because SPC involves both statistical (numerical) and process control systems, it is ideally suited as a framework to analyze and evaluate the cost impact of almost anything quantifiable, once the levels of quality are established and proper tools are developed to maintain these levels.

Briefly, establishing an SPC system in printing involves four sequential steps:

1. Collect and analyze data or information related to the product or process;
2. Establish a group of desired results from this information;
3. Develop "tools," e.g. actions or methods, to ensure these results will be consistently met, and;
4. Train shop-floor workers in the use of these "tools" so that mistakes are minimized, quality of product is improved, and errors are reduced.

Just-In-Time (JIT) management actually began as an inventory concept, and has also been practiced in Japan for some years. JIT is a method by which the business is operated on an "immediate need" basis, whereby paper is ordered today for use tomorrow, or a customer orders printing in the morning for afternoon delivery. With customers demanding an increased level of printing service and more printed products in a reduced time, the JIT concept fits well into the printing industry. Not only does JIT require that the printer become more responsive to the customer, which may lead to increased sales volume, but costs can be reduced due to increased production in less time.

Many progressive printers are combining SPC and JIT so that the printed work they produce is done at an enhanced quality level (the SPC portion) in a faster turnaround environment (the JIT portion). Most printing industry experts predict that both SPC and JIT will be fairly common in the printing industry by the mid-1990s, since each addresses critical areas which the printer must control if he is to remain competitive.

2.4 A Marketing View of Printing

Product and Process Specialization in Printing

The 1966 "McKinsey Report to PIA" is considered to be a cornerstone investigation of markets and pricing strategy for commercial printing. In general, the report stated that profit variation between printing companies was not always related to cost, volume, or price—even though each of these factors played a part in the profit picture—but to the market demand, services provided by the printer, and factors of competition experienced by companies vying for work in those markets. The study stressed that printing management should consider marketing as well as production manufacturing for enhanced profitability since the industry was market driven. Almost 20 years after the "McKinsey Report" was released, its findings are still considered relevant to commercial printers since there has been little change in the competitive factors at work in the industry.

In terms of products, the "McKinsey Report" found that the demand for tailormade printed products was diversified, but that no printer could reasonably expect to provide all products and services for all customers, even though many seemed to try. In terms of procedures used for manufacturing printed products, McKinsey reported that there was extreme diversity in process: Some equipment was automated and high-speed, while other equipment was slow and required the individual operator's attention. In fact, the study observed that printing companies were beginning to branch out with the addition of nonimaging processes. Conventional printing techniques of ink-on-paper still were used during printing, but extra equipment, such as special wrapping equipment, was being installed to provide additional services as an enticement for the customer's business.

The "McKinsey Report" concluded that market diversity and process variation both contributed to the significant and intense competition in the printing industry. Thus, if printers practiced specialization of product or process (or both), the intense competition between them for the same types of work would be lessened. If a printing company could enter a specialized market and control it, they could then control prices in that market. If such prices were under control, profits would be greater since price exerts tremendous leverage on profitability. Specialization is a way out of the intense, price-cutting free-for-all that most top managers dislike.

Essentially, product specialization is a marketing concept, while process specialization emphasizes the manufacturing component. The "McKinsey Report" stated that this dichotomy—these contradictory segments—in the printing business was an important element for consideration by printing management. McKinsey found that the significant majority of printing company top management who considered marketing and service as their principal lines of business and manufacturing as secondary were high-profit printers. Those top managers who considered manufacturing as first and marketing and service as secondary lines of their business demonstrated an inclination toward low or average profit levels.

Product specialization in printing is the production of an item difficult to produce economically by most other printers. Process specialization relates to the application of a unique process to accomplish production output of a printed product. An example of a specialized product is the manufacture of theatre tickets. An example of a specialized process is the use of shrink-wrap to seal packages for delivery after printing and finishing are completed.

It is possible to specialize in a product line using nonspecialized printing equipment, as many printing plants do. It is also possible to specialize in a process area such as web printing and produce a nonspecialized product. In recent years, analysis has shown that a combination of both product and process specialization yields the most profitability; it provides the company with excellent market leverage and high production output.

Advantages of Specialization. The following list outlines some advantages of product and process specialization:

1. Competition, in terms of large numbers of companies doing the same type of work and producing the same kind of product, is significantly reduced.
2. Specialized products and processes allow for definite cost advantages such as higher volume and reduced preparatory costs, thus providing price leverage over the competition.
3. Production during manufacturing is usually streamlined.
4. Production control and quality control, which assure a consistency of product and may be vital facets of the product produced, are simplified.
5. Employees working with specialized products and processes usually have narrower job assignments. This situation can mean reduced job errors and less on-the-job training.
6. There is a tendency to have less equipment, even though it is more specialized.

Disadvantages of Specialization. There are also disadvantages associated with specialization, as the following list indicates:

1. The more specialized the process or product, the greater the built in-inflexibility. Printers cannot serve all markets and all customers' needs. Thus, there is a tremendous dependence upon the market demand of their particular product or process.
2. While the number of competitors is reduced, those who remain can provide tremendous competitive pressure. If the market for the specialized product or process is not broad enough, there may eventually be too few customers for even a limited number of firms. The inevitability of price cutting then results.
3. Obsolescence can be a problem. One major technological advance can destroy a specialized product or process. In addition, the developed skills of craftpersons that have been specialized must change along with technology, perhaps necessitating expensive retraining to run the new specialized equipment.
4. In many cases, specialization requires dependence upon a few suppliers. While it may be beneficial to deal with this limited number, suppliers may also have specialized markets to control their pricing structure. In the long run, supplies and materials may be more, not less, expensive.
5. There is a human tendency to resist change, even if the product or process specialization is not longer profitable. Firms locked into a specialty may have a tendency, because of the initial very profitable years, to continue to do the

same thing over an extended period of time and may pass up changes in technology and products. Over this time period, profits and sales may begin to decline as the competition shifts to a more modern or different process or product specialty. Eventually, it could become necessary for the company to file for bankruptcy or to merge because of this resistance to change.

The chart in Figure 2.1 identifies the relationships between customers, product assortments, and markets when considering specialization. Marketing investment costs increase as the company changes from present customer/present market to new customer/new market. Investment costs in new equipment and technology increase as present products are changed to modified products, which later become new products.

	Present Products	Modified Products	New Products
Present customer in present market	present customers present products present markets	present customers modified products present markets	present customers new products present markets
Present customer in new market	present customers present products new markets	present customers modified products new markets	present customers new products new markets
New customers in present market	new customers present products present markets	new customers modified products present markets	new customers new products present markets
New customers in new markets	new customers present products new markets	new customers modified products new markets	new customers new products new markets

Increase in technological investment costs →

Increase in marketing investment costs ↓

Figure 2.1

Comparison of customers, products, and markets with respect to costs

It is critical to note that each industry segment—quick printer, commercial printer and high volume printer—already product-specializes within their own competitive area. Thus, Quick Printer A competes with Quick Printer B since each produces fast-turnaround, plain-paper copies, even though A may provide certain customer services not found with B. However, Quick Printer A doesn't compete head-on with Commercial Printer A because each has identified different and specific market niches or specialties, and each is geared up to serve those defined markets.

Of course, it is possible that some overlap between industry segments, by product specialty, could occur. For example, a commercial printer might have identified annual reports as a market specialty and a high volume printer might be pursuing the same market. In such a case, intense price competition is a likely result, even though the costs incurred by the commercial printer and the high volume printer may be quite different. The result is that one company will be forced to leave the annual report market over time, or each company will identify a particular market niche within the general annual report market best suited to their company production processes.

It should be noted as well that market specialization naturally leads to manufacturing standardization. For example, quick printing is geared to smaller press sheet sizes, with such reduced product sizes leading to similar paper handling techniques, common sizes of all products, etc. In sum, specialization and standardization go hand-in-hand, providing additional impetus for any printer to consider market specialization.

An Overview of Printing Sales, Sales Representatives, and Customer Service

More than ever before, printing management is aware of the importance of effective sales and focused customer service efforts. In an industry that has a history of strong production management—in fact, a production management orientation—the industry today is an integration of modern marketing management and sophisticated production management and control. The evidence is increasingly clear that the printer in the 1990s—quick printer, commercial printer or high volume printer—must deal effectively with marketing, sales, and customer service issues for even average business success.

The general appearance, personality, and drive of the printing sales representative are vital to his or her success. He or she is the company representative outside the plant and, in some cases, is the only person to whom the customer or much of the public talks. Because the sales representative must embody the company in a positive, honest, and friendly manner; the selection and training of one should include emphasis on these aspects of the job, as well as on a clear understanding of technical and production elements of printing manufacturing.

The intervention of customer service representatives (CSRs)—common with higher profit printers who generally produce complex products and desire customer closeness on a "customer as family" basis—has lead to somewhat of a changing role for the sales representative. In sum, while the job duties of CSRs vary from company to company, the sales representative is relieved of having to be extremely sophisticated in printing technology matters, of having to know complex job details, and of having to work closely with customers on a daily basis. Thus, the sales representative is released to focus efforts on selling and generating new business, as opposed to selling and maintaining accounts, with each client at a different manufacturing step for its printing. Also, with the CSR handling details and nuances of production, and working closely with the customer, the 1990s high profit printing sales representative can concentrate on marketing issues, generally done in concert with top management.

The normal sales sequence begins with the sales representative contacting a selected client, commonly known as a print buyer. It is vital that the sales representative learn as quickly as possible what person in the selected client company has buying authority. Selection of customers—determining the markets to approach or who to call on—is essential to effective sales as well as enhanced company profitability. Smaller printing companies may follow a pattern where the sales representative decides the markets to be approached, or the owner/operator of the company may make this decision with input from the sales staff. Medium size and larger printers typically have a sales manager, upon whom marketing decisions largely rest.

In most high profit printing operations, selection of markets and clients is based on the development of some type of customer/product/profitability profile through careful analysis by the sales manager, the owner/operator of the company, or top management. Such a system is emphasized in the 1966 "McKinsey Report to PIA" and has become even more emphasized through the 1980s as printers have become more familiar with identifying market niches instead of providing "total printing services" to every client. Some important factors for selecting customers are those who pay promptly, have good credit, and submit work that fits the printer's developed production niches. The actual development of a marketing plan—represented through a client list—is not easily executed yet represents one of the most important jobs of top management in the successful printing company.

For a printing sales representative meeting the customer, e.g. print buyer for the first time—perhaps in a "cold call" situation—the sales rep introduces the company's line of products and the print buyer identifies the type or kinds of products they desire. If the buyer and sales representative agree on a product, discussion follows as to the job specifications needed for estimating. Included in the job specifications are such factors as type of paper, size of the final product, quantity, colors, and artwork production; it is vital that the sales representative determine any customer-supplied parts of the job. Furthermore, it is essential that the printing sales representative obtain complete job specifications prior to leaving the print buyer's office.

At this point the sales representative returns to the printing plant and requests an estimate and price proposal, based on the job specifications. When this step is completed, the sales representative returns to the print buyer and presents a price proposal outlining all aspects of the potential printing order. The print buyer may then bargain, accept, or reject the price proposal. After the selling price has been agreed upon, a quotation is issued, and when accepted by the customer, the job begins production. Once the job is finished and delivered, an invoice is sent to the customer. The sales representative normally receives a commission for services provided, which may be based on the sales price of the product, the profitability of the job to the company, or on the value added basis of the job.

As a general rule, print buyers purchase printing on the basis of three overriding criteria: service, quality of work, and price. From the print buyer's viewpoint, service relates to meeting delivery dates on time, as well as to what types of help and aid accompany the printing in the form of services provided by the printer. For example, the sales representative or the customer service representative (if the printer has one) might work with the buyer to procure camera-ready artwork, arrange for color separations, help specify typesetting, photographic or other special production services, coordinate the design of the product so that a group of matching products can be produced—practically anything accompanying the production of the printed product. The meeting of delivery dates—perhaps the most critical element in customer service—is important because such established dates, usually determined around the time the job is estimated, directly relate to the customer's own planning base into which the printed product must fit at the appropriate time.

The 1966 "McKinsey Report" stressed the importance of detailing the specific services provided when the printing job is sold. McKinsey noted that many customers purchased printing initially, and did significant amounts of repeat work, based on the quality and quantity of services provided and not just the printing. During the 1980s various informal studies of print buyers have come to the same conclusions: give the print buyer the kind of quality and promptness he wants and price becomes a secondary consideration. In sum, deliver service and acceptable quality, and price is a distant third in the purchasing process.

Quality is the second purchasing consideration made by most customers. When the desired level of quality varies from that normally provided by the printing company, it becomes an important element of the work. For example, if the customer wants work quickly and is not concerned about the visual appearance to any great extent, or if, in contrast, the customer wants to order a process color annual report on coated stock and desired a polished product, these factors should be discussed during initial sales contact. It must be noted that since quality level is a matter of relative judgment, the customer and printer should agree in writing as to what standards are applicable. If this is not done, disagreements can easily occur after the job is printed and delivered. Depending on the use of the product, some customers will purchase printing based on the quality of the work alone and disregard price and service aspects.

The selling price quoted to the customer offers perhaps the most debated and continually discussed controversy of the three buying criteria. Numerous studies over the past 25 years have shown that the selling price quoted on a printed product—estimated from the same sample and for the same quality and quantity levels—can vary tremendously. It is obvious that some printing companies would be making a substantial profit if they were awarded the work, while others would not even be recovering costs if they were given the job. The net result is that pricing is not a simple matter. Pricing philosophies vary among printing company management, sales managers, sales representatives, and print buyers. For the most part, the sales representative and customer discuss price, many times in terms of price related to the buyer's budget allocation. Negotiation of price can be done exclusively by the sales representative, in consultation with no one else; however, most printers—perhaps the majority—allow the sales representative to serve as a go-between in coming to a final price, with the buyer on one side and the sales manager or company owner/operator on the other. Price becomes very important in times of economic inflation, when dealing with a contract where the price is set for a longer term, with government agencies who purchase printing, and, of course, when the buyer must work within a certain budget.

2.5 Pricing and Profit in the Printing Industry

Defining Selling Price and Profit

The *selling price* is the dollar amount assigned to the printed product that the customer will pay. It should include all fixed and variable costs, all material costs and buyouts (plus markups), and the addition of profit. If all costs are not fully recovered, if materials and buyouts are not added in, and if a suitable profit is not included, the selling price of the product will be too low and the company will lose money. When and if such losses occur on a consistent basis, assuming no additional sources of revenue are available, the company will be forced out of business through bankruptcy or coerced into merging with another company that will serve as a new revenue base.

Profit, the dollar amount added to costs, is vital to the business. Profit dollars are the incentive that the investors in the business receive for their involvement and risk. Profits are used for the growth of an operation. If there is little or no profit, usually signifying that the risk is not worthwhile, there will be no growth and investors and stockholders (if any) will leave the business. Profit dollars are shared by owners and stockholders. In addition, they are distributed indirectly to all employees in the business, not just top management, through improved working conditions, higher wages, and increased fringe benefits.

In the selling price formula, profit is the most flexible of all the inputs. Thus, while costs are defined as precisely as possible, the variable when pricing is how much profit to add to the summed costs. This subject creates considerable controversy in all segments of the printing industry. In fact, profit margin is a matter of concern to all businesses, not just printing establishments.

Numerous authors writing on the subject of printing management have clearly stated the importance of pricing for profit for printing management. Spencer Tucker in *Creative Pricing in the Printing and Allied Industries* (North American Publishing Company, 1975) states that "pricing is the sole profit generator, the function that provides the profit opportunity which is enhanced or diluted in the printing company's internal operation." The late Thomas Hughes, a graphic arts consultant who wrote the book *Profit Leadership in Printing* (North American Publishing Company, 1976), states: "The proper policy for an effective level of pricing comes about to make or continue a level of profitable operations, or to aid in the continued growth of the firm." The key to profit is pricing.

It should be noted that some businesses neither desire nor are outwardly motivated by profit. Such enterprises as churches, social groups, foundations, and charities expressly state that they are nonprofit organizations. Both state and federal law require that the function of the group be clearly stated prior to the use of the nonprofit classification.

A final point should be made clear. Determining price and profit are the major duties of the top management of any printing company. No one else—accountant, estimator, salesperson, or any other individual—should be making such important decisions. If top management abandons the function of pricing, they have removed themselves from a most vital element of the business. They have neglected one of their prime directives and functions.

Many variables enter into the establishment of price on the printed product. Since the overriding subject of this book relates to accurate cost determination, it must be stressed that the general costing procedure in the printing industry—practiced by quick printer, commercial printer, or high volume printer—is to determine costs prior to pricing, then use these costs as a minimum basis for price. In effect, costs—fixed, variable, material, and buyout—added together serve as a bottom line figure in making a price decision.

A cardinal rule for survival in any area of printing is that no printer should ever accept any order that will not cover his or her own cost to manufacture the product. To follow this rule, of course, the printer must have the ability to accurately establish such costs.

Management of any printing establishment must make every attempt to maximize selling price and thus increase profits while maintaining a suitable volume of work. If the price is too high, customers will elect not to pay it (even if it is "at cost" to the company). If the price is too low, the company will be swamped with orders but will make no profit. One vital key to success in the printing industry—quick printer, commercial printer or high volume printer—is knowing precisely the costs to produce the product, then using an effective pricing/profit strategy based on customer wishes, needs, and ability to pay

Pricing Determinants in the Printing Industry

The price for printed goods is a function of many factors. Some of these factors include the market demand for the product, manufacturing costs, product mix of the printer, price offered by the competition, quality of the product, services provided in addition to printing, quantity required, current volume of business in the printing company, and profit desired. These and other factors will be discussed below.

Managers of printing companies use varying combinations of such determinants to set price. Some managers consider cost and competition as primary factors when pricing their work, some may apply volume data and quality level when establishing a price, and some may use various, selected criteria on a job-by-job basis, such as the customer's ability and willingness to pay for the work.

It is vital to note that price is a customer-driven, market-related factor. The selling price charged for printing becomes the cost of goods to the customer. Such costs must fit into the customer's normal business plans and budgets. Dollars as selling price to the printer translate into dollars as business costs to the customer.

Market Demand for the Product. Market demand for printing products may be either elastic or inelastic. *Elastic demand* means that as price increases, the volume sold decreases, and as price decreases, the volume sold usually increases. *Inelastic demand* signifies that price is not necessarily related to volume: Prices may rise or fall and there may be no direct reduction or increase in volume. The printing industry characteristically has an elastic demand structure. However, when a printing establishment specializes in a particular product line or item not easily obtained from other printers, the market becomes inelastic. Typically, the more inelastic the market, the greater the control of prices.

This determination in effect, was a major conclusion of the 1966 "McKinsey Report to PIA." Specializing provides entrance into markets wherein prices can be controlled. If such prices are then raised, there will be limited volume loss. McKinsey noted that when prices were increased 3% in an elastic market, volume decreased by 10%, yielding a profit loss of 50% (Figure 2.2). For the printer in an elastic market who makes 5% profit, an attempt to increase prices by even a small amount could be disastrous. In the same study, it was noted that price has tremendous leverage over profit, in fact, three times more than other variables, assuming the market served is inelastic. As Figure 2.3 shows, in an inelastic market, if material costs were reduced 10%, then profits increased 66%; if labor costs were reduced 10%, profits increased 56%; if volume was increased 10%, profits increased 33%; but if price was increased 10%, profit increased 180%. This finding is significant because it clearly demonstrates the relationship of marketing to price and profitability.

Additional factors enter into the market demand category. The most important factors here are the customer's ability to pay for the product and willingness to pay the desired price. Such factors are difficult to determine with respect to the customer but are certainly important considerations.

Manufacturing Costs. The necessity of accurate cost estimating, which provides the total cost to manufacture the printed product, has been sufficiently stressed. As stated already, total manufacturing cost must serve as a bottom line price for the goods produced. In addition, it must be noted that there is no established relationship between total manufacturing cost and the selling price of the order from one company to another. For example, one plant may decide to double the total manufacturing cost figure to calculate the selling price of an order, while another may add a very small amount to the total manufacturing cost for final selling price determination. Thus, there is no defined procedure or formula for this cost-price relationship: it is determined on an individual plant basis. In fact, if printing companies in a geographic region attempt to establish prices together at a fixed level, which reduces the competitive relationship, unlawful price collusion may be charged.

Product Mix. With changing technology and manufacturing processes, there have been tremendous modifications in product mixes available through specific printers. Product mix follows the elastic-inelastic market relationship to pricing as discussed

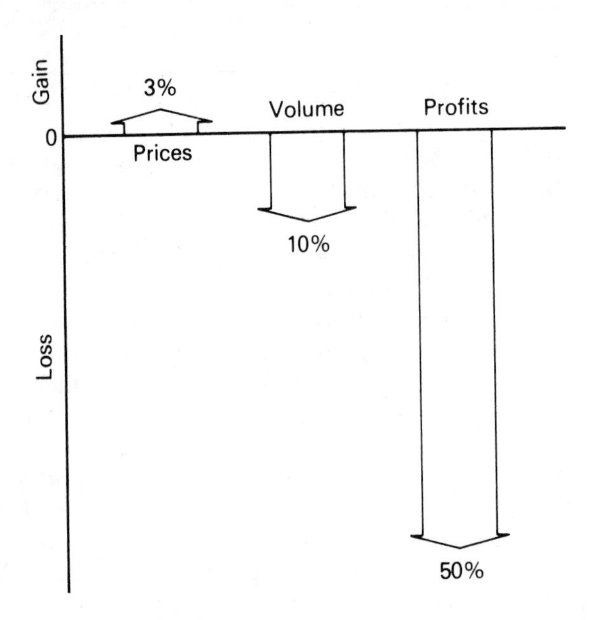

Figure 2.2

A price increase with a resultant volume loss for the commercial printer in an elastic market (Reproduced from the "McKinsey Report to PIA," with permission from Printing Industries of America, Inc.)

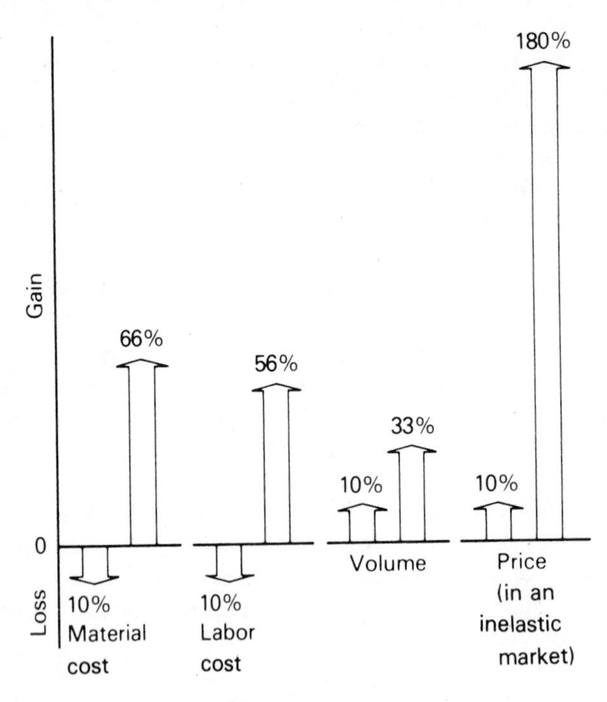

Figure 2.3

The leverage of price over profit for the commercial printer in an inelastic market (Reproduced from the "McKinsey Report to PIA," with permission from Printing Industries of America, Inc.)

previously. But product mix has another variable. Even with a narrowed, specialized product array, each particular product may contribute differently to the overall profit of the firm. Because of this fact, it is wise for management to investigate and determine the profitability of all products in a specific line or mix. Once this determination is done—and it is neither a precise nor simple procedure in many cases—the products that provide the greatest profitability should be emphasized through sales efforts. The less profitable products, depending upon ease of production and level of profitability among other factors, may continue to be produced, may be dropped, or may be purchased as a buyout from another printer.

Competition. For too many years, and even to a great extent today, printers priced their work by what the person down the street was charging. As previously discussed, this method is a poor way to price for two reasons. First, it is difficult to determine accurately the prices charged by the competition. Second, and more important, there is generally little similarity between companies. They vary in terms of type and size of equipment, labor skills and rates, management and sales salaries, and production orientation. Even minor aspects of two apparently similar companies may be different. Printing management, when pricing by competition, essentially uses an inaccurate basis—one that does not represent their own business costs and values. They fundamentally hurt their own profitability while they simultaneously reduce the profit level of surrounding printing establishments.

Quality of Product. The desired level of quality of the printed product, which the customer determines, may have a significant bearing on his or her willingness to pay for that product. The greater the value the customer places on quality, the greater the selling price relationship. Quality is, of course, a matter of judgment, and the determination of what is good, excellent, or poor must be decided from initial discussions between sales representative and client. The higher the quality, the more production time normally required, which increases manufacturing costs. 'The lower the quality, the less production time needed, resulting in a reduced cost to manufacture.

Services Provided to the Customer. As noted in the "McKinsey Report" the quantity of services made available in addition to actual printing provides a tremendous lever on price and profit. Such services relate to two areas:

Those provided by the sales force and affiliated plant personnel in design and other preliminary work on the order,
The additional services during the manufacture of the product.

For example, the sales representative may redesign an internal business form for the customer as a requisite service for printing that same form. In addition, after the form has been printed, it may be collated in a special manner or wrapped in special materials for shipment or storage.

There is no doubt that services provided at the beginning, during, and after production related to the final product will provide a significant price lever for the printing company. In fact, the "McKinsey Report" indicated that providing services to the customer was a key to being a high-profit printer. As demonstrated in Figure 2.4, the preponderance of company presidents who viewed their business as a service were high-profit printers, while most of the company presidents who viewed their business orientation as manufacturing were low-profit printers.

Quantity of Work to Be Printed. Production of any printed product—that is, the actual manufacturing of the product—requires a high fixed cost at the outset, even before the first piece is printed. After the order is put into production, costs for preparatory work begin to mount. Artwork, typesetting, photographic work, image assembly, and platemaking all must be completed prior to the actual production of the first copy. Then, as the job goes on the press and the initial copy is printed, that single copy bears the total preparatory cost burden, plus the variable cost for labor and materials (in this case, one sheet of paper and press operator's labor cost.) With the production of the second press sheet, the fixed prepatory cost is cut in half, but the labor and material cost—that is, variable cost—increases by another sheet of paper and the associated time for labor. As the number of copies continues to increase, the fixed cost per copy decreases as it is spread over the larger quantity, while the variable costs increase in proportion to the quantity produced.

The net effect is that the total cost to manufacture a printed product is not in direct proportion to the volume produced. For example, assume 1000 single-color letter-

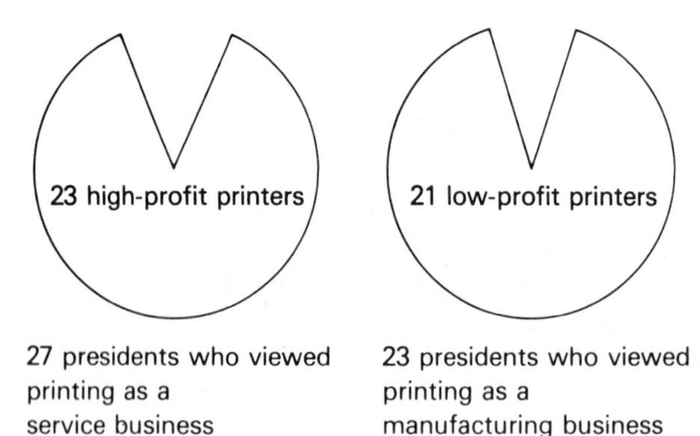

Figure 2.4

Relationship of service or manufacturing orientation to profit (Reproduced from the "McKinsey Report to PIA," field interviews, with permission from Printing Industries of America, Inc.)

heads cost $100 to manufacture. If this order is inreased to 2000, the cost does not double (to $200) but may be only $175. Even though the quantity was doubled, the total cost was not doubled because fixed costs were spread out over this larger quantity. Of course, material costs for paper and labor do increase in a direct proportion to the quantity produced.

To study this concept—especially with printing presses where equipment costs are tremendous and press size sheets are so changeable—printing management utilizes the break-even or changeover point concept. This concept is explained in detail in Chapter 12, where the selection of printing presses is discussed, but Figure 2.5 is presented here to demonstrate graphically the changeover point method.

Break-even or changeover analysis is a common procedure for comparison of profitability between different pieces of equipment. In Figure 2,5, a four-color Harris sheetfed press and a four-color perfecting web offset press are compared. The sheetfed press offers dollar savings up to a quantity level of about 60,000 units of production. After that level, the web press provides greater savings in costs and, thus, greater profitability. Actual chart development is based on accurate prepatory (setup) cost calculations and establishment of a uniform basic unit of product.

The effect of volume on price, with cost related to price, is an important concept for printing management. Analysis should be made on quantity levels for each customer's products. Generally, the greater the quantity desired, the greater the price, with a reduced cost per unit. This statement means that while the customer pays more for a greater quantity, the price paid is actually less per unit as quantity increases.

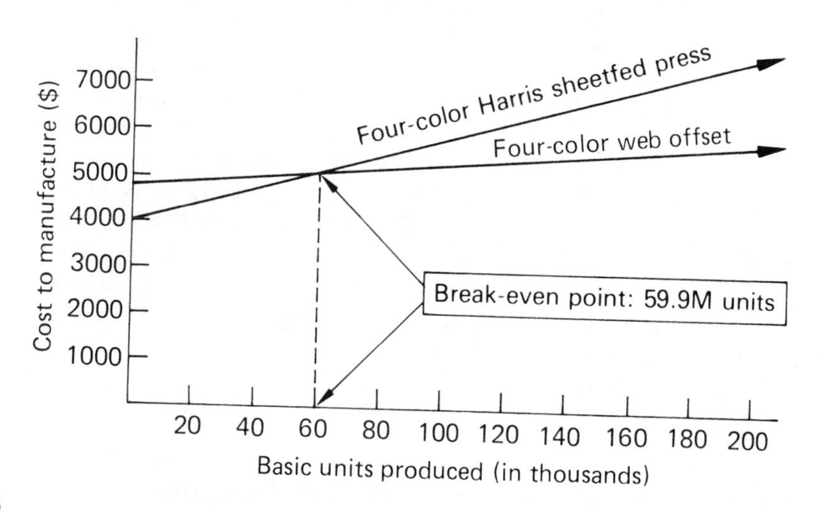

Figure 2.5

Example of break-even/changeover point analysis

Volume of Work in the Plant

This relates to how busy the printer is at the time the job is estimated and the quotation generated for the customer. If the printer is exceedingly busy, the tendency is to increase the price since the work will represent production anticipated to be beyond the normal capacity of the plant; if the printer's volume in the company is low—business is slow—then the price may be lowered to encourage the client to purchase.

When volume is a factor, the timing as to when the job will arrive and begin production becomes critical. For example, the estimate/quote may be completed when the company volume is low, and the price dropped to encourage the work. However, the job for which the price was dropped may not actually enter production for perhaps weeks or even a month after the proposal is accepted, and when it finally begins production, the company may be running at capacity. It is therefore important that dates of production and delivery be factored into the pricing equation should volume in the company be considered a major factor in setting price.

Customers Willingness and Ability to Pay for the Job

Based on a number of informal studies with printers across the country during the 1980s, pricing printing by "what the market will bear"—gauged by the customer's ability and willingness to pay for printing—is fairly common. This is particularly true as related to smaller commercial printers and quick printers, where the entrepreneurial philosophy of business operation is used. In sum, the entrepreneurial printer does not attempt to maximize profit—only to get the job and attempt to produce it at a financial gain. Thus, this type of pricing procedure fits well in the smaller printer's philosophy.

How much the customer is willing to pay is, in many cases, difficult to determine without some input from the customer; negotiation of price may represent the most common method to determine this aspect. Furthermore, whether the customer has the ability to pay a particular dollar amount for a job can only be determined as the printer and customer work together over time. Thus, whether a customer has the "willingness and ability to pay" is best evaluated with repeat customers, which commonly represent more than half of the client base of the typical smaller printer or quick printer.

It is fair to say that if a printer believes the customer has both a willingness and ability to pay for a particular product or job, the printer might decide to charge the highest possible price he can achieve. If doing this, the good or service—in this case the printed product—must be tailored specifically to the customers needs, since various state and federal laws may preclude the printer from charging different prices for exactly the same good or service.

Potential Future Business from the Customer

Similar to "what the market will bear" pricing, adjusting the price based on new or anticipated other printing is fairly common. Usually the customer informs the printer that he has a growing business, and if the printer will keep the price lower on the current work, future printing will go to the printer. Again, this pricing process is more common with smaller printers and quick printers.

The disadvantages of pricing by this method are obvious. First, the printer may be doing the current work at cost or even at a loss, simply to encourage future business. By the time that work is available—if it ever is—the printer may be out of business. Second, promises are typically not a basis for a legal contract, therefore the customer's statements are only promises, not facts. Finally, there is no reason for the printer to believe the customer will ever return to place those future orders.

It is true that encouraging business through astute pricing is one way to grow the business. This can be done through sales promotions or other special pricing systems that are offered on an infrequent basis. Another way to address the "future business" issue is to enter into a longer-term contract with the customer, locking up a specific type of product or quantity of work in a signed agreement.

Value Added. To many in the printing industry, value added is the amount of dollars a product increases (in worth) due to the various manufacturing and service components which improved it. Some define value added more quantitatively: total dollar value of the product less material costs and other directly chargeable costs. While it is not a defined measure of profit performance—as are ROS and ROI—value added does represent an important point of view for printing management in that the profit is achieved from sales of products manufactured by one's own company. Sometimes value added is called inside sales since it represents sales dollars generated inside the plant, as opposed to sales dollars used for materials and other outside-the-plant requirements.

Printers should always try to maximize their value added by working to reduce material costs, chargeable costs, and buyout charges (perhaps by beginning to do more trade work inside their own plant when practical). For example, Printer A sells a job for $100, $38 of which can be directly traced to paper ($23), film and plates ($8), and buyouts ($7). The value added for this job is then $100 minus $38, or $62. Printer B sells the same job for $100 with $30 traced directly to paper ($20), film and plates ($6), and buyouts ($4). Printer B's value added is $70. Thus, Printer B has $8 more than Printer A to contribute to overhead costs, other costs, and profit. In all likelihood, Printer B is a more profitable printer than Printer A.

With the value added concept, it is possible to have lower total sales dollars but greater profit than the competition. Company A has total monthly sales of $175,000, which produces $90,000 in value added. Company B has monthly sales of $125,000 and a value added of $100,000. Even though the total monthly sales of Company A is $50,000 higher than Company B, Company B has $10,000 more than Company A to cover overhead and for profit. Company B appears more profitable than Company A, which should reevaluate its material and buyout cost relationships. Company A might then decide to buy paper and other materials in larger quantities to take advantage of quantity discount or perhaps begin to do more work directly on some jobs, as opposed to sending the work to trade houses as outside purchases. Keep in mind that the value added figure is improved anytime printers do work inside the plant with their own equipment and personnel instead of sending it outside the facility. Value added banks on increasing internal production and reducing buyouts when quality or technical factors are not a problem.

Some printers have begun to use value added for salesperson compensation. This procedure provides for sales representatives to be compensated in direct relationship to the value added of the jobs they sell. The more the value added, the greater the commission; the less the value added, the less the commission. Since it is reasonably easy to determine the cost of paper, other chargeable materials, and buyout costs—one of the advantages of the value added concept—the procedure of determining value added allows the sales representatives to have a good idea when the job is estimated how much commission they will earn. For this reason, it is easy for sales representatives to direct their efforts at jobs that have a high value added, meaning improved profit to the company as well.

Profit Desired by Top Management. Profit provides incentive for the owners of any business. It is the reward for risks taken with invested capital. It allows for the growth of the business and for technological changes and advances. It provides for a general sense of security in the company that is shared by employees and management alike. Continual profit means continued stability in the operation of any printing business.

In raw figures, profit data may be categorized into gross and net figures. *Gross profit* is the amount of dollars made by the business before the subtraction of administration and sales expenses and salaries (and taxes). It reflects manufacturing costs for the product and not the cost of sales commissions or costs necessary for administration of the product. Gross profit in the printing industry normally averages between 8% and 24% of sales. *Net profit*, also known as *income*, is the amount of dollars remaining after all costs, including sales and administrative overhead, have been deducted. Net profit is the dollar amount on which taxes to both state and federal governments are paid and is sometimes referred to as *net profit before taxes* or *income before taxes*. Normally, the net profit in the printing industry falls between 3% and 7% of annual sales.

Two primary methods are used to measure profit in the printing industry: income (profit) as a percentage of sales and income (profit) as a return on investment. Income as a percentage of sales, also termed return on sales (ROS), is popular because it is a simple concept to understand and has traditionally been used. In 1988, the average net profit (before taxes) for all 874 respondents of the PIA Ratio Studies was 3.53% of sales. This statistic means that for every dollar the average printing company sold in products, it kept 3.53 cents. Thus if an average printing company had a total annual sales volume of $1,000,000 in 1988, and averaged 3.53% as a return on sales, it had before-tax profits of $35,300. It is important to emphasize that taxes have yet to be paid on this amount—which vary, but may take up to 50% of the dollar figure; the remainder would then be distributed to the owners or shareholders of the business.

While the percentage of sales method is a simple concept and easy to execute mathematically, it does not allow for a crucial element in the printing industry—the cost of investment. Because a printing company is so capital intensive—that is, because it requires a considerable amount of money to establish and operate—the return on investment concept of measuring profits has grown in popularity. Return on investment (ROI) is not quite as simple a measure of profitability as the percentage of sales concept. However, it is considered to be a more accurate measure because it relates net income (profit earned before taxes) to the assets (investment) required to produce that income.

Thus:

ROI = net income ÷ gross or net assets

There are two methods to calculate ROI: gross asset and net asset measures. Essentially, the difference between the two relates to the *depreciation* of the piece of equipment under consideration—that is, the loss in dollar value from use. *Gross asset* ROI makes no adjustment for depreciation, treating each piece of machinery as if it were a new investment, regardless of the year of use. *Net asset* ROI subtracts depreciation loss. Consider the following examples for a small lithographic duplicator.

The gross asset ROI method uses the following formula:

gross asset ROI = net income ÷ gross assets

In this method, the investment cost remains the same from year to year, but profits diminish due to reduced equipment efficiency and wearout. Thus, ROI is reduced from 12.5% to 5% over the eight year period of use, as shown by the following data:

Age of Duplicator	Income (profit)	Investment Cost	ROI (percent)
1 year	$1000	$8000	12.5
3 years	800	8000	10.0
8 years	400	8000	5.0

The net asset method uses the following formula:

net asset ROI = net income ÷ (gross assets – depreciation)

Thus, with the net asset method, depreciation reduces the investment, while income falls in the typical manner:

Age of Duplicator	Income (profit)	Investment Cost	ROI (percent)
1 year	$1000	$8000	12.5
3 years	800	6000	13.3
8 years	400	2000	20.0

However, for this example, the investment return on the duplicator in the eighth year is better than any previous year, including when the press was new.

On the one hand, it must be noted that there are numerous depreciation methods, making precise comparisons of net asset ROI data inaccurate. On the other hand, since depreciation is not included in gross asset ROI, comparison of data using this ROI method is more common.

Figure 2.6 is a ten-year comparison chart showing net income (before taxes) for "All Firms" and "Profit Leaders." Data in this chart is taken from the PIA Ratio Studies and is reproduced here with PIA permission. A profit leader is defined as a printing company making more than 8% return on sales. As will be noted in Figure 2.6, profit leaders were consistently making 11% or more ROS, while "All Firms" made between 3% and 6% ROS.

Figure 2.7 provides a comparison of net income (before taxes) to return on investment for the period 1978–87, which clearly mirror each other. Data for this chart is also taken from the PIA Ratio Studies and represents "All Firms" included the studies over that ten year period.

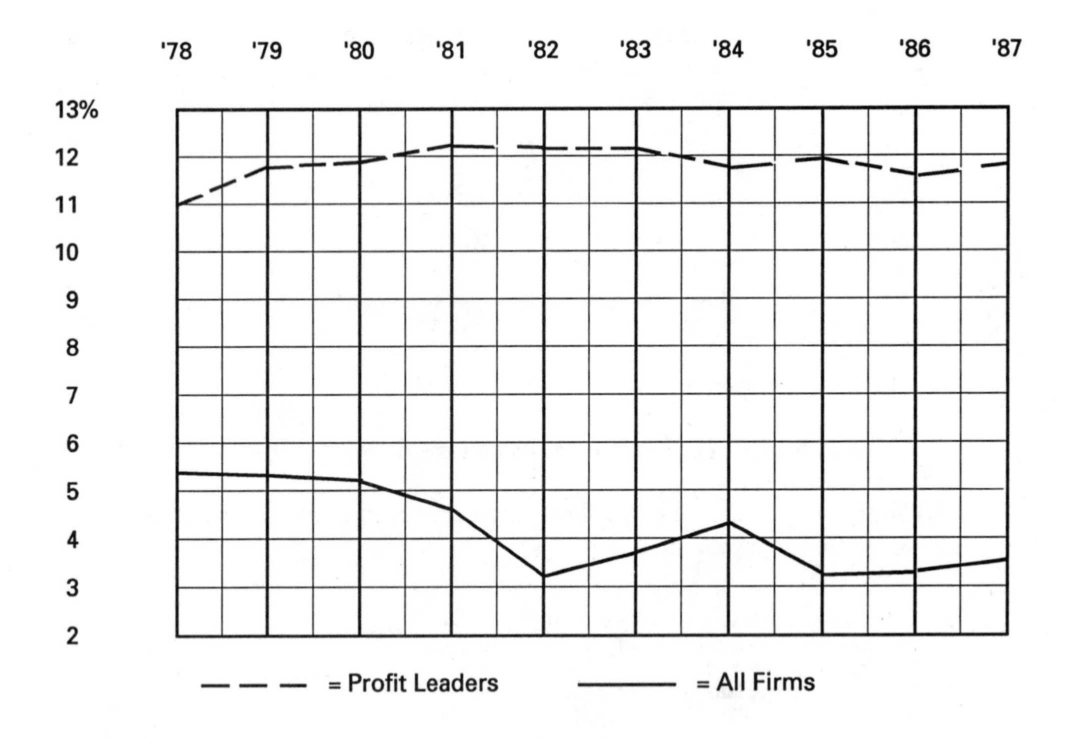

Figure 2.6

Ten-year trend comparison of net income (before taxes) for "All Firms" and "Profit Leaders" (Reproduced from PIA Ratio Studies, 1978-87, with permission from Printing Industries of America)

The *PIA Ratio Studies* are a valuable tool for printing management. They provide factual, yearly data and financial information related to costs, prices, and profits for the printing industry. The *Ratio Studies* are produced annually and provide complete comparative data through the fourteen volumes broken out by return on sales, return on investment, and value added cross-referenced with company size, and types of products and processes used by the responding printers. *The Ratio Studies* have been produced for more than 60 years, providing long-term financial guidelines for the printing and allied industries.

Another pertinent publication to the printing industry is NAPL's quarterly Business Indicator Report (QBIR). Figure 2.8, from a first quarter 1989 QBIR publication, shows zoned results for sales growth (first quarter 1989 over the first quarter 1988) and pretax profits (first quarter 1989). Figure 2.8 represents only one of many charts presented in a typical QBIR publication, which accurately tracks the performance of the printing and allied industries with respect to quarterly profitability, sales growth,

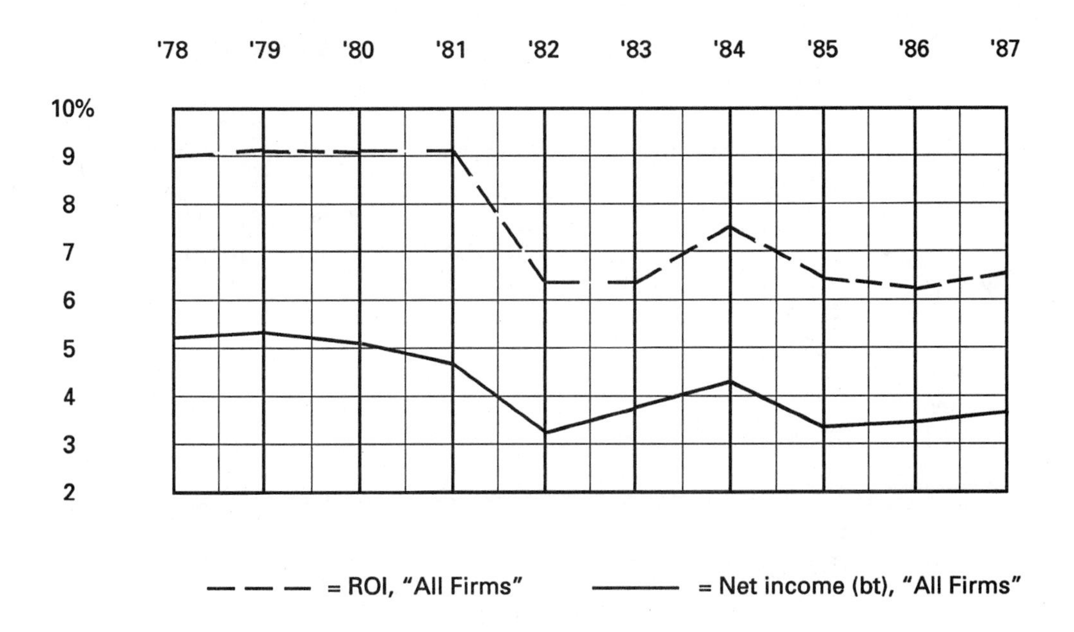

Figure 2.7

Ten-year trend comparison of net income (before taxes) for "All Firms" to return on investment for "All Firms" in the study (Reproduced from PIA Ratio Studies, 1978-87, with permission from Printing Industries of America)

inflation, paper, and other vital economic issues. Data is broken out by geographic region, company size, and product and process orientations. Each issue typically provides future forecasts along with an "Ideas for Action" segment which is always interesting reading.

While various measures of profit have been discussed, the actual addition of profit is a decision that normally rests with top management, such as the company president, and is best done on a job-by-job basis. The person making such a decision must consider the array of pricing determinants described herein.

Other pricing determinants, which vary in terms of application on a printer-to-printer basis, include:

Location of the Printing Company to the Customer. It is said in real estate that the three most important criteria for buying a house or piece of property are "location, location, and location." It is not unusual for printers to experience much the same attitude from their customers, in that being physically close to customers provides unequaled convenience, speed, and versatility for the printer. The location issue affects all types of print purchasing, more dramatically for quick printers and smaller commercial printers and somewhat less for larger commercial printers. Extremely convenient location appears as less of an issue for buyers from high volume printers, although location in the same city seems to be a factor even then.

Table 4 — Results by Zone

Zone	Sales*		Profitability †	
	Average (in %)	Median (in %)	Average (in %)	Median (in %)
USA	7.9	7.1	5.0	4.8
Pacific	7.3	6.4	3.7	3.2
Western	6.7	5.9	4.2	4.0
Central	7.3	7.1	5.1	5.0
North Central	10.2	9.5	4.8	4.7
South Central	4.2	3.5	3.2	3.0
Southeastern	8.6	8.4	5.8	5.5
Mid-Atlantic	6.1	6.9	4.8	4.7
Northeastern	5.2	5.6	5.0	4.7
Canada	9.1	8.2	5.8	6.0

Figure 2.8

Table from NAPL's quarterly *Business Indicator Report* (QBIR) providing results by geographic region for sales growth (first quarter 1989 over first quarter 1988) and pre-tax profitability data (first quarter 1989). (Reproduced from NAPL's first quarter QBIR publication with permission.)

The Customer or Print Buyer's General Attitude. This can be evaluated as an "ease of working" relationship and may represent an important pricing factor in that most printers don't wish to work for customers who are continually dissatisfied with every job, who continually complain about the service, price, or quality or who are simply difficult to get along with. Charging such "difficult" clients a premium price for printing goods and services thus compensates the printer for difficult, awkward, or pressured working situations.

The Technical Difficulty of the Job in Terms of the Printer Meeting the Service or Quality Requirements. For example, it would be unwise for a printer to attempt—even at a high price—to manufacture screen printed T-shirts. In such cases, the job might be sold by the printing sales representative and brokered to a suitable manufacturing facility or the printing sales representative might arrange for a sales representative from a screen printing company to visit the customer.

The Ego Satisfaction Gained by Printer Desiring the Customer or Customer Wanting the Printer. Let us say that Printer A desires to be the printer for Customer A and therefore reduces his price to become Customer A's printer. The printer's ego drive has affected price. Or the converse might occur: Customer B decides that Printer B is the best printer in the city/state/nation and is therefore willing to pay any price that Printer B charges. In this case the customer's ego has compromised his acceptance of any price the printer wishes to charge.

Both conditions exist fairly commonly and may play into the pricing equation.

Printer's Exclusive Production Capability. If a printer is the only manufacturer or source for a particular product or service, there is no doubt that a definite price lever exists. For any printer to have such a unique position would be unusual, yet this is sometimes evident in remote geographic areas, where the capital investment is prohibitive, or when patent or other legal barrier provides such exclusivity In general, if the preceding do not prevail, the intensely competitive environment of printing results in very few printers having exclusive production capability over the others.

Customer's Resale Value of the Printed Product. Some printers factor the value to the customer into the product they produce for the customer This is usually a subjective determination, since most printing is done for mass distribution and the cost per printed unit is not an issue. Instances where the customer's resale value might be important are for very specific printed goods of limited run, such as serigraphs and artwork, special short-run books, posters, or certain types of packaging products.

High-Profit Printers

Over the years, PIA has engaged various consulting firms to determine the relative profitability of the industry and to attempt to determine the differences between those companies making a large profit and those making less. To do this task effectively, PIA has defined two profit groups: high profit printers who make 8% or more ROS and those printers who make less than 8%. In 1988, based on PIA Ratio Study findings, there were 158 high profit printers averaging 11.83% profit on sales and a grand total of 874 firms averaging 3.53% profit on sales. In other words, approximately 18% of the PIA Ratio Study respondents were in the high profit category.

What distinguishes high-profit printers from all the rest? Do high-profit printers maintain this position year after year? Are there any significant differences in the manufacturing processes, management staff, pricing policies, and labor force? These questions are difficult to answer completely. However, PIA has attempted to distinguish reasons why the high-profit printer makes so much more money than the industry average; there do seem to be some general characteristics that separate the high profit printer from others in the printing industry.

Important Conclusions from the 1966 "McKinsey Report." The 1966 "McKinsey Report," which essentially directed printers to analyze markets with respect to price leverage and profit, made public two most interesting findings regarding profitability in printing. The first, as demonstrated in Figure 2.9 indicates that out of five printers, one loses money, three average about 6% return on assets (which is considered about average for the industry), and one printer makes three times more than all the rest. Interpreted another way, of five printers, 80% make average or less return on assets, while 20%—or one printer in five—make three times more return on assets than the others. This "80-20 rule" has been consistent in the printing industry for the past 20 or more years.

A second important finding, which has also been confirmed over many years, is that the size of the printing firm has no relation to the profitability of that firm. That is, a small printer can be equally or more profitable than a larger printer, or the converse. It is this fact that makes the printing industry attractive as a new business venture today, assuming, of course, appropriate initial start-up capital and marketing knowledge and orientation.

Six Characteristics of High-Profit Printers. In 1973, PIA provided the industry with a study titled "Management Practices and Philosophies Survey Used by High-Profit Printers." It contained a significant amount of valuable information and concluded by stating the following six characteristics that separated high-profit printers from all the others:

1. High-profit printers concentrate on one or two specific product categories (specialization). They tend not to be general commercial printers serving a large number of product groups.
2. High-profit printers have organized their businesses to serve the requirements of a specific product category.
3. High-profit printers tend to have nonunion or partially unionized plants.

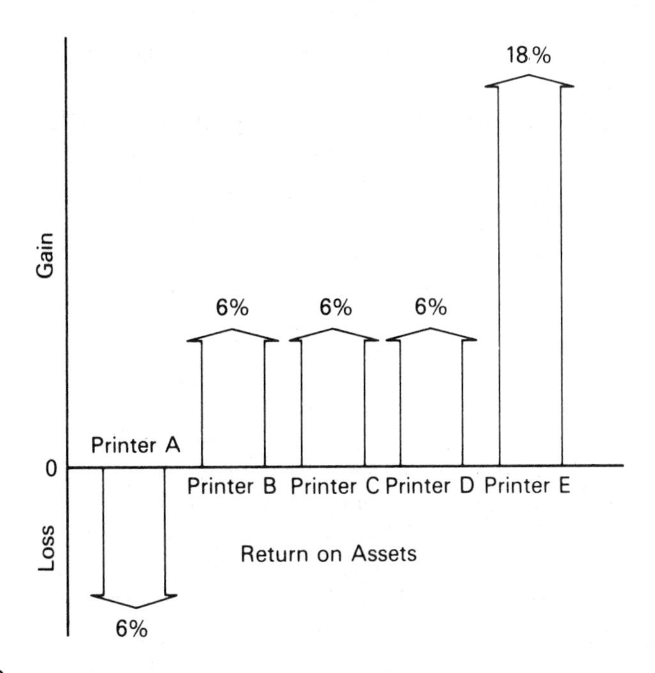

Figure 2.9

Graph demonstrating return on assets (reproduced from the "McKinsey Report to PIA," with permission from Printing Industries of America, Inc.)

4. High-profit printers realize higher productivity from their employees and equipment.
5. High-profit printers take a professional approach to sales and marketing functions.
6. High-profit printers make effective use of long-range planning.

Many of these findings reaffirmed conclusions of other PIA studies and results drawn by numerous consultants and management personnel in the industry.

An Update of High-Profit Characteristics. In 1976, PIA, along with the Harris Corporation, produced a study that looked again at the changing face of the industry with respect to technology and profits. It was titled "A New Look at Management and Profits in the Printing Industry." The study was a product of McKinsey & Company, the consulting firm that had produced the 1966 "McKinsey Report to PIA." The 1976 study was packed with valuable information for top management. For example, it noted the following distinguishing characteristics for successful printers:

1. High-profit printers develop a successful business niche. (Almost 80% of all study respondents mentioned this item as the vital key to success.)
2. High-profit printers have up-to-date equipment.
3. High-profit printers gear up the entire operation by integrating sales, production, and all other elements of the business.
4. High-profit printers work on personnel motivation and the improvement of personnel relationships.
5. High-profit printers are nonunion.
6. High-profit printers effectively manage funds throughout the business.

In another segment of the same study, the conclusion was drawn that both small printing companies and large operations will have no difficulty surviving in the future. Each operates on a different financial basis with different goals. The major squeeze, the study stated, would be for the medium-sized printing plant with annual sales between $1.5 and $8 million. For these medium-sized firms to survive in future years in the printing industry as high-profit operations, it was indicated that they must consider one of the following alternatives:

1. Find and develop a superniche for current equipment that will provide protection from eroding profit margins as other printers begin to specialize in their current product lines.
2. Retrench the business by cutting back the size and volume of work. This alternative means adopting a smaller-but-happier philosophy.
3. Consider merging with another printing company. This alternative would require combining sales efforts, personnel, equipment usage, and all other aspects of the business.
4. Redesign the company around the web lithographic process, which will provide greater production output.

Analysis and thought about such studies, along with appropriate action on the part of printing management are necessary for any printing business to become a high-profit enterprise.

Some Difficult Pricing Areas for the Quick Printer, Commercial Printer, and High Volume Printer

Management of any printing establishment is faced with the task of pricing company products effectively. The following discussion outlines some of the more difficult problems faced by printers in pricing certain types of work. It would be wise for the printing company management to discuss and develop a strategy or philosophy to deal with any of the following situations that arise frequently.

Filler Work. Filler work is printing that is used to keep the company busy in times of normal slowdown. The reason filler work is accepted is to at least recover manufacturing costs without employee layoffs. But this policy requires thought because, even though costs are recovered, little effort is made to obtain a profit. At the same time, equipment is being used (and worn out) for no financial gain.

Smaller printers and quick printers, whose volumes fluctuate on a day-to-day basis, don't typically experience the problems of larger commercial and high volume printers in dealing with filler work. Many printing companies that do aggressively seek out filler work cut prices to obtain it. Other plants attempt to schedule production in such a manner that an average output is consistent. Still others track customer needs closely and arrange for "last like period" printing to fill these slower periods. For example, if a customer ordered a particular job in March, the printing sales representative might contact the customer for a repeat run of the job—with or without changes—to be done in a slack period the company is currently experiencing.

Repeat Work. Many printers, particularly commercial and high volume printers, experience repeat orders for printing—the rerun of an existing job—with no changes. In such cases, since all preparatory work has been completed already (perhaps all that needs to be done is to make new plates), should the printer pass along such prepress savings to the customer? On the one hand, if the printer does pass along such savings, the customer will expect this savings on all other repeat work and, in fact, may try to reduce job changes in the future to enjoy this discount. On the other hand, if the printer does not share such savings with the customer, the profit on the job will appear larger, since the dollars paid in by the customer for prepress manufacturing have not actually been spent for this item. For the most part, printers faced with this situation do not pass along such savings, or do so only under specified circumstances or with certain customers.

Buyins/Trade Work from Other Printers. While buyout work is purchased from an outside printing company or trade house, a buy-in is an order placed into the plant by another printing establishment. The frequency of buy-ins is related to certain process and product specialty areas that the printer has available to the industry, such as perfect binding, special folding capabilities, or certain specialized copying or prepress operations. Normally price reductions to the trade are reciprocated between trade houses. When the specialty operation is truly streamlined, considerable cost savings are possible, and the savings may then be passed along to the purchasing company.

There appears to be an increasing amount of trade work between printers as market specialization has accelerated in the 1980s. For example the quick printer's, immediate production capacity is sometimes needed by a larger commercial or high volume printer, necessitating a trade relationship between the two. With the advent of color copying and the breakout of prepress companies supplying printers with fully prepared jobs for presswork, trade work is likely to continue to increase in the 1990s.

Rush Work. Rush work—printing done on a fast-turnaround basis—is an obvious market specialty for the quick printer, yet may represent major difficulties for both the commercial printer and the high volume printer. There is no question that the typical lead time—the time needed to manufacture the printed product inside the printing company—has been compressed during the 1980s, regardless of type of printing operation. In the early 1980s the typical lead time for a commercial printing job was about 21 calendar days; by the late 1980s this was reduced to a lead time of about 12 days.

For any printer not geared up to work overtime or not working on a two or three shift basis, rush work represents a difficult problem. On the one hand, providing customers with superb customer service necessitates that the printer do everything in his power to produce the customers work as desired. If not, the customer will find another printer who will do so.

There is no question that literally all customers need rush printing completed from time-to-time or even frequently, and no question that market factors and competitive issues require the printer doing the work to respond accordingly. However, rush work can be costly and may require overtime for employees, special supplies or processes to expedite production, increased paper waste, and other increased costs. Since these are valid, job-related costs, they should be borne by the customer.

In many cases, customers and print buyers are willing to pay extra for rush work, yet how much varies in proportion to how badly the printing is needed and the buyer's budget constraints. Rush charges may range from printer's additional costs only, with no markup, to a multiplier of the original price—perhaps double or triple—used by some printers to discourage fast-turnaround jobs. When rush work is promised by the printer it is essential that it be clear there will be additional costs for the customer and it would be prudent to provide him with an estimate of them.

Rush work has also become an important issue between trade houses supplying the printer with various prepress items, such as color separations. For example, the customer's job contains a series of four-color images, to be obtained via a color separation trade house through the customer's printer. If the work is rush, then the trade house must literally do the separations overnight, with stripping, platemaking, and printing to follow quickly. This makes pricing issues between printers and trade suppliers a difficult problem to resolve, particularly for price-conscious customers.

Two final points need to be made about rush work. First, there is no question that the printer of the 1990s must provide excellent customer service to remain in business and profit, this means that fast-turnaround work will remain as a fact of life and that customers will likely be requesting rush work on a more frequent basis. Second, since fast-turnaround work is a market niche, it is likely there will be an increasing number of printers—quick, commercial, and high volume—who will encourage this type of work. This adds a new competitive dimension to an industry already intensely competitive in many other ways.

Charity Work. It is safe to state that literally no printing plant exists that has not been approached by some church, foundation, school, or other nonprofit service organization requesting a free job or one at a substantially reduced price. Both the advantage and disadvantage of accepting charity work relate to reputation. If the work is done at a reduced price or for free, the company will benefit from a good reputation within the community. As this good reputation develops, however, additional organizations may request the same benefits. Saying no may then be difficult. Some printing plants take charity work at a time and materials rate and make no profit; others do small charity jobs for free but will not accept large jobs; and still others will accept charity work and run it using normal production practices, price it as regular work, then donate it to the charitable organization and write off the full cost as a donation (providing there are no legal entanglements). Whatever procedure is used, management must establish a policy regarding how to handle work for charities.

Work for First-Time Customers. Pricing work for new customers presents a particularly thorny problem to management, especially in printing companies wishing to keep equipment busy or to impress the new client with the dependability and service the printing company can provide. There is a tendency, as well, for new customers to indicate that they will place more work in the future if they are given a price break on the first job. While it is typical of some printers to reduce prices initially to get a new customer, this practice can be dangerous. After the first order, can prices be raised without question? How much compromising on price will be necessary in the future if the printer begins by cutting prices—will the customer expect this as a practice all the time? It is probably best not to put much stock in the customer's promises. It is also good to remember that as much as one-half of the volume of some commercial printers comes from repeat customers. Thus, if price cutting is done initially, it may become more of a pattern and practice than the printer would like because its long-term effects can be costly.

Work for Friends and Family. Friends and family seem to request discounts, or even free printing, more than the general public. The advantage of reducing prices, perhaps to time and materials, for a friend or family member is that the relationship will be maintained in a normal manner. The disadvantage is that if the price is reduced, the friend or family member may expect similar treatment in the future and may have a more significant printing request next time. Various philosophies exist with respect to dealing with this matter. Some printers use the time and materials method; others ask for a favor of equal value in return; others do the printing as a regular job, with the understanding that the friend or family member will provide something of like value through a barter trade-off; and some printers refuse to do printing for family or friends, but may aid the person in obtaining printing from a reliable source. Whatever policy is used, it should be one that will result in fair treatment to both parties in the relationship.

Rerun Work Because of a Customer Error. Assume that the customer has approved a press sheet and the job is printed, finished, and delivered. Later that day, the customer notices a major error in the work due to his or her own oversight. The job must be reprinted immediately, it is dated work, and the customer desires to have the job price refigured even though he or she is terribly sorry. If the concept of service is important to the printing plant manager, the job may be rerun at a reduced price figure. If the company philosophy is to make a profit on all work, the job will be rerun at the original price and the customer will pay for the work twice. Some plant managers consider the quantity of other work done for this customer as an important factor in making a pricing decision of this type.

If the job is rerun at a reduced price, it is possible that such errors may occur with future jobs and, should that happen, the customer may expect the same treatment then. If the customer must pay twice for the same job, he or she may never return as a customer or may refuse to pay the second bill (or pay it very slowly). One usual rule of thumb is not to rerun the order without payment of the first job in full. Then, if any misunderstandings arise, the company at least has received initial payment and will not be paying twice for materials with its own capital. The best advice is to work diligently to reduce customer errors by thoroughly checking all possible elements of the job during production and double-checking important facts such as times and dates with the customer prior to printing.

Contract Work. Most printing plants bid on contract printing at one time or another. Sometimes the reason for becoming involved in contract work relates to filling in a production scheduling gap. In other instances, that job might be appealing or the company might have a specialty in that area, and bidding on the contract may be the only procedure by which the work will be awarded. Perhaps the most important consideration is for whom the work will be done. Normally, state and federal government bids are accepted on the lowest dollar amount, assuming job specifications are met fully. Criteria for selection of other types of contract work vary with regard to the type of company, product use, and so on. Some printing companies cut prices on contract work simply to stay busy; others refuse to even bid on contract work because it normally requires exacting specifications that cannot be accurately cost estimated and priced.

Price Discrimination

The intention of the Robinson-Patman Act of 1936 (actually an amendment to the Clayton Act of 1914) is to eliminate price discrimination. The act essentially prevents the seller of a product, good, or service from offering the same kind of goods, "goods of like grade and quality," to two or more buyers at different prices.

Proof of discrimination under Robinson-Patman is not a simple procedure. Because it is federal law, interstate commerce must be involved. The buyer must prove that "an unjustified special price, discount, or special service was provided to the customer," while the identical price, discount ,or special price was not provided to the complainant. For the most part, purchasing agents—those persons who purchase printing consistently for larger companies—understand that Robinson-Patman is complicated and difficult to interpret. Robinson-Patman is enforced by the Federal Trade Commission through initial use of cease-and-desist orders from a court of law. If such price discrimination is actually proven in court, the offender pays treble damages. Most states have their own laws concerning price discrimination for interstate commerce, but such laws vary in terms of their application and intent.

Robinson-Patman and various state laws do have an effect on the establishment of prices for printed products. However, since the products are generally tailor-made, it is difficult under most circumstances for the buyer to prove such discrimination exists. Of course, if discounts are used or services are charged for at different levels and interstate commerce applies to the product, Robinson-Patman may be in force. Should any questions arise concerning the Robinson-Patman Act or your own state law with respect to price discrimination, contact an attorney.

2.6 Information Resources

This chapter has covered marketing, sales, customer service, general management, costing and pricing procedures for the quick printer, commercial printer, and high volume printer. Chapters 14 and 15 respectively will deal with estimating and pricing for the quick printer and high volume printer. The following sources are recommended to the reader who desires greater depth of discussion or information in the marketing and pricing area.

Printing Industries of America (PIA)

PIA, the largest member association serving the printing and allied industries, offers a diversified array of publications and services for the industry. Because the list is so extensive, contact Printing Industries of America, 1730 North Lynn Street, Arlington, VA 22209, for a full listing of publications and services, and for the address of the PIA regional office serving your area. Telephone 703/841-8110.

The following list provides general categories of information through PIA Publications:

A diversified array of publications are available in each of the following categories: (1) design and copy, (2) financial management, (3) general business management, (4) government, legislative, and regulatory information, (5) human resources management, (6) production/technical/operational management, and (7) sales and marketing management.

The cost of publications is reasonable and further reduced for members. Of these published materials, the most significant long-term publication may be the PIA Ratio Studies.

Meetings, Seminars, and Conferences

PIA offers numerous workshops in financial management, general business management, government and regulatory issues, human resources, production and operational management, sales and marketing, and sales management. Major conferences include the annual President's Conference, Web Offset Annual Meeting, Executive Development Program, and the President's Roundtable. PIA also works with a number of affiliated offices throughout the U.S. to sponsor regional trade shows on an annual or biannual basis.

The cost of meetings, seminars, and conferences varies as to the length and type of program.

Products and Services

PIA offers numerous video and cassette tapes covering a wide array of different subjects including sales management, human relations, technical and production management, and financial management. There are numerous special sections available to member firms, including Graphic Arts Employers of America, Master Printers of America, Graphic Arts Marketing Information Service, Graphic Communications Association, National Composition Association, PIA National Bookstore, Web Offset Section, and Association of Graphic Arts Consultants.

National Association of Printers and Lithographers (NAPL)

NAPL is an active, progressive, and dynamic national printing association serving all areas of the printing and allied industries. They offer a growing list of impressive studies and reports too numerous to list. NAPL can be contacted at 780 Palisade Avenue, Teaneck, NJ 07666 for a complete list of publications and services, some of which are discussed below. Telephone 201/342-0700.

One of NAPL's most popular and significant contributions to the printing industry is represented in their "blue books," which detail production center costs for all areas of a printing company. This "cost study" series includes sheetfed press, phototypesetting and automated composition, web press, lithographic preparatory operations, and binding, finishing, and mailing operations. Also available is a computer-assisted program so that purchasers may develop their own hour rates using NAPL's model.

NAPL's *Quarterly Business Indicator Report* (QBIR) is one of the most important economic tools available to the industry because of its timeliness and quality of data and information presented. The QBIR provides four-times-per-year information on the performance of the U.S. economy, paper costs and paper supply issues, quarterly data on printing profitability, and sales growth broken out by geographic region, company size, product, and process. It is derived from a two-part database designed to monitor general economic trends in the North American printing industry; one part consists of historical data on over 20 broad economic indicators, and the other part is a detailed economic survey of North American printers, designed to accurately reflect both the geographic and company-size distribution of the industry. Both mail survey and telephone survey techniques are used.

NAPL provides services in the following additional areas:

Cost and financial management
Plant and production management
Sales, marketing, and customer relations
Safety, health, and environment
Computers
Membership library
Human resource management
Administrative management
Audio cassettes
On-site consulting services
Programs and seminars
NAPL Management Institute
Recognition programs

Of particular interest in the printing estimating area include NAPL's services in cost and financial management, plant and production management, sales and customer relations, and computers. Their audio cassette programs are numerous and well done; their bimonthly *Printing Manager Magazine* is "must" reading for any printing manager. NAPL's numerous traveling seminars and workshops—one- or two-day programs held frequently in larger cities on many vital printing subjects including estimating—have become quite popular for both those new to the industry as well as seasoned professionals.

Also of interest are NAPL's on-site consulting services, using both resident and nonresident consultants, generally well-known industry experts. NAPL's Management Institute has become a popular continuing education program, offering intense coursework in business planning, marketing, and production management.

National Association of Quick Printers (NAQP)

NAQP was founded in 1975 by George Pataky and has about 4,500 member companies representing about 6,000 quick printing and copying shops throughout the United States, Canada, and several foreign countries. In 1987 it was estimated that there were about 24,000 quick printers in the United States and Canada. Gross 1988 sales for United States quick printers is estimated to be roughly $6 billion, with NAQP members representing about $2 billion—or one third—of that sales figure.

The purpose of NAQP is to provide its members with pertinent, up-to-date information on all facets of the quick printing business. The association releases numerous reports and other information such as:

- Member Shop Profiles (1987) which provide data and information on NAQP members with respect to current equipment, services provided, and demographics.
- Quick Printing Industry Operating Study (1987) summarizes in detail the financial and other operating characteristics of NAQP respondent firms.
- Desktop Publishing Studies (1988) tells about the methods and procedures used by NAQP members in providing desktop publishing services to customers and client.
- 1988 Pricing Study/Final Report provides information to members on profit leaders, owner's compensation, sales per employee, and regional pricing differences.

The focus of NAQP is to help their members in all areas of business operation. Since the typical NAQP member is a small business owner—1988 annual sales for franchised quick printers was $434,000 while independents had sales of $360,000—the services provided covers a wide array of information and issues. Many members offer both printing and copying services, desktop publishing, and related graphic services. NAQP began their "Print Expo" in 1978, which is held annually in a major metropolitan location.

Graphic Arts Technical Foundation (GATF)

GATF is the most technically-oriented of the groups serving the printing industry and is located at 4615 Forbes Avenue, Pittsburgh, PA 15213. Telephone 412/621-6941.

Member benefits to GATF includes a monthly mailing containing a newsletter, a broadside covering recent developments at the foundation, and various reports produced on an intermittent basis including subjects such as productivity and training, research, environmental, technical services, and teachers. Also the monthly *Graphic Arts Abstract*, provides a quick look at abstracted articles and other published materials.

GATF provides the following other member products and services:

Free telephone consulting
Textbooks
Learning modules
Audiovisual materials
Career materials
Research findings including
—Annual research reports
—Unique research reports

—Niche reports
Technical information
—Quality control devices
—Color communicator
—Various printing control devices
Educational
—Special programs
—Productivity training
—Training assistance
Technical services
—Plant audits
—Troubleshooting
—Product evaluation
—Plant layout and workflow evaluation
—Production measurement analysis
—Quality control program

GATF is committed to providing the printing and allied industries with current educational and training materials and helping the industry focus its educational efforts. The Education Council of the Graphic Arts Industry encourages interest in printing technology and printing education at the high school and college level; the National Scholarship Trust Fund offers scholarships to students enrolled in accredited printing and graphic arts college programs throughout the U.S. and Canada. GATF also sponsors teacher development programs, directed specifically to help graphic arts teachers update their skills.

Production Standards and Budgeted Hour Cost Rates in the Printing Industry

3.1 Introduction

This chapter provides detailed information on the development of two critical aspects of cost estimating: standard production data and budgeted hour cost rates. As indicated in Chapter 1, these two bodies of data work together to provide estimated production times and production costs for a customer's job. The estimated production costs will be used by management when pricing the job, and the estimated production times will be used to schedule each production step to be completed on the job during the manufacturing of the product.

It is essential that printing estimators or anyone attempting to deal accurately with determining the costs to manufacture printed goods have a thorough knowledge of BHR development and the production standards. These two groups of data form the core from which accurate cost estimating is completed. For printing companies considering shifting to computer-assisted estimating (CAE) systems, many such computer systems are driven by these two cores of data, thus ensuring their accuracy.

Production standards are basic time units required for each production operation on a job. For example, the production standard for line exposures for a process camera might be 10.1 hour (6 minutes) and for halftone exposures, 0.15 hour (9 minutes). If a job requires 12 line exposures and 4 halftones, the estimated time for production is (12 × 0.1 hour) + (4 × 0.15 hour), which equals 1.8 hours total.

Budgeted hour cost rates (BHRs) are essentially an accounting procedure by which all costs are identified for each production center of the plant—such as a process camera—then broken down into hourly charges assessed against the customer's job. If a process camera has a BHR of $31.50 per hour, then the 1.8 hours of estimated production time × $31.50 per hour equals $56.70 as the estimated production cost of the job. Material costs—in this example, the cost of film and developer—must be added to the $56.70 cost to arrive at an estimated time for process camera production for this job.

Estimators must have a thorough knowledge of BHR development and the production standards used in their plant. Without such important groups of data, accurate cost estimating is impossible. Also, shifting to computer assisted estimating (CAE) will be very difficult since CAE systems typically require establishment of these two cores of data for system initialization.

3.2 Definition and Use of Standard Production Data

Accurate cost estimating mandates that production standards, individualized for each particular printing plant, be developed and used. The terms "production standards" and "standard production data" are used interchangeably. A *production standard* is defined as an hourly unit value that represents the average output of a particular operating area producing under specified conditions. Output can be measured as a quantity unit of material, such as sheets of paper, number of signatures, pounds of paper, or sheets of film; the hourly value then becomes a function of the number of units produced. For example, the output of a printing press may be measured in the number of press sheets produced per hour, the number of press sheets per shift, or any other output per unit of time. Both output and time relationships vary with respect to the type of product produced and the relative speed and operating conditions of the equipment and personnel.

Two Methods to Apply Production Standards

Two common and interrelated methods of applying standards of production are used in the printing industry: as a measure of output per unit of time and as a value of time per output quantity. As an example of a measure of output per unit of time, 6000 sheets printed per hour might be the production standard for a lithographic sheetfed press. That is, in one 60 minute time period, 6000 sheets of paper will be printed. The second production standard, time per output quantity, relates production on a decimal hour base. The given example of 6000 sheets printed per hour may be converted to 0.1667 hour per 1000 sheets, or 0.1667 hour per M sheets. Calculation of this production standard is derived as follows:

$$1000 \div \text{number of pieces per hour} = \text{number of hours per 1000 pieces}$$

Then, substitute:

$$1000 \text{ sht} \div 6000 \text{ sht/hr} = 0.1667 \text{ hr/M sht}$$

If this measure is used as the production standard, the number of sheets (in thousands) multiplied by the hours per 1000 sheets (which is the production standard) will yield the estimated production time. Thus, if 4000 sheets were to be printed on this lithographic press (including throwaway, or spoilage, sheets) at the standard of 0.1667 hour per M sheets, the following calculation would apply:

$$4000 \text{ sht} \times 0.1667 \text{ hr/M sht} = 0.6668 \text{ hr}$$

This answer translates to 40 clock minutes (0.6668 hour × 60 minutes per hour).

If the production standard as a value of time per output quantity is known—in this case, 0.1667 hour per M sheets—the following formula can be used to convert to a measure of output per unit of time:

$$1000 \div \text{number of hours per 1000 pieces} = \text{number of pieces per hour}$$

Then, substitute:

$$1000 \div 0.1667 \text{ hr/M sht} = 6000 \text{ sht/hr}$$

It is vital to remember that standard production data are an average of production and that such data are based on all the operating conditions for a specific piece of equipment or work area. Averages are necessary because of various differences in equipment (much equipment sold in the printing industry contains optional devices that cause differences in output), because of differences in the skill levels of the various plant production employees, and because of the varying criteria specifically affecting the work produced on the equipment. Far too many small and unaccountable differences exist, even on a day-to-day basis with production employees, to allow such figures to be anything other than averages.

Categories of Production Standards

Four categories of production standards generally applicable to any printing company are procedural, machine, manual, and integrated. These standards are discussed in the following paragraphs.

Procedural Standards. Sometimes referred to as "technical standards," these standards state specifically the recognized technique or procedure for a given job, as well as the expected outcome. They are not time standards but serve as a framework for development of accurate time assignments. For example, the procedural standard for the proper exposure and development of a lithographic plate would include a step-by-step analysis of job components; the technical standard for opaquing a line negative would include all component operations needed to complete that job successfully. Procedural standards define the job duties and expected outcomes as thoroughly as possible.

Machine Standards. These standards are time standards established specifically for automated production operations wherein the significant portion of output is completed by machine operation and not manually. Examples of equipment that would apply for machine standard rates are stand-alone units such as film processors and phototypesetting equipment of some kinds.

Manual Standards. These time standards are developed specifically for operations done entirely by hand with no aid or assistance from automated equipment. Many bindery operations, paste-up, and stripping would be examples of operations to which manual standards would be applied.

Integrated Standards. Because printing manufacturing is essentially a combination of manual and automated operations, these standards apply more frequently than either the machine or manual types. These time standards are developed after procedural standards have been delineated. Integrated standards are by far the most difficult type of standard to develop accurately and require continual monitoring and reassessment on the part of production management.

Standardization of Printing Production

Printing manufacturing requires standardization of product and process. The establishment of thorough procedural standards for product and process provides for more efficient production. Efficiency reduces the cost of manufacturing the product and, eventually, gives management a flexibility of price not enjoyed by the competition (unless they do the same). The more efficient the manufacturing operation, the less the manufacturing cost and the greater the competitive advantage for the business. Standardization yields efficiency.

Process Standardization. Essentially, process standardization relates to the equipment and employees who execute production. In recent years, the printing industry has been fortunate to have suppliers and equipment manufacturers who have developed and marketed hundreds of new products and equipment, most of them standardized for improved efficiency. Thus, when a printing company decides to purchase an automatic film processor, they, in effect, standardize film sizes, types of film to be used, exposure times, and operational procedures. When in production, the processor not only speeds up film processing in the photographic area, but also requires that standard procedures be used for handling and exposing all film images. Such standardization has made photographic workflow simple and efficient with resultant film images of excellent quality. Many other examples of standardized processes, with accompanying standardized materials, can be found in the modern printing plant.

Product Standardization. Process standardization is common, but product standardization is developing rapidly. Some product standardization is due to equipment factors, and some to the increasing marketing interests of printing management with respect to product specialization. In terms of equipment, for example, web presses, which have fixed cutoffs, mandate that products remain within common size formats. Otherwise, extreme costs from high paper waste may result. Some bindery equipment is also manufactured only for specific product sizes. With regard to specialized marketing, wherein a certain type of product is manufactured for the customer, standardization fits in well. In fact, the more the product line is specialized, the greater the likelihood that extreme standardization of production is possible.

It must be noted that neither specialization nor standardization is dependent on or directly related to the other. It is possible to specialize in a particular product or process and utilize little standardization or to standardize greatly while attempting to meet every customer's total printing needs.

3.3 Methods to Establish Production Standards

There are six methods used in the printing industry to establish production standards. The comparison chart in Figure 3.1 shows the relative merits of each. It is advisable

	Accuracy	Relative Cost	Time Needed	Outside Experts or Consultants	Impact on Employees/Product
Standards based on historical plant data	excellent (over time)	expensive (over time)	considerable	optional	may be extensive
Standards based on intuition	poor-terrible	low	modest/little	none	little
Standards based on published information and catalogs	fair-good	low-moderate	little	none	little
Standards borrowed from competition	varies (depends on production similarity)	low	modest	none	little
Standards based on equipment manufacturers data (new purchase)	initially poor	low (when considering purchase)	little	yes	little
Standards based on time and motion study (plant engineering)	excellent	expensive	considerable	usually	may be extensive

Figure 3.1

Comparison chart for various methods used to establish production standards in printing

that the methods be used in combination with one another when time and cost permit; in this manner, cross-checking to ensure accuracy of data will be possible.

In some printing plants, the establishment of production standards will be completed by one individual in estimating or production management, in others, by a group or committee. For the most part, the persons involved will be middle management staff with experience in the production aspects of the business. Such persons have the best judgment regarding procedural and time standards.

The use of the historical plant data procedure, perhaps the most popular of all those listed, has evolved into what is today broadly termed *job costing*. Job costing is essentially the development of production data on jobs that have been done in the plant by keeping accurate records as when the job is being completed, thus allowing management to review manufacturing details of the job and the role these elements played in making or losing money on that job. Job costing also allows the company to maintain exacting and precise production standards for the plant. Since historical data/job costing is vital to generating accurate future estimates, many firms considering computer implementation may initially establish a manual job costing procedure to provide relevant historical production data upon which the new CAE system will be based.

It is important to note that historical data collection, or job costing, requires a continuing effort. If production standards have been manually established, a job time/cost summary form would be used (Figure 3.2) upon which actual production times are compared to estimated times. If a computer system is used, which may include computerized factory data collection, job costing and establishing production standards are much more quickly and accurately completed.

Standard Production Data Based on Historical Plant Performance

Developing production data based on historical plant performance can be done using either manual or computerized procedures. It is generally agreed that whichever procedure is used, the data developed are quite accurate and reliable. Disadvantages may include the time needed by the production worker to record the data (taking away from production time) and also the time needed to tabulate and analyze data, especially if the procedure is executed manually.

The use of computerized or electronic factory data collection (EFDC) has been an important direction for job costing and the establishment of production standards based on historical job performance. Essentially, electronic factory data collection begins with a "user-friendly" terminal located close to the production center where the work is being done. The employee working on a specific job enters his or her name or employee number, the job number, activity being completed, materials used, as well as any other required information (such as down time, waiting for a material, etc.) on the terminal. This information is immediately transferred electronically to the computer elsewhere in the building. At the end of a day, week, or other time period, the data is

Job Cost Summary	Customer		Job Number
Estimator	Dept./Organization		Date
Costed By		Quantity Delivered	Date Wanted
Job Description		Finish Size	Date Delivered
		Colors	Repeat [ ] Repeat W/Changes [ ]

Paper

Type	Weight	Color	Size	Quantity	Unit Price	Total

Flat Material

	Type	Quantity	Size	Unit Price	Total
Film					
Plate					

Ink

Type	Quantity	Color	Unit Price	Total

Labor/Time Cost

Flat	Act.	Est.	Diff.	Press	Act.	Est.	Diff.	Bindery	Act.	Est.	Diff.	**Total**
Hand/Ludlow				Make Ready				Cutting				Flat
Linotype				Run				Trimming				
Make-up				Make Ready				Fold Set-Up				
Proofs/Repros				Run				Fold Run				Press
Tape Perf.				Make Ready				Gather				
Photo Set-Up				Run				Stitching				
Photo Run				Make Ready				Drilling				Bindery
Paste-Up				Run				Padding				
Camera				Wash-up				Wrapping				
Stripping												
Platemaking												
Total				**Total**				**Total**				

Recapitulation

Item	Actual	Est.	Diff.
Matl's			
Flat			
Press			
Bindery			
Overhead			
Buy-outs			
Total			

Remarks

Figure 3.2

Job time/cost summary (Developed by Jake Swenson; reproduced with permission)

retrieved, sorted and analyzed by computer, allowing precise determination of center productivity, employee output, production problems, material overages, inventory, and so on.

Electronic factory data collection has become popular for larger commercial printers and high volume printers since it was first introduced in 1984. EFDC saves countless hours of time otherwise spent in manually collecting and analyzing such data. Collection is fast and generally the process is simple to learn and easy to do continuously. Yet such electronic data collection systems may offer certain disadvantages too, among them the time-consuming problem of entering information after the fact, such as when production employees do not enter materials or other needed information at the time they were used, which must then be input later by a secretary or staff employee. Also, since employee productivity is very easily monitored, it is possible for management to scrutinize individual employees who are required to use the system, setting up a "snooper-vision" relationship. This can lead to disgruntled workers and, ultimately, reduced or inaccurate data input over time. Chapter 5 provides a more detailed discussion of computer use in the industry and electronic factory data collection.

The following step-by-step procedure is recommended for companies setting up a manual historical production standards data collection system:

Step 1: Establish a defined data collection procedure.
Step 2: Develop forms and train employees in the use of the data collection system.
Step 3: Use the data collection procedure.
Step 4: Review collected data and modify standards.
Step 5: Repeat the procedure (steps 1–4) over time.

Establish a Defined Data Collection Procedure. Data collection requires accurate tabulation of information in concise form, generally using a coding procedure. Such codes describe, accurately and in limited space, such things as the area of work (production center), the specific employee doing the work, the type of work being done, and any additional information to specify type of work stoppage (for example, because of lack of materials or lunch breaks) or problem. As an example, a process camera (production center code 003) is used to expose line negatives (code 110) by employee Fred Smith (code E2160). Any numerical, alphabetic, or combination system can be easily developed as a code procedure. When work interruption codes are also provided, management can easily determine production times that are chargeable to the customer, as distinguished from those times that are nonchargeable and must be borne by the company.

Generally, the more detailed the coding system, the more confusing it becomes to production employees who actually must work with the codes daily. When the codes are kept to a minimum, however, data collected may not reflect production situations that actually occur. Management should develop a data collection procedure adequate to allow for meaningful comparison of actual versus estimated production times and costs.

Develop Forms and Train Employees in the Use of the Data Collection System. After the coding system is developed, management must design data collection forms for daily reporting of such production data. Form development should be tailored to the coding system. Some plants utilize manual employee record keeping wherein the employee completes the form with pencil or pen. In other cases, forms may be developed for time clock use: Production times are automatically recorded by the use of a time clock, with appropriate coding manually recorded by the worker. This system saves employees time since they do not have to keep track of time continually and generally improves the accuracy of data. More sophisticated systems may design forms and collection procedures for subsequent computer analysis.

It is vital that the employee understand that management does not intend to use the system punitively with respect to individual production output. Thus, when a new data collection procedure system is installed, management should notify employees in advance. Under no circumstances should such a procedure be established without consultation with the workers who will actually record the data. During initial discussions, employees should be fully informed about the intent and purpose of the system, including affirmation that no individual or group will be singled out for detailed analysis of production output. It should not be management's intention to use "snoopervision" with such a procedure because employees will not cooperate fully and data collected may not be accurate.

Employees should be trained in the proper procedure to complete the collection forms. Explanation of code applications should be provided, especially where interpretation of codes might be unclear.

Use the Data Collection Procedure. Management must decide initially how often the data collection system is to be used. For example, a plant with only a few jobs in process may record data on a job-by-job basis. A large printing facility may select, randomly or individually, only a percentage of jobs to be evaluated. Some plants may decide to collect data on all jobs done in the facility. The more jobs for which data is collected, the greater the number of comparisons to be made, yet the more time required in analysis.

All production data should be collected as accurately and completely as possible. Data collection is a task of the production employee, who must take appropriate working time to code the job category, production center, production time, and all other essential information required by management.

Review Collected Data and Modify Standards. Sometimes after a particular job is completed, the comparison of actual versus estimated production times is made. Usually, an individual assigned by management does this comparison, perhaps a production control employee or estimator. The comparison can be completed either in hours of production, or dollars of production cost, or both. If the comparison is made in terms of hours, the comparison is called a *job summary*. If the comparison is in terms of dollars of production cost, then it is called a *job cost summary*. If both time and cost are included, then it is a *time/cost summary* (see Figure 3.2).

Even though comparison procedures vary from company to company, the outcome translates into modification of existing production standards used for estimating and scheduling. Essentially, the times are modified to relate to actual production. The more accurate the data collection process and the comparison procedure are, the more in line future estimates and the more precise production scheduling will be.

It is wise not to allow the collected data to accumulate. Job summaries are best completed on a biweekly or monthly basis. The reason is that the procedure requires constant monitoring to maintain accuracy. In addition, when manual comparisons are made, especially with a large number of jobs, the procedure can become a laborious task.

Repeat the Procedure (Steps 1–4) over Time. The production standards generated through comparisons of estimated and actual production times must be constantly monitored to keep estimating and scheduling data current and accurate. This task is accomplished by following the indicated procedure consistently over time.

Many printing managers who use the historical production standards system, as described, make variations in the technique to suit their individual work force and plant operation requirements. For example some managers allow a copy of the time estimate to travel with the job, using it as the data collection form for actual production. Thus, the production employees see the estimated time during the recording of the actual time for production. In some plants, the work force is limited to a yearly review of times instead of a monthly or weekly review. Whatever the modifications of the procedure are, it must be noted that accuracy of collected data, with resultant updating of production standards, insures that estimated and actual times (and costs) will be in alignment with one another. This time alignment is the ultimate goal of such a historic production standards system.

Standard Production Data Based on Intuition

Using intuition is a common procedure for establishing production standards in the printing industry. This is not to say that the technique provides accuracy in the developed standards. In fact, the opposite is sometimes the case. But this method is low in cost, very easy to complete, and has little impact on employees. It is not unusual for estimators and production planners to use intuitive standards for production when none have been developed for a particularly complex manufacturing procedure or a new piece of equipment. Small printing plants also develop standards by intuition.

The major problem with basing production standards on intuition is that human nature gives each person a different outlook. People tend to assess quantifiable data in different ways and with different standards. The individual's personality has a significant bearing on the standards he or she establishes. While it may be possible for a production manager to watch an employee carefully during a normal working situation and then establish standards based on such observations, the standards are not necessarily accurate or consistent.

Use of the intuitive method as a cross-check, or judgment check, with other procedures is acceptable, especially if a standard appears grossly out of line. The observer should be thoroughly familiar with the procedural standards required for the job and should attempt to overlook personal feeling and bias while watching the individual performing the task. The working conditions should be as normal as possible. It is unwise to attempt to judge and assign a time value for a production sequence at the beginning or end of any shift or close to a lunch break. It is best to let employees know that they are being evaluated but that such results will have no future impact on their jobs.

Standard Production Data Based on Published Information

Using standards based on published information is a valuable and inexpensive way to obtain such data. Published sources require little investigative time, no outside consultants (unless desired), and offer no impact on production or employees in the normal working situation. The single biggest problem with such published standards is that they may not reflect the exact procedures used in a specific company, thus making extreme accuracy almost impossible. Regardless, published standards are a good cross-checking source for standards verification.

There are four major sources for published information. The first source is offered through national associations serving the printing industry, including Printing Industry of America, National Association of Printers and Lithographers and the National Association of Quick Printers. PIA's published sources include "Production Profit Targets" and "Production Benchmarks"; out-of-date publications from PIA, still found in some printing plants are PIA-Production PAR and PIA Sim-PAR. Some of NAPL's published materials which provide production standards include "Increase Your Profits with Production Standards," "Controlling Sheetfed Waste and Spoilage," "Controlling Web Press Waste and Spoilage," "An Overview of Waste and Spoilage in the Bindery," and "Simple Scheduling Methods." NAQP offers publications that relate exclusively to quick and smaller commercial printing production. See Chapter 2 for addresses of each of these associations.

The second source for published standards is offered in books and textbooks serving industry and education. The major advantages of such books are they are inexpensive and fairly easy to obtain; the primary disadvantage is that they sometimes address older technology and may not effectively cover current machines and newer technological processes. *Estimating Standards for Printers* by Fred Hoch has been popular for many years, yet may be too out-of-date with current technology. Gerald Silver's two books—*Introduction to Printing Estimating* (Kendall/Hunt Publishing Co., Dubuque, IA) and *Professional Printing Estimating* (North American Publishing Co., Philadelphia, PA)—both contain production standards and related estimating data. Jack Klasnic's *In-Plant Printing Handbook* (GAMA Publications, Salem, NH) contains a fair amount of estimating information including production standards. Don Piercy's *Simplified Estimating and Pricing for Greater Profit* (Printing Industries of the Gulf Coast, Houston, TX) covers both estimating and pricing theory, largely oriented to the smaller commercial printer or quick printer. Another book for the quick printer or smaller commercial printer addressing productivity and costs is Larry Hunt's *Keys to Successful Quick Printing* (Larry Hunt, Dunedin, FL 34698).

The third source of published information, one that offers various amounts of standard production data plus articles on estimating and related topics, is through trade magazines serving the graphic arts. Major monthly periodicals serving the broad-based commercial industry include *Graphic Arts Monthly, American Printer* and *Printing Impressions.* Two magazines—*Quick Printing Magazine* and *Instant & Small Commercial Printer*—serve the quick printer with excellent articles including many addressing price list development, estimating and marketing issues, while *High Volume Printing* directs it's editorial and advertising efforts to the needs of larger web and sheetfed printer.

The fourth source is represented by the published catalogs and regional studies that contain both production standards and cost rates. Perhaps the catalog in widest use is the *Franklin Catalog* (Porte Publishing Company, Salt Lake City, UT), described in Chapter 1. Franklin is a system for pricing printing and contains various production standards as well. A number of PIA regional offices produce standards that are developed for their own geographic membership, equipment, and staffing considerations. Printing Industries of Southern California (PIA-SC) publishes the *Blue Book of Production Standards and Costs for the Printing Industry*, updated annually and a very popular reference for estimators throughout the commercial printing industry. Other PIA regional offices, such as Association of the Graphic Arts (New York City), Printing Industries of the Carolinas, Printing Industry of Ohio, Pacific Printing Industries, and Printing Industries of Illinois provide similar information for their membership.

Published information on printing estimating, including production standards, must be closely scrutinized. The type of publication and the anticipated user have a great deal to do with the accuracy and reliability of information. However, even though the accuracy may vary, published data is very modest in cost and, for the most part, requires little time to obtain. Also, published sources are an excellent method to cross-check existing production standards and other estimating data.

Standard Production Data Borrowed from the Competition

The problems that exist when pricing by competition (Chapter 2) are very much the same as those encountered when borrowing production data from the competition. Inaccuracies of data are possible because equipment and production techniques may not be the same. Also, gathering and verifying of such data are sometimes awkward.

While the cost for obtaining such information may be modest, and there will be little impact on plant production employees, the time needed to gather the data may fluctuate tremendously.

In regions where large numbers of printing establishments exist, exchange of such information is not uncommon. Typically the information is shared as employees move from one printing company to another, i.e., they take to their new employer the standards and other production information they learned from their previous employer. With quick printers and smaller commercial printers, the equipment may vary somewhat but the time and hour rates may be very similar. In larger commercial plants, which have many different avenues of production and different equipment between companies, the exchange of production data and hour rates is less similar and therefore less interchange of data is likely. Of course, when competitive pressures in a market niche are intense, sharing of production standards and methods between competing firms is much more closely scrutinized by printing management.

Opinion varies significantly with respect to how much production data should be shared among printers. Certainly one of the functions of trade associations—PIA, NAPL, NAQP and others—should be to encourage exchange of such information for the betterment of the industry. The International Association of Printing House Craftsmen (IAPHC), a social club for printing personnel from all phases of the industry with affiliate branches across the country, openly promotes such "sharing of knowledge."

Standard Production Data Developed by Equipment Suppliers

Equipment suppliers are a potentially good source for production information when the purchase of a new piece of equipment is being considered. Small equipment items require a modest amount of standard data. The sellers of large printing equipment, however, usually offer extensive production standards for the purchaser to compare with competitive equipment. Sometimes equipment manufacturers are willing to develop a comparison study between existing equipment in the plant and the equipment under consideration for purchase. Usually, this study will be done by a sales representative, a sales team, or, in the case of a very large purchase, a group of staff consultants working for the equipment supplier. Because this study is undertaken by the equipment manufacturer, the cost to the printing plant is very small.

The data generated and used by these suppliers are usually accurate and provide a good base from which to begin estimating once the equipment is operational. Of course, there may be a tendency on the part of the supplier to present the standard data under optimum or exaggerated conditions. For this reason, standard data generated by equipment suppliers should be reevaluated following the break-in period of the equipment. The supplier also will not include in the analysis of standard production data those less determinable conditions affecting the equipment, such as changes in markets, varying degrees of quality of product, and training and skill level of personnel operating the equipment. These factors may have a tremendous bearing on production standards, and the manufacturer of a piece of equipment cannot possibly include such assessments.

Standard Production Data Based on Time and Motion Study

This procedure is the most exacting and precise method for determining production standards. Time and motion study is considered a plant engineering function and is normally not done in small or medium-sized printing companies because of the extreme expense. Large printing companies have a tremendous array of equipment and the capital necessary to complete time and motion analysis.

Motion study is a procedure by which a particular job is broken down into its component elements and analyzed in terms of the hand and body movements involved. It leads to work simplification procedures or methods engineering. Motion study makes use of charts, graphs, and photography to break all operations into detailed specific components. Once the analysis is completed, other methods to produce the same product are investigated, with emphasis on increasing efficiency. Such operations analysis is particularly useful in manufacturing areas that require considerable handwork and the establishment of manual standards, such as the binding and finishing area of a printing plant. It is also used for the establishment of integrated standards. The overall intention is to streamline procedures and techniques, making them more efficient.

Once the best, most efficient method has been developed and tested, *time study* may be used to establish standard production times. That is, standard times are developed for each specific hand or body movement or for an entire sequence of movements together. A good time study engineer is usually a good methods engineer and vice versa, even though the job duties of each are different. In many large printing plants, one engineer may establish both standard methods and standard times.

If a specialized small printing facility considers motion and time analysis, they should consider hiring a consulting engineer; rarely does the printing manager have the tools, skills, or time necessary to effectively perform this task. Because cost can be very high, consideration should be given to the expected cost savings over time in relation to the full cost of the engineer or consultant. Sometimes this cost-effective determination is difficult to make.

Motion and time study may not be popular with plant employees. Generally, the attitude is against such techniques as micromotion analysis, flowcharting, and photographing employees performing their jobs. Thus, it is desirable that plant employees be informed well in advance of any such investigation, with perhaps incentives provided for their cooperation. If such notification is not given, a work stoppage or slowdown might result. When the announcement is made, it should be pointed out that no one individual will be singled out for study. The methods/time engineer, with the aid of management and employees, will select and study a sample of sufficient size to develop reliable standard data for the company.

Motion and time study is one of the most exacting methods by which plant production standards can be established. It is also expensive, time-consuming, and may have undesirable effects on the company labor force.

3.4 Keeping Production Standards Current

It is common for changes in production techniques, equipment, and personnel to occur as time passes, requiring modification of standard production data. The following alterations represent the usual range of changes that necessitate revision of standards:

New equipment or labor-saving devices added to existing equipment that provide for increased output in the plant;
Adoption of new materials that expedite production or modify the quality of the product;
Any physical change or modification of a product;
Revision of procedural standards because of greater specialization;
A change in the flow of production because of any revision of the physical plant—for example, a new layout in the existing or new building, the sale of a specific process to another printer, and changes that modify production lines and procedures;
A change in the mix of materials on the buyout list, thereby reducing output of that product in the plant;
Increased experience levels of shop personnel through additional training or improved on-the-job competence.

Updating production standards is necessarily an ongoing task in plants that use historical standards and job summaries. When standards are based on intuition, derived from published sources, borrowed from the competition, based on manufacturer's data, or developed through motion and time study, periodic updating must still occur and should especially take into account the range of changes first listed.

Opinions vary as to how frequently production standards should be updated. Some plants, especially those with computers and job costing in place, have an ongoing procedure. Others, such as firms that complete manual job costing, modify their production standards annually. Of course, production standards should be frequently reviewed and updated when new, faster equipment is put into use by the company or when technological changes or work flow modifications are made by the firm.

3.5 Development of Budgeted Hour Cost Rates

The development and use of budgeted hour cost rates (BHRs), sometimes termed machine hour rates (MHRs) or all-inclusive hour rates (AIHRs) is a reasonably common procedure in the commercial printing industry and higher volume printers, and less common with smaller commercial printers and quick printers. The intent of the technique is to accurately recover all costs incurred in the production, sales, and administration of the printed product. The procedure must include all elements of cost for which the customer must pay: equipment costs, costs for rent and heat, the cost of labor, and support labor costs, among others. At the same time, the customer must not be overcharged for cost items not actually required in the production of the product. Thus, cost recovery is a "fine-line" procedure. If too much is charged, prices will be too high, threatening to reduce volume in the plant, which might lead to layoffs and other negative actions; if too little is recovered, prices will be too low and perhaps no profit will be made.

It is essential that BHRs be as accurate as possible, reflecting company costs without inflating or minimizing them.

As detailed in Section 1.5, the usual cost-estimating procedure includes breaking the job into operational components and then assigning standard production times to each individual operation. Since each manufacturing portion of the production plan requires different equipment and employee skills, a separate cost rate must be developed for each portion to reflect these differences. If the estimated time is accurate and the BHR fully provides for all operational costs incurred in the production area, then the multiplied result of the two figures represents the manufacturing cost of the product. Along with paper, ink, film, and plates, which are substantial in cost, materials whose use can be noted and quantified with regard to the manufacture of a particular product are added on. Those materials that are more difficult to identify specifically with any one job, such as solvents, industrial towels, and lithographic blankets, are included in the BHR recovery system. For a process camera center, the following formula would apply:

estimated time to expose and process one half tone × budgeted hour cost rate
+ film and chemistry cost = total cost for halftone

The initial establishment of a BHR system should be completed by someone familiar and experienced with variations of the procedure—a company accountant, a production management staff member or estimator, a team of plant personnel with each knowledgeable in a certain area, or outside consultants. Various industry associations—specifically NAPL and PIA—have worked extensively with BHRs and have personnel experienced in the installation of such procedures in a printing plant. In fact, some progressive local associations, such as Printing Industry of Central Ohio and Printing Industries of Southern California, have developed complete BHR systems for their regional membership. The technique presented in this book was initially developed for members of Printing Industry of Central Ohio.

There are some important general rules for proper use of BHRs. It is most important that each plant apply its own individual cost figures when starting the BHR procedure because costs can vary extensively due to many factors that act on production, such as company finances, purchasing methods for equipment, depreciation, and so on. For example, Plant A and Plant B have the same lithographic press. Plant A has purchased it outright with inherited capital (no interest charges), has located it in a building with low rent (but a bad area of town, which increases fire and sprinkler insurance costs), and pays an operator $12.50 per hour (no fringe benefits) to run it. Plant B, on the other hand, is financing the press over 3 years, has located it in a building with high rent (but excellent fire protection), and pays an operator $9.25 per hour (with paid-up health insurance). Thus, each company will have different cost values for the same press, so the same BHR will not apply equally to each. The point is that accuracy in BHRs is possible only when such rates are tailor-made to fit the company. Costs simply vary too much to be applied generally.

The BHR sheet in Figure 3.3 clearly identifies equipment specifications and employee costs, fixed and variable costs, and (at the bottom of the sheet) manufacturing costs and BHR values. The rest of this section will provide a discussion of each cost individually. All dollar amounts (to be written in the right-hand column) should represent annual costs; they may be rounded to even dollar figures if desired. The technique is one in which costs are built from the four fixed cost items, through the variable costs, to total factory cost and total annual center cost. Some percentages apply as a sum of all preceding costs.

The basic hour value is *chargeable hours*—that is, those production hours that will be paid for by a customer or client. In effect, chargeable hours are derived by determining total hours annually, subtracting vacations and holidays (which are paid days off), and then applying a utilization percentage to adjust the hour amount to what normally is productive time. Both cost and time values are developed on a yearly basis.

Once the system is in operation, there may be a tendency to use the same cost rates for too long a period of time. It is suggested that BHRs be reviewed at least once a year. Of course, in times of steeply rising costs, such reevaluation could be performed even more frequently. When completing the updating program, individual costs should be reviewed line by line. Chargeable hours should also be reinvestigated and revised based on actual plant scheduling history. Management should assign this task to plant personnel who will monitor the program continually.

Many current CAE packages contain ready-to-run programs that quickly and accurately calculate BHRs. For the printer who purchases such computer capability, all that must be done is to identify those areas that require individualized data—that is, data specific to the printer's own company. Once given this variable information, the computer rapidly develops the BHRs, by center, for the printing facility. It is thus possible for the printer to update BHRs on a frequent basis—perhaps monthly—and adjust such hour rates for variations in chargeable hours, wages, and so on. The computer, of course, saves considerable time for company staff who would normally

BUDGETED HOUR COST RATE

Title of center__________________

SPECIFICATIONS:

Cost of equipment (incl.install) . . .	______	Wage scale	______per hr.
No. employees in center	______		______per hr.
Sq. ft. working area(incl.aisles) . .	______	Paid vacation	______days
Total motor horsepower.	______	Holidays,etc	______days
Arc lamp wattage.	______	Weekly hours	______

	ONE SHIFT
FIXED COSTS	
1. Fixed overhead (rent/heat/general utilities, based on $3.25/sq. ft).	
2. Insurance on equipment ($5.00 per each $1,000 valuation)	
3. Property taxes applicable to center ($1.25 per $100 valuation).	
4. Depreciation (10% straight line)	
Total fixed costs (lines 1-4)	
VARIABLE COSTS	
5. Wages - direct labor at straight time hours	
6. Wages - indirect labor (24.5% direct labor payroll)	
Total: direct and indirect labor (lines 5-6)	
7. Light/power (10¢/KWH): Light______ Power______Arcs ______	
Total: fixed costs+labor cost+light/power/arc (lines 1-7)	
8. Departmental direct supplies & expenses (10.3% lines 1-7)	
9. Payroll taxes: SocSec, Unemploy & Workman Comp.(10.5% total payroll)	
10. Employee benefits: Group insurance & pension (9.2% total payroll).	
11. Repairs to equipment (2% gross investment)	
Total: lines 8 - 11.	
DIRECT FACTORY COST (total lines 1-11)	
12. General factory expense (17.1% of Direct Factory Cost).	
TOTAL FACTORY COST	
13. Administrative and selling overhead (48% of Total Factory Cost)	
TOTAL ANNUAL CENTER COST.	

Manufacturing cost per chargeable hr. (Total factory cost ÷ chargeable hrs)

80%:__________ annual hours . . . . . $__________ per hour

70%:__________ annual hours . . . . . $__________ per hour

60%:__________ annual hours . . . . . $__________ per hour

Budgeted Hour Cost per chargeable hr. (Total center cost ÷ chargeable hrs)

80%:__________ annual hours . . . . . $__________ per hour

70%:__________ annual hours . . . . . $__________ per hour

60%:__________ annual hours . . . . . $__________ per hour

Figure 3.3 Budgeted hour cost rate sheet (Percentages and dollar amounts are for example only.)

develop such BHRs manually. Also, CAE allows for quick and easy storage of such hour rates; no one needs to write them down for manual storage in a book or other company document. Both BHRs and standard production times are essential to utilization of CAE programs.

Title of Center

The production *center* is a specifically defined area for production of some portion of the printed product. The center must be properly identified, defined, and given a title. It may contain one piece of equipment operated by one employee, one piece of equipment with a complement of workers, or perhaps one employee working essentially without a piece of equipment.

Specifications

The specifications of a production center establish the parameters of the center and provide the basis on which the BHR is established. The specifications should be carefully determined and tailored to the specific center and the specific printing company. The following information provides details with respect to each specification item needed, referenced to Figure 3.3.

Cost of equipment: This should include all installation costs and the cost of any specific peripherals. In general, three dollar figures could be used here, each representing a different view of how the company wishes to recover its costs:

1. The dollar amount paid by the company when it originally purchased the equipment, which continues to be the figure used each time the BHR is recalculated.
2. The depreciated cost for the equipment, which is a continuously declining cost figure each time the BHR is recalculated.
3. The anticipated replacement cost of the equipment, which would be a higher-than-normal cost and would tend to inflate the BHR yet provide a cushion for replacement of the equipment item when necessary. Given the intensely competitive nature of the printing industry, inflating job costs typically inflates the price of the job, thus reducing volume in a printing firm. For this reason replacement cost data is not usually used in building BHRs.

Equipment that is usually used together in production, such as an automatic film processor and process camera, should be combined together when building BHRs. In the case of equipment that is sometimes shared and other times used as "stand-alone," or serves more than one other cost center continuously, separate BHRs for each piece of equipment may be required. Analysis of plant production operations is needed to determine these relationships.

Number of employees in the center: This indicates the number of employees who productively work in the center. There may be a fraction of a person when multiple pieces of equipment are operated simultaneously by one employee or there may be more than one person if a crew is necessary to operate the center.

Square feet of working area: The working area measurement should include all productive space and aisles. Space which is indirectly productive to the center, such as storage area, is not normally included in this figure.

Total motor horsepower: Determined from equipment specifications actually operating in the center. If more than one motor is used in the center, or if some motors for a center operate at different rates or at different times than others, a weighting factor should be used to evaluate the horsepower demand.

Arc lamp wattage: Includes the input wattage required for high-intensity light sources used to expose photographic materials, including cameras, contract frames, platemakers, typesetting, and scanning equipment.

Wage scale: This should indicate the hourly dollar amount paid to each employee working directly in the center. If the center requires a crew, then listing each wage paid, or the sum of these wages, is necessary. Using the BHR process detailed in Figure 3.3, a crew is considered as one operating group, as if it were a single employee operating the center.

Paid vacation: This states the number of vacation days the person or crew will be paid yet be unavailable to work in the center.

Holidays: This indicates the number of holidays the person or crew will be paid yet be unavailable to work in the center.

Weekly hours: This is the number of weekly hours the center will operate, typically the same as the number of hours per week week company operates.

It is important to note that the BHR recovery process described here and demonstrated in Figure 3.3 does not include overtime worked in the center, nor the cost impact of two-shift and three-shift operations. It also assumes that when a person or crew is on paid leave (vacations and holidays), the center will shut down and not operate using a substitute employee or crew. Each of these may offer significant cost impact on the center and the company and should be addressed by modifying Figure 3.3 to accommodate such variations.

Fixed Costs

Fixed costs are those costs that do not change with changes in production output in the center. In all cases, the values assigned must be determined for the individual plant. The base values indicated in the chart may be representative but are not intended as established standards throughout the industry.

Fixed Overhead (Line 1). Fixed overhead provides for recovery of all fixed costs necessary to maintain the environment of the production center. Included are rent, heating/air conditioning, and general utility costs, such as water, sewer, and gas. Any electrical power costs used to maintain the general environment should be included. However, power requirements for specific equipment operation, considered a variable cost, will be recovered separately on line 7 of the form. Cost basis is $3.25 per square foot per year, which is determined by summing all environmental costs as specified above over a yearly period and then dividing them back on a center-by-center square footage basis.

Insurance (Line 2). This item covers all insurance paid on equipment and materials in the building. Types of policies vary, as do amounts for protection. Normally, fire and sprinkler insurance is minimum coverage; some policies include business interruption coverage also. The individual plant can determine this cost by summing the insurance premiums for the year and apportioning this figure back over the center covered by the policy.

Property Taxes Applicable to Center (Line 3). Property taxes fluctuate widely across the country, but must be included in center costs. The best procedure is to find out how such taxes are paid and use that system, center by center, for totalling tax costs. Since rates change annually in many sections of the country—in the past few years some areas of the country have experienced large increases—this element of fixed cost must be carefully watched.

Depreciation (Line 4). Depreciation is the allowance for normal wearout of equipment. Through the depreciation process, the cost of existing equipment or replacement equipment is recovered. The dollar figure used on this line may reflect, or use as a base, the replacement cost of a new piece of equipment, the current full cost of the existing equipment, or the current depreciated cost of the existing equipment. Top management must decide which base to use when setting up the BHR procedure in the plant and apply that figure as if this dollar figure represented the anticipated depreciation of the existing piece of equipment currently in the center.

If the printing company leases the equipment instead of owning it, the full annual lease cost would be written on this line. In sum, whether the cost basis is in depreciated dollars (from ownership) or out-of-pocket dollars paid for leasing, the amount on this line represents a dollar amount to be paid back to the owners or lessors.

Variable Costs

Variable costs are those costs that fluctuate as production output changes. The major variable cost incurred in printing manufacturing is the cost of labor and the associated support labor charges. Of the seven variable cost categories on the BHR sheet in Figure 3.3, four are directly related to labor costs. The other three costs are for light and power (line 7), departmental direct supplies and expenses (line 8), and repairs to equipment (line 11).

Direct Labor Wages (Line 5). This figure is determined by multiplying the straight time hourly rate—for example, the wage scale in the specifications—by the number of annual hours worked by production personnel in the center. Since employees are typically paid for holidays and vacations, these hours also must be included in the number of annual hours worked.

Indirect Labor Wages (Line 6). This item covers all support persons working in the plant and having some relationship to the center. This group might include working supervisors, miscellaneous employees such as custodial help, porters, delivery persons—that is, employees who support some segment of the production center, but are not directly tied to its production. The best procedure for determining the percentage of indirect to direct labor is to categorize all support labor personnel and then sum up the costs for such indirect labor and divide it by the direct labor cost for the center. An expected range of indirect to direct labor is between 15% and 25% of direct labor.

Light and Power (Line 7). Light and power costs are most accurately apportioned to each cost center by summing annual lighting and electrical power costs, then allocating these costs back to each working area on a square footage or usage basis. Another procedure, strictly a mathematical technique, provides for approximate lighting and power cost determinations and is presented in the examples that follow.

The cost of electric power used in the following examples is $0.10 per kilowatt hour (kWh), but the rate varies throughout the country, contingent upon geographical location and demand usage for the individual company. While electrical power is the typical energy source for a printing plant, natural gas for press dryers and other energy systems are also used. Because natural gas is used in a limited number of centers, it should be prorated according to approximate usage from the annual bills.

Example 3.1

There is a total of 150 square feet in a duplicator press center, and the lighting level is 3 watts of coverage per square foot. Power cost is $0.10 per kilowatt hour. Determine the cost annually to light this center operating one shift, at 40 hours per week.

Three variables must be known to make cost of lighting calculations:

Number of annual hours worked,
Lighting coverage in watts per square foot,
Cost per kilowatt hour for power.

The steps in this example are represented by the following formula that can be used for lighting cost problems in general:

lighting cost = [(number of annual hours worked × wattage per square foot) ÷ 1000 watts per kilowatt)] × cost per kilowatt hour × number of square feet in center

1000 watts per kilowatt is a constant.

Solution

Step 1 Determine the number of kilowatts per square foot:

3 W/sq ft ÷ 1000 W/kW = 0.003 kW/sq ft

Step 2 Determine the number of annual operating hours:

40 hr/wk × 52 wk = 2080 hr/yr

Step 3 Determine the number of annual kilowatt hours per square foot:

0.003 kW/sq ft × 2080 hr/yr = 6.24 kWh/sq ÷ ft/yr

Step 4 Find the annual lighting cost per square foot:

6.24 kWh/sq ft/yr × $0.10/kWh = $0.624/sq ft/yr

Step 5 Find the lighting cost per year for the press center:

$0.624/sq ft/yr × 150 sq ft = $93.60/yr

Example 3.2

Determine the cost annually to operate a 3000 watt platemaker center (3000 watt are lamp) operating one shift at 40 hours per week. The platemaking unit is in operation 50% of the time, with the arc lamp in use. Power cost is $0.10 per kilowatt hour.

Four variables must be known to make cost of arc lamp calculations:

Number of hours worked per week,
Arc lamp input wattage for the specific platemaker or camera,
The cost per kilowatt hour for power,
The usage adjustment factor.

The steps in this example are represented by the following formula that can be used for arc lamp cost problems in general:

arc lamp cost = (number of rated input watts of source ÷ 1000 watts per kilowatt) × number of annual hours worked × cost per kilowatt hour

Solution

Step 1 Determine the number of kilowatts per hour:

3000 W ÷ 1000 W/kW = 3 kWh

Step 2 Determine the number of annual operating hours:

40 hr/wk × 52 wk = 2080 hr/yr

Step 3 Determine the number of annual kilowatt hours:

2080 hr/yr × 3 kWh = 6240 kWh/yr

Step 4 Make usage adjustments:

6240 kWh/yr × 50% = 3120 kWh/yr

Step 5 Find the arc lamp cost per year for the platemaker center:

3120 kWh/yr × $0.10/kWh = $312.00/yr

Example 3.3

Determine the annual cost of electrical power for one 4 horsepower (hp) motor used to power a single-color offset press center. The press operates a 40 hour week. Cost for electrical power is $0.10 per kilowatt hour

Three variables must be known to make cost of power for motors calculations:

The number of annual hours worked in the center,
Aggregate total of all motor horsepower in the center,
The cots per kilowatt hour for electricity.

The steps in this example are represented by the following formula that can be used for cost of power problems in general:

power cost = total horsepower of motors × number of annual hours worked × 0.746 kilowatt per horsepower × cost per kilowatt hour

0.746 kilowatt per horsepower is a constant.

Solution

Step 1 Determine the number of annual operating hours:

40 hr/wk × 52 wk = 2080 hr/yr

Step 2 Determine the number of annual horsepower hours (hp-hr):

2080 hr/yr × 4 hp = 8320 hp-hr/yr

Step 3 Determine the number of annual kilowatt hours fusing 0.746 kilowatt per horsepower as a constant):

0.706 kW/hp × 8320 hp-hr/yr = 6207 kWh/yr

Step 4 Find the annual cost of electrical power to run the motor:

6207 kWh/yr × $0.10/kWh = $620.70/yr

Departmental Direct Supplies and Expenses (Line 8). This item is a catchall charge to recover costs for materials used in the center that are not specifically traceable or quantifiable to any one job but are necessary to produce the products in the center. Some examples of such direct supplies are solvents, lithographic blankets, cotton wipes, shop towels, writing tools, opaque, and opaquing tools. Equipment repair costs are not included, nor are materials that can be charged directly to the customer, unless those charges would represent a minor cost. For example, some companies might include ink used for a job or masking materials for stripping as a departmental direct supply. The best way to determine the percentage figure for departmental direct supplies and expenses is to sum the costs of those items to be included in the category and divide that dollar figure by the sum of all fixed charges (lines 1–4), plus all wages (lines 5 and 6), plus light and power (line 7). Generally, departmental direct supplies and expenses, as a percentage of labor, power, and fixed charges, are between 5% and 15%.

Employee Payroll Taxes and Insurance (Line 9). This item is a major category of support labor costs sometimes hidden or unknown to employees. Included are social security taxes, unemployment insurance, and workman's compensation insurance. Calculation of the percentage that applies is completed by summing social security unemployment, and workman's compensation costs or premiums paid annually and dividing that figure by the total straight time wages paid to the company's total factory payroll, covering both direct and indirect employees.

Social security taxes are mandated by the federal government, which sets the percentage of contribution and the ceiling. In recent years, both have been increasing dramatically. The system requires that both the company and the employee contribute separate dollar amounts. The company's portion is recovered through the BHR costing procedure, while the employee's is deducted from his or her paycheck in proportional amounts over the year.

Unemployment compensation insurance is paid by the employer to support the unemployment system. The rate of contribution varies from state to state and with respect to the number of persons drawing on the system at any given time.

Workman's compensation insurance is a procedure used to compensate employees for work-related accidents. Normally, the company carries an insurance policy for this contingency based on their own safety rating or one for the industry. The methods and dollar charges vary from state to state, and a company should use. their own premium cost or contribution rate, as applicable.

Employee Benefits (Line 10). This category lumps together two important and expensive fringe benefits typically provided by an employer: group health insurance and pension/retirement benefits. The percentage figure used can be determined by summing together total premium costs and pension/retirement contributions that are employer-paid and dividing that figure by total factory payroll, covering both direct and indirect personnel. Typically, the percentage figure falls between 5% and 15% of total factory payroll.

Group health insurance covers all company-paid insurance premiums offered to the employee for health care. Generally, such policies include major medical coverage and hospitalization. Dental and optical plans may also be included if provided by the employer. If a group life insurance plan is offered to factory employees, it should be included here also.

Pension/retirement benefits include any dollar amounts paid by the employer for pension/retirement of factory personnel (those working in production centers). If the company pays only a portion of such costs, that amount should be recovered. Generally, pension/retirement approximates 4% of the factory employee's gross wage, but this percentage may vary depending on the age of the employee at the time of hire and other factors.

Repairs to Equipment (Line 11). The average cost for equipment repairs is 2% of the investment cost throughout the industry. However, the age and condition of equipment, usage level, and preventive maintenance procedures all play a part with respect to equipment repairs. If possible, the actual annual repair cost should be used. When such information is not available, discussion with local suppliers may help to establish a realistic figure to use for repair costs.

General Factory Expenses (Line 12). These expenses are those that cannot be specifically allocated to another cost area. Examples are expenses for uniforms, building maintenance, painting, lawn care, and trash pickup. Any expense related to the general operation of the printing plant should be included. To determine the dollar amount for costing, the annual charges for all items should be totalled and then allocated back to each production center.

Administrative and Selling Overhead (Line 13). This item is actually a fixed cost. This cost recovers for all administrative salaries and expenses (principally for top and middle management and management staff) and for all sales salaries, expenses, and commissions. When actual dollar figures can be obtained, they should be used. If percentages are used, they are based on total factory cost as a function of sales and administrative expenses.

Manufacturing Costs per Chargeable Hour

This figure is determined by dividing the number of chargeable hours into the total factory cost (which does not include administrative and selling charges). Chargeable hours are calculated by determining the number of total annual hours in the particular center, subtracting the sum total of vacations, holidays, jury duty, and similar items, and then multiplying that result by an estimated percentage of utilization for that center.

Manufacturing costs are typically used when a printing firm wishes to evaluate the direct cost to make or manufacture the product, without administrative and selling overhead. For example, if a printer was required to restrip a job due to an error made in the stripping department—commonly termed a "house error" (or HE)—the actual cost to the company would be based on the manufacturing cost of the product, not the budgeted hour cost. There are also instances when management wishes to look at the direct cost to make a product. This might include jobs that do not involve significant administrative or selling efforts such as work sold by printing brokers or from over-the-counter sales.

Example 3.4

One employee works 37.5 hours per week in a platemaking center. He receives 3 weeks paid vacation and 10 paid holidays. Production management has evaluated his level of utilization (actual working time) at 70% based on studies recently made in the plant. How many chargeable hours will this employee work annually?

Solution

Step 1 Find the total annual hours:

37.5 hr/wk × 52 wk = 1950 hr/yr

Step 2 Determine annual time-off hours:

25 day/yr × 7.5 hr/day = 187.5 hr/yr

Step 3 Find the maximum annual chargeable hours:

1950 hr/yr − 187.5 hr/yr = 1762.5 hr/yr

Step 4 Determine the actual annual chargeable hours (70%):

1762.5 hr/yr × 70% = 1233.75 hr/yr

To determine manufacturing costs per chargeable hour for the platemaking center in Example 3.4, the total factory cost is $54,758 annually. If the annual cost is divided by the number of chargeable hours, then:

$$\$54{,}758 \div 1233.75 \text{ hr} = \$44.38/\text{hr}$$

which is the manufacturing cost per chargeable hour. This figure means that for each hour this platemaker is used for making a plate, brownline, or blueline, it will cost $44.38 to cover all employee and equipment charges. The administration and selling of the product has not been included, and the cost of the metal plate or proofing material has not been recovered.

Budgeted Hour Cost Rate per Chargeable Hour

BHRs recover for all costs incurred in the manufacture of the product, as well as the administrative and selling overhead. In Example 3.4, the Total Factory Cost was $54,758. If we add 48% for administrative and selling overhead, the annual center cost then becomes $81,043. If this cost is then divided by the number of chargeable hours at 70% utilization (actual hours worked), which is 1233.75 hours, the result is a BHR of $65.69. Thus, for one hour of platemaking time and to recover all costs from the client, we would have to charge the customer $65.69. If any costs increase or the utilization fluctuates downward, the cost per hour will increase. If utilization is improved or the costs go down, the BHR will be reduced accordingly.

Figures 3.4 and 3.5 are two completed BHR sheets. The first (Figure 3.4) is for a NuArc platemaker and the second (Figure 3.5) is a Heidelberg SORM single-color press. Both figures are presented for example only and are not intended to reflect costs or hour rates for any particular printing operation.

The appendix of this chapter provides all-inclusive BHRs representative of the Southern California geographic region. These rates are based on prevailing 1988 wages for Southern California in a 40 hour work week. They have been reproduced with permission from the Printing Industries Association, Inc. of Southern California from their *1988-89 Blue Book of Production Standards and Costs for the Printing Industry*. Copies of the complete book, which is revised annually, may be obtained by contacting the Management Services Department, PIA-SC, P.O. Box 91-1151, Los Angeles, CA 90091.

It is essential to note that BHRs are not prices. They represent a composite of all costs to produce work in a particular location in the plant with defined personnel and equipment and at a specific rate of production. In fact, BHRs multiplied by the number of estimated hours anticipated for a job equals the expected breakeven cost if the company received the order, with no profit addition or other markups. In many printing plants, the total job cost serves as the basis upon which the final price of the job is established (see Section 2.5 for pricing information).

BUDGETED HOUR COST RATE

Title of center NuArc Platemaker

SPECIFICATIONS:

Cost of equipment (incl.install) . . .	$ 5,100	Wage scale	$14.20 per hr.
No. employees in center	1		per hr.
Sq. ft. working area(incl.aisles) . .	175	Paid vacation	15 days
Total motor horsepower.	0.75	Holidays,etc	10 days
Arc lamp wattage.	3,000	Weekly hours	37½

	ONE SHIFT
FIXED COSTS	
1. Fixed overhead (rent/heat/general utilities, based on $3.25/sq. ft).	$ 568.75
2. Insurance on equipment ($5.00 per each $1,000 valuation)	25.50
3. Property taxes applicable to center ($1.25 per $100 valuation).	63.75
4. Depreciation (10% straight line)	510.00
Total fixed costs (lines 1-4)	1,168.00
VARIABLE COSTS	
5. Wages - direct labor at straight time hours	27,690.00
6. Wages - indirect labor (24.5% direct labor payroll)	6,784.05
Total: direct and indirect labor (lines 5-6)	34,474.05
7. Light/power (10¢/KWH): Light $102.38 Power $109.10 Arcs $292.50	503.98
Total: fixed costs+labor cost+light/power/arc (lines 1-7)	36,146.03
8. Departmental direct supplies & expenses (10.3% lines 1-7)	3,723.04
9. Payroll taxes: SocSec, Unemploy & Workman Comp.(10.5% total payroll)	3,619.78
10. Employee benefits: Group insurance & pension (9.2% total payroll).	3,171.61
11. Repairs to equipment (2% gross investment)	102.00
Total: lines 8 - 11.	10,616.43
DIRECT FACTORY COST (total lines 1-11)	46,762.46
12. General factory expense (17.1% of Direct Factory Cost).	7,996.38
TOTAL FACTORY COST	54,758.84
13. Administrative and selling overhead (48% of Total Factory Cost)	26,284.24
TOTAL ANNUAL CENTER COST.	$ 81,043.08

Manufacturing cost per chargeable hr. (Total factory cost ÷ chargeable hrs)

80%: 1410 annual hours . . . . . $ 38.84 per hour

70%: 1233.75 annual hours . . . . . $ 44.38 per hour

60%: 1057.5 annual hours . . . . . $ 51.78 per hour

Budgeted Hour Cost per chargeable hr. (Total center cost ÷ chargeable hrs)

80%: 1410 annual hours . . . . . $ 57.48 per hour

70%: 1233.75 annual hours . . . . . $ 65.69 per hour

60%: 1057.5 annual hours . . . . . $ 76.64 per hour

Figure 3.4

BHR for a NuArc platemaker

BUDGETED HOUR COST RATE

Title of center: Heidelberg SORM 1/C Press (20½x29 max. pss)

SPECIFICATIONS:

Cost of equipment (incl.install) . . .	$ 95,800.00	Wage scale	$ 15.50 per hr.
No. employees in center	1		per hr.
Sq. ft. working area(incl.aisles) . .	275	Paid vacation	15 days
Total motor horsepower.	5.6	Holidays,etc	10 days
Arc lamp wattage.	0	Weekly hours	40

	ONE SHIFT
FIXED COSTS	
1. Fixed overhead (rent/heat/general utilities, based on $3.25/sq. ft).	$ 893.75
2. Insurance on equipment ($5.00 per each $1,000 valuation)	479.00
3. Property taxes applicable to center ($1.25 per $100 valuation).	1,197.50
4. Depreciation (10% straight line)	9,580.00
Total fixed costs (lines 1-4)	12,150.25
VARIABLE COSTS	
5. Wages - direct labor at straight time hours	32,240.00
6. Wages - indirect labor (24.5% direct labor payroll)	7,898.80
Total: direct and indirect labor (lines 5-6)	40,138.80
7. Light/power (10¢/KWH): Light $171.60 Power $868.94 Arcs 0	1,040.54
Total: fixed costs+labor cost+light/power/arc (lines 1-7)	53,329.59
8. Departmental direct supplies & expenses (10.3% lines 1-7)	5,492.95
9. Payroll taxes: SocSec, Unemploy & Workman Comp.(10.5% total payroll)	4,214.57
10. Employee benefits: Group insurance & pension (9.2% total payroll).	3,692.77
11. Repairs to equipment (2% gross investment)	1,916.00
Total: lines 8 - 11.	15,316.29
DIRECT FACTORY COST (total lines 1-11)	68,645.88
12. General factory expense (17.1% of Direct Factory Cost).	11,738.45
TOTAL FACTORY COST	80,364.33
13. Administrative and selling overhead (48% of Total Factory Cost)	38,584.48
TOTAL ANNUAL CENTER COST.	$118,968.81

Manufacturing cost per chargeable hr. (Total factory cost ÷ chargeable hrs)

80%: 1504	annual hours	$ 53.45	per hour
70%: 1316	annual hours	$ 61.08	per hour
60%: 1128	annual hours	$ 71.26	per hour

Budgeted Hour Cost per chargeable hr. (Total center cost ÷ chargeable hrs)

80%: 1504	annual hours	$ 79.10	per hour
70%: 1316	annual hours	$ 90.40	per hour
60%: 1128	annual hours	$ 105.47	per hour

Figure 3.5

BHR for a Heidelberg SORM single-color press

3.6 Overtime Considerations and Cost Savings for Additional Shifts

One of the critical production management decisions that must be made from time to time relates to the startup of a second or third shift. Typically, those cost centers that are the most labor and capital intensive are the ones that require this decision. One vital guideline used is the amount of overtime currently being worked in the center for which employees are paid time and one-half or double time. A general rule is that when overtime consistently approaches 40% of the straight time hours worked, a complete new shift should be considered. For example, if an employee running a single-color press consistently works 3 hours over each 8 hour day, which is a 37.5% increase in center usage, a complete additional shift should be seriously considered.

There are advantages to running second and/or third shifts. Essentially, additional shifts save money for the company by reducing overtime and spreading fixed costs for the center over a larger time block, thus reducing center costs. With added shifts, the company can significantly increase the volume of work without additional equipment purchases.

There are some other important considerations to be made, however, prior to the startup of an additional shift. First, the company should plan to increase the volume of the plant through increased sales efforts or other methods to bring in more work. Second, additional shifts—second shifts and especially third "graveyard" shifts—typically present both employee and supervisory staffing problems. Since many employees do not wish to be moved to a late afternoon/evening work time or do not wish to work at night (even though a shift differential pay increase from 5% to 20% over straight time pay will be earned), it may be hard to find competent personnel to work during the second or third shift. Third, employees working overtime make one and one-half or double their straight pay scale This situation will essentially end completely with the addition of a new shift, making these employees disgruntled at the loss of such extra income. Finally, additional shifts represent increased equipment wearout. For example. a press that normally might last 9 years on a one-shift basis would probably last only 3 years on a three-shift operation.

In terms of the major advantage of additional shifts—cost savings to the plant—some general rules can be made. Because the center's fixed costs are spread out over a larger working time and do not increase with increases in production, a significant savings in hour cost results. For example, such savings may range from 10% to 20% for a second shift BHR over the first shift and from 5% to 15% in BHR savings for the third shift over the second shift. Thus, for a center working on all shifts at 75% production and with a first shift BHR of $50.00, the second shift BHR would reduce to $42.50 per hour (a 15% savings), and a third shift would reduce the BHR to $39.10 per hour (an 8% savings over the second shift). It is important to note that the savings percentages may vary with respect to the type of equipment, operating factors, labor rates, and fringe benefits paid to employees, among some of the more important conditions.

3.7 Effects of Automation on BHRs and Chargeable Hours

As previously discussed, BHRs are used in conjunction with standard production times to determine estimated selling costs. One particularly important aspect to study relates to the effects of automation on BHRs, especially comparing new, automated equipment to older, less automated units. To make this comparison, refer to Figure 3.5. which is a BHR for a Heidelberg SORM press. Note the two "bottom line" components in Figure 3.5: the total annual center cost of $118,968.81 and the chargeable hours at 70% utilization, which is 1316 hours annually. The BHR for this press is then $118,968.81 divided by 1316 annual hours, which equals a rate of $90.40.

Now let us assume that this Heidelberg press is retrofitted with electronic controls to enable it to be 25% more productive. Consider the following examples.

Example 3.5

After a line-by-line analysis redoing the BHR sheet, we find the additional electronic gear increases our total annual center cost by $10,000—from $118,968.81 to $128,968.81—with no reduction in labor costs. Chargeable hours are reduced by 25%, too, since there has been no increase in volume in the center, which makes the chargeable hours (at 70% utilization) 987 hours. Determine the BHR for the retrofitted Heidelberg press.

Solution $128,968.81 ÷ 987 hr = $130.66/hr

Example 3.6

Total annual center cost remains at $128,968.81, with no reduction in labor costs. Chargeable hours are increased by 25% through increased sales efforts, putting them at 1316 hours (70% utilization). Determine the BHR for the Heidelberg press.

Solution $128,968.81 ÷ 1316 hr = $98.00 hr

Example 3.7

We find because of our nonunion environment and added electronic gear that we can save $10,000 in labor costs, reducing our total annual center cost to $108,968.81. However, with no increase in volume in the center, chargeable hours are 25% less, or 987 hours. Determine the BHR for the Heidelberg press.

Solution $108,968.81 ÷ 987 hr = $110.40 hr

Example 3.8

Our total annual center cost is $108,968.81 because of our $10,000 savings in labor costs. With a 25% increase in volume through increased sales, the chargeable hours increase to 1316 (at 70% utilization). Determine the BHR for the Heidelberg press.

Solution $108,968.81 ÷ 1316 hr = $82.80/hr

Even with arbitrary figures, the preceding examples demonstrate that the impact of automation on hourly costs can be significant. Both top management and the estimating staff have a responsibility to work out such annual cost and productivity values, which should include increasing sales volume to keep the automated equipment busy, so that BHRs are current and reasonable for the company.

3.8 Cost Markup

Cost markup is an additional dollar amount added to the cost of the job as profit for the company, or to the cost of materials, outside purchases (items not included in the BHR cost), and labor/overhead costs for various reasons. Thus, the term *markup* has two somewhat different applications in the industry—a markup on the estimated cost of the job—also called a profit markup—and markups on materials, buyouts and internal labor costs, usually completed during the estimating process.

In a major nationwide study by NAPL (December 1985 Special Report) based on an *American Printer* survey, markup percentages in all of these areas were determined. In absolute terms, there were 897 usable responses from more than 900 responses, with commercial printers representing 68% of the study respondents. Quick printers totaled 7% of the respondents, "other" such as screen printers represented 10% of the sample and scattered smaller percentages for business forms printers, trade services, book printers, and magazine printers. Almost 60% of the study respondents reported gross sales of under $1 million, with 24% of the participants having yearly sales of less than $200,000. In sum, this was a study of smaller printers, likely a fairly significant mix of smaller commercial printers and quick printers.

What Specific Areas are Marked Up?

Literally all printing companies—smaller commercial printers, quick printers, larger commercial printers and high volume printers—mark up costs in three general areas: (1) profit markup on the entire job based on an estimated cost, (2) material markups based on the individual cost of paper, ink, film, and plates, and (3) labor/service markups for buyouts and internal production of a job where labor costs represent the primary cost factor.

	Percent Estimating Markup by:					
Speciailty	Out-of-Pocket Hourly Rates	Manufacturing Hourly Rates	All-inclusive Hourly Rates	Price Lists	Other	Total*
General Commercial	7	15	61	16	7	106%
Quick Printing	8	6	38	57	9	118
Advertising Specialists	15	27	52	10	2	106
Business Forms	13	19	35	35	10	112
Trade Services	12	15	62	19	—	108
Book Printers	8	8	71	—	13	100
Magazine, Printers	—	7	71	28	—	106
Total %	8%	15%	58%	20%	7%	108%

*More than 100% indicates several respondents use more than one method.
All percentages are rounded to the next highest number.

Figure 3.6

Cost markup by specialty (Reproduced from NAPL's *Special Report* titled "Cost Markup in the Printing Industry," December 1985.)

Companies By Annual Sales	Percent Estimating Markup by:					
	Out-of-Pocket Hourly	Manufacturing Hourly Rates	All-Inclusive Hourly Rates	Price Lists	Other	Total*
$0-$200,000	17	13	35	31	11	107%
$200-$500,000	4	15	48	32	12	111
$500-$1 million	6	12	67	18	3	106
$1-$2 million	4	14	72	14	3	107
$2-$3 million	4	12	77	8	5	106
$3-$5 million	3	21	73	8	—	100
$5-$10 million	7	20	68	5	5	105
$10+ million	4	28	61	11%	—	104%

*More than 100% indicates several respondents use more than one method.
All percentages are rounded to the next highest number.

Figure 3.7

Cost markup by annual sales (Reproduced from NAPL's *Special Report* titled "Cost Markup in the Printing Industry," December 1985.)

Profit Markup

The profit markup is based on the estimated cost of the job using the cost estimating formula and other estimating/pricing systems discussed in Chapter 1 of this book. The accuracy of the estimating system is a critical factor in determining profit markup additions, since the more accurate the estimated (base) cost the more accurate and competitive the final price may be.

As noted in Figure 3.6 reproduced from the study, 58% of all respondents used budgeted hour rates (also known as all-inclusive hourly rates) to determine the cost basis of printed work, with 20% using price lists and 15% using manufacturing costs. Broken out by individual company category, 57% of the quick printer respondents used price lists as their first-choice costing base, with 38% of the quick printers using BHR rates. Commercial printers chose all-inclusive rates as their first costing system 61%, with price lists in second place at 16%.

When broken out by annual sales volume as indicated in Figure 3.7, all-inclusive hourly rates and price lists dominate the two selected systems, with manufacturing hourly costs and out-of-pocket hourly costs indicated less. One conclusion evident from this figure is that a higher percentage of small volume printers tend to use price lists, while higher volume printers move to all-inclusive hour rates and manufacturing hourly rates. Regional differences were also studied, with the most noticeable variation evident with New York printers in an almost even three-way split with 30% using manufacturing hourly costs rates, 33% using all-inclusive hourly cost rates, and 36% using price lists.

Markups on Paper, Ink, Film, and Plates

As noted in Figure 3.8, the average markup for materials by printing industry specialty was 31% for paper, 30% for ink, 44% for film, and 44% for plates. Only one percent of the study respondents indicated that they added no markup on paper, two percent indicated they did not mark up film or plates, and six percent said they did not mark up ink.

Based on Figure 3.8, quick printers marked up all materials higher than any other single specialty: paper at 43%, ink at 45%, film at 64%, and plates at 65%. Lowest markups by specialty was paper for magazine printers with a 20% markup, ink for book printers with a 17% markup, film for magazine printers with a 26% markup, and plates for trade services with a 23% markup.

Markups for paper decreased almost directly to the size of the company: firms with less than $200,000 in annual sales had an average markup of 45%, while firms with sales over $10 million used an average markup of about 16%. When paper represented a disproportionately large part of a specific job cost, 57% of the respondents indicated they would reduce their normal markup.

As noted in Figure 3.9, paper markups vary by geographic region with the western states using the highest at a 40% markup, and the southeastern states the lowest with a 26% markup.

Printing Specialty	Average Markup for:			
	Paper	Ink	Film	Plates
General Commercial	30%	31%	44%	44%
Quick Printing	43	45	64	65
Advertising Specialists	36	30	48	43
Business Forms Houses	36	24	35	38
Trade Services	23	27	29	23
Book Printers	21	17	47	33
Magazine Printers	20	21	26	29
Overall Average	31%	30%	44%	44%

Figure 3.8

Average markup for materials by specialty printing group (Reproduced from NAPL's *Special Report* titled "Cost Markup in the Printing Industry," December 1985.)

Region	Average Markup For Paper	States Included
Pacific	35%	WA, OR, CA, AK, HI,
Western	40%	AZ, NM, CO, UT, NV, ID, WY, MT, ND, SD
Central	37%	NE, IA, KS, MO, OK
North Central	29%	MN, WI, MI, IL, IN, OH
South Central	39%	TX, AR, LA
Southeastern	26%	FL, MS, AL, GA, TN, KY, NC, SC
Mid-Atlantic	30%	WV, MD, VA, Wash. DC
Northeastern	31%	PA, NY, NJ, CT, RI, MA, NH, VT, ME

Figure 3.9

Average paper markup by geographic region (Reproduced from NAPL's *Special Report* titled "Cost Markup in the Printing Industry," December 1985.)

Labor and Services Markup

As noted in Figure 3.10, markups for labor and services are separated into inside and outside categories. The figure notes that book printers reported the largest internal labor and services markup at 61%; quick printers were second with a 46% markup on internal labor and services. The overall average of internal markups for labor and services was 41%.

Markups on outside labor and services—common in the commercial printing industry—were highest for quick printers at 45% and lowest in the book printing specialty with 22%. The commercial printing segment was roughly at the 30% average for the entire sample.

Lifting Charges, Storage Charges, and Filler Work

Printers were asked to respond to questions about charging customers for "lifting" work from a press for another job, whether customers were charged for storage and practices on pursuing filler work. When asked "If Customer B forces you to lift Customer A's job off the press, is Customer B charged for the additional expense?" A significant 70% of all respondents answered "yes" to this question. When analyzed by process, the numbers showed that 90% of all heatset web respondents were likely to charge for lifting; in contrast, quick printers were least likely to charge for lifting of all specialty areas, split evenly 50/50 on their responses.

Do printers charge for storing customers finished printing when held for later delivery? Of all respondents, 28% said "yes" and 72% said "no." The highest industry segment to charge was book printers with 58% who charge, while only 3% of quick printers charge; 29% of the commercial printers said "yes" to charging. For those who did charge, the basis for calculating charges varied from skid per month to square footage or cubic footage to flat rates, percentage of the job's total price, interest computed at prime rate plus, length of storage time coupled with space required, and at a poundage rate. Generally the charges were based on comparable area warehouse charges or derived from the company's own records; some firms added a fee for insurance.

The final question: "Is it your practice to take on 'filler' work during slack periods? If yes, how does the markup differ?" Overall, filler work was taken on by 49% of the respondents with book printers, trade services, and advertising printers leading the list with 58%, 54%, and 56% respectively. General commercial printers responded "yes" 44% and only 35% of quick printers took on filler work. Analysis by type of press showed that "filler" work is sought by 44% of all sheetfed printers, 50% by heatset web plants, 60% by non-heatset operations, and 69% of the companies with web offset business forms presses.

In terms of markup on filler work, the data was hard to quantify with a variety of approaches indicated. The most interesting figure here was that 6% of those printers responding to the study said they made no change in markup on filler jobs. Some printers mentioned they reduced the markup on paper, others cut labor and overhead

	Average Markups for Labor & Services	
Specialty	Internal	Outside
General Commercial	40%	30%
Quick Printing	55	45
Advertising Specialists	46	41
Business Forms Houses	46	27
Trade Services	34	27
Book Printers	61	22
Magazine Printers	31	24
Overall Average	41%	31%

Figure 3.10

Average markup for labor and services by specialty printing group (Reproduced from NAPL's *Special Report* titled "Cost Markup in the Printing Industry," December 1985.)

to cost or below cost, others eliminate markup only on manufacturing costs, and still others reduced markup across the board on all materials, to cost if necessary.

3.9 Establishing Prorated Costing

Introduction

Prorated costing may have particular appeal to smaller commercial printers and quick printers who do not wish to establish full budgeted hour costs but may wish to cost recover for a general area of the plant. For example, a smaller commercial printer or quick printer may have a small stripping department which is used somewhat infrequently or which has "roaming" employees who complete many other plant duties including stripping, platemaking, press operating, and bindery. Since prorated costing is more of a "global" costing system for a printing firm, it more easily accommodates these varied types of production situations.

It is also important to point out that prorated costing allows for separate development of fixed and variable costs, also on a more global basis. This provides the smaller printer or quick printer with a view of the total fixed or variable cost picture, not encumbered with specific, intricate details of cost as derived by the BHR process. This is not to say that the BHR system will not work for the smaller printer or quick printer. Prorated costing simply provides a less specific costing system, although its accuracy will approximate the BHR process if time is taken to be certain the data used and the process followed is correct.

System Overview and Description

Different types of allocation (or prorated) units may be used depending upon available information. The most popular allocation base is square footage as physically measured in the plant. Sometimes the number of employees or the amount of capital investment in a working center may be the allocation base. It must be noted that the criteria for selection of the prorated base should be such that the base for the units used is common to all centers and reasonably reflects plant production capacities in each center. For example, Example 3.9 uses square footage as the allocation base, with the assumption that the plant is utilizing floor space efficiently.

As with most types of costing systems, different approaches and philosophies can be taken. For this discussion of prorated costing, we investigate two different methods. The first, called the *all-inclusive center method* (AIC) allows the administrative areas of the plant to represent a cost center, just as if administration were a production oriented portion of the plant. The administrative center will then have a calculated cost base. It is not desirable to use the AIC method for estimating purposes because printing production is normally not completed in administrative areas. Of course, the AIC method will provide a specific hourly cost value that may be helpful for comparison of the cost of administrative centers over time or with other segments of the plant.

The second method, called the *production center method* (PC), subtracts the administrative units from the total number of units. Thus, fixed and variable costs are then prorated over only those centers specifically dedicated to production. The PC method is preferred by the estimator because it more accurately reflects the true costs that must be paid for each order moving through the plant. Administrative variable costs will be prorated over production centers.

Database Needed to Establish Prorated Costing

The five groups of information that follow are needed to complete prorated costing.

Employee demographics: This information includes (1) number of hours worked per week per employee, (2) the average number of days of vacation per employee, and (3) the average number of paid holidays per employee per year.

Plant square footage by working area or center: Aisle and storage space should be included as they apply. If an area is not dedicated to production, it should be classified as a "special" area or administrative area.

Utilization time: This information includes the percentage of time (based on the annual total of chargeable hours) that each center utilizes space or equipment.

Annual fixed costs: These costs include fixed overhead (rent, heat, and general utilities), insurance on equipment, property taxes applicable, and depreciation.

Total annual variable (labor) cost: This information should include the cost of all labor in the plant whether salaried or waged. All fringe benefits should be included as a dollar amount, as should support labor costs such as social security, retirement, unemployment compensation, and medical insurance. In the administrative category, costs should include administrative and selling expenses paid by the company, as well as sales representatives' commissions paid annually.

The more accurate the collected data, the more precise the results. Keep in mind that such information gathering may be the most difficult task in establishing costs precisely. In many cases, getting the array of proper information collected and formatted in a manner sympathetic to the system is more difficult than the actual mechanics of the system itself.

Example 3.9

Let us assume that we have gathered the following information:

Employee demographics: All work 35 hours per week, take an average of 15 days paid vacation per year, and receive an average of 10 paid holidays annually.

Square footage and *utilization time*, by center:

Center	No. Units (sq ft)*	Utilized Time (percent)
Administration	25 (250)	100
Camera	15 (150)	70
Stripping and platemaking	5 (50)	65
Pressroom	30 (300)	80
	75 total units	

*Any equivalent ratio system may be used, such as the number of employees or capital investment in dollars.

Annual fixed costs: $41, 600
Total annual variable (labor) cost, by center (number of employees indicated in parentheses):

Administration (6): $88,600
Camera (2): $31,500
Stripping and platemaking (4): $14,200
Pressroom (4): $58,500
Total labor cost: $192,800

Determine the prorated cost per hour for each center using the AIC method.

Solution See Figure 3.11.

On an annual basis, if costs and hours hold as determined and we "charge out" to customers at the indicated dollar values, cost recovery should be effective. Of course, many tangential variables enter to cause this status to change, which implies that any such system must be monitored to maintain consistency and accuracy. The AIC method does generate a cost for the administrative areas of the plant; the PC method, presented next, does not.

In the PC method, variable and fixed costs will be prorated over only those centers that are actually dedicated to production. In addition, the variable costs of administration will be prorated over the other production centers. The total number of units will not include administration. The same example problem data will be used so comparisons can be made. The total of annual production hours and chargeable hours used in the AIC method will apply with no changes to the PC method with the exception that administrative hours will not be used.

Example 3.10

Use the data in Example 3.9 to determine the prorated cost per hour for each center according to the PC method.

Solution See Figure 3.12.

Determining Annual Production Hours and Chargeable Hours

35 hr/wk × 52 wk/yr:	1820 hr
subtract vacations and holidays (25 da × 7 hr/da):	− 175 hr
annual max production hr:	1645 hr

Center	Maximum Production Hours		Utilization		Number of Chargeable Hours
Administration	1645 hr	×	100%	=	1645 hr
Camera	1645	×	70	=	1152
Stripping and Platemaking	1645	×	65	=	1070
Pressroom	1645	×	80	=	1316

Fixed Cost Proration (AIC Method)

annual fixed cost ÷ total units = annual fixed cost/unit:
$41,600 ÷ 75 units = $554.67/unit/yr

Center	Units		Cost/Unit		Annual Cost/Center		Number of Chargeable Hours		Fixed Cost/Center/Hour
Administration	25	×	$554.67	=	$13,867	÷	1645 hr	=	$ 8.43
Camera	15	×	554.67	=	8,320	÷	1152	=	7.22
Stripping and Platemaking	5	×	554.67	=	2,773	÷	1070	=	2.59
Pressroom	30	×	554.67	=	16,640	÷	1316	=	12.64
	75				$41,600				$30.88

Variable Cost Proration (AIC Method)

Center	Annual Wage and/or Salary		Number of Chargeable Hours		Variable Cost/Center/Hour
Administration	$ 88,600	÷	1645	=	$ 53.86
Camera	31,500	÷	1152	=	27.34
Stripping and Platemaking	14,200	÷	1070	=	13.27
Pressroom	58,500	÷	1316	=	44.45
	$192,800				$138.92

Totals Summation

Center	Fixed Cost/Center/Hour		Variable Cost/Center/Hour		Total Cost/Center/Hour
Administration	$ 8.43	+	$ 53.86	=	$ 62.29
Camera	7.22	+	27.34	=	34.56
Stripping and Platemaking	2.59	+	13.27	=	15.86
Pressroom	12.64	+	44.45	=	57.09
	$30.88		$138.92		$169.80

Figure 3.11

Solution (AIC method) for Example 3.9

Fixed Cost Proration (PC Method)

annual fixed cost ÷ total production units* = annual fixed cost/unit
$41,600 ÷ 50 units = $832.00/unit/yr

Center	Units		Cost/Unit		Annual Cost/Center		Number of Chargeable Hours		Fixed Cost/Center/Hour
Camera	15	×	$832.00	=	$12,480	÷	1152 hr	=	$10.83
Stripping and Platemaking	5	×	832.00	=	4,160	÷	1070	=	3.89
Pressroom	30	×	832.00	=	24,960	÷	1316	=	18.97
	50				$41,600				$33.69

Variable Cost Proration (PC Method)

administrative variable cost ÷ total production units = annual variable cost/unit
$88,600 ÷ 50 units = $1772.00/unit/yr

Center	Units		Administrative Cost/Unit		Administrative Cost/Center		Center Annual Wage		Annual Labor Cost		Chargeable Hour		Variable Cost/Center/Hour
Camera	15	×	$1772	=	$26,580	+	31,500	=	58,080	÷	1152 hr	=	$ 50.42
Stripping and Platemaking	5	×	1772	=	8,860	+	14,200	=	23,060	÷	1070	=	21.55
Pressroom	30	×	1772	=	53,160	+	58,500	=	111,660	÷	1316	=	84.85
	50				$88,600		104,200		192,800				$156.82

Totals Summation

Center	Fixed Cost/Center/Hour		Variable Cost/Center/Hour		Total Cost/Center/Hour
Camera	$10.83	+	$ 50.42	=	$ 61.25
Stripping and Platemaking	3.89	+	21.55	=	25.44
Pressroom	18.97	+	84.85	=	103.82
	$33.69		$156.82		$190.51

*75 total units minus 25 units for administration

Figure 3.12

Solution (PC method) for Example 3.10

When comparing method results, note that the high utilization of administration, as well as the averaging of administrative costs over working production centers, causes differences between solutions. Remember that as chargeable hours increase, hourly costs are reduced. It is this fact that has stimulated printing management to automate production techniques. Such automation also reduces the long-term cost of labor, which has been rising considerably.

3.10 Keeping and Using a Manual for Production Standards and Costs

A problem in many printing plants. regardless of size, is that they do not keep their production standards and BHRs updated and current. Updating should be done annually at a minimum and twice annually is preferable. Another problem is that many companies do not keep their production standards and BHRs in written form. These problems are related because it is impossible to revise production standards and BHRs if there is no record of what the standards are or have been. Thus, it is strongly recommended that each printing company develop, in written form, a production standards and costs manual.

This written record will be valuable to production managers who schedule and plan work for production and to plant estimators and customer service representatives. It will provide a fixed standard from which deviations in production can be plotted and identified, and it will provide a starting point from which standards can be updated as technology and equipment change. It will give a new employee in the production management area or an estimating trainee access to current written standards and costs, expediting the training process. It will also protect the company if a key estimator or production management employee is sick, is on an extended vacation, or has left to work for another employer. A company cannot rely on information stored only in an employee's head.

Maintenance and updating of production standards and costs should be an ongoing concern. For those printing firms with computers, the job-costing procedure provides a comparison of actual versus estimated times and costs, allowing for a review and revision of standards. For plants without CAE and job costing, a manual job summary or job cost summary can be completed. This summary provides for the manual comparison of actual versus estimated times and costs and, ultimately, for revision of production standards and hourly costs.

It is possible to password information considered confidential to the printing company using some CAE systems. Production standards and costs can be held in storage in the computer where they are accessible only to a certain number of persons familiar with the required code to retrieve this information. Thus, instead of a written manual of production standards and BHRs, the information is conveniently and confidentially stored for use as needed.

Appendix to Chapter 3 All-Inclusive Hourly Cost Rates

The figures in this appendix were reproduced with permission of Printing Industries Association, Inc., of Southern California from their *1988-1989 Blue Book of Production Standards and Costs for the Printing Industry* Note that all rates are based on prevailing wages in Southern California and a standard 40 hour work week.

ALL-INCLUSIVE HOURLY COST RATES

BASED ON PREVAILING WAGES IN SOUTHERN CALIFORNIA AND A STAND 40 HOUR WORK WEEK

NAME OF COST CENTER	SIZE	APPROXIMATE INVESTMENT IN EQUIPMENT	CREW	COST PER HOUR AT PRODUCTIVITY OF 60%	70%	80%
COMPOSITION						
Proofreading	N/A	1000	1	46.13	39.53	34.61
Hand Composition	N/A	13000	1	48.41	41.49	36.32
Photo-typositor	N/A	6420	1	53.77	46.08	40.35
Pasteup and copy prep	N/A	1000	1	47.63	40.82	35.74
Linotype Meteor Mod. 5	15 fonts	21400	1	50.17	43.00	37.64
CRTronic	N/A	19207	1	58.12	49.81	43.61
CRTerminals	N/A	9630	1	57.55	49.32	43.18
CRTronic 150	N/A	22417	1	61.42	52.64	46.08
CRTronic 200	N/A	26697	1	62.22	53.33	46.69
CRTronic 300	N/A	30977	1	62.57	53.63	46.95
Typeview 300	N/A	10647	1	58.39	50.04	43.81
Typeview 202	N/A	43870	1	65.06	55.76	48.81
Typeview 202 N	N/A	55640	1	67.28	57.66	50.48
Typeview 202 W	N/A	66287	1	71.55	61.32	53.69
Typeview 101	single drive	23487	1	61.83	52.99	46.40
Typeview 101	dual drive	24557	1	62.54	53.60	46.92
Typeview 101 / S	N/A	21347	1	61.68	52.86	46.28
Linoscreen Composer II	N/A	32047	1	62.95	53.95	47.24
MycroComp 2000	N/A	34775	1	62.96	53.96	47.24
CG System 1 keyboard	10 fonts	19207	1	60.14	51.55	45.13
CG System 2 keyboards	20 fonts	28730	1	73.98	63.40	55.51
Compugraphic 7300	10 fonts	19367	1	66.96	57.38	50.24
Compugraphic 7500	20 fonts	23968	1	67.65	57.98	50.76
Compugraphic 8600	45 pica	51146	1	73.37	62.88	55.05
Compugraphic 8600	68 pica	61846	1	75.93	65.07	56.97
Compugraphic 8600 Imagesetter	N/A	72546	1	78.37	67.16	58.80
Digitek PTW w IBM XT	N/A	36375	1	57.43	49.22	43.09
Digitek P/W w IBM AT	N/A	37445	1	57.88	49.61	43.43
Digitek 4000 for IBM XT	N/A	29955	1	56.72	48.61	42.56
Digitek 4000 for IBM AT	N/A	31025	1	56.67	48.57	42.52
IBM Selectric Tyewriter	N/A	1177	1	42.22	36.18	31.68
IBM Electric Composer	N/A	10058	1	43.77	37.51	32.84
Penta Full system 8 input terminals 2 output drives	2 vdts 500 MB	200000	1	106.91	91.63	80.22

Figure 3A-1

NAME OF COST CENTER	SIZE	APPROXIMATE INVESTMENT IN EQUIPMENT	CREW	60%	70%	80%
LITHOGRAPHIC PREPARATORY						
Line and halftone camera with dark room—small shop	14 x 20	10700	1	72.34	62.00	54.28
Line and Halftone camera Dark room and xenon	24	26750	1	83.40	71.48	62.58
Litho Film Processor Rapid access	24	26750	1	82.02	70.30	61.55
Line and Hlftone w/ processor and xenon	24	53500	1	90.90	77.90	68.20
Stripping-Small presses		1284	1	66.52	57.01	49.91
Stripping-Color and Large presses		3210	1	70.68	60.57	53.03
Platemaker-Small presses		3210	1	68.20	58.45	51.18
Platemaker-Large Presses		8025	1	72.36	62.01	54.29

Figure 3A-2

ALL-INCLUSIVE HOURLY COST RATES

BASED ON PREVAILING WAGES IN SOUTHERN CALIFORNIA AND A STAND 40 HOUR WORK WEEK

NAME OF COST CENTER	SIZE	APPROXIMATE INVESTMENT IN EQUIPMENT	RATED SHEETS PER HOUR	CREW	COST PER HOUR AT PRODUCTIVITY OF 60%	70%	80%
DUPLICATORS & SMALL PRESSES							
Multi, Chief, chute del.	10 x 15	12412	9000	1	47.68	40.86	35.77
Multi, Chief, chain del.	10 x 15	14873	9000	1	48.14	41.26	36.12
Multi, Chief, chute del.	11 x 15	12840	9000	1	47.76	40.93	35.83
Multi, Chief, chain del.	11 x 15	15194	9000	1	48.20	41.31	36.17
Multi, Chief, chute del.	11 x 17	14552	9000	1	48.31	41.40	36.25
Multi, Chief, chain del.	11 x 17	16692	9000	1	48.67	41.71	36.52
Multi, Chief, 2/c chain del.	10 x 15	20972	9000	1	49.78	42.67	37.35
Multi, Chief, 2/c chain del.	11 x 17	21935	9000	1	50.13	42.97	37.62
Multi, Chief, chute del.	14 x 18	17869	9000	1	49.43	42.37	37.09
Multi, Chief, chain del.	14 x 18	19421	9000	1	49.73	42.62	37.31
Multi, Chief, 2/c chain del.	14 x 18	28141	9000	1	51.22	43.89	38.43
Ryobi 17	11 x 17	16050	9000	1	49.21	42.18	36.93
ATF Davidson 501	11 x 15	16050	9000	1	48.60	41.65	36.46
ATF Davidson 502P	11 x 15	17334	9000	1	49.09	42.07	36.83
ATF Davidson 701	15 x 18	18083	9000	1	49.32	42.27	37.00
ATF Davidson 702P	15 x 18	29933	9000	1	51.98	44.55	39.00
ATF Davidson 901	15 x 20	21133	9000	1	49.82	42.69	37.38
Heidelberg TOM	11 x 15 ½	19900	10000	1	49.28	42.24	36.98
Heidelberg TOK	11 x 15 ½	20900	10000	1	49.57	42.48	37.19
Heidelberg GTO 52	14 x 20 ½	43500	8000	1	60.70	52.02	45.55
Hamada 500 CDA	10 x 15	14500	9000	1	48.11	41.23	36.10
Hamada 550 CDA 2/c	10 x 15	19800	9000	1	49.30	42.25	36.99
Hamada 600 CD	11 x 17	15100	9000	1	48.95	41.95	36.73
Hamada 660 CD 2/C	11 x 17	20400	9000	1	50.07	42.92	37.57
Hamada 611, 612 CD	11 x 17	15400	12000	1	48.90	41.91	36.69
Hamada 661, 662 CD 2/c	11 x 17	21100	12000	1	50.10	42.94	37.60
Hamada 700 CD	14 x 18	15400	9000	1	48.78	41.80	36.60
Hamada 770 CD 2/c	14 x 18	25300	9000	1	51.02	43.73	38.28
Hamada 800 DX	14 x 20	24600	9000	1	51.02	43.72	38.28
Hamada 800 CDX	14 x 20	27400	8000	1	51.67	44.29	38.77
Hamada 880 DX	14 x 20	33000	9000	1	52.90	45.33	39.69
Hamada 880 CDX	14 x 20	35800	9000	1	53.42	45.79	40.09
Oliver 52	14 3/16 x 20 ½	54300	12000	1	56.75	48.63	42.58
Imperial 2200 Maxim	12 x 18	16350	10000	1	48.82	41.84	36.63
Imperial 3200 Maxim	12 x 18	17350	10000	1	49.01	42.00	36.77
Imperial 4200 Maxim	12 x 18	17850	10000	1	49.23	42.19	36.94
A.B. Dick 369	11 x 17	17800	9000	1	48.88	41.89	36.68
A.A. Dick 369 T *	11 x 17	44100	9000	1	53.84	46.15	40.40
A.B. Dick 8810	11 x 17	10000	10000	1	47.45	40.67	35.60
A.B. Dick 9840	17 ¾ x 13 ½	17400	10000	1	49.23	42.19	36.94
A.B. Dick 385	17 ½ x 22 ½	21000	8000	1	50.59	43.36	37.96
A.B. Dick 385 w/T head	17 ½ x 22 ½	28000	8000	1	52.29	44.81	39.24
Oliver 58	17 ½ x 22 ½	79000	10000	1	61.10	52.36	45.84

Figure 3A-3

OFFSET PRESSES

One Color Presses

NAME OF COST CENTER	SIZE	APPROXIMATE INVESTMENT IN EQUIPMENT	RATED SHEETS PER HOUR	CREW	60%	70%	80%
Solna 125/80	18 x 25 3/16	66000	8000	1	70.30	60.25	52.75
Heidelberg KORD	18 ⅛ x 25 ¼	49950	6000	1	67.05	57.46	50.31
Oliver 66	19 x 26	77500	12000	1	72.22	61.90	54.19
Royal Zenith R2 26	19 x 26	63665	12000	1	70.01	60.01	52.53
Solna 164	19 x 26	79000	10000	1	72.92	62.49	54.71
Heidelberg MO	19 x 25 ½	83600	11000	1	73.62	63.09	55.24

* Denotes Perfector

Figure 3A-4

ALL-INCLUSIVE HOURLY COST RATES

BASED ON PREVAILING WAGES IN SOUTHERN CALIFORNIA AND A STANDARD 40 HOUR WORK WEEK

OFFSET PRESSES

YOU MUST COMPUTE YOUR OWN COST FIGURES FOR TWO-, FOUR-, FIVE- AND SIX-COLOR PRESSES. THE HOUR RATE FOR THE 2ND PRESSMAN AND FEEDER MUST BE ADDED TO THE COST CENTER HOUR RATE WHERE APPLICABLE BEFORE COMPUTING THE HOURLY COST FOR MAKE READY, IMPRESSIONS PER THOUSAND OR WASH-UP.

NAME OF COST CENTER	SIZE	APPROXIMATE INVESTMENT IN EQUIPMENT	RATED SHEETS PER HOUR	CREW	COST PER HOUR AT PRODUCTIVITY OF 60%	70%	80%
One Color Presses (Cont.)							
Komori 126 Sprint	20 x 26	95000	11000	1	75.85	65.01	56.91
Oliver 72	20 x 28 5/16	79000	10000	1	72.72	62.32	54.57
Heidelberg SORM	20 1/2 x 29 1/8	95800	11000	1	76.50	65.57	57.40
Miehle Roland 201	20 1/2 x 28 3/8	96100	12000	1	77.09	66.07	57.84
Miller SC 74	20 1/2 x 29 1/8	81000	13000	1	79.72	68.32	59.81
Heidelberg SORD	24 1/4 x 36	133700	11000	1	90.50	77.56	67.91
Miehle Roland 36	25 x 36	172500	10000	1	98.62	84.52	73.99
Miller SC 95	25 5/8 x 37 1/2	134000	12000	1	90.24	77.34	67.71
Komori 140 Lithrone	28 x 40	175000	13000	1	98.75	84.63	74.09
Heidelberg SORS	28 3/8 x 40 1/8	144300	11000	1	93.64	80.25	70.26
Miehle Roland 40 low pile	28 3/8 x 40 3/16	197400	10000	1	103.48	88.69	77.65
Miehle Roland 40 high pile	28 3/8 x 40 3/16	222700	10000	1	108.26	92.78	81.23
Miehle Roland 40 Perfector	28 3/8 x 40 3/16	374000	10000	1	143.74	123.19	107.85
Two Color Presses							
Heidelberg GTO ZP	14 x 20 1/2	102800	8000	1	82.42	70.63	61.84
Oliver 252 RP	14 3/16 x 20 1/2	148200	10000	1	90.84	77.85	68.16
Oliver 258	17 1/2 x 22 1/2	116100	10000	1	85.63	73.39	64.25
Oliver 258 RP	17 1/2 x 22 1/2	140800	10000	1	90.51	77.57	67.91
Solna 225	18 1/8 x 25 3/16	130000	10000	1	89.02	76.29	66.79
Fuji Perfector	19 x 25	184040	10000	1	99.22	85.03	74.45
Heidelberg MOZP	19 x 25 1/4	195900	11000	1	101.62	87.09	76.25
Royal Zenith 26	19 x 26	137495	12000	1	90.64	77.68	68.01
Royal Zenith 26-2CP	19 x 26	155925	12000	1	94.12	80.66	70.62
Solna 264	19 1/16 x 26	149265	10000	1	92.90	79.62	69.71
Komori 226 Sprint	20 x 26	179000	11000	1	98.43	84.36	73.86
Komori 226 Lithrone	20 x 26	219000	13000	1	106.28	91.09	79.75
Komori 226 Sprint*	20 x 26	215000	11000	1	105.53	90.44	79.18
Akiyama 228	20 x 28	229000	13000	1	108.17	92.70	81.16
Oliver 272	20 x 28	116000	10000	1	86.13	73.82	64.63
Oliver 272 RP HP	20 3/8 x 28	202500	10000	1	102.75	88.06	77.10
Miehle Roland 28 2/c	20 1/2 x 28 3/8	236800	10000	1	104.55	89.60	78.45
Heidelberg SORMZ	20 1/2 x 29 1/8	172600	11000	1	99.11	84.94	74.37
OMSCA H226-NP 2	20 7/16 x 29 1/8	187250	10000	1	104.27	89.36	78.24
Miller TP 74	20 1/2 x 29 1/8	243000	12000	1	115.01	98.57	86.29
Akiyama 232	23 x 32 1/4	325000	13000	1	130.65	111.97	98.03
Miller TP 84*	24 x 33	251000	10000	1	116.85	100.15	87.68
Miehle Roland 36	25 x 36	272600	10000	1	123.42	105.77	92.61
Miller TP 95*	25 5/8 x 37 1/2	301000	10000	1	127.01	108.85	95.30
Heidelberg SORSZ	28 3/8 x 40 1/8	227000	11000	1	114.13	97.81	85.63
Komori 240 Lithrone	28 3/8 x 40 1/2	395000	13000	1	146.05	125.17	109.58
Akiyama 240	28 5/16 x 40 1/8	395000	13000	1	146.13	125.24	109.64
Miehle Roland 40 2/c	28 1/8 x 40 3/16	308700	10000	1	130.43	111.78	97.87
Miller TP 104*	28 3/4 x 41	360000	11500	1	138.44	118.65	103.88
Miehle Roland 800 50	39 3/8 x 50	572600	10000	1	190.76	163.49	143.14
Miehle Roland 800 55	39 3/8 x 55 1/8	598700	10000	1	196.53	168.43	147.46
Miehle Roland 800 63	43 5/16 x 63	681000	10000	1	212.69	182.28	159.59
Feeder					40.02	34.30	30.03
Second Pressman					60.03	51.45	45.04

* Denotes Perfector

Figure 3A-4 cont

ALL-INCLUSIVE HOURLY COST RATES

BASED ON PREVAILING WAGES IN SOUTHERN CALIFORNIA AND A STANDARD 40 HOUR WORK WEEK

OFFSET PRESSES

YOU MUST COMPUTE YOUR OWN COST FIGURES FOR TWO-, FOUR-, FIVE- AND SIX-COLOR PRESSES. THE HOUR RATE FOR THE 2ND PRESSMAN AND FEEDER MUST BE ADDED TO THE COST CENTER HOUR RATE WHERE APPLICABLE BEFORE COMPUTING THE HOURLY COST FOR MAKE READY, IMPRESSIONS PER THOUSAND OR WASH-UP.

NAME OF COST CENTER	SIZE	APPROXIMATE INVESTMENT IN EQUIPMENT	RATED SHEETS PER HOUR	CREW	COST PER HOUR AT PRODUCTIVITY OF 60%	70%	80%
Four Color Presses							
Heidelberg GTOV-52	14 x 20 1/2	164400	8000	1	105.34	90.28	79.04
Solna 425	18 1/8 x 25 3/16	203300	10000	1	115.00	98.56	86.29
Heidelberg MOV	19 x 25 1/2	303500	11000	1	135.03	115.73	101.32
Heidelberg MOVP	19 x 25 1/2	329500	11000	1	140.02	120.00	105.06
Solna 464	19 x 26	310300	10000	1	136.40	116.90	102.34
Komori Lithrone 426	20 x 26	439000	13000	1	161.10	138.07	120.88
Komori Lithrone 426**	20 x 26	529000	13000	1	178.08	152.62	133.62
Akiyama 428	20 x 28	400000	13000	1	153.91	131.90	115.48
Heidelberg 72V	20 1/2 x 28 3/8	433900	11000	1	160.77	137.78	120.63
Miehle Roland 28	20 1/2 x 28 3/8	476500	10000	1	170.84	146.41	128.18
Miller TP 74	20 1/2 x 29 1/8	472000	12000	1	167.18	143.28	125.44
Akiyama 432	23 x 32 1/4	475000	13000	1	168.50	144.41	126.43
Miller TP 84*	24 x 33	519000	10000	1	179.56	153.89	134.73
Miehle Roland 36	25 x 36	566200	10000	1	190.81	163.53	143.17
Miller TP 95*	25 5/8 x 37 1/2	553000	10000	1	187.18	160.42	140.45
Akiyama 440	28 5/16 x 40 1/8	675000	13000	1	210.70	180.58	158.10
Heidelberg 102V	28 3/8 x 40 1/8	575000	11000	1	189.26	162.20	142.00
Miehle Roland 40	28 1/8 x 40 3/16	625000	10000	1	203.91	174.75	153.00
Komori 440 Lithrone**	28 3/8 x 40 1/2	795000	13000	1	235.32	201.67	176.56
Miller TP104	28 3/8 x 41	699000	10000	1	214.70	184.01	161.10
Miehle Roland 50	39 3/8 x 50	1070100	10000	1	296.81	254.37	222.70
Miehle Roland 55	39 3/8 x 55 1/8	1129200	10000	1	308.86	264.70	231.74
Miehle Roland 63	43 5/16 x 63 3/4	1214900	10000	1	325.57	279.03	244.29
* Denotes Perfector							
** Remote Inking Console							
Feeder					**43.66**	**37.42**	**32.76**
Second Pressman to 29"					**60.03**	**51.45**	**45.04**
Second Pressman over 29"					**60.94**	**52.23**	**45.73**

Figure 3A-4 cont

ALL-INCLUSIVE HOURLY COST RATES

BASED ON PREVAILING WAGES IN SOUTHERN CALIFORNIA AND A STANDARD 40 HOUR WORK WEEK

NAME OF COST CENTER	SIZE	APPROXIMATE INVESTMENT IN EQUIPMENT	RATED SHEETS PER HOUR	CREW	COST PER HOUR AT PRODUCTIVITY OF 60%	70%	80%

OFFSET PRESSES

YOU MUST COMPUTE YOUR OWN COST FIGURES FOR TWO-, FOUR-, FIVE- AND SIX-COLOR PRESSES. THE HOUR RATE FOR THE 2ND PRESSMAN AND FEEDER MUST BE ADDED TO THE COST CENTER HOUR RATE WHERE APPLICABLE BEFORE COMPUTING THE HOURLY COST FOR MAKE READY, IMPRESSIONS PER THOUSAND OR WASH-UP.

NAME OF COST CENTER	SIZE	APPROXIMATE INVESTMENT IN EQUIPMENT	RATED SHEETS PER HOUR	CREW	60%	70%	80%
Five Color Presses							
Heidelberg GTOF-S2	14 x 20 ½	208000	8000	1	120.34	103.14	90.29
Heidelberg MOF	19 x 25 ½	370400	11000	1	157.79	135.23	118.39
Heidelberg MOFP	19 x 25 ½	396400	11000	1	162.69	139.43	122.07
Komori 526 Lithrone**	20 x 26	625000	13000	1	205.83	176.40	154.44
Akiyama 528	20 x 26	495000	13000	1	180.45	154.65	135.40
Miehle Roland 28	20 ½ x 28 ⅜	608800	10000	1	207.41	177.75	155.62
Miller TP 74	20 ½ x 29 ⅛	586000	12000	1	198.74	170.33	149.12
Akiyama 532	23 x 32 ¼	595000	13000	1	201.19	172.43	150.96
Miller TP 84*	24 x 33	645000	10000	1	210.96	180.80	158.29
Miller TP 95*	25 ⅝ x 37 ½	692000	10000	1	224.73	192.60	168.62
Akiyama 540	28 5/16 x 40 ⅛	805000	13000	1	247.22	211.87	185.49
Heidelberg 102F	28 ⅜ x 40 ⅛	698000	11000	1	228.46	195.80	171.42
Komori 540 Lithrone**	28 ⅜ x 40 ½	965000	13000	1	278.84	238.98	209.23
Miller TP 104 *	28 ⅜ x 41	1001000	11500	1	284.14	243.52	213.20

* Denotes Perfector
** Remote Inking Console

NAME OF COST CENTER	60%	70%	80%
First Feeder	35.91	30.78	26.95
Second Feeder	43.10	36.93	32.34
Second Pressman to 29"	59.26	50.79	44.46
Second Pressman over 29"	61.05	52.32	45.81

Figure 3A-4 cont

ALL-INCLUSIVE HOURLY COST RATES

BASED ON PREVAILING WAGES IN SOUTHERN CALIFORNIA AND A STANDARD 40 HOUR WORK WEEK

OFFSET PRESSES

YOU MUST COMPUTE YOUR OWN COST FIGURES FOR TWO-, FOUR-, FIVE- AND SIX-COLOR PRESSES. THE HOUR RATE FOR THE 2ND PRESSMAN AND FEEDER MUST BE ADDED TO THE COST CENTER HOUR RATE WHERE APPLICABLE BEFORE COMPUTING THE HOURLY COST FOR MAKE READY, IMPRESSIONS PER THOUSAND OR WASH-UP.

NAME OF COST CENTER	SIZE	APPROXIMATE INVESTMENT IN EQUIPMENT	RATED SHEETS PER HOUR	CREW	COST PER HOUR AT PRODUCTIVITY OF 60%	70%	80%
Six Color Presses							
Komori 626 Lithrone**	20 x 26	725000	13000	1	225.67	193.41	169.33
Akiyama 628	20 x 28	550000	13000	1	193.86	166.14	145.46
Miehle Roland 28	20 ½ x 28 ⅜	673500	10000	1	221.83	190.11	166.44
Miller TP 74	20 ½ x 29 ⅛	695000	12000	1	222.47	190.66	166.92
Akiyama 632	23 x 32 ¼	650000	13000	1	215.17	184.40	161.45
Miller TP 84*	24 x 33	765000	10000	1	241.00	206.55	180.83
Miller TP 95*	25 ⅝ x 37	819000	10000	1	261.07	223.74	195.89
Miehle Roland 40	28 ¼ x 40 3/16	888800	10000	1	280.74	240.60	210.65
Akiyama 640	28 5/16 x 40 ⅛	958000	13000	1	291.55	249.87	218.76
Heidelberg 102S	28 ⅜ x 40 ⅛	835000	11000	1	268.09	229.76	201.15
Komori 640 Lithrone**	28 ⅜ x 40 ½	1125000	13000	1	322.39	276.30	241.90
Miller TP 104*	28 ⅜ x 41	1132000	11500	1	322.88	276.72	242.27
Miehle Roland 50	39 ⅜ x 50	1570900	10000	1	414.89	355.58	311.31
Miehle Roland 55	39 ⅜ x 55 ⅛	1621200	10000	1	426.91	365.87	320.32
Miehle Roland 63	43 5/16 x 63	1711100	10000	1	447.99	383.94	336.14

* Denotes Perfector

** Remote Inking Console

	60%	70%	80%
First Feeder	**43.10**	**36.93**	**32.34**
Second Feeder	**35.91**	**30.78**	**26.95**
Second Pressman to 29"	**59.26**	**50.79**	**44.46**
Second Pressman over 29"	**61.95**	**53.09**	**46.48**

Figure 3A-4 cont

LETTERPRESS

Heidelberg, Stamp & emboss	10 x 15	26215	5500	1	50.60	43.36	37.97
Heidelberg, Stamp & emboss	13 x 18	37878	4000	1	53.17	45.57	39.90
Kluge, Stamp & emboss	14 x 22	33812	4000	1	52.58	45.07	39.46
Platen, Hand-fed*	8 x 12	1605	1800	1	45.45	38.95	34.10
Platen, Hand-fed*	10 x 15	1605	1800	1	45.45	38.95	34.10
Platen, Hand-fed*	12 x 18	1605	1800	1	45.45	38.95	34.10
Heidelberg Platen	10 x 15	23380	5500	1	49.75	42.64	37.33
Kluge Automatic	11 x 17	13375	4500	1	47.86	41.02	35.91
Heidelberg	13 x 18	36380	5500	1	52.33	44.85	39.26
Kluge Automatic	13 x 19	16050	5000	1	48.79	41.82	36.61
Heidelberg*	15 x 20 ½	9630	5000	1	47.81	40.98	35.87
Cylinder Press*	14 x 20	10700	5000	1	48.01	41.15	36.03

* Reconditioned Equipment

Figure 3A-5

ALL-INCLUSIVE HOURLY COST RATES

BASED ON PREVAILING WAGES IN SOUTHERN CALIFORNIA AND A STANDARD 40 HOUR WORK WEEK

BINDERY

THE RATES FOR THE BINDERY EQUIPMENT INCLUDE ONLY THE OPERATOR SOME OF THESE COST CENTERS MAY REQUIRE ADDITIONAL PERSONNEL. BELOW ARE THE ALL-INCLUSIVE HOURLY COST RATES FOR HAND BINDERY, HEAVY AND LIGHT. THESE RATES MUST BE ADDED TO EACH COST CENTER WHERE APPLICABLE TO GET THE CORRECT HOURLY COST FIGURE.

NAME OF COST CENTER	SIZE	APPROXIMATE INVESTMENT IN EQUIPMENT	CREW	COST PER HOUR AT PRODUCTIVITY OF 60%	70%	80%
M-M "Apollo" Counter-stacker	N/A	23326	1	62.13	53.24	46.62
Harris Collator 106B-8	N/A	39697	1	65.22	55.89	48.93
Didde Graphic Gather All 12 station Collator—Flat sheets and signatures	N/A	73028	1	72.61	62.23	54.48
Rosback 2 head Autostitcher model 201	N/A	8995	1	59.93	51.36	44.97
Rosback 2 head Autostitcher model 203R	N/A	14700	1	61.19	52.44	45.91
Harris Booklet Binder 50	N/A	23380	1	62.25	53.35	46.71
M-M "Fox" Saddle Stitcher 4 pocket	N/A	100901	1	77.49	66.41	58.14
Rosback Bindery System 203 6 pocket feeder-stitcher-trimmer	N/A	49650	1	68.40	58.62	51.32
Didde Graphics Bindall-9 station stitcher, folder and face trimmer	N/A	60455	1	70.72	60.61	53.07
Harris Saddlebinder II, 6 pockets, stitcher, 5 knife trimmer and stacker	N/A	147179	1	87.09	74.64	65.34
Rosback Perfect Binder model 880	N/A	14200	1	61.26	52.50	45.96

Figure 3A-6

CUTTERS—WITHOUT SPACERS

NAME OF COST CENTER	SIZE	APPROXIMATE INVESTMENT IN EQUIPMENT	CREW	60%	70%	80%
Imperial Semi-Automatic	18 ½	3595	1	56.48	48.41	42.38
Imperial Automatic	19	3995	1	56.56	48.47	42.44
Imperial Semi-Automatic	24	4895	1	56.73	48.62	42.56
Imperial Automatic	24	5895	1	56.91	48.78	42.70
Imperial Automatic	26	5995	1	56.93	48.79	42.72
Imperial Standard	30	12400	1	58.50	50.14	43.90
Polar 76 SD-P	30	16900	1	59.35	50.87	44.53
Challenge 305 MCPB	30 ½	11500	1	58.33	49.99	43.77
Baumcut 3050	30 ½	15000	1	58.99	50.56	44.26
Imperial Standard	36	21950	1	60.62	51.96	45.49

Figure 3A-7

ALL-INCLUSIVE HOURLY COST RATES

BASED ON PREVAILING WAGES IN SOUTHERN CALIFORNIA AND A STANDARD 40 HOUR WORK WEEK

NAME OF COST CENTER	SIZE	APPROXIMATE INVESTMENT IN EQUIPMENT	CREW	COST PER HOUR AT PRODUCTIVITY OF 60%	70%	80%
CUTTERS—WITH SPACERS						
Imperial Microprocessor	30	17495	1	59.34	50.85	44.52
Polar 76 EM	30	18900	1	59.41	50.92	44.58
Baumcut 3050	30 ½	18000	1	59.73	51.19	44.81
Wohlenberg Regent MCSTV-2	36	41800	1	64.05	54.89	48.06
Imperial Microprocessor	36	33950	1	62.97	53.97	47.25
Royal Zenith S-92 PMC	36 ¼	32450	1	62.69	53.73	47.04
Polar 92 EMC	36 ¼	35800	1	63.32	54.27	47.51
Imperial Microprocessor	40	42500	1	64.90	55.62	48.70
Imperial Microprocessor	45	48000	1	66.03	56.59	49.55
Wohlenberg MCSTV-2	45	55900	1	67.58	57.92	50.71
Polar 115 EMC	45 ¼	44910	1	65.55	56.18	49.19
Baum-Lawson MC	52	49000	1	66.96	57.39	50.24
Polar 137 EMC	54	58200	1	68.76	58.93	51.59
Wohlenberg MCSTV-2	54	68500	1	70.74	60.63	53.08
Polar 155 EMC	60	69100	1	70.96	60.82	53.25
Wohlenberg MCSTV-2	61	78900	1	72.90	62.47	54.70
Wohlenberg MCSTV-2	73	124700	1	81.75	70.06	61.34

Figure 3A-8

ALL-INCLUSIVE HOURLY COST RATES

BASED ON PREVAILING WAGES IN SOUTHERN CALIFORNIA AND A STANDARD 40 HOUR WORK WEEK

NAME OF COST CENTER	SIZE	APPROXIMATE INVESTMENT IN EQUIPMENT	FEET PER MIN	CREW	COST PER HOUR AT PRODUCTIVITY OF 60%	70%	80%
BINDERY							
Folders							
Baum Ultra 714 friction	14 x 20	2600	9	1	44.79	38.39	33.61
Baum Ultra 714 airfeed	14 x 20	4000	9	1	45.06	38.62	33.81
Imperial Whisperfold	14 x 20	2495	9	1	44.77	38.37	33.59
Baumfolder 517 friction feed 2 pl.	17 ½ x 22 ½	6000	50	1	45.89	39.33	34.43
Baumfolder 517 friction feed 2-4 pl.	17 ½ x 22 ½	12000	75	1	47.18	40.43	35.40
Baumfolder 518 pile feed 4 pl.	18 x 25	11000	75	1	59.68	51.15	44.78
Baumfolder 518 pile feed 4-4 pl.	18 x 25	17000	75	1	60.82	52.12	45.63
Stahl B-19 4-4 pl.	19 x 25	18600	175	1	61.97	53.11	46.50
MBO T-94 4-4 pl.	19 x 27	19367	175	1	62.12	53.24	46.61
Baumfolder 520 pile feed 4 pl.	20 x 26	13000	100	1	60.26	51.64	45.21
Baumfolder 520 cont. feed 4 pl.	20 x 26	17000	100	1	61.01	52.29	45.78
Baumfolder 520 pile feed 4-4 pl.	20 x 26	19000	100	1	61.45	52.67	46.11
Baumfolder 520 cont. feed 4-4 pl.	20 x 26	23000	100	1	62.21	53.31	46.67
Baumfolder 520 pile feed 4-4-4 pl.	20 x 26	25000	100	1	62.58	53.64	46.96
Baumfolder 520 cont. feed 4-4-4 pl.	20 x 26	29000	100	1	63.34	54.28	47.52
Stahl C-22	22 x 35	29800	200	1	64.79	55.53	48.61
MBO T-55 4-4	22 x 36	22684	200	1	63.02	54.01	47.29
Baumfolder 523 pile feed 4 pl.	23 x 36	17000	200	1	61.95	53.09	46.48
Baumfolder 523 cont. feed 4 pl.	23 x 36	21000	200	1	62.71	53.74	47.05
Baumfolder 523 pile feed 4-4 pl.	23 x 36	23000	200	1	63.08	54.06	47.33
Baumfolder 523 cont. feed 4-4-4 pl.	23 x 36	27000	200	1	63.84	54.71	47.90
Baumfolder 523 pile feed 4-4-4 pl.	23 x 36	29000	200	1	64.21	55.03	48.18
Baumfolder 523 cont. feed 4-4 pl.	23 x 36	33000	200	1	64.97	55.68	48.75
Baumfolder Liberty pile feed 4 pl.	23 x 40	24000	250	1	63.57	54.49	47.70
Baumfolder Liberty cont. feed 4 pl.	23 x 40	26000	250	1	63.95	54.81	47.98
Baumfolder Liberty pile feed 4-4 pl.	23 x 40	36000	250	1	65.84	56.43	49.40
Baumfolder Liberty cont. feed 4-4 pl.	23 x 40	38000	250	1	66.22	56.75	49.68
Baumfolder Liberty pile feed 4-4-4 pl.	23 x 40	48000	250	1	68.10	58.37	51.10
Baumfolder Liberty cont. feed 4-4-4 pl.	23 x 40	50000	250	1	68.48	58.69	51.38
Baumfolder Liberty pile feed 4-4-4-2 pl.	23 x 40	56000	250	1	69.61	59.66	52.23
Baumfolder Liberty cont. feed 4-4-4-2 pl.	23 x 40	58000	250	1	69.99	59.98	52.52
MBO T-65 4-4-4 pl.	26 x 41	50611	250	1	68.85	59.00	51.66
Stahl B26 4-4-4-2 pl.	26 x 41	58000	250	1	70.66	60.55	53.02
Stahl C-30 4-4-4-2 pl.	30 ¾ x 45 ¼	39500	300	1	67.47	57.82	50.62
Stahl B-30 4-4-4-2 pl.	30 ¾ x 45 ¼	64200	300	1	72.13	61.82	54.12
Baumfolder Liberty pile feed 4 pl.	26 x 50	29000	350	1	65.44	56.08	49.10
Baumfolder Liberty cont. feed 4 pl.	26 x 50	31000	350	1	65.81	56.40	49.38
Baumfolder Liberty pile feed 4-4 pl.	26 x 50	41000	350	1	67.70	58.02	50.80
Baumfolder Liberty cont. feed 4-4 pl.	26 x 50	43000	350	1	68.08	58.35	51.08
Baumfolder Liberty pile feed 4-4-4 pl.	26 x 50	53000	350	1	69.97	59.96	52.50
Baumfolder Liberty cont. feed 4-4-4 pl.	26 x 50	55000	350	1	70.34	60.29	52.78
Baumfolder Liberty pile feed 4-4-4-2 pl.	26 x 50	61000	350	1	71.47	61.26	53.63
Baumfolder Liberty cont. feed 4-4-4-2 pl.	26 x 50	63000	350	1	71.85	61.58	53.91
Baumfolder Liberty cont. feed 4 pl.	30 x 50	36000	400	1	67.12	57.53	50.36
Baumfolder Liberty cont. feed 4-4 pl.	30 x 50	50000	400	1	69.77	59.79	52.35
Baumfolder Liberty cont. feed 4-4-4 pl.	30 x 50	61000	400	1	71.84	61.57	53.90
Baumfolder Liberty cont. feed 4-4-4-2 pl.	30 x 50	69000	400	1	73.35	62.86	55.04
Stahl C-37	37 x 52	49900	400	1	69.91	59.92	52.46
MBO T-101 4-4-4-2 pl.	40 x 57	81106	500	1	76.64	65.68	57.50
Stahl B-42 4-4-4-2	42 x 58 ¼	76520	500	1	75.77	64.94	56.85
MBO T-112 4-4-4-2 pl.	44 x 65	87472.5	500	1	77.84	66.71	58.40

Figure 3A-9

Printing Orders from Sales through Invoice

4.1 Introduction

The purpose of this chapter is to provide an accurate general description of the typical procedure for orders produced in a commercial printing plant. The discussion begins with direct sales contact with a customer or print buyer via a printing company sales representative, then moves in sequential order beginning with completing job specifications, processing the estimate, customer service involvement, checking the customer's credit, completing the price proposal and quotation, producing the order, invoicing and cash flow, and finally job costing. A discussion of printing brokers is included in this chapter, as are the distinctions between customers and printing (or print) buyers. Also, the general profile of the customer service representative will be provided.

The flowchart in Figure 4.1 indicates the usual operating procedure for a commercial printing plant, from sales through completing the job costing process. It should be clear that deviations from the general order as presented in the flowchart are not unusual. For example, not all work that is produced and billed is estimated, not all printing firms have customer service representatives, not all check the credit of an existing or repeat customer with a good payment history. Also, shortcuts in the described procedures—that is, removing or combining steps—may be common, as deemed appropriate by company management.

The procedures followed by smaller commercial printers and quick printers are not discussed in depth here but left for Chapter 14. This is because the quick or smaller commercial printer differs from the larger commercial printer and high volume printer in two critical areas: first, the quick printer operates a printing business with a focus on the customer and markets served, and much less on a production or manufacturing orientation needed to produce the product. Second, the typical owner of a quick or smaller commercial printing firm operates the business as an entrepreneur, providing an openness—an operating freedom—at all levels of the business. Chapter 2 covers a general discussion of entrepreneurial versus investment community philosophies and how they affect printers of different sizes and types in the industry.

4.2 Selling the Printing Order

Sales Representatives, Customers, and Print Buyers

For many printing companies, sales representatives are the company's only contact with the customer or print buyer. These salespersons are a key link in the functioning of any printing establishment because they bring in work from outside the plant, serving as the company's public personna. Sales representatives should be thoroughly trained in the elements and processes of printing, especially paper, prepress techniques including color variations between proof and press sheet, and finishing processes; they should understand the essentials of printing manufacturing and be able to articulately explain the products and services they sell.

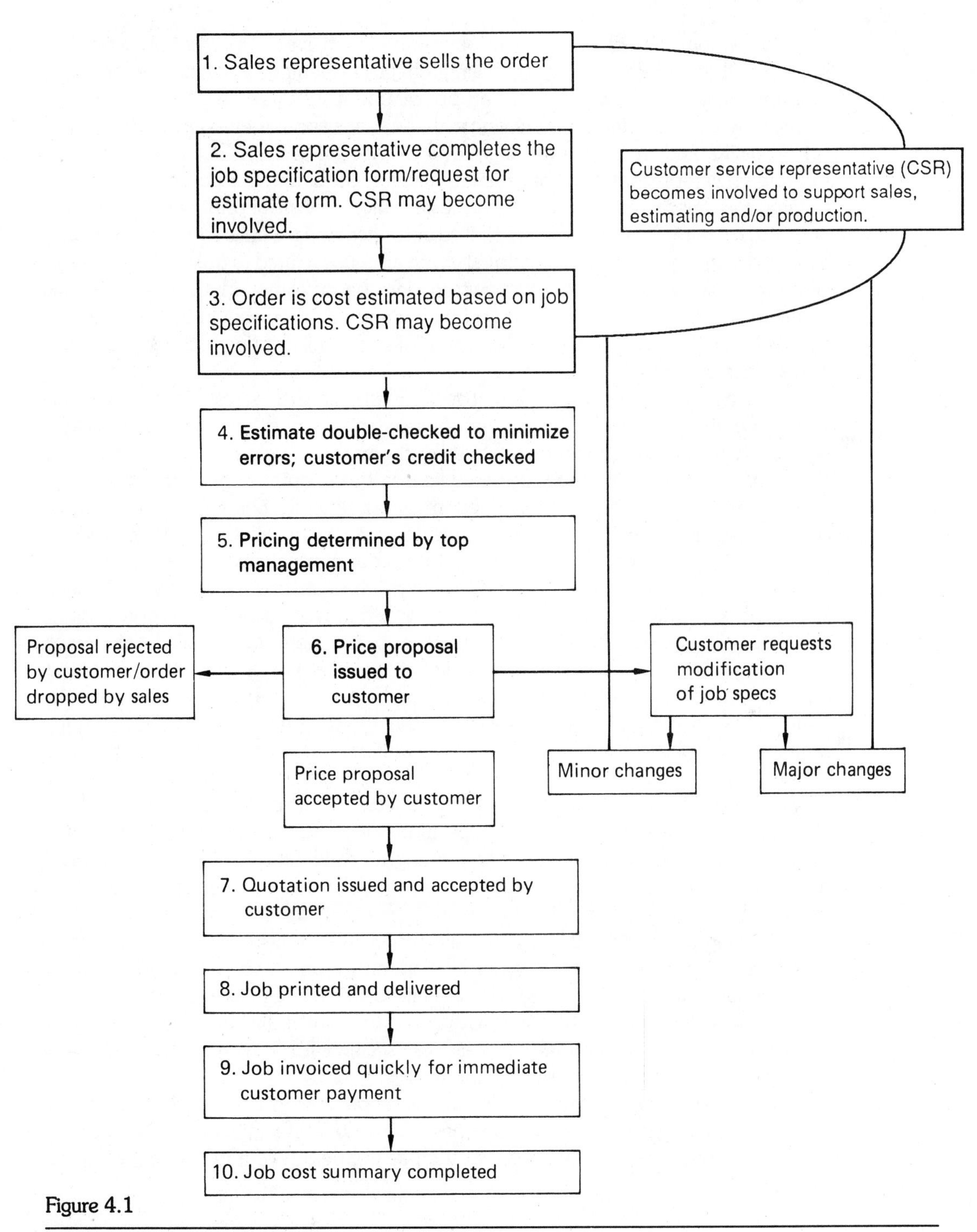

Figure 4.1

Flowchart for selling, estimating, quoting, producing, and billing a printing order

In the competitive environment of printing today, sales representatives must be familiar with the elements of good selling: effective openings and closings for a new customer or buyer, how to obtain an appointment with the client, providing fast response to customer inquiries, dealing with questions about their company's service/price/quality, and maintaining a high level of account service. The most successful printing sales representatives have unique personalities, and are particularly focused to meet customer needs quickly and professionally. Successful printing sales reps are not "order-takers," but rather are skilled intermediaries with customers and print buyers, dealing openly and professionally with all issues related to their printing needs. Because printing sales representatives provide the customer or buyer's first—and sometimes continuous—view of the printing company, they should be keenly aware that their public image speaks for the firm. While they sell printing to a customer, they also sell the company.

The customer service representative (CSR) plays a quasi-sales role in most plants which use them. Because of their absolutely critical position between the sales representative outside the company and the internal production of a customer's job inside the company, CSRs are discussed later in this chapter in greater detail.

As intense competition has forced many printers into becoming more market-oriented and more aware of the marketing environment in which they provide products and services, the difference between a customer and print buyer has become more distinct, even though the lines are still somewhat unclear. Also, as most sales representatives quickly learn, customers usually require a different sales approach and sales support than do print buyers. A customer can be generally profiled as a person representing his or her company who purchases printing on a fairly infrequent basis and who has a somewhat limited knowledge of the printing process. A print buyer, by contrast, is an experienced professional—the typical print buyer knows printing manufacturing processes thoroughly, understands the variables of the printing process, knows finishing techniques, and can distinguish otherwise insignificant differences between press sheets or finished goods. Print buyers are commonly found with printing clients who purchase large quantities of printing, such as advertising agencies and large corporations. Many print buyers buy printing as a full-time job or as the primary function of their duties; they represent a very large dollar amount of printing purchased in the U.S. In some cases, print buyers are former printing sales representatives.

Printing sales representatives develop their own individual selling procedures based on their personalities. They usually work on a commission basis. *Commission* is the amount earned by the sales representative on a job and may be calculated on the basis of the selling price of the job or perhaps on the profitability or value added of the job. Compensation and commission payment systems for sales representatives vary throughout the commercial printing industry.

It is only natural that salespersons who work on a commission try to increase the amount of printing their clients will purchase. Therefore, in addition to the customer's order, they may suggest other services such as the design of a new company logo, development of a house publication, or a new form or other product. However, regardless of the types of printing orders sold or the market served by sales representatives, the sales representative and the customer service representative should work to see that the customer's best interests are served. Both should be committed to providing the client with the best product for the money spent, delivered on time and with minimum client hassle.

Sales representatives also should be concerned about the ultimate use of the product being ordered. If a product is not easily produced by the printing company that the salesperson represents, he or she may arrange for printing from another company as a buyout, although the customer will be billed by the sales representative's company. Generally, the customer is willing to pay a premium rate for printing if the services provided by the sales representative are difficult to obtain elsewhere.

Printing Brokers and Printing Consultants

A printing broker is a person who sells printing goods and services to customers and print buyers, then arranges for printing manufacturing through any printer willing to complete the job. In the past ten years print brokering has become more common since the number of brokers has increased, thus increasing the competition in certain market niches. Brokers sell "head-to-head" with printing sales representatives, and have the latitude of selecting any cooperating printing firm to do the manufacturing, while the printing sales representative is locked into his one company, with little latitude to consider options that may provide a more competitive advantage. Some printing brokers were formerly sales representatives for a printing company and chose to work independently.

A printing broker generally works by selling customers and print buyers a printed product then finding a cooperating printing company to produce the work. All clients of the broker remain his and the printer serves only a manufacturing role. The good printing broker provides excellent customer services—arranging for art boards, photographs and other job elements, providing design ideas, and so on—and charges a fee for this above the printing costs. The broker invoices the client and pays the printer so that the customer does not deal directly with the printer. Those clients who work with successful printing brokers find the process of designing and obtaining printed goods simplified and with minimum hassles; brokers sell services as the most important part of their operation.

Depending on the printing company's experience with brokers, they seem to fit into the extremes of being either liked or disliked. Those printing manufacturers who have found brokers to provide additional volume without detracting from current customers, who have found the brokers to be easy to work for, and who have found that printing bills are paid in a timely and professional manner, work with brokers without any problems. However, when a broker practices a "hit-and-run" procedure whereby the printer has been used to produce brokered printing on a marginally profitable basis for a period of time, then receives no orders at all from that broker, relations can become strained. Also, brokers who provide slow payment for printing, meanwhile taking current brokered work to other printers, can cause angry feelings with printers.

The Printing Brokerage Association (PBA, 1700 N. Moore Street, Arlington, VA 22209) represents professional printing brokers across the U.S. and may be contacted for more information, if desired.

Printing consultants are also an active part of the printing industry. Instead of selling and coordinating printing needs for a customer or print buyer, the printing consultant is retained by the customer or print buyer to ensure consistent quality, coordinate printing projects at different stages, and even buy printing for the customer on an as-needed basis. Printing consultants are very valuable to large purchasers of printing when the customer or buyer must purchase printing that requires complex manufacturing, such as gravure or flexographic products.

Walk-in Clients

While larger commercial printers and high volume printers use sales representatives and brokers for most if not all of their orders, walk-in customers are common to smaller commercial printers and particularly quick printers. Generally "walk-ins" are geographically close to the printing company, need "on-demand" printing which is mostly short-run and fast-turnaround, pay cash for the completed printing, and expect a high level of customer service, rendered immediately. Walk-in customers may or may not be familiar with the printing or copying processes, and therefore artwork or the copy to be reproduced may be poorly prepared.

There is no question that many quick printers and smaller commercial printers gear their business operation toward walk-in customers. For this reason most of these types of printing firms do not have full-time sales representatives. The bulk of the orders are placed over-the-counter, necessitating very customer-oriented, pleasant counter personnel. In many cases, as the smaller commercial printer or quick printer grows, they begin to encourage brokered work or hire a part-time sales representative; as the business expands, one or more full-time sales representatives works for the company. Chapter 14 addresses in detail customer issues for the quick printer or smaller commercial printer.

4.3 Customer Service Representatives

Customer service representatives (CSRs) have become popular with larger commercial printers and high volume printers primarily due to an increasing need for customer sensitivity and customer focus. The CSR of the 1980s and 1990s serves as a liaison between the customer, sales representative, and the production area of the plant; in past years a CSR was more commonly known as a "production coordinator," whose focus was largely on internal production matters, not directly customer-related.

The four primary duties of a CSR are:

1. Coordinate orders and handle details of a particular job between customers and sales representatives.
2. Coordinate and work with estimators, customers and sales representatives.
3. Coordinate timely delivery of the job at the expected quality level with the production side of the printing firm.
4. Continually sell the company.

The job duties of a customer service representative may vary from company to company. For example, some printing firms ask their CSRs to complete estimating in addition to coordinating production. Other firms have a separate estimating staff for completing this function, but they want the CSR to be the primary contact between the customer and estimator, instead of the estimator contacting the customer directly. Some companies ask the sales representative to transfer the customer to a CSR as soon as the quotation or proposal is accepted, giving the CSR primary customer contact very early in the process, and freeing the sales representative to find more customers. Still other CSRs continue to handle mostly production coordinating duties, with varying amounts of direct customer contact.

How the customer service representative is paid appears to relate to the relative time an employee spends as a CSR. Some printing firms compensate their CSRs on a straight salary basis, yet ask them to participate in a fairly significant amount of sales-service including direct support of the sales representatives; these same firms pay their sales representatives on a commission, value added, or profit basis, driven by the motivation and dynamic nature of the sales rep. Thus, the CSR sometimes does a large portion of what might be considered "sales," yet is paid on a straight salary basis only. This inevitably leads to CSRs moving from customer service duties to positions as sales representatives—from straight salary to a compensation package dictated by their own drive and motivation—since they have functioned largely as a sales representative anyway. Some CSRs also move from the customer service role into full-time estimating, which is typically compensated on an roughly equivalent straight salary basis.

The CSR function appears essential for the very survival of larger commercial printing firms and high volume printers, just as the educated, friendly counter person serves the quick printer or smaller commercial printer. In sum, with an inherent overcapacity in our industry, thousands of printing firms competing for the same business, increasingly intense foreign competition, and artificially low ceilings on prices, there is no question that enhanced customer sensitivity and greatly improved customer satisfaction play key roles in keeping customers. The CSR ensures that customer happiness and customer satisfaction are continually monitored and maintained at a high level.

4.4 Preparing Job Specifications and Request for Estimates

The completion of a job specification and the follow-up request for an estimate is done when the printing order is complex, difficult, or exceeds the obvious economic range of quick copy capability. Larger printers and high volume printers complete the job specification and request for estimate process routinely, with most jobs.

For the smaller commercial printer or quick printer, the printing order is typically reproduced directly from customer-supplied art, passed across the counter, printed on-the-spot, with the customer paying for the work immediately and leaving. Artwork for the quick printer may come from a computer-driven system such as a Macintosh with dot matrix or LaserWriter output, from paste-up mechanical art containing images from various sources, or from anything in between. In sum, for fast-turnaround work there is no need to provide the job specifications, a request for estimate, a cost estimate, or continue with the proposal/quotation process as will be described. Typically price lists are the primary method used by quick and smaller commercial printers to charge their customers for printing services.

For larger printing jobs, the job specification form is completed by the printing sales representative, usually in the customer's presence. It details the specifics of the order to be produced. The type of form and its development vary considerably from plant to plant and are usually modified for the types of products produced by a particular company. In some cases customers may provide written specifications on their own forms, which the sales representatives will use to complete the job specifications form when they return to the plant.

After the job specification form has been completed, the sales representative may return to the plant and complete a request for estimate form that reviews and clarifies the original job specifications. It is during the completion of the request for estimate form that the sales representative may wish to ask the customer service representative, estimator, or job planner specific questions concerning matters such as production of the job, buyouts, and artwork. In fact, the job may be transferred to a customer service representative, estimator, or retained by the sales representative for final job specification completion, depending on the company.

Some plants combine the job specification and request for estimate forms, thus saving time in completing two separate input sheets. The major disadvantage of combining the forms is that the specifications ultimately used to estimate the job may be incomplete, which slows down the estimating process. A sample request for estimate form is shown in Figure 4.2.

The sales representative or customer service representative is the usual contact for job specifications from customers. Sometimes clients—especially professional print buyers—will telephone the plant with a request for estimate. In such cases the estimator, or customer service representative should carefully record the information. It is also possible that customers will request an estimate or printing quote by letter, or may personally appear at the plant to place the order. Regardless of the method used,

Request	Customer				Request Number
Salesman	Organ./Dept.				Phone
	Quantity 1	Quantity 2	Quantity 3	Size 1	Date of Request
Job Description				Size 2	Date Wanted
				Color	Rerun ☐ Rerun W/Changes ☐

Paper / **Ink**

No. Type	Weight	Color		Color	#Colors

Composition

Camera Ready ☐ UGS sets ☐ See Copy ☐ Proofs: Customer O.K.________

	Type Face	Size	Line	Approx. # Words	Other
Text					
Heads					
Other					

Camera

	Number/Pages	Original Size	Size Wanted	Other
Halftones				
Line/Art				

Stripping/Platemaking

# Colors/Side	Register	#Sides	Brownlines ☐	Other
	Tight ☐ Loose ☐	One ☐ Two ☐	Bleeds ☐ Screens ☐	

Press Special Instructions

Bindery

Cut & Trim ☐ Folding ☐ Saddle Stitch ☐ Side Stitch ☐	Gather in sequence ☐ Collate in sequence ________ ________ ________ Pad in Pads of________'s Numbering Beg.________ End________	Perforate ☐ Drill ☐ Rounding ☐ Wrap ☐ Box ☐

Additional Notes From:

Competition Notes

Previous Supplier

Date Quantity

Price

Marketing Information

Walk-in ☐ Cold call ☐ Telephone ☐ Referred By
Previous Customer ☐ Advertising ☐

Quantity/Color	Price	Quote

Customer Approval

Administration Approval ________________ Date ________

Authorized Purchaser ________________ Date ________

Figure 4.2

Request for estimate form (Developed by Jake Swenson; reproduced with permission)

the customer should receive a copy of the complete job specifications and/or written job proposal that includes the specifications. These documents will serve as written clarification and will provide the client with complete job information to compare estimates from other printers.

When the sales representative handles a standardized product line, the specifications will be related only to the variables offered for that standard product, such as different colors of ink, changes in format size, and modest changes in typeface. If a sample portfolio of standard products is available from which the customer can select, the selling effort, especially the difficulties of providing complete job specifications, is simplified. Standardization of product and the accompanying reduction of complex job specifications expedite the sales effort and reduce mistakes and misunderstandings throughout production.

Any set of complete job specifications should include the following: the customer's name, address, and telephone number; sales representative's name; general description of the job; quantity levels; flat sheet and folded dimensions (with possible alternate sizes); bleeds and margin information; number of colors; complete details regarding all artwork (who is to supply it and whether the work will be camera-ready); typesetting required; complete paper specifications; binding and finishing requirements; type of packaging to be used; and type of delivery method. An area on the form should be provided for additional notes and information related to activity of the competition.

Obtaining complete job specifications prior to estimating is an essential requirement of the customer service representative, sales representative, or estimator. Because of the frustration of incomplete job specifications, some printers have made the estimator the key collector of job specifications through direct telephone contact with the customer, even though this is an accepted part of the sales representative's job. Also, customer service representatives have begun to actually complete estimating duties, whereby the CSR gathers the job specs and estimates all as one operation. Regardless of who obtains the job specifications, inaccurate estimates are a result of poor detailing of job specifications in a great many cases.

4.5 The Cost Estimate Based on the Job Specifications

The estimating function includes the determination of a plan of production to produce the customer's order as detailed in the specifications and assignment of time and cost values to determine final cost to manufacture the product. In large plants, the planning function is completed specifically by a production planner who obtains the most efficient production sequence. In many medium-sized and small plants, the estimator will complete both tasks. A sample cost-estimating form used for this purpose is shown in Figure 4.3.

Estimate	Customer			Estimate Number
Salesman	Dept./Organization			Phone
Estimator	Quantity 1	Quantity 2	Quantity 3	Date of Estimate
Job Description			Size	Date Wanted
			#Colors	Rerun Rerun W/Changes

Paper							
No.	Type	Wt.	Color	Size	Quantity	Unit Price	Total

Flat Matl's					
	Type	Size	Quantity	Unit Price	Total
Film					
Plate					

Ink/Chemicals				
Type	Color	Quantity	Unit Price	Total

Flat Costs					Running									
Number of Colors	1	2	3	4	Press	M	M	M	Bindery	M	M	M	Labor Costs	
Hand/Ludlow					Make Ready				Cutting				F/1	
Linotype					Run				Trimming				F/2	
Make-up					Make Ready				Fold Set-Up				F/3	
Proofs/Repros					Run				Fold Run				F/4	
Tape/Perf.					Make Ready				Gathering					
Photo/Set-Up					Run				Stitching				R/1	
Photo/Running					Make Ready				Padding				R/2	
Paste-Up					Run				Wrapping				R/3	
Camera					Wash Up								Materials Cost	
Stripping													Mtl/F	
Plate Making														
Total Flat 1					**Total** Press 1				**Total** Bindery 1				Mtl/R1	
Total Flat 2					**Total** Press 2				**Total** Bindery 2				Mtl/R2	
Total Flat 3 & 4					**Total** Press 3				**Total** Bindery 3				Mtl/R3	

Recapitulation										
	Flat			Running			**Total**		Quantity	Price Quote
Matl.										
Labor										
Total										
Overhead										
Buy-out										
Total										
Quote										
Tax										
Total										
Price										

Notes

Figure 4.3

Cost-estimating form (Developed by Jake Swenson; reproduced with permission)

Production Planning

Production planning begins the complex process of minimizing manufacturing costs and should be done carefully. It should include:

An analysis of the job by breaking it into operational segments,
Determination of buyouts,
Preparation of internal plant planning forms such as dummies and rough sketches.

Estimators and job planners learn by their mistakes and by experience. It is important that the persons doing this work understand thoroughly relative costs of production and the many additional costs that can erode profits considerably if not included. Trainees should be carefully educated with respect to the many production variables and contingencies possible.

Cost Estimating

After job planning is completed, cost estimating begins. It includes the application of:

Time assessments for all operational components, based on established production standards,
Application of accurate BHRs, by production center, to determine manufacturing costs.

The estimator should investigate product and cost alternatives beneficial to the customer. These alternatives might include modification of job specifications that provide a better cost to the customer. Such modifications might be the selection of a less expensive paper stock or changes that will result in an improved product more suitable to the customer's ultimate needs.

The cost-estimating form (Figure 4.3) contains the following information: customer's name and company organization, the names of the sales representative and estimator, date of the estimate, assigned estimate number, quantities desired, and date of delivery. Space is provided for a brief job description, even though the estimator will be working from a list of job specifications supplied by the sales representative.

Materials are specifically indicated and might include paper, lithographic "flat" materials such as film and plates, and ink and chemical requirements. Costs (using estimated hours and BHRs) are then determined for all production segments at requested quantities. Space at the bottom of the form is provided for recapitulation of data, notes, and price quote information. Pricing normally follows this estimating function and is done by top management. A copy of this form is made and forwarded from the estimator to the appropriate top management person.

The cost estimate should be kept on file with the original job specification form supplied by the customer and any pages of calculations. Sometimes estimators use the back of the estimate form for such calculations. Generally, the more clear and readable the calculations retained are the more expeditious production scheduling and purchasing will be if the company is awarded the job. The estimator should build a file for each customer and cross-reference it by type of job for future use. There should be ample filing cabinet space provided for this purpose in the estimator's office.

Hundreds of different types of estimating forms exist. Each printing plant sooner or later develops one that best fits its particular type of work and production sequencing. The layout of the estimating form should be carefully planned, with the use of trial forms prior to adoption of the final format. Some plants combine the estimating form with production scheduling and purchasing forms by preparing the estimate in triplicate. The top copy may be used as the official estimate, the second copy (different color but same form) may be retained by purchasing as a record of materials needed (if the company gets the order), and the bottom copy may be a kraft envelope that serves as a job ticket that will move through the plant with the job. Some controversy exists over this type of system because estimating data then travel with the job through production. It may be considered undesirable to allow production employees to see times and costs associated with their particular operation.

It is essential that the sales representative, customer service representative, and estimator each have a clear understanding of the responsibilities of each to the other. As previously discussed, obtaining the job specifications must be carefully and meticulously done, regardless of who collects the data. An important rule is that the estimator who receives complete job specifications owes a quick turnaround of the estimate; one day or less is common for many printers today, particularly those with reliable CAE systems.

The use of computers certainly speeds up the estimating process, ensuring accuracy and consistency of estimates as well. Another possible technique to be certain the relationship between sales representatives and estimators is coordinated might be to place estimating staff and sales staff under the same supervisor. This person, a vice president of sales or director of sales, should be familiar with both sales and estimating, preferably having done both as past work experience. This technique would allow for mediation of problems between the two sometimes adversary groups in a fair and equitable manner. Also, since estimating is essentially a sales support position, working directly under a sales-related supervisor seems natural.

It is essential to note that the actual estimating process is typically fast-paced and quickly completed. Thus, while printing estimating may appear slow and laborious, the reality is that most estimators become very skilled at job analysis, production planning, and cost calculations. In the daily routine of estimating, many jobs are similar or can be categorized into groups that allow analyzing and estimating them to be completed very quickly. Speed of estimating is also impacted by other factors such as the estimator's experience and whether estimating is done with computer aid or manually.

4.6 Double-Checking the Estimate and Checking the Customer's Credit

Printing estimating is a complex procedure that unfortunately has a large margin of error. Consider this situation: The estimator must develop a detailed plan of production with perhaps dozens of manufacturing steps, must include buyouts and material costs. and then must accurately assign time and cost values to determine total cost to manufacture the product. Errors can, and do, occur anywhere in the estimating sequence.

By far the most important consideration with respect to errors is the relative cost impact on the business. If an error causes a significant cost loss to the company, profit dollars must be used to make it up. However, if the error causes too high a price for the product, the customer will perhaps not authorize the order and the company will lose a potential sale. In essence, a printing company cannot afford to price work at a loss in profits, nor can it afford to overprice work and lose volume in the plant.

Double-Checking Cost Estimates

The best way to determine the number and impact of mistakes in estimating is to consistently review estimates at two levels.

At the estimator level: Here a second person would double-check the estimate after it has been initially completed and prior to pricing the job. This might be done by another estimator or perhaps a customer service representative who also does estimating. In some larger printing firms, one person—such as the estimating supervisor—might complete the checking process. Also, some printing companies establish guidelines for checking estimates which may include estimate checks only for jobs that are above a set dollar value, checking estimates for jobs which are unique or new to the company, and/or checking estimates for a specific category of jobs that have a history of being particularly troublesome in production or not financially rewarding.

At the job costing level: Job costing, which is usually done after-the-fact, is essential in providing feedback with respect to estimating errors. Some printing companies job cost every job which they have produced, while other companies review only certain types of jobs, those above a certain dollar level or those which appeared to be troublesome during production. With electronic factory data collection becoming more common, job costing prior to invoicing is becoming popular. Because it is computer-driven, this provides a high level of accuracy and allows the printer to charge for all customer-approved alterations tracked during production. Job costing is discussed in Chapter 3 and also later in this chapter.

Double-checking estimates is a most effective procedure to eliminate estimating errors. Usually, such reviews are completed by the estimating supervisor (in a large plant) or by the exchange of estimates between estimators. Some printing companies require that estimates be double-checked only when the production plan is reasonably complex or when the cost appears significant for the proposed job. Other plants may require double-checking all estimates. A major disadvantage to such reviews is that extra personnel and time are required, which are not available in many companies. Sometimes spot-checking a selected number of estimates for errors saves time. Spot-checking will not eliminate all mistakes, but a large percentage should be found and corrected. The greater the number of estimates that are double-checked, and the more thorough the checking procedure, the more accurate the estimates produced.

Categories of Estimating Errors

Estimating mistakes generally fall into four categories: mathematical errors, assumption errors, policy errors, and mistakes in communications. These categories are discussed in the sections that follow.

Mathematical Errors. This area is perhaps the most common for mistakes in estimating. Today, electronic calculators and CAE hardware provide a greater capacity for accuracy, but sometimes estimates prepared using these tools are less accurate. Estimators may depend upon calculators for the correct answer without substantiating mentally whether the final figures are correct. While calculators aid in faster estimating with the potential for considerable accuracy when they are used carefully, sometimes too much confidence is placed in their displayed numbers, and subsequent estimates contain more mistakes than those prepared with slower hand methods.

The use of CAE removes literally all mathematical errors in estimating, certainly one of its strongest selling points. But CAE, mathematically error-free, is not errorless because input into the system is done by people. As with calculators, if there is too much of a rush during inputting, mistakes will inevitably result.

Assumption Errors. Some estimators make too many assumptions during the estimating process. Essentially, the fewer the assumptions that have to be made during estimating, the more accurate the estimate. Assumption errors may be divided into two categories: job specification assumptions and production planning assumptions.

In the case of job specification assumptions, major errors usually result from:

Not knowing what the customer will supply,
Not fully understanding exactly what the customer wants produced,
Assuming the use of certain materials that have not been specifically requested by the customer, such as paper or type.

The less information and fewer specifications available prior to estimating, the less precise the estimate.

Regarding production planning assumption errors, the most common are:

Assuming one knows the most economical production sequence when actual comparisons should be made,
Assuming certain capacity and production levels that may in reality be different when the order begins production or may not be generally representative of the plant,
Assuming and using a delivery date that is unreasonable.

Keep in mind that there are many other assumption errors that can occur during production planning, but those listed are some of the most common and must be carefully watched.

Policy Errors. Established policies allow the estimator to work under known conditions and procedures. There must be established methods for submitting job specifications, completing the estimate, writing up the proposal, and quoting on work. There should be rules regarding all policy areas such as buyouts, material purchasing, modification of production standards, development and interpretation of BHRs, relations between production planning and estimating, and so on. It is the job of top management to establish and oversee such policy areas. The fewer established policies, the sloppier the estimating system and the greater the capacity for errors.

Mistakes in Written and Verbal Communications. These errors usually occur because of poor or misunderstood communications between customer and sales representative or sales representative and estimator. To minimize such problems, all parties should communicate on the same level—that is, on the same knowledge and language plane. Every attempt should be made to put as much as possible in writing between all parties. Samples and visual images should be included whenever possible. Verbal agreements and "okays" should not be used. Also, a conscious effort to communicate slowly and accurately should be made. The more efficient the communications, the fewer the misunderstandings.

Everyone makes mistakes. Doing so is a part of the human condition. No matter how much effort is made to eliminate mistakes, they will continue to occur. Yet all attempts should be made to reduce estimating errors to a minimum. Today's printing managers understand this situation and know that they must maintain a tolerance level as to the type, frequency, and number of errors that occur during the estimating process. Most professional printing estimators work continuously to insure accuracy.

Checking the Customer's Credit

Sometime during the estimating/pricing process the customer's credit may be checked. Opinion varies among printers as to when this is best completed. Some companies check credit for jobs that are in estimating which appear to have a likelihood of becoming orders. Other firms may not check credit until the job has actually been won by the company, while some printers check credit during the pricing phase of the process.

Opinion also varies among printers as to whose credit should be checked. Some printers check credit for all new customers and depend on the payment history of established customers to indicate credit and payment problems. Other printers check credit of all customers on a periodic basis, regardless of credit history with the company. Still others rarely, if ever, check credit; or they check credit randomly or only when a job is large and represents a significant impact on the company relative to production and cash needs.

There are essentially three ways to check on a potential customer's credit. The first is to have the customer complete an application for credit with the printing company, listing all the customer's creditors and credit references. The application can be checked with routine telephone calls at little cost to the printer.

The second procedure is to join or contact a credit association to investigate the credit of a particular customer. Generally, if the printer is a member of the credit association, there are fees only for extra services when doing a credit check on a potential customer. If the printer is not a member but wishes to use the credit association services, there are typically initial access charges plus additional charges as they apply.

The third procedure for checking credit is to utilize the credit referral service offered through most PIA regional offices. While available only to PIA members, this service is typically efficient, inexpensive, and accurate. Contact your PIA regional office for particulars.

Checking the credit of customers avoids bad debts and slow paybacks, which adversely affect the company's cash position and cash flow in general. (Later in this chapter, effective cash flow is discussed, along with the split invoice technique by which a customer pays a certain amount on a job before production begins—as a deposit — and the remainder after the job is delivered.)

4.7 Pricing Determination by Top Management

Pricing decisions are the single most important aspect of profitability. The more prices can be controlled, the more leverage exerted over company profits. As detailed in Chapter 2, controlling prices in printing requires that markets be controlled; specialized product and process applications were mentioned in this regard. Other determinants of pricing include manufacturing costs, product mix of the printer, competition's prices, quality of the product, services provided to the customer, volume of work, and desired profit.

The intention here is not to spell out in detail the application of each pricing determinant but to indicate that pricing is a vital function of top management. Thus, it is up to top management to establish a procedure by which all jobs are priced, once the total cost to manufacture the product is known. In some printing plants, pricing is left up to the estimator; in others, to the sales representative or sales manager. Some companies provide the estimator with the mechanics of price determination, usually a percentage markup to add to the total manufacturing cost of the job. When this percentage is the same for all types and kinds of work done in the plant, it is not wholly accurate: Each product or market may have different levels of profitability. Thus, when using such a percentage markup system, different percentages may be considered for different customer segments and major product groups. Such a percentage should be based on factual data developed by top management with respect to the customer, the product, and the desired profit.

Pricing is a crucial matter and must be given thoughtful consideration. Both federal law (Robinson-Patman Act) and state laws in most states require that pricing patterns be consistent for like goods and services provided to the customer or within the major product group. Of course, the customer needs the work, and the printing company has the available skills, personnel, and equipment to provide it. There is no process by which pricing can be simplified in the printing industry because it is solely a company-by-company, top management decision.

Before a printer makes the final commitment to print an order, the customer's good credit should be verified first. Top management should consider carefully the aspect of customer credit—that is, the customer's ability to pay for the work done—prior to extending a proposal to the client. The more substantial the dollar amount of the job, the more thorough the credit check must be.

4.8 The Price Proposal

The proposal is a typewritten or computer-generated document that specifies prices for a printing order. It may be sent by mail or personally delivered to the customer by the sales representative or company representative. A proposal describes completely the work to be done, specifies quality and quantity levels, and suggests selling prices to the customer based on estimating and pricing considerations.

Because proposals represent suggested prices, one of three actions will occur. First, the customer may accept the proposed price, in which case the proposal may serve as a quotation—the legal agreement between the printer and customer. In some instances, a separate quotation will be issued after acceptance of the proposed price. (Because of the legal implications of the quotation, it will be discussed in greater detail in Section 4.9.)

The second action, rejection of the proposal, may occur because customers find the price too high for their present budget allotment or because they decide they do not want the job for other reasons. Sometimes customers have made arrangements for printing with another company because it took the printer too long to complete the proposal. Regardless of the cause, at this point, the salesperson will drop the order and inform the printing company about the customer's decision.

The third action, and perhaps the most frequent, especially with large orders, is for the customer to modify the job in one way or another. When extensive modifications are indicated, it is best for the sales representative to write new job specifications and submit the work as if no previous proposal had been issued. For minor changes in the order, such as a different type of paper stock, only that segment of the job should be recalculated. Such minor revisions are normally completed by the estimator on the original estimate or with a supplementary estimate form attached to the original estimate. After this revision is completed, a new proposal, or final quotation, will be issued to the customer.

Price negotiations are handled differently from company to company. In some cases, management will allow the sales representative to serve as the final judge for prices. In other cases, management may decide that final decisions will be made by a certain top management individual or group, with the sales representative serving as a go-between. Most companies normally have established procedures to handle such negotiations.

4.9 The Price Quotation

A price quotation, offered by the printer and accepted by the customer, is a legal and binding contact between the two parties. It is therefore important that the quotation specify exactly the job to be produced, with firm price, quantity, and quality levels indicated. In many cases, it will contain additional specifications, such as the date of shipping or delivery, type of wrapping for the final packages, special conditions of the final product, and so on.

The quotation should be issued in duplicate and delivered or mailed to the customer. Upon receipt and review, the customer must sign and return one copy to the printer. Good business practice dictates that no quotation be modified or changed unless both parties agree and initial such changes. It is most important that no verbal agreements regarding any aspect of the quotation be made between customer and salesperson or customer and printer. Such verbal agreements can lead to many misunderstandings.

Because the quotation is a legal and binding contract, it should be accompanied by a copy of the Printing Trade Customs (Figure 4.4). *Printing Trade Customs* establish the parameters and working conditions between customer and printer. Many times, they are printed on the back of each quotation. The Trade Customs cover a wide array of job elements that have in past years been controversial. They are based on legal fact, so their inclusion with the quotation provides communication between customer and printer about ownership of preparatory materials, alterations, overruns and underruns, and fourteen other contingencies. The sales representative is wise to point out the Trade Customs to all customers when the quotation is issued.

The signed quotation is a vital ingredient in the chain of printing sales and production. It should specifically "lock up" all elements of the job. Once signed and returned to the printer, production on the customer's order may begin.

PRINTING TRADE CUSTOMS

Trade Customs have been in general use in the printing industry throughout the United States and Canada for more than 60 years.

1. **QUOTATION** A quotation not accepted within sixty (60) days is subject to review. All prices are based on material costs at the time of quotation.
2. **ORDERS** Orders regularly placed, verbal or written, cannot be cancelled except upon terms that will compensate the printer against loss incurred in reliance of the order.
3. **EXPERIMENTAL WORK** Experimental or preliminary work performed at the customer's request will be charged at current rates and may not be used until the printer has been reimbursed in full for the amount of the charges billed.
4. **CREATIVE WORK** Creative work, such as sketches, copy, dummies, and all preparatory work developed and furnished by the printer, shall remain his exclusive property and no use of same shall be made, nor any ideas obtained therefrom be used, except upon compensation to be determined by the printer, and not expressly identified and included in the selling price.
5. **CONDITION OF COPY** Upon receipt of original copy or manuscript, should it be evident that the condition of the copy differs from that which had been originally described and consequently quoted, the original quotation shall be rendered void and a new quotation issued.
6. **PREPARATORY MATERIALS** Working mechanical art, type, negatives, positives, flats, plates, and other items when supplied by the printer, shall remain his exclusive property unless otherwise agreed in writing.
7. **ALTERATIONS** Alterations represent work performed in addition to the original specification. Such additional work shall be charged at current rates and be supported with documentation upon request.
8. **PRE-PRESS PROOFS** Pre-press proofs shall be submitted with original copy. Corrections are to be made on "master set," and returned marked "O.K." or "O.K. with Corrections" and signed by customer. If revised proofs are desired, request must be made when proofs are returned. Printer cannot be held responsible for errors under either or both of the following conditions: if the customer has failed to return proofs with indication of changes, or if the customer has instructed printer to proceed without submission of proofs.
9. **PRESS PROOFS** Unless specifically provided in printer's quotation, press proofs will be charged at current rates. An inspection sheet of any form can be submitted for customer approval, at no charge, provided customer is available at the press during the time of makeready. Lost press time due to customer delay, or customer changes and corrections, will be charged at current rates.
10. **COLOR PROOFING** Because of differences in equipment, processing, proofing substrates, paper, inks, pigments, and other conditions between color proofing and production pressroom operations, a reasonable variation in color between color proofs and the completed job shall constitute acceptable delivery.
11. **OVER-RUNS AND UNDER-RUNS** Over-runs or under-runs not to exceed 10% on quantities ordered, or the percentage agreed upon, shall constitute acceptable delivery. Printer will bill for actual quantity delivered within this tolerance. If customer requires guaranteed exact quantities, the percentage tolerance must be doubled.
12. **CUSTOMER'S PROPERTY** The printer will maintain fire, extended coverage, vandalism, malicious mischief and sprinkler leakage insurance on all property belonging to the customer while such property is in the printer's possession; printer's liability for such property shall not exceed the amount recoverable from such insurance. Customer's property of extraordinary value shall be insured through mutual agreement.
13. **DELIVERY** Unless otherwise specified, the price quoted is for a single shipment, without storage, F.O.B. local customer's place of business or F.O.B. printer's platform for out-of-town customers. Proposals are based on continuous and uninterrupted delivery of complete order, unless specifications distinctly state otherwise. Charges related to delivery from customer to printer, or from customer's supplier to printer, are not included in any quotations unless specified. Special priority pickup or delivery service will be provided at current rates upon customer's request. Materials delivered from customer or his suppliers are verified with delivery ticket as to cartons, packages, or items shown only. The accuracy of quantities indicated on such tickets cannot be verified and the printer cannot accept liability for shortage based on supplier's tickets. Title for finished work shall pass to the customer upon delivery to carrier at shipping point or upon mailing of invoices for finished work, whichever occurs first.
14. **PRODUCTION SCHEDULES** Production schedules will be established and adhered to by customer and printer, provided that neither shall incur any liability or penalty for delays due to state of war, riot, civil disorder, fire, labor trouble, strikes, accidents, energy failure, equipment breakdown, delays of suppliers or carriers, action of government or civil authority and acts of God or other causes beyond the control of customer or printer. Where production schedules are not adhered to by the customer, final delivery date(s) will be subject to renegotiation.
15. **CUSTOMER-FURNISHED MATERIALS** Paper stock, inks, camera copy, film, color separations, and other customer-furnished material shall be manufactured, packed, and delivered to the printer's specifications. Additional cost due to delays or impaired production caused by specification deficiencies shall be charged to the customer.
16. **TERMS** Payment shall be whatever was set forth in quotation or invoice unless otherwise provided in writing. Claims for defects, damages, or shortages must be made by the customer in writing within a period of fifteen (15) days after delivery of all or any part of the order. Failure to make such claim within the stated period shall constitute irrevocable acceptance and an admission that they fully comply with terms, conditions, and specifications.
17. **LIABILITY** Printer's liability shall be limited to stated selling price of any defective goods, and shall in no event include special or consequential damages, including profits (or profits lost). As security for payment of any sum due or to become due under terms of any agreement, printer shall have the right, if necessary, to retain possession of, and shall have a lien on, all customer property in printer's possession including work in process and finished work. The extension of credit or the acceptance of notes, trade acceptance, or guarantee of payment shall not affect such security interest and lien.
18. **INDEMNIFICATION** The customer shall indemnify and hold harmless the printer from any and all loss, cost, expense, and damages (including court costs and reasonable attorney fees) on account of any and all manner of claims, demands, actions, and proceedings that may be instituted against the printer on grounds alleging that the said printing violates any copyrights or any proprietary right of any person, or that it contains any matter that is libelous or obscene or scandalous, or invades any person's right to privacy or other personal rights, except to the extent that the printer contributed to the matter. The customer agrees, at the customer's own expense, to promptly defend and continue the defense of any such claim, demand, action, or proceeding that may be brought against the printer, provided that the printer shall promptly notify the customer with respect thereto, and provided further that the printer shall give the customer such reasonable time as the exigencies of the situation may permit in which to undertake and continue the defense thereof.

Originally formally promulgated, Annual Convention, United Typothetae of America, 1922. Revised and updated and repromulgated, Annual Convention, Printing Industries of America, Inc., 1945 & 1974. Updated and adopted by the Graphic Arts Council of North America, 1985.

Printing Industries of America, Inc.

Figure 4.4

4.10 Production and Delivery of the Job

Job Ticket

As the job begins the production sequence, a *production docket*, sometimes called a *job ticket*, is completed (Figure 4.5). While the form of the job ticket may vary from plant to plant, it is usually printed on a kraft envelope that holds various articles pertinent to the job—for example, paper-ordering documents, the job dummy, and requisition forms for buyouts. Production information is recorded on the form on the envelope relating to the sequence of production through the plant. The docket will have the customer's name, purchase order number, order date, due date, a job ticket number, and quantity, finish size, and colors to be run. In addition, all production segments will be identified with respect to the contribution each will make for the job. The production docket is normally completed by production control personnel. However, other plant personnel may be assigned this duty by top management.

Time and Materials Record

With larger commercial and high volume printers, tracking of employee operations and activities, inventory and chargeable time during the actual production of the job, is essential to maintaining accurate production standards and budgeted hour cost rates. As discussed in Chapter 3, the process of collecting such data—the first step in the job costing process—may be done manually or by computer. While both systems find common usage, manual data collection tends to be laborious and sometimes inaccurate while electronic factory data collection (EFDC) is convenient and has become widely accepted in the latter 1980s in the industry.

When completing manual data collection, a time and materials record accompanies the job through all steps of production. Sometimes printed on the back of the job ticket, this document is used to record actual production times and materials used, recorded accurately by the employees doing the work. A manual time and materials record is provided in Figure 4.6. Chapter 5 provides further information on electronic factory data collection.

There are some general rules that should be followed during production, rules designed to resolve problems before they occur. First, every attempt should be made to minimize the errors and mistakes in all work completed. Production personnel should double-check work when conditions permit. The customer should be consulted for approval on artwork, press proofs, process color press run sheets, and any other areas wherein difference of opinion might occur. Customer approvals should be in writing.

Second, the customer should be informed of and should approve any modifications or changes in the job specifications. For example, if the paper selected is no longer available, the customer should choose the substitute; if a typeface is not available in the size proposed, any substitution should be approved by the customer. Of course, the

Production Docket	Customer		Job Number
Salesman	Dept./Organization	Phone	Date Ordered
P.O. Number	Deliver To	Quantity	Date Due
Job Description		Finish Size	Litho Letterpress
		Colors	Rerun Rerun w/Changes

Stock

Type	Weight	Color/Finish	#Parent	#Out	Size

Other Materials

Ink	Film	Plates	Buy-Outs

Composition

Machine	Type Face	Size	Line Length	Other instructions
				Proofs ☐ Out ☐ Ok'd ☐

Camera

Line Shots	Size	Halftones	Screen	Size	Other

Stripping/Platemaking

Flats	Size	#Up/Flat	Plates	Burns/Plate	Other
					Brownlines Out ☐ Ok'd ☐

Press

Press	Press Size Sheet	#Finished Sheets	#Spoil	Imposition	#Up	Bleeds

Numbering Beginning ¦ Ending	Score ☐ Perforating ☐ Embossing ☐	Other

Binding

Cutting From ¦ To	Finish Trim Size	Stitch Side ☐ Saddle ☐ Gather ☐ Collate in sequence
Folding From ¦ To	Perforating ☐ Drilling ☐ Rounding corners ☐	Padding ☐ in pads of__________'s Delivery Box ☐ Wrap ☐

Other Instructions

Figure 4.5

Production docket (job ticket) (Developed by Jake Swenson; reproduced with permission)

Time Record

Operations Completed

Composition ☐ Camera ☐ Stripping ☐ Plate ☐ Press ☐ Bindery ☐

Employees Name	Date	Start	Finish	Total	Initial	Operation

Materials Usage

Paper Type	Parent Size	Quantity	Press Size	Number	Ink Color	Pounds Used

Film Type	Size	Quantity	Plate Type	Size	Quantity

Other Materials

Figure 4.6

Time and materials record (Developed by Jake Swenson; reproduced with permission)

printer does have some flexibility within the guidelines of the Printing Trade Customs. Even so, it is best to resolve necessary modifications directly with the customer. It is sound business practice to be certain that the client will be pleased with the work produced by the printer.

The third rule is that the specified delivery date should be met unless extraordinary conditions do not permit it and prior arrangements have been made between both parties. Even though it may not be possible to meet each job deadline. every attempt, including overtime, should be made to meet deadlines as much as possible. Many printing plants have excellent production scheduling and control systems. Some systems are even computerized, backed up with accurate production standards and estimating procedures. Meeting complete job specifications, within the deadlines as agreed, increases the quantity of repeat orders for any printing company.

4.11 Invoicing and Cash Flow

Cash Flow Sequence

An important element in the operation of any printing business is the manner by which capital moves through the company. In elementary form, this sequence of *cash flow* events follows a circular pattern (Figure 4.7). It begins with cash on hand in the business. This cash is used to meet all out-of-pocket and fixed expenses included in the preparation of the customer's printing order: artwork, typesetting, photographic work, image assembly, and platemaking. In the next step—actual production on the press and binding operations—additional variable costs, such as labor and paper expenses, are incurred. At this point, the cash outlay represents the total manufacturing cost for the job. After the work is delivered, the customer is invoiced for payment, and the bill represents what is termed an *account receivable*. Once the customer pays this bill, usually with a check or sometimes in cash, the capital starts through the cycle again. The pattern is one of continual repetition.

The ideal situation for a company is to have enough capital to meet all of its obligations and anticipated growth yet not have too much cash tied up in noninterest accounts, such as checking accounts, wherein the money is providing no financial return. In any one given time period, the total cash received should be greater than the total disbursements to be paid out. Thus, if total cash paid in over the month of March was $65,000, and the disbursements (total of all incurred expenses) paid out totalled $67,000, the company must borrow or obtain from some other source an additional $2000 to meet all March commitments completely. The cost of borrowing money is high and not a desirable solution for most businesses. Consistent borrowing inevitably leads to greater indebtedness and will, in the long term, cause a forced bankruptcy or liquidation of the business.

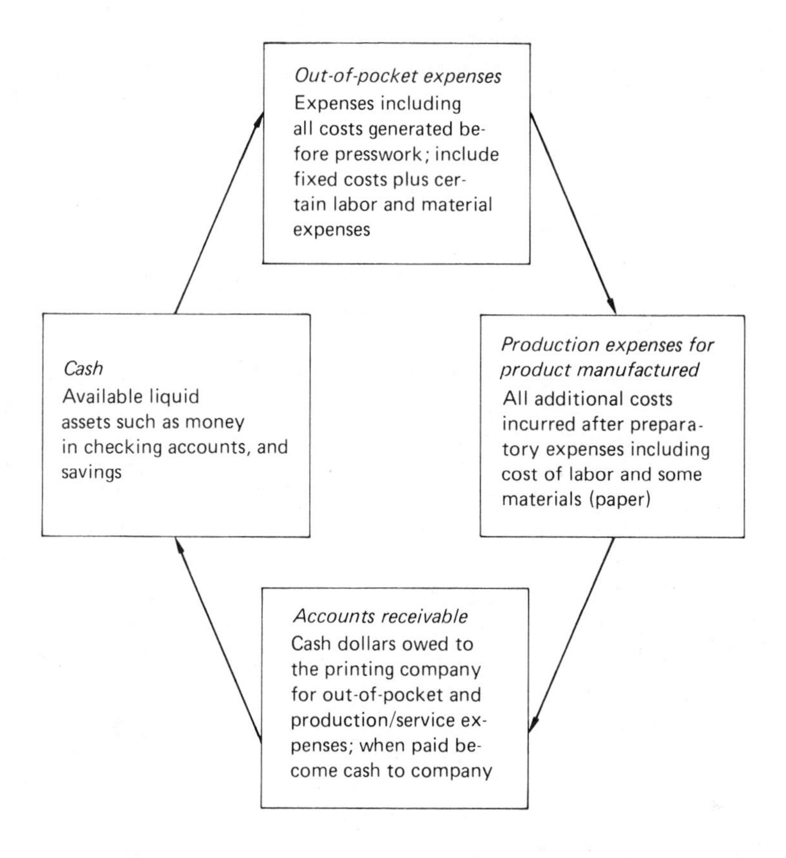

Figure 4.7

Diagram of cash flow in a printing company

Accounts Receivable Collection Period

One major problem with many printing companies is that their flow of cash is limited by the amount of time it takes for individual customers to pay their bills—the *accounts receivable collection period.* For example, when billing for all work done over a 1 month period is completed at the end of the month, as much as 4 weeks has elapsed between delivery and billing for some jobs. In addition, the customer will probably not pay the bill immediately but may wait 30 or even 60 days before paying. The net result is that the customer is using the printed goods, paid for with the printer's capital, for perhaps as long as 3 months. As tightening credit requirements by graphic arts suppliers and paper retailers becomes common, many material costs must be paid quickly. In addition, the cost of labor for plant personnel must be paid within a very short time of completion of work. The net effect, unfortunately, is that the printing company is indirectly financing the customer's business while at the same time incurring an unfavorable cash position by being forced to meet its own obligations quickly. When printing management allows this squeeze to occur, the effect can be devastating on every aspect of the business.

Effective Cash Flow

Cash flow is a very important aspect of the printing business for reasons already discussed. It is a matter with which the top management of the company must deal carefully and effectively.

The key to effective cash flow is twofold. First, the customer should be invoiced as soon as possible after delivery of the printed goods—within 2 working days is recommended. Thus, the customer receives the goods and immediately thereafter the invoice, after which he or she will then have 30 days to make payment. Sometimes discounting for early payment—such as 2% net 10 days—will be used by the printer. In other cases, a finance or interest charge will be levied for any unpaid balance on the amount owed after the initial 30 day period. Both of these techniques receive varied reaction on the part of printing management and customers alike.

Besides cash discounting and addition of interest charges on the unpaid balance, the split invoice technique is used by printing management to improve the cash situation of the business. This procedure is usually followed by commercial printers. Basically, the *split invoice technique* requires customers to pay a certain amount of the quoted price on deposit prior to any work on a prospective order. The deposit can vary from 10% to 100%, with 50% the amount normally used. This procedure allows the printers to pay the incurred out-of-pocket costs of the work with the customers' money and not tie up their own capital. Once the job is printed and delivered, the customers will be invoiced immediately for the remaining amount due, which they should be obligated to pay within 30 days. The split invoice technique is effective because it commits customers to the work before any actual production is completed, reduces the final payment by the customers, and allows printers to use less of their own capital in the production of the customers' orders. Of course, the customers' cash flow position will be adversely affected, which they may quickly point out. The split invoice technique is most helpful when printing management desires a good cash position; some customers may not like the procedure, and some printing orders may be lost because of it.

A second key to effective cash flow is for the printer to establish some type of cash forecasting system. Using such a procedure to project income and disbursements for a 6 month or 1 year period based on forecasted sales and anticipated collections of receivables is a wise planning tool. In effect, the printer can "roadmap" cash movement through the business for the selected time period, identifying those months where excess cash will exist and others where the cash balance will be unfavorable. Cash forecasting is desirable for printing firms wishing to tighten or define income and disbursement patterns or to project available excess cash for investment or other purposes.

4.12 The Job Cost Summary

After the job has been completed, it is necessary to review the actual time and materials data collected during the production process, then compare this actual data to the estimated times and costs originally developed. This procedure—commonly known as job costing—can be accomplished either manually or by computer, depending on how the data was collected or available computer tools to complete the process. For example, if the company used electronic factory data collection, analysis is likely to be completed entirely by computer. If the data was collected manually, as described in Chapter 3, it is possible to build a computer spreadsheet to analyze this same data. This would be done by inputting the manually-collected data into the spreadsheet program and then use available spreadsheet tools to complete the analysis. Spreadsheets are discussed in Chapter 5 in further detail.

Job costing provides the company with feedback in three important areas:

- Estimating, showing estimating errors and highlights of jobs relative to actual versus estimated times and costs.
- Production, showing deviations from existing production, equipment utilization and unexpected modifications in production from actual to estimated
- Planning deviations or planning errors

The result of job costing is typically to point out variances between actual and estimated times and costs allowing the company to modify estimating production standards, budgeted hour costs and planning procedures. While the process is always completed at the conclusion of production, some progressive printers with electronic factory data collection complete the job costing process immediately after the job is finished and before invoicing. This allows the printer to bill for the addition of all author's/customer's alterations, which can be a significant source of lost revenue if not included. Other printers job cost only after certain (perhaps larger volume) jobs, a random selection of jobs, or only those jobs which appeared to have many author's alterations.

Computers for Printing Estimating and Management

5.1 Introduction

During the 1950s and 1960s there was limited use of computers in the printing, publishing, and allied industries. With the development and acceptance of phototypesetting in the early 1960s, followed by rapid and major advances in electronics in industry, the computer entered the printing plant permanently. As we approach the last decade of this century, computers and electronic systems utilizing microprocessors are in wide use in the printing industry. Technological development has continued to improve these tools and is likely to move us forward to even more sophisticated systems in the next century. During the 1980s some significant changes have occurred to facilitate computer integration throughout the printing industry:

1. There has been a slow but obvious diminishment of computer "fear," largely brought on by the explosion of microcomputers, simplified operating systems which can be easily learned for programming purposes, and improved training. Computer literacy—a general understanding of electronic computing and the general technology and application of computers to solve problems—is common.
2. Hardware has greatly improved through electronic advancements, downsized to save space, and improved to reduce maintenance problems and repairs.
3. A variety of software programs covering areas such as printing estimating, inventory control, production management, and BHR development have become available at reasonable cost.
4. The microcomputer—or micro—has become a very popular tool for many printing firms, completing dozens of operations including estimating, production control duties, and inventory management. Microcomputer hardware has continually dropped in price and is therefore easily available to any printing company wanting to computerize.
5. Inexpensive mass storage devices including hard disks and magnetic tape units have become easy to obtain and provide fast, convenient operation. Laser disks are the next—and even faster—mass storage medium.
6. Printing industry computer vendors selling turnkey systems—complete packages of integrated hardware with specific application software—have largely stabilized with a number of vendors offering nationwide sales and full, professional services to support their installed systems.

This chapter will focus on management and information systems available to the printer. An overview of electronic and computer devices for production will be covered briefly in this chapter with specific details later in the text. The step-by-step process to select and implement a computer for estimating and information processing will be provided along with a complete 1990 listing of printing industry computer vendors in Appendix C.

While there have been numerous magazine articles covering the subjects of computer estimating and management information systems over the past five years, the major source providing guidelines and pertinent information in this subject is available through NAPL's *GraphPro* (for larger computer and MIS systems) and *MicroGraphPro* (covering microcomputer systems). Throughout this chapter the 1983 and 1987 *GraphPro* and 1983 and 1987 *MicroGraphPro* editions will be cited as appropriate.

5.2 Computer Applications in a Printing Company

Mainframes, Minicomputers and Microcomputers

In past years, computers have been generally categorized into three basic units, by size, from large to small: mainframe computers, minicomputers, and microcomputers. Today distinctions between each are much less defined. While mainframes and minicomputers still exist, they have been downsized to fit into very small areas, and have thus begun to appear like larger versions of the microcomputer.

NAPL's *GraphPro* (1987) defines a mainframe as a computer which costs in excess of at least $100,000, probably requires environmental controls such as air conditioning, and requires dedicated support. *GraphPro* goes on to define a microcomputer as frequently desktop (or lap) in size, which costs as little as "hundreds" of dollars and is frequently "single user." A minicomputer, then, is whatever falls in between the other two. Generally the cost of a single microcomputer system with hard disk media is between $1,000 and $2,000 dollars.

Information Processing and Printing Production Applications

It is important to distinguish between computer systems for management, used largely for processing information of various kinds, and computer systems developed to aid in the manufacturing of printed goods. While these two different general groups of computer systems exist with fairly clear distinctions, there is yet an overlap or blurring in some cases. Perhaps the most obvious duality of roles—a computer system which serves both production and information functions—is the microcomputer. The uses of "micros" are so varied, and the machines so versatile, that they can easily fit into both information and production uses without difficulty.

The quick printer and smaller commercial printer has seen the rapid evolution of a various number of computer systems, lead by the revolutionary Macintosh, as well as computerized peripheral attachments including laser printers, black-and-white copiers, and color copiers with electronic controls. For the quick printer or smaller commercial printer, the Macintosh represents the first truly "front-end" computer system economically feasible, yet versatile enough to handle both graphics and information processing. In fact, piggybacking both information processing and graphic production duties on a "Mac" is common. For example, during normal working hours a single Macintosh, running desktop publishing (DTP) software such as PageMaker or Ready-Set-Go, can be used as a saleable piece of equipment on which customers generate original artwork. In the evening and "off" hours, the same Macintosh computer can be used to complete various business and management duties including invoicing, inventory control, and monitoring work-in-process. Other uses of the Macintosh as well as IBM and the IBM-compatibles will be discussed as appropriate to the prepress areas later in this text.

Larger printers and high volume printers have embraced the computer with equal enthusiasm, although with much more focused, stand-alone equipment. These larger firms have hundreds of customers, dozens of jobs in process at any given time, and literally thousands of pieces of information to track and monitor. Without a computer to aid and expedite the information flow, details would be forgotten, too many mistakes would be made in jobs, and there would be a plethora of unhappy customers. The larger printer has not been as directly affected by the Macintosh revolution as his small printer counterpart, but more affected by development of specific computer systems that serve particular functions. For example, information processing, including estimating, might be completed on a turnkey computer system developed specifically to meet the larger printers' information uses and needs. In terms of computers for production, electronic stripping would be done on a Gerber AutoPrep, also a turnkey system, but made specifically to complete this task.

Typesetting and Word Processing

The process of typesetting has lead the evolution of computers in the printing industry. From hand set type and hot metal in the 1940s to phototypesetting in the 1950s and 1960s, typesetting today is done almost completely by capturing of keystrokes on literally any computer system which converts them into digital form, then transmits these digital signals to an electronic typesetting unit, or printer, able to render final typeset material.

Perhaps the most dramatic change in typesetting is desktop publishing (DTP). While figures vary, it is estimated that between 250,000 and 300,000 DTP packages have been sold since the products were first introduced in the mid 1980s. Briefly, desktop publishing software provides user-integration of text and graphic images so that the computer can be used as both a design and typesetting device. Microcomputers input (or digitize) characters, and inexpensive, laser-driven printers are used for output. The development and adoption of Postscript code as the accepted common language for printing any digitized image (type or graphic image), coupled with the parallel introduction of Apple's LaserWriter in 1985, played a major role in the acceptance of microcomputers and laser imaging systems.

While critics indicate that the final laser typeset image at 300 dots per inch (dpi) is not of acceptable quality, adoption of such systems has been nothing short of phenomenal, particularly for less critical end-use requirements such as newspapers and certain types of commercial design and typesetting work. Also, use of higher resolution typesetting output units such as Linotype's Linotronic L300 provide final quality typeset output at a a very high level.

With the advent of front-end typesetting systems, dedicated word processors are fading from the scene. The front-end typesetting system includes a microcomputer, laser printer or phototypesetter, a good word processing program, and DTP software. Hard disks and at least one megabyte of internal memory are general machine requirements for the micro driving the system. The near future will likely bring more sophisticated word processing and DTP programs, improvements in scanners for inputting graphic images into the micro, better mass storage devices, and improved-resolution laser and electronic typesetting units for output. Networking—connecting systems together so that they can share common programs, output devices and other peripherals—appears likely to continue.

There are three significant technological advancements which have also aided in the acceptance of the microcomputer as a front-end typesetting system: inexpensive hard disks for mass storage of electronic information, improved microprocessor design allowing for very large memory capacity at very affordable prices, and the continued improvement of output resolution and speed. Hard disks and improved microprocessors also speed up the microcomputer's operation significantly, making the user wait much less during the actual word processing operation.

Other Prepress Systems

Of all the areas affected by electronics and computers, prepress—design/art production, typesetting, photography, film assembly, proofing and platemaking—has been perhaps the most drastically changed. Prepress includes those areas of printing production necessary before any ink-on-paper printing is actually completed.

It is impossible to quickly summarize all the electronic changes which have occurred in the prepress area since the 1980s. Typesetting, which began the revolution in the 1960s, followed with word processing systems in the 1970s, has already been briefly discussed. The remaining production areas—design and art production, photography, image assembly, and platemaking/proofing have each been affected in different ways. Furthermore, as one success has lead to another—such as the scanner for color separation production—"steppingstone" development of related electronic units has been the rule.

Noncomputerized prepress production, done without computerization or electronic aid, represents very labor-intensive production, and tends to become a major bottleneck for many commercial printers. Without computers much of the work is man-machine oriented, with a high degree of skill required by the craftsmen doing the production; the equipment is fairly slow since it is paced to the operator's skills.

Thus high labor costs, unavailability of skilled labor, and slow production output stimulated computerization in the prepress area. Also, as the quick printing industry matured in the 1970s and began to provide good-quality printing on a fast-turnaround basis, pressure on the commercial printer and high volume printer to do the same increased. Other developments also had an impact: the natural evolution of computer-aided design/computer-aided manufacturing (CAD/CAM) systems in other industries "tracked over" into the printing industry, and the parallel development of affordable microprocessors and microcomputers—particularly the inexpensive Macintosh—to effectively interface graphic images, typesetting, and other graphic duties which previously were each stand-alone operations.

Chapters 8, 9, 10 and 11— respectively on estimating production components of printing including artwork, copyfitting and composition, photographic procedures and image assembly, and platemaking—will address computerization of each area and provide information on estimating.

Data Collection and Database Systems

Printing industry data collection—also known as shop floor data collection—uses electronic units to track work-in-process production. An installed electronic (or automated) factory data collection (EFDC) system—typically found in larger commercial printers and high volume printers—consists of small, user-friendly terminals at strategic production points throughout the plant, connected to one single computer to which all information is sent and stored. Employees doing the production work log all activities performed, materials used, production notes, and any other vital production data into this data collection terminal; the employee is also required to log "on" and "off" to begin and end the workday, for lunch, breaks, etc. Ultimately the employee's workday can be traced and reviewed to the jobs on which he or she worked, times spent in various activities on a job-by-job basis, times which are chargeable and those which are nonchargeable to particular customers, and so on. EFDC is an ideal system for immediate job cost review and is discussed in some detail in Chapter 4.

Databases for microcomputers are marvelous tools for processing vast amounts of data quickly and accurately, and for building and maintaining even the most complex numerical files. To control inventory, the user purchases a microcomputer database program, inputs into the program his on-hand inventory, then updates this inventory package weekly based on depleted items noted on job tickets. Database software may be relational in nature, where data in one field can be associated with data in another field, or nonrelational where the data is generally not intended to be interfaced. Thus, instead of a fairly expensive EFDC system, the printing company could track all job costing data manually during production, then build a microcomputer database to analyze this information quickly and accurately.

Number Crunching and Spreadsheets

As with data collection and processing, computers also speed up number-crunching (numerical analysis) duties in the printing plant. Because they are mathematically error-free, computers are valuable tools for a diversified series of important chores.

Perhaps the most significant number-crunching duty is estimating. CAE essentially is used in two different ways. The first, discussed in theory in Chapters 1 and 2, utilizes the application of cost-estimating data via computer. These data include the development of BHRs (by computer), the application of standard production times, the addition of material costs and buyouts, and finally, the addition of company profits (which may or may not be a part of the CAE system). Doing cost estimating manually is a time-consuming, laborious task prone to mathematical errors. When cost estimating is computerized, not only are math errors completely eliminated but also the process is generally fast and convenient. The majority of CAE programs available from vendors serving the printing industry utilize the cost-estimating model.

Another direction for CAE is the development and use of electronic spreadsheet programs, typically developed in-house by the estimator to fit company needs. A *spreadsheet* is a row and column arrangement of numerical data that can be related by the establishment of formulas upon which the cell information is generated. Initially, to develop a spreadsheet, a general-purpose spreadsheet program would be purchased for the micro and loaded into the computer. Spreadsheet base programs such as VisiCalc, ® Multiplan, ® Excel,® and Lotus 1-2-3™ are available for purchase at most computer stores to run on an array of microcomputers. The micro then becomes a stand-alone estimating system for the company

To create a spreadsheet, the estimator develops a series of blocks of information, called *cells*, via the computer monitor. Each cell is cross-referenced by a letter for the row and a number for the column, so that "A-l" would be the first cell to accept data. For example, the "A" row might be built to contain titles of desired data, such as "Press Number" (A-1), "Average Imp/Hr" (A-2), "Quantity Needed/Job" (A-3), "BHR $/Hr" (A-4), and "Total Press $/Job" (A-5). The "B" row directly below the "A" row would then receive standard estimating data such as "6000 shts/hr" (B-2) to represent "Average Imp/Hr" in cell A-2 or variable data such as "25,000 pss" (B-3), which would vary from job to job. Thus, the spreadsheet forms a model by which "what if" questions can be asked and answered. For example, changing "Quantity Needed/Job" from 25,000 pss to 35,000 pss in cell B-3 will change the resultant "Total Press $/Job" figure in cell B-5. The spreadsheet system allows for very fast calculations that are done automatically when the variable data are changed. Cell data can be based on formulas or on an identified piece of data. The spreadsheet can ultimately be developed to estimate all segments of a normal printing job using both a reasonably responsive, simple matrix of fixed plant production standards and hourly costs and variable data that pose the "what if" questions. Because each company develops its own spreadsheet data based on individual company information, the system can be tailored for the company.

Some of the other obvious number-crunching duties that can be quickly and accurately done by computer—and easily on spreadsheet programs—include BHR development, "what-if" comparisons of all kinds, machine loading, payroll services, accounting, purchasing, forecasting sales, production, cash flow and budget preparation.

Telecommunications, Electronic Transmission, and FAX Systems

The most common and inexpensive telecommunications unit is the modem, an electronic interface between the computers, via a normal telephone line. The modem changes the sending computer's digital code into transmittable electronic signals, which are then picked up by the other computer and converted back to digital form. The speed—or baud rate—at which modems transmit these electronic signals is commonly 1200 or 2400 baud, which is considered fairly slow when the data to be transmitted is extensive. Since digitized graphic images such as color separations may require considerable electronic space, modems represent a less frequent source of transmission. When transmitting limited data, or repairing or debugging software, telecommunications via a modem is a popular medium.

For transmitting large data groups such as color separations or fully digitized pages, including desktop publishing materials, either direct electronic transmission or shipment of transportable floppy disks is common. Electronic transmission using satellites or microwave systems represent the fastest method, yet these systems are typically available to only the largest and most sophisticated prepress trade shops, printers, and publishers. More common and available to most printers is the overnight transporting of formatted floppy disks containing data representing final or finished images in digitized form. Federal Express, United Parcel Service, and other carriers provide such services.

Facsimile (FAX) machines have also become popular for transmitting images on paper by telecommunication. The FAX is a tabletop electronic input/output (I/O) unit which "reads" graphic images or text on paper, converts these images to electronic signals, and sends these signals over normal telephone lines to an accepting FAX on the other end. FAX units can either send or receive and typically operate at between 4,800 and 9,600 baud, which is considerably faster than modems used for telecommunications. Many quick printers sell FAX services along with their other customer services they have available.

Electronic Press Systems

The printing process, specifically lithographic sheetfed and web printing, has been significantly affected by computerization and electronics in the 1980s. To the casual observer this statement may not appear accurate for two reasons: the final printed piece physically looks much the same compared to a printed image ten years ago, and the presswork manufacturing process still involves the same essential production steps: makeready, press run, and washup.

Electronic and computer changes in presswork can be divided into two broad categories: (1) makeready and press preparatory improvements, and (2) process controls during printing. Together these improvements have had significant effect on the internal structure of the pressroom, the quality and cost of the final product, and the speed with which the product is delivered.

When completing a makeready for a specific job, the time taken from the beginning of the makeready period to the point in time when the first sheet is approved for printing has always been a key productivity issue for printing managers. The longer a makeready takes, the higher the fixed costs to produce the first saleable sheet and the more costly the product to both the customer and the printer. Changes to thus reduce makeready times—and thus, costs—include plate scanning devices which electronically read image densities for ink adjustments prior to startup, accurate plate registration controls, image movement via electronic consoles, and various feeding and delivery controls allowing both simplified makeready and higher running speeds.

During the printing process, most newer sheetfed and web presses are operated from an electronic console that provides the operator with instantaneous controls over literally every aspect of the image being printed on the sheet. In addition, computer devices read ink densities and automatically adjust ink controls to provide precise, preset values that are consistent through the pressrun. The result of these changes has improved the quality and speed of production while concurrently reducing the number of skilled employees required in the pressroom to operate the equipment.

Chapter 12 provides information on press improvements and estimating sheetfed presswork, and Chapter 15 investigates web production and web estimating in detail.

5.3 Computer Management Systems Available to Printers

Computer systems serving the printing and allied industries can be classified into two general groups: services provided by outside service suppliers, including service bureaus and timesharing, and systems that are purchased by the printer to he used in-house, generally broken down into turnkey, microcomputer and customized system segments.

Outside Service Suppliers

Service Bureaus. Service bureaus are organizations that offer complete data processing, using computer hardware and software in the service bureau's offices. The printer provides whatever source materials are needed—for example, timecards for payroll. The service bureau processes the timecards and provides the printer with payroll checks and all related financial information with respect to payroll transactions. Service bureau functions, unlike CAE, typically are not tailor-made to fit the printer's needs precisely, but can be made available by the bureau if desired. Service bureaus usually charge monthly, and the printer pays a flat fee for all processing completed during the last transaction period.

Timesharing. With timesharing, printing companies rent time on large mainframe computers that hold their programs. Communication with the computer is typically through a terminal located in the printing plant, where employees enter information, translated via a modulator-demodulator (modem) and telephone line to the computer at another location. The information is then processed and returned to the printing company via modem either as a display on the computer monitor or in printed form. Sometimes report documents can be printed at the timesharing facility and mailed to the printer.

Fees for timesharing are divided into access charges for each time the computer is accessed via the modem and running charges for actual time spent on the computer. Program modification may be provided by an in-house programmer for the timesharing firm or by a programmer, consultant, or firm hired by the printer to handle program changes.

In-House Systems

There are three ways for printers to computerize the information processing portion of their company. Each is described in some detail below with major advantages and disadvantages indicated.

Turnkey Systems. Turnkey systems have been designed, programmed, produced, and sold as a complete package of hardware and software, with the purchaser "turning the key" to begin operation of the system. Most turnkey vendors purchase hardware from major manufacturers and assemble the computer system from various selected components. Programming is completed by individuals knowledgeable in both computer programming for the configured hardware and a knowledge of printing. The vendors sell the complete package under their company name and are called OEM (Original Equipment Manufacturers) vendors.

Figure 5.1 provides a view of a typical modular turnkey approach to computerizing a printing company with a full management information system (MIS). There are two distinct layers for the system—production and accounting. Installation of a typical MIS turnkey system, as indicated in Figure 5.1, would begin with order entry/job costing/production analysis, followed with BHR development and then quotation control/estimating. This may take from 30 to 90 days to implement. Sometime after the system is used for estimating, raw material inventory and purchase order control would go on-line, and sometime after they were fully operational, job loading and scheduling would be installed. Using the modular approach, system implementation is focused and training of operators oriented to one function at a time.

The accounting functions could be installed in parallel fashion with estimating, since estimating requires the entering of customers' names and other customer information used by various accounting modules. The interface of such information between modules determines if the turnkey system is integrated, meaning that information or data within the systems is shared from one module to another. For example, entering a customer's name and having an assigned job number or estimate

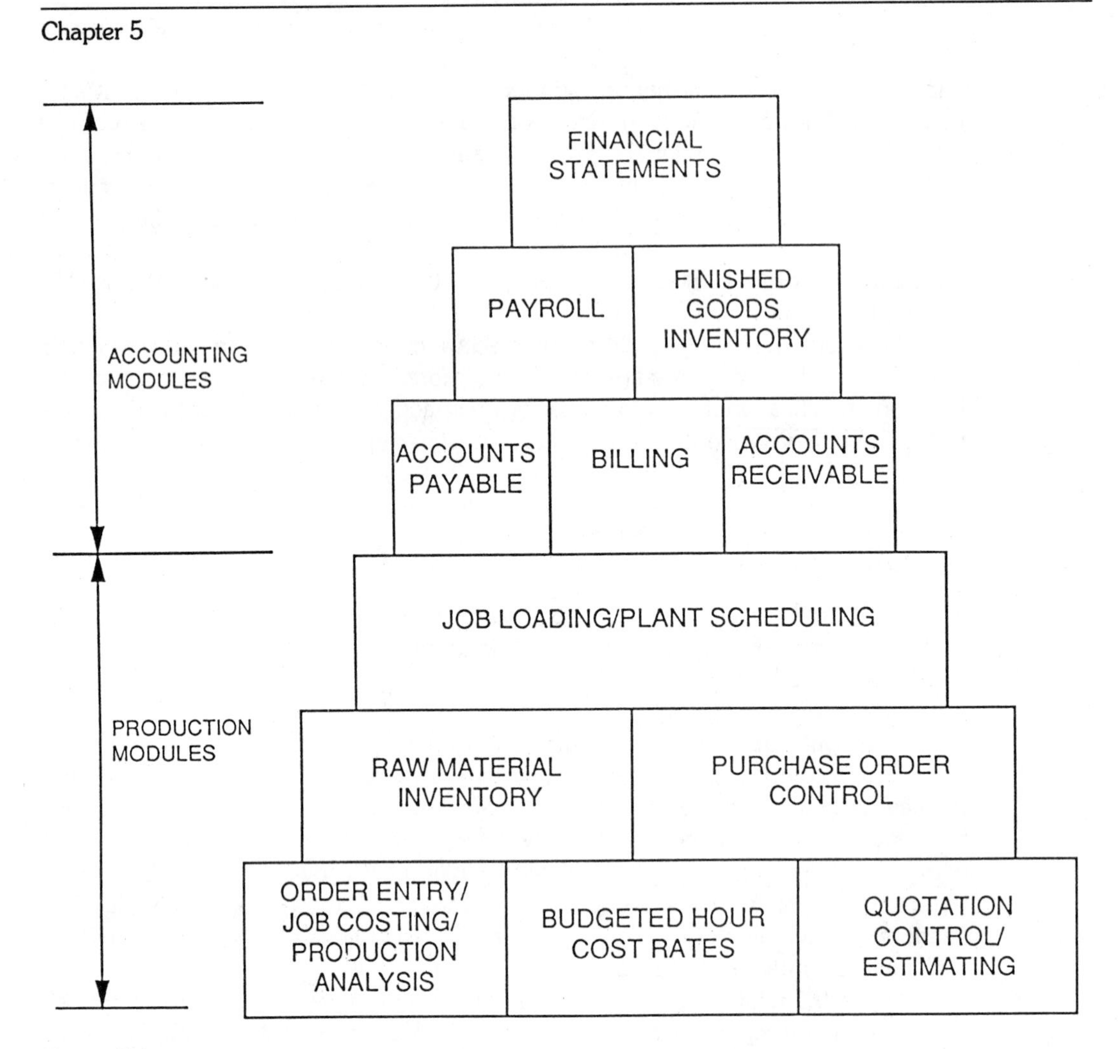

Figure 5.1

A complete management information system (MIS) for a printing firm, presented in modular form. (Reproduced with permission from Leo Salerno)

number means that all other parts of the system will accordingly recognize and respond to that name or number.

Turnkey systems represent the easiest way for a typical printing company to computerize, since the selected vendor has already made choices in both hardware and software and established a foundation upon which the system will function. Most turnkey vendors who have been in the industry for five or more years have an established history of users and system installation, and attempt to upgrade systems and installation procedures on a continuous basis. Figure 5.2 provides a chart of typical

Accounting

Typical improvements
- Better control
- Improved cash flow management
- Reductions in bad debt collections
- Reductions in accounts receivable collections
- More timely financial reports

Typical labor savings
- 50% reduction in accounts payable payment preparation
- 75% reduction in preparing and typing statements
- 75% reduction in check preparation and posting

Job Costing/Production Control

Typical improvements
- Better control over costs
- Faster billing
- Timely information on production performance
- More information with less people
- Better controls over waste and spoilage

Typical labor savings
- 50% reduction in job cost preparation and billing

Inventory

Typical improvements
- Improved inventory utilization
- More controls over inventory levels and purchases
- Aids in controlling waste and spoilage

Typical labor savings
- 25% reduction in posting and reporting

Estimating

Typical improvements
- More accurate and consistent estimates
- Faster estimate turnaround
- Allows more time for planning
- Better cost and production control

Typical labor savings
- Up to 50% reduction in estimate cost

Note: The percentages shown are based on typical savings in a range of service/manufacturing businesses. The improvements shown also have tangible benefits in some cases such as reductions in inventory costs, reductions in loads (due to better cash flow), and so on.

Figure 5.2

Chart of typical benefits when using a computer for accounting, job costing and production control, inventory, and estimating in a printing company (Reproduced from NAPL's *GraphPro,* 2nd ed., 1983, with permission) benefits when using an integrated turnkey system which includes accounting, job costing, production control, inventory, and estimating for a printing company

Advantages. Turnkey systems major advantages include:

1. A complete package is purchased and available for immediate use or modular integration into the printing plant.
2. Documentation is usually available from the time the system is installed in the company.
3. The vendor likely will provide support through a hotline which allows for fairly immediate resolution of glitches and programming problems.
4. Usually the system has been fairly well debugged through field testing, e.g. beta site, with selected printing companies.
5. System integration provides a fairly complete MIS.
6. Electronic factory data collection is usually a simple add-on.

Disadvantages. The disadvantages of turnkey systems are:

1. Turnkey systems can be moderately to extremely expensive.
2. Once a turnkey system is on-line, it is very difficult to shift to another computer system or a different turnkey vendor.
3. Any packaged program may require compromise or may be inflexible to the specific needs of the user.
4. Software support costs can be expensive, yet are essential to be certain the system is always functional.
5. Software for the system may be more than desired by the user, yet it may be impossible for the vendor to unbundle these undesired modules or parts, meaning that they must be purchased with the system even though they won't be used.
6. The user is almost entirely dependent upon the vendor for support.

It is important to note that most of the established printing industry turnkey vendors encourage, work with and/or support user groups, who usually meet from one to four times per year to discuss shared experiences, problems, and solutions. As the number of users with any turnkey system increases, their collective ability to have a specific, system-wide problem attended to is usually quite good. This could be considered an advantage of selecting a turnkey system

Microcomputer and Application Software Systems. The second way any printing company can computerize is to purchase a stand-alone microcomputer from a local computer store, then separately purchase an application software package ("canned" software) from a printing industry computer vendor. This "micro and software system" (M&SW) package generally appeals to quick printers, smaller commercial printers, and other larger printers who are purchasing a first-time computer system for management purposes or wish to utilize an existing micro now dedicated to typesetting or some other internal production function.

The microcomputer world is generally divided into two major hardware and operating system camps: IBM and IBM-compatible systems, and the Macintosh system. IBM and the compatibles typically run the popular Microsoft Disk Operating Systems (MS-DOS) or OS/2; less popular are UNIX, CP/M-80 (Control Program for Microprocessors), and others. The Macintosh runs with a system specific to the Macintosh known as a System File, coupled with a Finder that locates and opens files for use. The Macintosh System File—which contains the operating program—may also contain utility programs, desk accessories, font data, and a message file.

Vendors writing application software for M&SW systems must write the program in a language compatible with the micro's operating system. Because programming for IBM and the compatibles has largely focused on the MS-DOS operating system—and because IBM or compatible equipment was heavily purchased in the mid-1980s—most printing software vendors originally developed programs using MS-DOS, and to a lesser degree, OS/2, UNIX or CP/M. At the same time, as the Macintosh computer became faster through microchip improvements, and economical hard disk capacity became more readily available, a few vendors began offering estimating programs to the industry. Then, with the parallel explosion of the Macintosh for DTP and graphics use, more vendors began to offer production and estimating software. In almost all cases the targeted market for these programs was the quick printer and smaller commercial printer.

Advantages. Major advantages of M&SW systems include:

1. Most M&SW systems are much less expensive than turnkey systems.
2. Installation time is usually very short, allowing almost immediate use of the system.
3. The microcomputer can be used to complete other tasks such as typesetting and DTP functions, when not being used for management functions.
4. Documentation from national software-only vendors is generally good.
5. If the change to a more sophisticated system is done before too much time and effort is invested to make the M&SW system fit growing demands, the financial and time commitment with the M&SW system will be minimized.
6. System expandability through networking is simply accomplished, allowing the system to grow with the user's growth.

Disadvantages. Some major disadvantages of M&SW systems are:

1. Most "canned"software has a built-in inflexibility, limited in part by the memory capacity of the microcomputer and in part by the orientation of the developer to write a generic program for many possible purchaser-users.
2. Vendors may go out of business, leaving the purchaser-user stuck with a system that won't be upgraded or supported over time.

3. Unless the system is fairly large—beginning to approach the size of a smaller turnkey system—electronic factory data collection is not economically feasible.
4. Documentation may be limited and the vendor may not have an established process by which software support is easily provided.
5. System files will be difficult to integrate due to size capacities and the fact that production and accounting functions may run with different background software.
6. Expanding the system as the business grows will likely point out the major system limitations and programming difficulties, forcing the user to a bigger—perhaps turnkey system—thus making the expenditures for the M&SW system to be wasted money.

Customized Systems. Customized computer systems for operating any size or type of printing company provide the most tailored, precise result, generally at greater initial expense, since the process of developing the system is borne completely by the printing firm. It is not unusual for any printing company to become frustrated with canned software or turnkey packages that they find do not address their own production, estimating, inventory, or other needs. This is particularly true of smaller printing operations and quick printers who have largely been ignored by vendors or given reduced vendor interest because both sales and support costs are higher for smaller firms. Why would a vendor sell and support small shop installations when selling a system to one larger firm represents the same essential income and is far easier to install and support?

This text will address the two extremes of customized systems with the understanding that there are intermediate customizing solutions between them. On the "low end" side, which addresses the needs of first-time users or smaller firms, a microcomputer and generic spreadsheet software are all that is required. On the other end, for the larger printer or "high end" user, customized programs, hardware and support are purchased through a company or person professionally selling such services.

For the smaller printer or quick printer, a customized estimating, inventory, or production control system can be accurately built using a microcomputer and an available spreadsheet package such as Lotus 1-2-3, Multiplan, Excel, or Visicalc. It is critical that the person building the spreadsheet understands the interrelationship between each cell and be familiar with the process he or she is attempting to computerize. For example, building a spreadsheet program to reflect specific budgeted hour costs rates for the company should be attempted once the person doing this work understands, in detail, how BHRs are manually calculated. Essential spreadsheet information was presented earlier in this chapter.

Advantages. Major advantages of spreadsheets as a number-crunching tool are:

1. The cost of the software is generally quite low.

2. The system can be tailored to meet the specific, particular needs of the person or company for which the spreadsheet is intended.
3. When not running the spreadsheet program, the microcomputer can be used to complete other tasks such as typesetting and DTP functions.
4. The commercial spreadsheet programs available are usually thoroughly debugged and have clearly written documentation.
5. Development time can be very short, allowing almost immediate use of the spreadsheet.
6. Modifications can be quickly and effectively completed without the aid or expense of any outside sources.

Disadvantages. Some major disadvantages of spreadsheets are:

1. Data and calculations built into one spreadsheet are not easily transferred to spreadsheet. This means that a "big" spreadsheet program might be required to handle integrated functions such as BHR rates and estimating.
2. There is a dollars vs. time component consideration. The printer is faced with balancing the time required to build a spreadsheet program versus purchasing a canned program that may do almost all of the functions desired of the spreadsheet.
3. It is not difficult to leave out or forget a specific item or element, and catching such an error may be difficult if the only person building the spreadsheet is also the primary one using it daily. By like token, spreadsheets don't provide any outsider's input with respect to what is being completed. Professionally software vendors check and recheck the software they sell.
4. Spreadsheet cells require that data be quantifiable or boiled down to clear mathematical facts. This is sometimes hard to do within the cell framework available.
5. Spreadsheets handle number-crunching duties easily, yet this is only one component of the information processing system needed to run even a small printing firm. Expansion of the business also may force the company to sooner or later consider a canned software package.
6. The person building the spreadsheet—the person who knows it best—might move to another job, taking his or her skills to modify the program with them.

While spreadsheets represent the smaller company's method of tailoring a computer package to their printing company, the larger printer is faced with different requirements. Larger printers and high volume printers, due to the intensely competitive cost/price/volume relationship they encounter, coupled with the hundreds of customers and minute details of each job in production, need a computer system that tracks everything occurring in the plant as well as with each specific customer's job. This amounts to literally thousands of detailed pieces of information daily, with minimum errors allowable.

The solution for the larger printer or high volume printer may be to employ a person or company specifically to develop a full MIS system tailored to meet the printer's needs. Vendors providing customizing services work closely with the printer to determine both present and future computer needs, then select hardware and write (or otherwise develop) software specific to the printer's requirements. Custom vendors usually also coordinate the installation of the system and train employees in system use. They may also provide an extended service contract to maintain the system.

Advantages. Major advantages of a customized MIS system are:

1. The system is tailored to fit the printer's needs exactly .
2. The system can be fully integrated, incorporating full production and accounting functions with minimum difficulty.
3. The printer can have complete say as to the way the system is developed and worked on a daily basis.
4. Expansion of the system is easily accomplished since a professional vendor would always include this in the scheme of system design.
5. The system can be easily changed or modified since no part of the system is proprietary to the user.
6. The printer can exercise more control over both hardware and software maintenance procedures.

Disadvantages. Major disadvantages of a major MIS system are:

1. The cost of initially developing and implementing the system will be high.
2. Training times and costs will be higher and must be fully paid by the user-printer, as opposed to turnkey system training costs which are factored into the cost of a typical system and thus recovered by the collective number of system users.
3. Software modification is not ongoing but completed only when the printer wishes it to be done, with the full cost borne by the printer.
4. Because the system is tailored to the user-printer, there is typically little documentation written to support the system.
5. System development and implementation can be a long-term process, taking more than one year to complete.
6. The user-printer is required to support the system once it is fully operational, using company employees. Some custom vendors provide software service support after the system has gone on-line, yet this can be expensive.

Between the low-end number-crunching spreadsheets for the smaller printer, and fully customized information processing MIS packages for the larger printer, there are a variety of programming languages (BASIC, COBOL, C, RPG, DBMS, etc.) and intermediate solutions for customizing a computer system for a printing company. Also, it is possible for a timesharing vendor to have a specific program written in a language and operating system to fit his equipment, much as a custom program would be written by a printing industry vendor, specific to one printing firm.

Shareware has also become a way to find a program which is largely oriented to a specific printer's needs. Many smaller printers, frustrated with canned package inadequacies, have built their own estimating, inventory, and other packages—usually for IBM (or compatible) microcomputers running with MS-DOS operating systems. These software authors feel other printers should have the opportunity to use their developed packages—literally customized for their company—and make this software available through shareware groups such as Printer's Shareware (5019-5021 W. Lover's Lane, Dallas, TX 78701).

5.4 Computers for Estimating and Management in a Printing Company

Introduction

The printing industry has slowly been recognized as an important vertical market by computer vendors, largely driven by their interest in capturing the thousands of untapped customers in the industry. While there have been no significant studies measuring the computerization of the printing industry since 1985, vendor sales and industry trade show interest throughout the U.S. has been intense. Computers as a management tool, for estimating, and for other production and accounting duties are now widely available in any kind and size and for any type of printing company.

Printing estimating has been a key area for computerization, since estimating intermeshes so completely with most other parts of the printing company including accounting, production management, purchasing, proposal/quotation generation, and inventory control.

From a global perspective, two general approaches to printing estimating computer systems have evolved: computer-assisted estimating systems (CAE) which require an estimator to directly input information as each estimate is completed, and computer-generated estimating systems (CGE) where the computer determines the best way to run a job using preset job standards.

Other computerized information processing services—in addition to estimating—include order entry, inventory management, machine loading and job scheduling, electronic factory data collection, and production management. As previously discussed, the computer has also contributed significantly as an aid in producing graphic images throughout the printing production process.

Consider the following reasons for computers to continue to be increasingly accepted for estimating and other information processing duties as we enter the 1990s:

1. Estimates and other databased duties can be turned around extremely fast when compared to manual methods.
2. Computers provide consistency when estimating since the same production standards, BHRs, and logic sequences are used for all jobs.
3. All number-related production and estimating duties are completed without mistakes since computers do mathematics and logic (as programmed) completely error-free.
4. Computers can be quickly utilized to answer what-if questions, allowing press selection and other production management and estimating duties to be easily resolved. Spreadsheets are an example of this use.
5. Management will have a tool to provide for tracking many different details of the business that might otherwise be overlooked with manual estimating and production systems. Management may then monitor employee and equipment productivity, accurately watch inventory and materials, and carefully review profitability by customer, product, and salesperson, among other important aspects. Management will have a tool to better monitor and control the business and to aid in making crucial business decisions.
6. Storage and retrieval of data are fast and do not require burdensome copying and filing of plain paper copies. Freed from some of the tedious office chores, employees have time for more productive work.
7. With a growing number of less expensive microcomputer systems, integration of data is possible, permitting interactivity of the database so that the computer can do numerous tasks, such as estimating, order entry, job costing, production and sales analysis, and many different accounting duties.
8. Quotations and proposals can be generated as computer printouts, which in terms of the customer, seems psychologically to add to the accuracy and logic upon which the quote and proposal figures are based.
9. When used for estimating, the computer allows for a greater percentage of jobs to be estimated. Computers free estimators from more routine estimating chores so that they can spend more time on larger estimating chores, job planning, customer service, production and cost controls, and other duties.
10. If the printer has a full MIS system, the computer system will handle all accounting functions, production scheduling, produce profit reports, and provide financial analysis. The computer becomes a vital business management tool to aid in making many decisions and help with the direction of the business.

However, there are still some problem areas to be addressed when considering computerization of a printing company.

1. Since a computer is not a production center that produces billable work that can be billed to a customer, it represents a significant expense that must be recovered through the firm's operating expenses. As noted in the discussion on the three ways computerization can be done in a printing firm, and while there is some overlap, M&SW represents the least expensive way to go, with turnkey systems in the mid-range, and customized systems on the higher end of the cost continuum. Computer system costs will be discussed later in this chapter.

2. The software support for many systems typically represents a major expense. While the cost basis varies among vendors, a typical turnkey vendor may charge 1% of the total cost of the system per month to maintain full system use, ensuring the printer-user minimum downtime. Such support is essential, for without it the system is literally unusable.

3. Management is rightly concerned that only certain personnel learn the system—which may be complicated—giving these employees undue leverage in salary matters, access to information that may be considered proprietary and opening the company to possible critical employee shortages if or when these persons move to other employers, go on vacation, or wish to move to other positions in the plant.

4. Obsolescence of both hardware and software represents a possible problem, even though hardware technology somewhat stabilized in the late 1980s. Certainly, major technological breakthroughs will still occur—particularly in mass storage devices and microprocessors—yet the general state of the computer technology has stabilized and obsolescence has become less of a consideration.

5. Some plant personnel may not look upon the purchase and installation of a computer system as a positive step but as a threat to their job security and also as an awkward tool to learn and use. In general, the fear of computer implementation and use in the printing industry remains, although it has been somewhat reduced with the growing availability of microcomputers. Job security is an issue when electronic factory data collection is used to monitor employee productivity and output in a negative way by management.

6. Management may purchase a computer system with the intention of implementing it throughout the company, yet provide only limited or partial support during the installation phase, break-in periods, or when running parallel manual and computer checks and for training support. It is essential that management provide full and complete support to employees attempting to learn a sometimes complex, unique, and new process. Management's support and dedication toward the integration of the computer system in the company must not be lip service, but backed with direct management action.

7. For companies without a computer system that wish to purchase one, or for those firms with a computer system that they have outgrown or no longer use, "what to buy" and "how to buy" and implement such a system are major issues. Determining which way to go—turnkey, customized, or M&SW—then selecting the software and hardware within the framework of cost restrictions, immediate versus longer-term needs, employee compatibility and willingness to work on the system, and many other attendant problems must be dealt with. In too many cases management's response is to "make do" with the old system for a while longer, to wait too long to make a decision or to purchase a system that is ultimately too much or too little for the company's needs.

Selecting a Computer System for Printing Estimating and Management

Of all the areas to be reckoned with regarding computerization of a printing business, none may be more frustrating or intimidating than deciding which computer system to buy. Overall this purchasing decision is harder now than even five years ago, since there are more vendors now and most have improved their software products significantly, running on faster and better hardware as well. Also, management has a tendency to look upon a computer system as the solution to many of the problems of the printing firm, instead of a tool to provide ways to analyze and repair the problems. The cost of a computer system for estimating and related MIS duties is also a very important issue, since there is no precise method to evaluate the cost-benefit payback to the company.

Figure 5.3 is a flowchart covering the typical steps necessary to computerize estimating and other information processing functions in a printing company.

Discussion of each point will follow in some detail. It is important to note that a proper, carefully reasoned decision will take time—valuable time—away from management and other company personnel. For a small printer, it may be especially difficult to allocate extensive time blocks or additional personnel as the selection process proceeds, but in the long term, the time will be very well spent. How much time is difficult to pin down. Overall, it might take hundreds of hours to decide on the system to buy, purchase it, learn the essentials, and break it in before cutting it over for permanent use. Undoubtedly, the invested time will be recovered with future savings in time and energy—plus enhanced and detailed controls to make better management decisions—but sometimes short-sighted managers may not be willing to make the initial time commitment. Time will also be needed simply to keep the computer system running, to revise hour cost rates, do job costing, and so on. The bottom line is that computerization of a printing company is a time-consuming and, thus, expensive process.

Buying a computer system is to an extent like purchasing other types of printing equipment. As with most equipment decisions, which printers make frequently and with generally good success, the decision must be carefully approached using logic and reason as primary inputs, with "gut level" feelings and reactions mixed in.

Step 1. Determine if a computer system is needed. The question of whether a computer system is needed is certainly the place to begin. For smaller printers and quick printers, comparing cost versus benefits may result in a final decision of making no purchase. Many signals point to a need for an MIS or computer system, which include noticeable estimating errors, the loss of important details during production

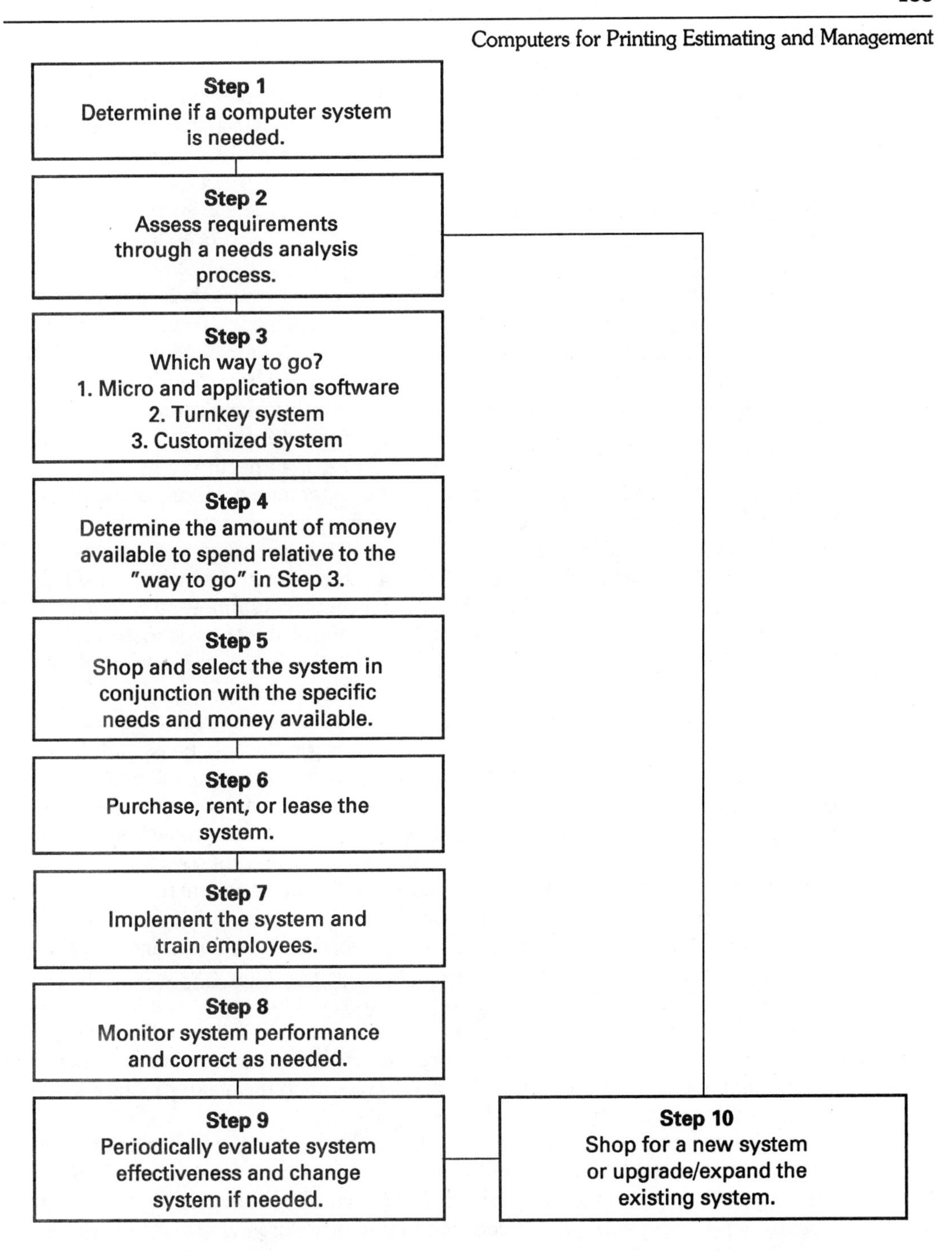

Figure 5.3

Flowchart for selecting, implementing and evaluating a computer system for a printing company

causing excessive remakes, inventory fluctuations, billing and invoicing errors caused by typing and math mistakes, frequent production delays which translate into missing customer delivery dates, or excessive costs for clerical services.

Step 2. Assess requirements through a needs analysis process. Once the decision has been made to computerize the information processing areas of the company, "what to computerize" becomes the issue. Essentially this relates to identifying detailed needs for each area of the company where the computer can aid or help resolve problems. Figure 5.4 lists some of the criteria upon which the computer system should be selected; the items contained in the list are not intended to be all-inclusive or to cover the selection factors for every printing operation. The list presents a starting point from which management can begin to make a careful and reasoned decision about what computer software and hardware to buy.

In general, management should assess six areas of need: production management, estimating and job costing, accounting, management, practical needs, and those factors relating to the implementation and maintenance of the potential computer system. The needs criteria list is essentially based on two general categories: criteria that are quantitative—for example, the number of estimates per day, which may help to determine the size and storage capacity of the computer system—and criteria that are qualitative—for example, the factors related to the types of reports ultimately to be produced by the database in the system. Figure 5.4 lists software and generated reports for computer systems currently in use in the printing industry.

Production management needs: Computer systems greatly aid production management through providing details on many production areas not normally reviewed. Production management needs include an assessment of the number of jobs beginning and ending production on a daily basis and the number of average jobs in process during a typical workday. Order entry reports begin the production cycle, with job tickets printed by the computer, accompanied by reports covering backlogs in production, order acknowledgement for customers, and a due date report that lists all delivery dates for current work-in-process.

Production analysis reports, which are valuable to production managers, provide excellent details about cost center usage, downtime, overtime, productivity, remakes, and spoilage. The relative size of inventory is also important—for example, the quantity of items. Inventory control reports should include material transactions, inventory usage reports, finished goods inventory, and reorder status reports. Production management is also very interested in reports regarding the status of jobs and work-in-process, generally on a daily basis.

As noted on Figure 5.1, once the lower level modules of BHRs, estimating, order entry, raw material inventory, and purchase order control have been installed, job loading and scheduling are ready to be implemented. Job loading is a process by which jobs to be produced on a specific piece of equipment are organized in the most efficient way for that specific piece of equipment. Scheduling, on the other hand, is the arrangement of all jobs in production in the most efficient manner possible, thereby ensuring maximum productivity and minimum downtime. In sum, machine loading

Production Management Needs

Number of average jobs in the plant per day; number of jobs beginning and ending production daily
Order entry reports
Production analysis reports
Number of items in inventory
Inventory control reports
Job status and work-in-process reports

Estimating and Job-Costing Needs

Anticipated number of estimates per day
General complexity of estimate
Detailed and summary estimate capability
Quotation and proposal generation
BHR development
Number of cost centers
Link with automated factory data collection
Job-costing reports

Accounting Needs

Number of employees, customers, and vendors
Invoicing and statement production (number of monthly invoices issued)
Payroll (number of checks written monthly)
Accounts receivable reports
Accounts payable reports and production (number of checks written monthly)
General ledger

Management Needs

Potential direction for company growth (long-range business plans)
Future company computer needs
Initial cost of the computer system; long-term support costs
Sales analysis, profitability, and ratio reports
Balance sheet
Income statement

Practical Needs

Physical requirements of the computer system
User-friendliness of the system
Computer system storage requirements and capacity
Support personnel for the computer system
Other possible uses of the computer system

Implementation and Maintenance Needs of System

Warranties on hardware
Software modification procedures
Projected cost of software support and hardware maintenance
Training programs (availability, cost, need, and effectiveness reports)
Clarity of documentation (manuals) supplied with the system

Figure 5.4

Needs criteria to be evaluated when selecting a computer system for printing estimating and management use deals with specific jobs on a specific piece of equipment, while scheduling is a much more complex, global process of production management.

Scheduling printing production is difficult since there are so many variables that many enter the rather complex printing manufacturing process. For example, a job can't be printed on press without plates, so the scheduling program must link platemaking before presswork. Another example: a job can't be printed without paper, so the scheduling program must be interactive with the computer's inventory system to be certain that the paper for the job is on-site and ready for production.

It is also important to note that production scheduling requires that all jobs to be scheduled by the computer be estimated on the computer, or be entered in the computer scheduling database so it exists as a record to schedule. However, while figures vary, a typical commercial printing company might use their CAE system for perhaps 70% of their estimates (this figure may vary from a limited percentage to 100%), making full production scheduling more difficult to accomplish than it appears initially.

Estimating and job-costing needs: Perhaps the biggest number-crunching duty of the computer is CAE and job costing. The size of the computer system should generally be based on the anticipated number of estimates per day and an assessment of the general complexity of estimates. Good CAE systems provide for estimates to be generated in both detail (item-by-item) and summary form and allow for the printing of proposals and quotations through the computer's printer. Identification of the number of plant cost centers will help define the general size of the production areas of the company and will be used to develop BHRs, a program needed for the computerized printing plant since hour rates are so vital to estimating and job costing.

Job costing can be handled easily by the computer system, especially if automated factory data collection (Chapter 3) is also available. Job-costing reports provide for a detailed assessment of actual versus estimated times and costs for management to review, pinpointing extreme differences between estimated times and costs and what actually occurred during production. When a consistent variation between actual and estimated times and costs can be noted, production standards can be modified accordingly.

Accounting needs: The number of employees, customers, and vendors used throughout the accounting process should be determined. The system should provide invoicing and statement production, and the general number of monthly invoices issued should be known. Payroll duties and the number of checks written for payroll purposes are an essential computer need. The system should monitor accounts receivable through an invoice list, cash receipts journal, aged trial balance, and similar receivables reports. The system should also handle accounts payable, including a cash disbursements journal, purchase journal, accounts payable summary, and vendor summary report. General ledger information should be able to be utilized as well, which includes a complete listing of a chart of accounts and trial balance.

Management needs: When making the decision about the size and expandability of the computer system, management needs should include assessment of long-range business growth linked to potential future computer needs. The initial dollar cost of the system is, of course, an important criterion for management, as is the long-term support cost after the system is installed and operational. In terms of analysis documents and reports, management will be most interested in sales and profitability reports and ratio data. Top management and stockholders will be vitally interested in balance sheet and income statement reports, which interface with the accounting needs previously discussed.

Practical needs: Practical needs relate to the physical requirements of the system: space allocation, electrical power requirements, environment (air conditioning, soundproofing, and so on), and any other observable matter. Another practical need—actually a working need—is that the computer system be user-friendly so that employees can feel comfortable working with it continuously. User-friendliness is essentially a software matter, determined primarily by the operating system used in conjunction with the application program. A practical need for computer networks is the location of terminals relative to the storage devices and the file server for the system. Support personnel are a practical need as well, since these people will ultimately be the backbone of the system. Here the practical needs relate to both the initial and longer-term numbers of people to be using the system, working space for them, and related personnel needs.

Implementation and maintenance needs of system: Putting the computer to work—implementation—and maintaining it over the long term are both important factors when considering computer purchase. Initial start-up hardware warranties, typically 90 days to 1 year, are important, as are the extensions that may be purchased from some vendors. Software implementation and modification of software are also absolutely critical matters, both in the start-up and in long-term usage, and should be definite items discussed during the purchasing stage. Training programs—their availability, cost, effectiveness, and necessity—are all factors to be weighed, as are the clarity and quality of system documentation—for example, manuals that provide the operator resource material to troubleshoot problems as they occur.

Step 3. Determine which "way to go." As previously discussed, the printing firm has three general approaches to computerizing: purchase an M&SW system (microcomputer and buy prewritten "canned" software from a printing industry vendor), purchase a turnkey system, or customize a system specific to the company on spreadsheets or through vendors who provide customizing services. Each of these choices has significant advantages and disadvantages which must be carefully addressed at this point, including the relative cost of the systems, system expandability relative to company growth, and overall impact on company employees in contact with the system.

Step 4. Determine the amount of money available to spend relative to the selected "way to go." The process of buying a computer for estimating and other management duties is a balancing act between the amount of money the company wishes to spend and the actual costs related to purchasing, installing, and operating the system.

There is no clearcut method to determine consistently how much money the printer should be prepared to spend for a system. NAPL's *GraphPro* (2nd edition, 1983) indicates that a good rule of thumb is 7% of the first million dollars or less of sales, plus 3.5% of the excess over one million dollars. Thus, for a printing company doing four million dollars annual sales, the amount of money for computerization of estimating and management duties would be $70,000 (for the first million in sales) and $105,000 (for the additional three million at 3.5%), which totals $175,000. Dividing this figure by three—representing a three year use and amortization period—results in an annual computerization figure of $58,334.

Other, more subjective, methods to determine the amount of money for a system include evaluating the relative cost savings to the company from computerization by reviewing the savings from hiring fewer staff, looking at costs saved through more effective inventory control, savings from faster internal processing of data, and improved cash turnaround through faster invoicing and better tracking of cash flow.

Looking at what the system will cost—not how much the company is willing to spend—seems to be a more common approach for making this purchase. For a beginning computer system, M&SW is generally low in cost ($2,500 to $5,000) depending on hardware and software packages. Customizing by using a spreadsheet and in a microcomputer is fairly low in cost—perhaps $3,000 for hardware and a standard spreadsheet program—yet time commitments might be very high to develop the program from scratch. Turnkey systems begin at roughly $30,000 and go up from there, with customized systems varying in cost based on specific user needs.

Of course, the cost of the system is relative to meeting the needs criteria addressed previously. These criteria are not easily quantifiable and may be summarized as follows:

Size and power of the computer system: These factors relate to the costs of microcomputers versus turnkey systems and the processing power inherent with each. Typically smaller printers purchase a small computer—usually a micro in the beginning—which is expanded as their business grows and develops. Hard disk media for electronic storage is common, with floppy disk or magnetic tape backup depending on the size of the system. Medium-size and larger printers move toward increasingly more powerful computer systems based on their identified needs, with additional peripherals and related software to enhance their capabilities.

Applications software costs: The cost of software for a computer system can be a major and significant expense if the software is to handle job costing, estimating, and other printing plant duties, plus provide suitable reports that will enable management to monitor carefully the production aspects of the business and complete accounting and management functions. It is important to carefully assess the needs criteria already covered and pick out the most important factors when specifically selecting software.

Start-up costs: These costs essentially are covered in the need criteria under practical needs and those requirements associated with implementation and maintenance costs. Start-up costs may exceed the initial hardware and software costs of the system in the first year of use.

Peripheral hardware: The peripherals selected when purchasing the system can have a definite cost impact. Considerations include buying a better printer (a letter quality over a dot-matrix) or purchasing multiple printers (as related to the number of invoices, checks, reports, and other documents to be output by the computer system). Modems will increase costs, as will the quantity and type of monitors (color versus black-and-white). Other peripherals might include hard disk storage devices or other expensive magnetic storage units.

Expandability factors and future needs: Cost increases when the computer initial purchase incorporates expandability of the hardware to meet possible future needs and company growth. Thus, while the printer may "overbuy" initially to have a greatly expandable system, future needs will be met and there will be little, if any, need to deal with system incompatibility at a future date. This consideration is an important one.

Long-term costs: These costs include all the warranty costs after the initial warranties expire, software maintenance, follow-up training costs, expansion of the system, and the projected cost of personnel to operate the system. Long-term costs can be very high, especially the software maintenance and related personnel costs expended simply to keep the system operational.

Completing a cost-benefit analysis may be an important factor in deciding what system to buy. Essentially, the cost-benefit analysis procedure is a measurement of required cost inputs compared to the anticipated benefits for money spent. Since these benefits cannot usually be given a specific cost value, the system is typically subjective in result but does provide a formula or method to begin to assess the relationship between cost and benefit when buying equipment.

Figure 5.5 provides one approach to determine the cost-benefit relationship when making a computer purchase. Inputs include annual company sales and costs for office, estimator, and executive. The recommended target maximum cost of the system can be calculated using 7% on the first million dollars of sales ($70,000) and then 3.5% per million for each remaining million dollars of sales. At this point, NAPL recommends that the company determine the cost savings of three (or more) applications of the computer system—perhaps estimating, job costing, and inventory, or some other mix—and compare the anticipated costs with the anticipated savings of the applications. It is important to note that the same applications should be selected when comparing systems from different vendors.

Management may decide to spend more money than needed for the first computer system, wisely providing a system with greater capacity and power so it will be expandable and flexible. If a company has decided to look at the longer term, generally the conclusion is that networked micros or a turnkey system represent the best approach. Larger customized systems major users are high volume printers and printers who specialize in specific products for which software is not available.

Step 5. Shop and select the system in conjunction with the specific needs and money available. The process used to shop for a computer system relates to the type of system desired and the time available to conduct the search. A smaller printer or quick printer may find that there are a limited number of systems to be considered, thus saving time and making the decision easier. Those printers looking at turnkey systems—because of the increased cost and commitment on the part of many company employees—must be more diligent in the shopping phase. If the printer has decided to go with a customized spreadsheet program written internally, employees who will do this must take a computer course to learn how spreadsheet development would be accomplished. Finally, if a custom system is desired, shopping for a skilled vendor may represent a major task.

Some industry experts recommend that software considerations take precedence over hardware selection at the shopping stage. This advice is generally true, with the exception that some vendors may sell a software package that will exceed the hardware

1. Annual sales ______
2. Average cost per hour—office ______
3. Average cost per hour—estimator ______
4. Average cost per hour—executive ______

Costs

Up to $1 million revenue × 7% ______
Over $1 million revenue × 3.5% ______
Total Cost ______

Benefits

Application 1
- Savings by improvements ______
- Labor savings ______
- Total ______

Application 2
- Savings by improvements ______
- Labor savings ______
- Total ______

Application 3
- Savings by improvements ______
- Labor savings ______
- Total ______

Total Annual Benefits ______

Total Net Benefits ______

Total benefits less total costs

Substitute actual costs or benefits, where known, for more accurate computations.

Figure 5.5

Cost-benefit analysis form for evaluating the costs and savings when deciding on a computer system purchase (Reproduced from NAPL's *GraphPro*, 2nd ed., 1983, with permission)

investment the printer wishes to make. Using a software-before-hardware approach is reasonable because the final reports and other needs of the computer system are generated as a result of software packages, and hardware becomes a secondary matter.

Appendix C contains a 1990 alphabetical listing of printing industry computer vendors. In general, the vendors know the needs of the industry and have developed turnkey systems or software specifically for the printing industry. Included in the list are vendors providing Macintosh programs, as well as those who provide software in MS-DOS and other operating systems. The vendors are usually responsive to printers' needs and knowledgeable about specific printing industry applications.

Some printing industry vendors may be geographically limited, necessitating a long trip to view a system that if purchased, may then require long-distance support. Trade shows seem to represent the best way to see a cross section of computer systems offered by industry vendors, yet some do not participate in every trade show and may offer a more appropriate package, for a better price, than other vendors that do display their systems consistently at all trade shows. Those vendors with greater industry exposure and systems in use may have a regional office where a typical system can be seen and demonstrated; sometimes vendors are willing to arrange for the client/printer to visit another printer who already has a system in operation.

As when buying any piece of printing equipment, tempering the statements made by the sales representative may be necessary. While printing industry computer vendors talk the language of printing and computers together and probably will be willing to take extensive time and effort to explain their systems in detail, computer salespersons with no knowledge of printing may be able to spend only a limited amount of time to discuss a system and will not know much about the printer's own specific needs, reports, or business.

During the shopping phase, the printer and vendor should talk openly about all details of the systems under consideration. Details should include software cost and support, hardware cost and support, speed and power of the computer for current and future use, networking of computers together to expand systems. compatibility of systems with other computers, and so on. Many of these items have already been discussed under the listed needs criteria in Figure 5.4. Resolution of these details at the initial purchasing stage may save the printer thousands of dollars in the future.

When viewing and working with the systems, try to have each vendor use common software applications, such as inventory, overtime reports, payroll, and so on (Figure 5.6). Actual printed reports from these different vendors can then be taken back and compared leisurely at the printing plant. Report format and information may vary extensively from vendor to vendor.

It is wise to have more than one person from the printing firm spend time with the system so that different details are noted during demonstrations. These persons should be involved in using the computer after it is purchased and probably should have some basic computer knowledge before the shopping begins. While it may be expensive to pay for extra personnel to attend a trade show for two or three days—taking valuable time away from work as well—proper computer selection requires that this procedure be followed.

Estimating
- Detailed estimate
- Summary estimate
- Quotation

Order Entry
- Job jacket/sheet
- Production backlog
- Order acknowledgment
- Due date report

Job Costing
- Job cost details with estimate comparisons
- Job status report
- Work-in-progress

Production Analysis
- Cost center details
- Labor details
- Downtime report
- Overtime report
- Productivity by cost center
- Spoilage and remake report

Sales Analysis
- Sales and profits by salespersons
- Sales by customer and product

Inventory
- Material transactions
- Usage reports
- Finished goods
- Re-order status

Invoicing
- Finished bill

Accounts Receivable
- Invoice list
- Cash receipts journal
- Aged trial balance
- Sales allowances
- Statements

Accounts Payable
- Purchase journal
- Cash disbursements journal
- Accounts payable summary
- Vendor summary

General Ledger
- Chart of accounts
- Trial balance

Financial
- Balance sheet
- Income statement
- Ratio statement
- Hourly rates

Figure 5.6

Chart of typical software and generated reports for computer systems in the printing industry (Reproduced from NAPL's *GraphPro*, 2nd ed., 1983, with permission)

As noted in NAPL's *GraphPro* (3rd ed., 1987) there are traditionally seven ways to evaluate computer systems for a printing company, each described briefly from that resource. The most used of these are subjective evaluation, literature review, and the weight-score/cost effectiveness ratio system.

- **Subjective Evaluation.** This process involves evaluation of computer systems in a nondefined and casual manner. Although commonly used, it can lead to expensive mistakes and should be avoided unless the outcome is unimportant or not costly.

- **Literature Review.** Magazine articles, books, and other published sources are an essential input when gathering information toward any major computer purchase. This is an important early step in choosing a system.
- **Cost/Value and Requirements Costing.** This evaluation process requires that a number of systems be chosen to study based on the identified needs analysis previously discussed, then specific parts similar to all systems be evaluated in terms of the cost requirements. The system with the least cost is the winner.
- **Weight-Score and Cost/Effectiveness Ratio.** Based on the needs analysis, selected system features are given an assigned weight, from highest or most important to features of less importance, which totals 100. Systems can then be compared in terms of those getting the highest marks relative to their evaluated effectiveness and cost. The evaluation process must be objective and carefully reasoned out.
- **Application Benchmark.** Some experts believe that benchmarking is the only way to effectively compare systems. To do this, two or more systems are compared point-for-point with each other, based on expected norms or quantifiable points of comparison. An example benchmark might be to compare estimating system A with estimating system B in terms of input time, estimate detail, and order entry characteristics. Actual printed reports from different systems can be analyzed for report format providing a benchmark for reports.
- **Cost/Performance Ratio.** To do this, values are assigned to specific hardware and software parameters which are then used to calculate performance ratios, and which in turn are used to determine an overall measure of cost to performance. This is a difficult and complex process.
- **Simulation.** This requires a simulation program. Data related to the user's job mix is fed into the simulator program, which calculates how different systems perform. This is expensive, hard to justify, and it requires expertise in simulator programs.

When comparing systems, it may be wise to use a benchmark process by which each vendor completes the same estimate, demonstrating how his system works. The printer should try to compare systems on an equal basis by looking at common system applications such as estimating, inventory, payroll, and so on. Figure 5.6 provides a look at some typical software and generated reports for computer systems serving the printing industry.

When shopping, do not be afraid to ask questions about specific needs. Use the terminal or monitor to get the feel of the system; discuss such financial matters as costs for hardware, software, and support; and discuss payment terms, delivery dates, and other critical considerations. Also, do not prolong the shopping and selection process. Spend a concentrated amount of time shopping, buy the system most appropriate based on all available input, and then work to put it in operation in the plant.

The shopping criteria is usually different for the printer who has no installed and operating computer system versus a printer looking to upgrade an existing system. Figure 5.7 provides a look at some of the typical considerations for first-time purchasers of computer systems versus the needs of the printer wishing to update or change to a second or newer system.

Selected criteria a typical commercial printer looks for with his first computer-assisted estimating system

1. A system that represents the type of production the company utilizes.
2. A system simple enough to be usable without extensive training.
3. A system that requires minimum maintenance for both hardware and software.
4. Expandability of the system (even though the printer doesn't always have a clear plan or proposed expansion direction in mind).
5. Trouble-free customer service from the vendor.
6. A system which is built to minimize input errors.
7. Maximum benefit at minimum cost.

Selected criteria a typical commercial printer looks for with his second or next computer-assisted estimating system.

1. Improved accuracy and reliability in all areas of the system.
2 Data collection.
3. Report generation.
4. Uniqueness that will show the company's differences.
5. Immediate fixes on software problems with minimum downtime.
6. No hassles from the vendor—vendor service is critical.
7. A system that will "do more" than what is being retired.
8. Expandability without building new files or requiring complex conversion between files.

Figure 5.7

Comparison of some criteria for first-time CAE purchase versus criteria for the purchase of a second or next CAE system.

One good way to find out about most turnkey systems is to ask the vendor to arrange a visit with a current user of his system, requesting that the vendor be absent during this meeting. This will allow the printer-user to talk frankly about the good and bad points of the system. Of course, the vendor will likely select one of the more successful installations. Another good way to investigate a particular system is to attend a user's meeting and listen to the discussions about the system, which tend to be frank and open and generally require the vendor to respond to the complaint or problem.

Step 6. Purchase, lease, or rent the system. The decision whether to purchase, lease, or rent the selected computer system relates to the company's cash position and

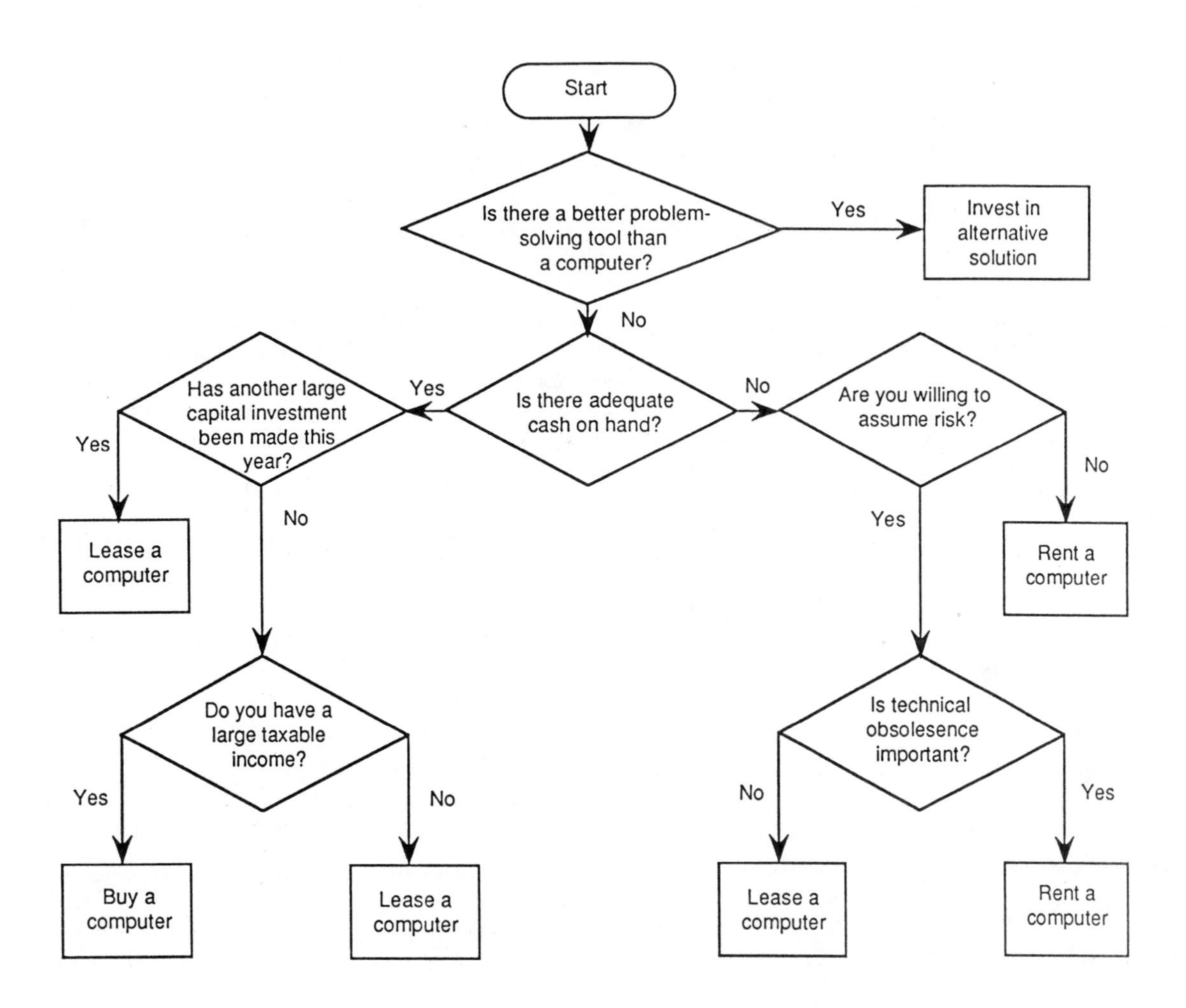

Figure 5.8

Rent, lease or buy decision flow chart (Reproduced from NAPL's *GraphPro*, 3rd Edition, 1987, with permission)

the amount of risk the company wishes to take relative to the benefit. Figure 5.8 from NAPL's *GraphPro* (3rd ed., 1987) provides a decision flow sequence in this regard.

Purchasing, which is common with smaller systems and some turnkey systems, holds fairly significant risk to the printer/buyer in that the maintenance, support, insurance, obsolescence, and other costs must all be borne by the purchaser. Many vendors require outright purchase of their systems, which can be a fairly expensive ordeal. Renting a system on a month-to-month basis is generally considered a non-risk situation to the buyer and is uncommon for the majority of printing industry computer vendors, who prefer either outright purchase or some type of lease arrangement.

Leasing a system, which is offered through some vendors, provides less risk and requires less immediate capital needs on the part of the printer-buyer. "Full-payout" leases spread the payments over what is typically the full life of the computer system, which lowers payments and usually terminates with an option to buy the system. "Operating" leases, another type of lease, have higher payments yet require a long-term, commitment—perhaps 10 or more years—and may be noncancellable.

Systems which are lower-cost—M&SW or smaller turnkey systems—typically require cash or quick payment terms, while turnkey vendors may be willing to arrange financing, provide a mechanism through which financing can be easily obtained, or work out a leasing arrangement. In sum, as the cost of the system increases, many vendors are willing to work with the printer-buyer to finalize the sale.

It is wise to have the vendor provide an itemized list of what will be included with the purchase price of the system and a breakdown of the costs of the items. The following suggests some areas that should be covered:

- Price of the hardware, including peripheral hardware such as printers, disk drives, backup systems, clock devices, modems, etc;
- Price of the software and specific names of software packages to be included in the purchase;
- Start-up costs which include nonchargeable on-site vendor time for initial installation and training and follow-up training costs;
- Maintenance and support costs including specifics on software support, hotline access and cost, hardware maintenance, and on-site costs for support, if required.

As with most printing equipment, purchasing a computer system involves a contractual relationship between seller and buyer. As the purchaser, it is best to know and understand the contract terms and conditions, and to have this provided in writing from the vendor.

Step 7. Implement the system and train employees. Installing the computer system in the plant can be a smooth, hassle-free process or one filled with frustration. Generally M&SW systems install with minimum difficulty, while turnkey systems and customized larger systems require much more time, concentrated effort, and training.

If implementing a turnkey system into a printing firm or building a customized system, the printer and vendor should work out a timetable to complete the process. Experienced vendors will have a good idea as to how long this will take, so vendor recommendations should be carefully considered. The timetable should be as realistic as possible, but does not have to be precisely followed during installation. Many vendors practice a modular installation process following the building block process covered in Figure 5.1.

The four most critical factors for a smooth transition once the timetable has been established are: clarity and interface of the present manual systems into the computer system configuration, the software requirements and software modification for inputting data and producing reports and other system documents, the training and documentation available from the computer vendor, and management dedication to making the system work. When any of these factors are neglected or ignored, the conversion process will be frustrating indeed.

Interface of manual systems into the computer database: There is truly no substitute for neat, clean manual records and files when transferring data from manual methods to computer systems. While each computer system will require specific information in a different way, essentially, the database for a specific application will not vary extensively. For example, a typical cost-estimating program will require BHRs, standard production times, material costs and markups, and other standard data. If this information is not available or not properly developed, implementation of the computer system will be more difficult. Accounting data, which is generally more standardized, is another example of a database that should be orderly in manual form to ease the transition into the computer system.

Software requirements and software modification: This factor may become a major problem at the implementation stage and may also represent long-term problems over the life of the computer system. Some industry vendors sell packaged software that essentially covers all areas of a commercial printing plant. This software may not have been written to be easily modified but may be offered at a price that is quite low compared to other programs. If this type of software has been purchased, it may have to be used as-is since modification may not be possible or is not cost justified.

Other printing industry vendors provide extensive software modification, allowing the programs to be closely tailored to the specific needs criteria of the printer. While these systems may be more expensive, they will allow the printer-user to generate the types of reports in the format he or she specifically wishes and use the computer system precisely for the purposes and applications desired. Some vendors offer hotlines for such software modification and troubleshooting as the applications programs are installed. Hotlines may or may not utilize a modem interface and telecommunications directly between the vendor's computer and the printer's unit.

It is inconceivable to think that a computer system and the accompanying software could be purchased by a printer with no software modification over the long term. Therefore, the software decision is particularly critical during the purchasing stage. For vendors that do provide software modifications, there may be an extra fee on a per hour or other consulting basis; some firms offer an annual contract for software support, providing many or all program changes for a single annual fee. The costs for this critical support should be carefully discussed when purchasing a system.

Training and documentation available with the computer system: At the installation stage, training and documentation from the vendor become critical. Most industry vendors provide some training as a part of the package price of their systems. Generally, the more costly the system, the more training that is provided as a part of that cost. Extra training from vendors, beyond that sold with the system, may be a factor to consider, especially with respect to the costs involved. During and after training, system documentation—manuals and books written to detail how the system works—becomes quite important since it may be the only source for immediate solutions to some problems. It is critical that documentation be clearly written, using plain English, with diagrams and other appropriate explanatory materials. Clarity is essential. Good manuals require a great deal of time to write, so there are many variations among vendor documentation; some is excellent and some not very good. Discussing and looking at system documents during the purchasing stage may be helpful.

As indicated previously, a basic knowledge of computers is essential during the shopping and selection stages, and this knowledge carries over into the installation stage. Depending upon the size of the system, it may be necessary to hire a systems consultant or person to work with plant personnel for a period of time to put the system into place and work out the bugs. This systems consultant, employed for a very brief period of 6 months or longer, can deal quickly and effectively with both hardware and software problems, provide programming support or hire short-term programmers to write company-specific software, and aid tremendously in training and breaking in operating personnel for the system. In general, small systems (under perhaps $100,000) do not require the employment of extra temporary specialists or consultants.

Management dedication: Perhaps the most important factor in the implementation of a computer system into a printing firm is management support. Management must remember, depending upon the size of the system and the other four previously mentioned factors, that it is difficult to select a specific date for the entire implementation process to be completed and for the company to be on the system. Some microcomputer systems may take only a few days to install and make operational, while other larger systems may take a year or more to implement. Some vendors recommend that different applications programs be put in place in a prescribed orderly fashion, beginning with perhaps job costing, then adding estimating, then inventory, and so on. This process may stretch out the implementation time yet allow for the most critical tasks to be initially met and carefully debugged—that is, errors are found and fixed—before moving to the next level of software.

During the implementation stage, management must remember that both manual and computer system duties are under way. Estimating, for example, will need to be completed manually at the same time the estimating program is being readied for computer use. Provisions should be made for this double-work aspect. Paying overtime for employees, hiring temporary persons to handle specific jobs to ease the workload of the permanent staff, and moving qualified plant personnel into temporary positions where they can provide help as the implementation process goes forward are some solutions.

Management must also remember that the implementation stage may necessitate extensive extra time and cost commitments and be prepared to pay for this. It is not in the best interest of management to push a system in place too quickly or demonstrate an insensitivity to the needs of implementing a high-technology piece of equipment with somewhat different skill requirements for company personnel. Unless management allows the implementation stage to go forward at a pace comfortable to the specific situation and personnel involved, time and money will be wasted and tremendous frustration and anger will result.

Step 8. Continuously monitor system performance and correct as needed. Once the system has been cut over and is operational—once the computer system has been turned on and manual systems have largely been abandoned—a resting period should follow. This period will be a time, perhaps 6 months to 1 year, when the system begins to settle into the plant routine and operating personnel become increasingly familiar with the good and bad aspects. Only modest changes, if any, should be made during this period.

Perhaps the most important review aspect relates to the types and format of reports generated from the computer database. These reports, some of which are listed in Figure 5.6, can take many forms and may initially appear confusing and too detailed—perhaps seem like overkill to the noncomputer-oriented manager. Those managers involved with the review of such reports should initially try to use the reports as-is since their development has typically been thought out by the vendor of the system. It is, of course, possible that some reports will not be needed. These reports should not be produced until such a need arises. If, over time, a report is unclear or unsatisfactory because of the format or style, the format should be changed if possible. Some vendors have a "report generator" that allows for reports to be customized to the databases available. Reports may then be produced using selected criteria determined by management, such as different reports of data by job number, then sorted by due date, and then sorted by profitability. Report generators expand the number of possible reports by providing variations from the database. In some cases, graphs are possible too.

Step 9. Periodically evaluate system effectiveness and change the system if needs dictate. As a printing firm grows and changes, or as customer demands change the structure of the business, computer system requirements may change. Precursors to investigating computer changes are evident when the company begins to employ more clerical staff, when manual processing of information becomes a burden, or when job details or customer service requirements can't be quickly and easily accommodated.

Step 10. Shop for a new system or upgrade/expand the existing system. If the printing company finds their current computer system cannot handle the additional burdens required, the options are to search out and purchase a new system, modify and upgrade the existing system, or remain partially computerized and develop manual procedures to cover those areas not effectively done by computer.

Opinion varies among experts as to how long a printing firm should keep the first system purchased and installed in the company. In general, the prevailing pattern seems to indicate that a printer's first microcomputer system will be used from three to four years. That initial experience, enlightening and instructional for the printer, will allow the next purchase—typically a larger micro- or minicomputer system that is more interactive—to more effectively meet the needs of the business for the quantity of dollars invested. It will be a more informed purchase as well since the printer will have a greater knowledge of computers as applied to the printing business and the advantages and disadvantages computerization provides.

Estimating Paper

6.1 Introduction

Paper is the most-used material in the printing industry. Based on trend patterns of PIA ratio studies over the past 15 years, paper generally accounts for between 20% and 25% of the cost of the average printing order. Because of its obvious importance, the estimator must thoroughly understand how to classify, order, and cost this material.

This chapter is divided into eight sections. Section 6.2 investigates paper manufacturing and introduces basic paper terms. Section 6.3 covers U.S. printing paper classifications, and Section 6.4 discusses basic sizes, weights, and the general mathematical concepts needed for estimating paper accurately (U.S. system). Section 6.5 gives an introduction to metrication of printing papers. Section 6.6 provides details about ordering paper, including how the paper pricing catalog is used. Section 6.7 covers the vital area of paper planning for sheetfed production and integrates much of the information from the previous sections. Bookwork impositions, Section 6.8, details the essentials of book production and imposition techniques. The last section of the chapter—Section 6.9—presents data for determining sheetfed press spoilage.

6.2 Paper Manufacture and Terminology

Essentially, *paper* is a thin mat of cellulose fibers, or fibers from cotton, wood, bamboo, grass, or other vegetables. The principal machine used in paper manufacturing is called a Fourdrinier machine, named after the Fourdrinier brothers, who held the patent. Papermaking generally involves the conversion of logs or other fibrous substances into a watery pulp known as *slurry*, which is then placed on the revolving wire screen of the Fourdrinier machine. After the water is drained off, the mat of overlapping fibers moves through a series of rollers that squeeze it dry and evaporate excess water.

Pulp

There are four classes of pulp used in papermaking: groundwood pulp, old paper pulp, rag pulp, and chemical pulp. *Groundwood pulp*, normally used for newsprint, is produced by grinding debarked logs into particle form, then refining the pulp in water, and pouring the solution onto the Fourdrinier screen. Substances such as lignin and resins, which are considered to be impurities, are not removed. Groundwood pulp is not strong and is usually quite inexpensive.

The second type of pulp is called *old paper pulp* and is made from old previously manufactured paper. The paper can be unprinted paper, called either *reclaimed paper* or *broke* by paper manufacturers, and is preferred if available, but printed paper (postconsumer paper), called *recycled paper*, is used in greater quantity because it is more readily available. Normally, old paper is dissolved in a lye bath (de-inked) and then bleached for whiteness.

The third type of pulp is obtained from cotton and linen rags and is known as *rag pulp*. Sometimes nylon or rayon can also be used. For the most part, unused white rags are preferred, especially in the production of rag content papers. This type of stock is very permanent and popular for letterheads and legal documents that must last for many years. Pulp content varies from 25% to 100%. Pulp with a content of 25% has one-quarter rag content and three-fourths wood fiber, while a pulp content of 100% has no wood content at all.

The last type of pulp, the one used for the majority of paper products, is *chemical pulp*. There are several types of chemical pulp based on the kinds of wood and chemicals used. Basically, however, the chemical process begins with the arrival of cut logs at the mill yard. A debarking procedure is used initially, after which the barkless logs are chipped into small pieces of wood. The chips are then transported to a large pressure-cooker tank, called a *digester*, where various chemicals, such as lime, sulphurous acid, and caustic soda, are mixed in. Over a period of 4 or 5 hours the wood chips are "cooked" apart, providing for the removal of lignin and resins and leaving the cellulose fibers. Following cooking, the fibers are washed to remove as much of the chemicals as possible and then bleached if printing papers are being made. Both hardwoods and softwoods are used for making chemical pulp.

Type of Wood and Fiber Length

The kind of wood used and the length of fiber contribute to the kind and type of paper manufactured. Softwoods, such as fir, pine, and spruce, provide the basic strength and durability for paper and are a major ingredient in most papers used for printing (as distinguished from those papers that have industrial applications). Hardwoods, such as poplar, beech, and aspen, are also added to printing paper and serve to increase bulk and improve surface uniformity. The proper mix of softwoods and hardwoods is determined by the type of paper desired.

Fiber length is related to paper strength. Generally, the hardwoods are the shortest of the fibers (1 to 3 millimeters) and contribute only slightly to strength. Softwoods are longer in length (1 to 8 millimeters) and usually provide the major strength component to papers used for printing. If cotton content is desired, which is accomplished by the addition of cotton fibers, the strength and durability of the stock produced, such as a rag content bond, are significantly increased. Cotton fibers may be as long as 30 millimeters. Fiber length can be shortened prior to papermaking with the use of special machines that chop the fiber as the slurry is passed through them.

Most papers used for printing are largely softwood composition, with an added amount of hardwood for improved printability. Different applications of paper require different strengths, so the fiber length of paper will vary with respect to the end use of the paper product.

Refining

All pulps go through a refining stage prior to being made into paper. The refining technique is known as *beating* and is a very important step in pulp preparation. The pulp is put into large tubs with water (and other ingredients), and the fibers are smashed against a metal or stone bed, fraying (fibrillating) and flattening the ends of the fibers so that they will hold together more firmly in paper form. The type and amount of beating have a great deal to do with the character of the resultant paper, as do fiber length and the type of wood used. Additional fillers, sizing ingredients, and colorants may be added during the beating procedure. The actual slurry sent to the papermaking machine may be from two or three different beaters, each with a different type of wood and amount of beating activity. Proper combinations are determined by manufacturing experience.

Papermaking

The slurry is transported to a *flow box*, or *headbox*, located at the beginning of the Fourdrinier papermaking machine through a piping system. At this point, the mixture is distributed, or *flowed*, across a moving belt of wire screen mesh that allows the fibers to mat together as the water drains off or is removed by vacuum. The speed of the moving belt, in fact the speed of the entire Fourdrinier, is extremely fast: as much as 1500 linear feet per minute. After the paper is formed on the screen belt, it moves into a long series of rollers that apply pressure and heat, thus squeezing the sheet smooth and compressing it while evaporating residual water. The papermaking process, as described, is a continuous procedure: Paper is made as one continuous ribbon, perhaps 100 inches wide or more. At the dry end of the Fourdrinier, the web (or roll) of paper stock is wound tightly, forming a *log* of paper. Here, also, it is sometimes slit into smaller roll widths for immediate wrapping and shipment to web printing plants.

Additives. Sizing compounds are added to paper to provide resistance to rapid moisture penetration. *Internal sizing* for the paper stock, which is added to the slurry, is called *engine* or *beater sizing*. If sizing is added to the surface of the sheet after formation, it is called *external* or *surface sizing*. Papers such as blotter and newsprint are poorly sized; they will accept ink, but it will spread quickly. Papers for writing purposes are normally sized using both internal and external processes. Rosin is the most common chemical used for sizing, but modern technology has provided other sizing ingredients as substitutes.

Fillers are added to the slurry prior to papermaking, normally to increase the opacity of the paper. Sometimes called *loaders*, fillers may be clay, calcium carbonate, talc, or titanium dioxide and generally constitute between 8% to 15% of the weight of uncoated offset book paper.

Coloring materials, *colorants*, which may be mineral pigments or organic or synthetic dyes, are used to color many kinds of printing papers. Most pigments are added during the mixing process when the paper is in slurry form. Dyes are applied after the paper has been formed and sized. For whiteness, titanium dioxide is the most common additive.

Additional special additives are also manufactured into paper for special considerations. Bleaches are added to increase brightness, and vegetable starches are used to bind the fibers together and increase surface smoothness. Resins may be added to increase the wet strength of lithographic printing papers.

Watermarks and Textures. *Watermarks*, or relief images, are put into the paper when it is wet. Immediately after the sheet has been formed on the wire belt, a *dandy roll* riding on top of the screen impresses a relief image into the wet paper mat. This procedure causes the fibers to be pressed tightly together, forming a visible image of the relief dandy roll form.

The dandy roll is also used for the manufacture of "laid" and "wove" papers in a manufacturing procedure very much like the watermark technique. Additional surface textures, such as linen, leather, and pebble-grain finishes, are produced using rotary embossing techniques after the paper is formed and dry.

Paper Surfaces

Coated Papers. *Uncoated paper* stock is paper as it is produced during the paper-making process. *Coated paper* has received a fine layer of mineral substance applied to the paper surface. Coatings may consist of clays, barium sulfate, calcium carbonate, or titanium dioxide-clay, which is the most common. Coatings may be applied by blades, rollers, or by passing the web of paper through a vat of solution. Paper can be *coated one side* (C1S) or *coated two sides* (C2S). Coating may be done *on the machine* during initial manufacture or done *off the machine* during a later stage of production.

Coated stocks represent the smoothest possible surface for the printing of halftones and process color art. Sometimes the coating is polished to impart an even shinier surface with greater reflectivity. For papers that will be used for textbooks. coated papers are usually *dull* to minimize glare and reflection of light. Papers used for other purposes may be *glossy* or *enameled* to provide increased light reflectance. Sometimes the term *enamel* is used when discussing either dull-coated or glossy-coated stocks.

Calendered Papers. *Calendering* is a polishing action of the paper surface without the addition of chemicals or coatings to the paper. The result is a smooth surface without the expense of additional coating procedures. Calendered papers, with their smooth surface, minimize lithographic printing problems and give a uniform base for printing inks.

Calendering may be done on the machine or off the machine, just as with coated stocks. Paper that receives a normal on-the-machine amount of calendering is called *machine finished* stock (sometimes abbreviated MF). Extremely smooth paper surfaces are possible with off-the-machine calendering, such as *supercalendered* papers that are calendered by running the web of paper through a series of *calendering stacks*. A brisk buffing action is provided to the paper with cotton and steel rollers. Sometimes paper that has been sized will be highly calendered also, producing a very smooth sheet surface called *sized and supercalendered* (abbreviated S&SC). In addition, some types of coated papers are calendered after coating, providing a very polished, highly reflective surface with excellent printability. Other kinds of coated papers, such as *antique papers*, receive no calendering at all during manufacture.

Many printers refer to paper stock by number, such as "No. 1 sheet." The numbering system is an index of the overall quality of the paper, relative to the paper ingredients and the amount of finish given the sheet during manufacturing. Better papers have higher numbers, while lower numbers represent less expensive stocks. The range for uncoated book papers is from No. 1 highly calendered, finely-made book stock to No. 5 which is groundwood paper with limited calendering. In general, the same numbering process used to identify coated book stocks and bond papers, even though ingredients and finishing processes vary with these papers.

Paper Grain

As the screen belt on the Fourdrinier turns, the fibers tend to align predominantly in the direction of the movement of the belt. This tendency causes what is known as grain—that is, the alignment of fibers in a sheet of paper in a common direction (Figure 6.1).

Paper grain has important production considerations. First, paper is stiffer in the direction of the grain. Thus, when paper must be folded, the direction of grain should be parallel to the fold or the sheet will crack along that edge instead of folding nicely. Second, paper tends to swell and shrink from humidity, and such changes are greater across the grain than with it. This problem can be significant when precise registration

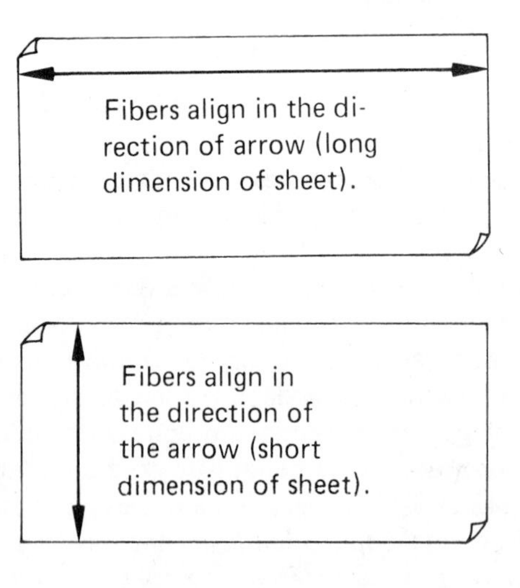

Figure 6.1

Examples of fiber direction in short and long grain papers

of images is necessary using the lithographic printing process. Third, paper tears more easily with the grain than against it. Fourth, papers used for lithographic production should, when possible, have the grain parallel to the axis of the printing cylinder. This is because paper swells much more against the grain than in the direction the grain travels. The following five tests are methods by which the grain of paper can be easily determined:

1. *Resistance test*: In heavier weights of paper, grain direction can be determined by curling the sheet in each direction and feeling the relative resistance to the curl. The less resistant direction is the direction of the grain.
2. *Fold test:* With literally all weights of paper, grain direction can be determined by folding the sheet in each direction. Paper will fold easier, straighter, and with less cracking and buckling parallel to the grain.
3. *Tear test* Grain direction can be quickly determined by tearing the sheet in each direction. Paper tears straighter in the direction of the grain.
4. *Moisture test*: Grain direction can easily be determined by moistening a piece of paper either with the tongue or a convenient water source. Paper will curl parallel to the direction of the grain. This test is good for lighter weight papers.
5. *Stiffness test:* Tear two 1 × 5 inch strips from opposite directions in a sheet of paper. Place the two strips together and hold flat between thumb and forefinger on one end. Observe that one strip sticks out almost straight while the other strip tends to droop and curl. Grain direction is long with the strip that tends to stick out and short with the one that droops.

Paper Thickness

Paper thickness is determined largely by the amount of fiber deposited on the moving screen belt during papermaking. Modern papers used for printing are fairly uniform in thickness. That is, if a sheet of paper measures 0.002 inch (2 thousandths of an inch) in thickness, it will be very close to that thickness throughout the sheet. The thickness of paper is also referred to as the *caliper*.

While most printing papers measure between 0.001 and 0.008 inch in caliper, cardboards are traditionally measured in *plies*. Cardboards are used for point-of-purchase displays and in instances where very stiff paper board is required. Use the following formula to convert ply thickness to caliper thickness:

$$\text{caliper thickness} = (\text{number of plies} \times 0.003 \text{ inch per ply}) + 0.006 \text{ inch}$$

Thus, a sheet of 6 ply board would be 0.024 inch thick:

$$(6 \times 0.003 \text{ in./ply}) + 0.006 \text{ in.} = 0.024 \text{ in.}$$

Sometimes points are used in conjunction with such measurements. A *point* is the equivalent of 0.001 inch. Thus, 24 point board is 0.024 inch thick.

Sheeting and Wrapping Paper

Once the paper has been completely manufactured, it moves to the sheeting area if it is going to be made into sheets. Here, large *sheeters*, feeding as many as 10 rolls at one time, cut the web stock into exact sheet dimensions. Since many of the webs have been slit during papermaking to the desired width, sheeting may determine only the length of the paper. Because sheeting is an additional manufacturing step, sheeted papers are more expensive than web stocks.

The final operation at the mill for both sheet and web paper is the wrapping and shipping of the paper products. Sheeted papers may be packaged in small amounts and then placed in cartons; paper may also be cartoned without packaging. Stock ordered in larger quantities is usually placed on *skids* or *pallets*, typically wooden or metal bases upon which the paper is carefully stacked. Skids require paper-handling equipment in the printing plant, but they are a convenient method for shipping large quantities of paper in sheet form.

The final wrapping of skids and rolls and the cartoning of paper are of tremendous importance. A completely airtight package is necessary to maintain the desired moisture relationship during shipping to the warehouse and ultimately to the printing plant. If moisture content is not controlled, many production problems may result during printing. The modern paper mill uses mechanized wrapping equipment with vacuum closure and excellent sealing compounds. Once paper is cartoned or wrapped. it is shipped by railroad or truck to the paper wholesaler, who in turn sells it to the printer.

Paper Supply and Demand

Paper is the most essential commodity to a printer, yet it is becoming increasing difficult to obtain. In the 1970s and 1980s the demand for paper outstripped production capacity at U.S. mills at various times, forcing the printing industry into unfortunate shortage situations. This problem has been further exacerbated by the export of printing papers and pulp to foreign countries beginning in the mid-1980s.

Paper manufacturers respond that not only is the use of paper on the rise, particularly for printing, but that an expanding economy is responsible for this greater demand. The apparent U.S. consumption of paper has reached an estimated 640 pounds per person, compared with an estimated worldwide average of slightly over 90 pounds; the total annual U.S. apparent consumption (production plus imports minus exports) reached 82 tons in 1988. The American Paper Institute (API), surveying data covering a period from November 1987 through November 1988, found that U.S. paper mills produced 4.3% more paper than in the previous year period, yet were still unable to meet demand. Manufacturers say that while there are new papermaking facilities going on-line, this expansion is not enough to keep up with the continually escalating demand.

History has shown that paper shortages invariably result in higher prices which usually remain somewhat escalated after the shortage is over. At the same time, paper shortages have a tendency to force printers to stockpile the major kinds of paper they frequently use, thus making the shortages even more pronounced and causing major cash outlays for excessive inventory held by the printer.

6.3 Printing Paper Classifications

There are five major classifications of printing papers. Each category has different physical characteristics and many are manufactured for specific end-use requirements. The major classifications are: business papers, book papers, cover stocks, cardboards, and miscellaneous papers. As shown in Figure 6.2, each major class has varying subgroups that relate to the desired end-use requirements for that category.

It should be understood that the paper sales representative will aid the estimator or estimating trainee whenever possible to understand the paper to be purchased for any given order. Of course, the more estimators work with the various categories of paper, the more familiar they become with the lesser-known differences between seemingly comparable paper stocks.

Business Papers

This classification includes bond, ledger, thin, duplicator/mimeographic, and safety papers. All have a basic size of 17 × 22 inches, although substance numbers vary with each type of stock (see Figure 6.2).

Bond. Bond papers are also called *sulfite bonds* or *sulfites* because they are manufactured using the sulfite chemical process. As a general rule, sulfite bonds are numbered with respect to their grade value. Number 1, the top and most expensive grade, may be readily recognized because it always bears a watermark; it is sized for excellent printing and writing qualities. Sulfite number 2 and number 3 are less expensive and of lesser quality, not watermarked, and normally not manufactured today. Numbers 4 and 5 sulfite are even less expensive and of commensurate quality, although they are suitable for many printing jobs, especially direct mail and other throwaway materials. The range of substance numbers applies equally to all sulfite bonds.

Premium grade bond stocks are made with cotton fiber, usually 25%, 50%, 75%, or 100% compared to cellulose fiber. Thus, using this system, a 50% rag bond is 50% cotton and 50% cellulose fiber; a 25% rag content paper has 25% cotton fiber and 75% wood products. The greater the cotton content, the more expensive the stock, with 100% cotton fiber the most expensive. Usually, rag content papers are used when permanence is desired with the printed matter, as with letters, legal documents. and map products.

All grades of bonds have generally good writing properties; have fair to excellent strength, depending on the amount of cotton, sizing, and cellulose fiber length; are sized to insure good printability by offset or letterpress processes; and are available in a wide array of colors. Bond stock is used in large quantity for business forms printing by letterpress; lithographic web, and flexography. Bond papers may be purchased in either sheet or web form from the paper merchant.

Ledger. Ledger papers are available in both sulfite and rag content types. Ledger stocks are strong and durable and are sized during manufacture to have excellent

Stock Classifications	Basic Size (in.)	Basis Weight (lb)
Business Papers	17 × 22,	
Bond		13, 16, 20, 24
Ledger		24, 28, 32, 36
Thin (Manifold and Onionskin)		7, 8, 9, 11, 12
Duplicator (Duplicator and Mimeograph)		16, 20, 24
Safety		
Book Papers	25 × 38	
Uncoated		40, 45, 50, 55, 60, 65, 70, 75, 80, 100
Coated		40, 45, 50, 55, 60, 65, 70, 75, 80, 100
Text		40, 45, 50, 55, 60, 65, 70, 75, 80, 100
Cover Papers	20 × 26	60, 65, 80, and 100 are common
Paper-boards,		
Index bristol	25 1/2 × 30 1/2	90 and 110 are common
Tagboard	24 × 36	100, 125, 150, 175, 200
Blanks	22 × 28	2-ply to 6-ply
Printing bristol	22 1/2 × 28 1/2	75, 90, and 110 are common
Wedding bristol	22 1/2 × 28 1/2	75, 90, 100, and 110
Miscellaneous,		
Blotter	19 × 24	100, 120
Label	25 × 38	60, 80
Newsprint	24 × 36	30, 32, 36

Figure 6.2

Printing paper classifications, basic sizes and basis weights

writing characteristics. Such papers are used for accounting and similar record-keeping functions. Ledger is available in many varieties conforming to machine tabulations and computer processing.

Thin. Thin business papers may either be manifold, which is the sulfite grade, or onionskin, which is the cotton content body. Some thinner papers are manufactured with thicker bulk while maintaining the same low substance weights. Thin papers include substance 12, which may sometimes be considered a bond paper. Owing to significant increases in mailing costs in recent years, perhaps the major use of thin papers is for direct mail. They are also popular as second sheets, the second copy or file copy of a typed letter. Thin papers are used when weight or storage is an element in the product application.

Duplicator/Mimeographic. Duplicator paper is manufactured to be used specifically with spirit duplication machines for the reproduction of limited-quantity carbon masters. Spirit duplication is popular in schools because of its extremely low per-copy cost. Mimeographic paper is made specifically for use with mimeographic machines. The mimeographic process involves the "cutting" of a paper master, through which ink is forced when the master is placed on the mimeographic machine. It is important to note that the printing techniques used with each process are completely different. Thus, paper is manufactured for each process with different sizing compounds and fillers. When purchasing either mimeographic or duplicator paper, be certain the correct stock is ordered. If such papers are switched, copies produced will appear poorly printed, with fuzzy, ragged images. Duplicator and mimeographic papers are generally purchased as "cut papers," which is usually 8 1/2 × 11 inch lettersize paper.

Safety. Safety paper is manufactured with an impregnated background pattern and is generally used for documents involving financial exchange, such as checks and letters of credit. The reason for the name "safety" is that once the paper has been written upon with ink, it is impossible to modify the document or the amounts indicated without removing the background pattern.

Book Papers

Book stock is used in the production of literally thousands of products: books, magazines, periodicals, trade journals, calendars, annual reports, advertising pieces, and labels, to mention only a few. Book paper is manufactured in three grade classifications:

Uncoated (or plain) book paper, which is very popular for less expensive books, textbooks, and other products;

Coated book paper, which is usually used when high-quality reproduction is desired, as in the printing of process color images;

Text papers, which are used in the production of books and other products when colored, special finish paper is desired.

Text papers generally have a deckled (ragged) edge, are available in many fancy colors (even fluorescents), and have definite surface textures.

Cover Papers

Cover stocks are thicker papers designed to be used as wraps for many types of books and other published materials. The advent of perfect binding for paperback books, telephone directories, and many other products has heightened the use of cover papers in book manufacturing. Cover papers generally have excellent printability, may be coated or uncoated, and are typically available in colors and with surface characteristics matching those of the text paper (book) category.

Paperboards

There are five major divisions of board stock, with a number of different basic sizes (see Figure 6.2). All board stocks have high bulk and are used in the production of thousands of packaging products, as well as for posters, announcements, point-of-purchase displays, and many other printed products. *Index bristol* is perhaps the largest single seller in this category and is used in the manufacture of index cards, business cards, and other similar products. *Tagboard*, another board classification, is a generally inexpensive stock used in the making of tags and other products where strength is important. *Blanks*, which may be purchased C1S or C2S, are used in the production of many types of displays; they are exceptionally thick and strong and print well when coated. *Printing bristols* are sulfite board stocks used for less expensive board items, such as business cards. *Wedding bristol* is a rag content board stock used for wedding invitations and other announcements of important events.

Miscellaneous Papers

The miscellaneous category includes three major types of paper. The first, *blotter* paper, is not used much today because of the popularity of ballpoint pens; however, blotter is still used in the production of checkbook backs and desk pads. *Label* paper, the second type, is somewhat similar to uncoated book paper and is used in the production of many types of food canning labels and other label products. Perhaps the largest seller in the miscellaneous category is *newsprint*, which is used as the substrate for newspapers, paperback books, and many other less expensive published products. The majority of newsprint today is sold in web (or roll) form and printed using rotary letterpress or web offset lithographic equipment. Sheet newsprint is not used to any great extent in the printing and publishing industry.

Most paper merchants provide the printing plant with sample swatches and books to aid the sales representative, estimator, and customer in making paper selections. These samples allow the customer and sales representative to discuss the different types of paper available for the job under consideration. The estimator uses them to verify the stock selections made. Sample books are supplied to the printing company on a free basis since they essentially stimulate sales of the paper merchant's products.

6.4 Printing Paper Sizes and Weights

The following information must be thoroughly understood by the printing estimator or estimating trainee to effectively estimate paper. It should be noted that the size and weight system used for printing papers can be confusing to the novice. Careful attention should be given to learn as much as possible about the sizes and weights of papers. While metrication of printing papers may someday become a reality, for the present it appears that the described system will prevail in the United States.

Basic Size of Paper

Basic size refers to the established size (in inches) of paper stocks upon which the basis weight of the stock is calculated. The basic size of most paper stocks is also recognized by buyers and sellers as one size that has common use requirements. For example, the basic size of business papers (see Figure 6.2, middle column) is 17 × 22 inches. As Figure 6.3 shows, both 8 1/2 × 11 inch letterheads and 7 1/4 × 8 1/2 inch invoices cut very efficiently from a 17 × 22 inch sheet. As another example, consider the basic size of book paper, which is 25 × 38 inches. When producing books, a very common standard page size, before final trimming and finishing, is 6 1/4 × 9 1/2 inches. (This size provides for a 1/4 inch trim on three sides of the book.) When 6 1/4 × 9 1/2 inches is divided into the 25 × 38 inch basic size (Figure 6.4), 16 pages can be positioned on one side of the press sheet.

The classifications cardboard and miscellaneous papers each have different types of stock with different basic sizes. However, the classifications of business, book, and cover papers each have one basic size that covers all types of paper offered in that category. Because there are so few basic sizes and because the estimator is working with them frequently, it is best for students and trainees to memorize them.

Basis Weight of Paper

By definition, the basic size of paper serves as the one size upon which basis weight is calculated. *Basis, or substance, weight* is defined as the weight of 500 sheets (one *ream*) of the basic size of a particular classification of paper. Figure 6.2 lists typical basis weights in the right-hand column. These weights are most common, but other less typical weights are also available. It should be noted that basis, or substance, weight is not directly related to thickness. Basis weight, substance, and poundage, or pounds, all mean the same thing—that is, the scale weight of 500 sheets of a particular type of paper in the basic size.

Using the basis weight definition and applying it to bond paper with a basic size of 17 × 22 inches, the chart indicates four basis weights: substance 13, substance 16, substance 20, and substance 24. Therefore, 500 sheets of 17 × 22 bond (basic size)

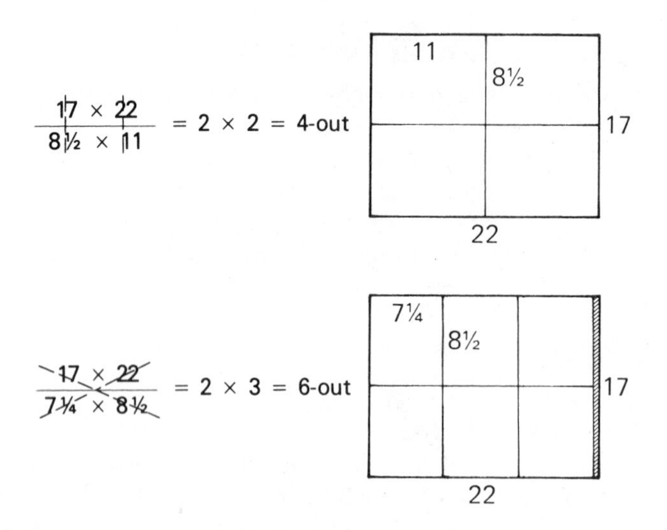

Figure 6.3

Cutting diagrams for basic size bond measuring 17 × 22 inches

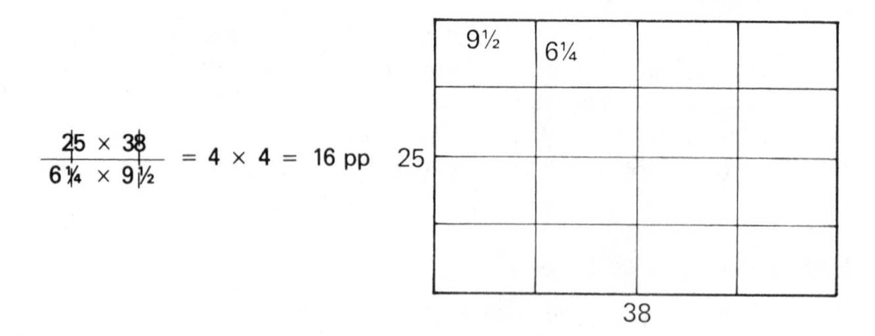

Figure 6.4

Imposition diagram for basic size book paper measuring 25 × 38 inches

of substance 13, 16, 20, and 24 will have scale weights, respectively, of 13, 16, 20, and 24 pounds. Thus, if we purchased 500 sheets of substance 20 (sub 20) bond in the 17 × 22 size, it would weigh 20 pounds; if we purchased 5000 sheets of substance 16, it would weigh 160 pounds.

Standard Sizes of Paper

Sometimes the basic size of paper is not the most suitable size for customer needs or production during printing. For this reason, paper merchants offer paper for sale in standard sizes other than the basic size that print, cut, or fold in a manner advantageous to the customer and printer. Thus, the paper merchant will sell bond paper in the basic size (17 × 22), as well as numerous other standard sizes conforming to printing equipment requirements or desired finish size of the customer's product.

To a certain extent, standard sizes of paper offered for sale vary in dimension from one paper merchant to another. Three general rules apply to the sizes selected to be standard:

1. The basic size of one type of paper is sometimes offered as the standard size of another. For example, book paper (basic size 25 × 38 inches) is also offered in the 17 × 22 inch size, which is the basic size for bond papers. Conversely, bond may be sold in the 25 × 38 inch standard size.
2. Standard sizes of paper are many times offered in exact mathematical ratio values of other sizes. For example, basic size book (25 × 38 inches) is also sold in the 19 × 25 inch size, exactly one-half of the longest sheet dimension. This same paper will also be sold in the 38 × 50 inch standard size, which is twice as big as the basic size. Bond, basic size 17 × 22 inches, is commonly sold in the 22 × 34 inch size (twice as large), the 34 × 44 inch size (twice as large again), and the 11 × 17 inch size (one-half the basic size).
3. Some standard sizes are offered for sale in unusual dimensions that fit well with various sizes of printing press equipment or correspond to standard product sizes popular with customers. Some of the more common standard sizes following this rule are: 16 × 21 inches (government size), 17 1/2 × 22 1/2, 17 × 28, 23 × 29, 23 × 35, 24 × 38, and 35 × 45 inches.

Basic and standard sizes are listed in any paper catalog, sometimes called a price catalog, which can be obtained from any paper merchant.

Comparative Weight of Paper

The weight system for paper works easily for basic sizes; however, standard sizes of paper are also sold in large quantities. The *comparative weight system* is used to determine the weight of the standard sizes of paper for shipping, handling, and manufacturing purposes. Because paper cartons indicate equivalent weight, it is important that the derivation of such information be known.

Equivalent weight is a comparison ratio of paper size to the weight of paper. The following comparative weight formula is used to determine the equivalent weight:

(area of desired paper size ÷ area of basic paper size) × basis weight = equivalent weight for 500 sheets of new size

where area is measured in square inches.

To find the paper poundage using the equivalent weight, use the following formula:

weight = (number of parent sheets ÷ 500 sheets) × equivalent weight

Example 6.1

We want to purchase a substance 80 cover paper (basic size 20 × 26 inches) in the standard size of 17 × 22 inches. Determine the weight of 3000 sheets.

Solution Use the comparative weight formula and substitute to find the equivalent weight:

$$\frac{17 \text{ in.} \times 22 \text{ in.}}{20 \text{ in.} \times 26 \text{ in.}} \times \text{sub } 80 = \frac{374 \text{ sq in.}}{520 \text{ sq in.}} \times 80 \text{ lb/500 sht}$$

$$= 0.72 \times 80 \text{ lb/500 sht} = 57.6 \text{ lb/500 sht}$$

Use the poundage formula to find the weight of 3000 (17 × 22) sht:

3000 sht ÷ 500 sht) 57.6 lb/500 sht = 345.6 lb

Example 6.2

We want to buy 8000 sheets of 24 × 36 inch bond, substance 16. (Find the basic size for bond in Figure 6.2.) Determine the 500 sheet equivalent weight and the total weight of 8000 sheets.

Solution Use the formula and substitute:

$$\frac{24 \text{ in.} \times 36 \text{ in.}}{17 \text{ in.} \times 22 \text{ in.}} \times \text{sub } 16 = \frac{864 \text{ sq in.}}{374 \text{ sq in.}} \times 16 \text{ lb/500 sht}$$

$$= 2.31 \times 16 \text{ lb/500 sht} = 36.96 \text{ lb/500 sht}$$

Use the poundage formula to find the weight of 8000 (24 × 36) sht:

(8000 sht ÷ 500 sht) × 36.96 lb/500 sht = 591.36 lb

Figure 6.5 is an equivalent weight table reproduced from a paper catalog; most paper catalogs contain a similar table. It provides a handy and quick reference for the estimator to determine equivalent weights without using the preceding mathematical formula.

M Weight of Paper

When cartons of paper are delivered to the printing plant, two weight measures are given on the carton label. The first is the substance number, or basis number, which by definition is the weight per 500 sheets of the basic size. The second is what is known as the *M weight*, or *1000 sheet weight*, and is defined as the weight of 1000 sheets of the size of paper in the carton. For example, if 3000 sheets of 17 × 22 inch substance 20 bond were delivered in one carton to a printing plant, the two weights on the carton label would be "substance 20," which, in this case, is the weight per 500 sheets of the basic size, and "40 M," which is the weight per 1000 sheets of 17 × 22 (which is the basic size in this case). For basic sizes of paper, the substance number may always be doubled to give the weight per 1000 sheets. To find the paper poundage using the M weight, use the following formula:

weight = (number of parent sheets ÷ 1000 sheets) × M weight

The paper in this example carton would weigh 120 pounds on a scale:

(3000 sht ÷ 1000 sht) × 40 lb = 120 lb

But standard sizes are different. Consider a carton of substance 60 book, containing 1500 sheets, purchased in the 24 × 36 inch size (basic size is 25 × 38 inches). On this carton label, the basis weight is 60, but because the paper is not in the basic size, a new equivalent weight must be calculated, then doubled, to represent the 1000 sheet weight for the 24 × 36 inch size. The following is the mathematical solution using the comparative weight formula:

$$\frac{24 \text{ in.} \times 36 \text{ in.}}{25 \text{ in.} \times 38 \text{ in.}} \times \text{sub } 60 = \frac{864 \text{ sq in.}}{950 \text{ sq in.}} \times 60 \text{ lb/500 sht}$$

$$= 0.909 \times 60 \text{ lb/500 sht} = 54.54 \text{ lb/500 sht}$$

EQUIVALENT WEIGHTS IN PAPER

While each type of paper uses a different basic size—book papers use 25 × 36—writings use 17 × 22, etc,—a chart like this brings out the fact that a 50# offset and a 20# bond are the same weight. Also a 65# cover and an 80# vellum bristol. The finish, bulk and strength could vary from one type of sheet to another. But on some jobs, because of size, availability or some other reason it might be advantageous to consider an equivalent sheet.

	Bond 17 × 22	Books 25 × 38	Cover 20 × 26	Vellum Bristol 22 1/2 × 28 1/2	Printing& Index Bristol 25 1/2 × 30 1/2	Kraft, News & Tag 24 × 36
Bond	**13**	33	18	22	27	30
Duplicator	**15**	38	21	26	31	35
Ledger	**16**	41	22	27	33	37
Mimeo	**20**	51	28	34	42	46
Writing	**24**	61	33	41	50	55
	28	71	39	48	58	65
	32	81	45	55	67	74
	36	91	50	62	75	83
Book	16	**40**	22	27	33	36
Offset	18	**45**	25	30	37	41
Text	20	**50**	27	34	41	45
	24	**60**	33	41	49	55
	28	**70**	38	47	57	64
	30	**75**	41	51	61	68
	31	**80**	44	54	65	73
	35	**90**	49	61	74	82
	39	**100**	55	68	82	91
	47	**120**	66	81	98	109
Cover	36	91	**50**	62	75	83
	43	110	**60**	74	90	100
	47	119	**65**	80	97	108
	58	146	**80**	99	120	133
	65	164	**90**	111	135	150
	72	183	**100**	123	150	166
Index &	43	110	60	74	**90**	100
Printing	53	135	74	91	**110**	122
Bristol	67	170	93	115	**140**	156
Tag	43	110	60	74	90	**100**
	54	137	75	93	113	**125**
	65	165	90	111	135	**150**
	76	192	105	130	158	**175**
	87	220	120	148	180	**200**

Figure 6.5

Equivalent weights (basis weights in bold; all weights are for 500 sheet reams) (Reproduced with permission of Unisource Corporation)

The M weight for this 24 × 36 book paper will then be:

(1000 sht ÷ 500 sht) × 54.54 lb/500 sht = 109 lb/1000 sht

This carton of 24 × 36 book paper will have a scale weight (paper only) of 163.5 pounds since there are 1500 sheets:

(1500 sht ÷ 1000 sht) ×109 lb = 163.5 lb

When paper is purchased, as will be explained later in this chapter, a *specified form* is used to detail all elements of the paper order. Both the substance weight of the basic size and M weight representative of the size of the paper purchased will be required.

It is important to remember the following definitions:

Basis weight is the weigh per 500 sheets of the basic size,
M weight is the weight per 1000 sheets of the size of paper purchased, which is not always the basic size.

These definitions hold throughout all discussions of paper weights used in the printing industry.

Bulk of Paper

The thickness of paper, or caliper, is a measurement in thousandths of an inch of the bulk of the sheet. The bulk of paper is important for press and folding requirements, especially during book production, and is often a marketing concern for the book publisher.

Most printing estimators have a *paper micrometer*, a hand-held device used to accurately determine paper thickness. A machinist's micrometer will work for this purpose also. When actually measuring paper thickness, a half-dozen measurements should be taken and the average figure used. Even though modern technology has assured a very flat and even paper sheet, slight variations in thickness still occur.

Figure 6.6 is a comparative bulking chart reproduced from a paper price book. It lists all classifications of paper, selects certain basis (substance) weights, and provides a caliper thickness measurement for these weights. It must be noted that such measurements are averages for many kinds of paper. Thus, basis 70 regular offset, which has a chart caliper of 0.005 inch, will vary with each manufacturer's basis 70 sheet. Some could average 0.0046 inch, others perhaps 0.0052 inch in thickness, and so on. The caliper of a sheet in a particular basis weight can vary but will not affect the weight at all.

AVERAGE CALIPER BULKING CHART

These are averages. Variations will be found if paper is run on light or heavy side of basic substance weight, depending on various mill runs. The thicker the caliper, the wider the variation.

	Basis	Thickness
Bond, Cotton Fiber (Flat Finish) 17 × 22	13	.0025
	16	.003
	20	.0035
	24	.004
(Cockle finish)	13	.003
	16	.0035
	20	.004
	24	.00475
Bond, Sulphite 17 × 22	13	.003
	15	.0035
	20	.004
	24	.0045
Ledger 17 × 22	24	.004
	28	.005
	32	.0055
	36	.006
Coated Book (Gloss) 25 × 38	60	.003
	70	.0035
	80	.004
	100	.005
	120	.006
Coated Book (Dull) 25 × 38	70	.004
	80	.0045
	100	.0055
Offset (Regular Finish) 25 × 38	50	.00375
	80	.0045
	70	.005
	80	.0055
	100	.007
	120	.008
	150	.010

	Basis	Thickness
Cover (Coated) 20 × 26	60	.0055
	80	.008
	100	.010
Cover (Antique)	50	.0065
	65	.0095
	80	.0115
	D.T.	.019
Index Bristol Sulphite 25 1/2 × 30 1/2	90	.007
	110	.0085
	140	.011
Printing Bristol (Antique) 22 1/2 × 28 1/2	100	.011
	120	.014
(Plate)	100	.009
	120	.011
Tag 24 × 36	100	.007
	125	.009
	150	.011
	175	.013
	200	.015

	Ply	Thickness
Board 22 × 28	3	.015
	4	.018
	5	.021
	6	.024
	8	.030
	10	.036

Figure 6.6

Comparative caliper bulking chart (Reproduced with permission of Unisource Corporation)

6.5 Paper Metrication

Introduction

While various efforts have been made to move the U.S. to the the metric measurement system, there seems to be little interest on the part of either the U.S. government or the American citizen to make the conversion. America is thus the only major industrialized nation in the world using non-metric measurements. Our English-speaking friends in Britain, Canada, and Australia began the conversion process in the early 1970s and are completely metric now. In 1975 President Gerald Ford signed into law a metric conversion bill, stating that the official policy of the U.S. will be to "coordinate and plan the increasing use of the metric system in the United States." The bill provided little money for this process and did not establish a timetable or mechanism by which the metric changeover would occur in the U.S.

In point of fact, it is slowly becoming a metric world in the U.S. As America becomes more internationalized in its business dealings—for example, as we import German, Japanese,or Swedish printing equipment—we will be forced into dealing with metric measurements, to the point that by the turn of the decade it is likely that we will be at least partially into the conversion process. While the American public may not yet feel comfortable with liters, grams, and meters, the integration of metric measurements into the products we buy will force us slowly to begin to work with this system.

To some, learning metric is like learning a second language. Just as in learning French by converting an English word to the French version with a two-way dictionary, there are conversion tables used to convert metric to U.S. measurements. For example, 1 meter equals 39.37 inches, and one inch equals 2.54 centimeters. Others find the best way to learn metric measurements is to learn the values independent of any comparison measurements, thus learning metric "straight," without conversion values. Regardless of the way metric is learned, most educators agree it is best to move slowly into the subject and to avoid trying to learn the metrication all at once.

The metric system includes measurements in length (meters), solid weight, (grams), liquid weight (liters), and temperature (Celsius). It is based on units of 10, 100, and 1000, allowing metric measurement components to be easily calculated. Thus a millimeter is one-thousandth of a meter and equal to 39.37/1000 or 0.03937 inches. For comparison purposes, the following basic chart is provided with a few common product examples:

1 meter	=	39.37 inches (slightly more than one yard)
1 millimeter	=	0.001 meter or 0.03937 inches (about the diameter of a paper clip wire)
1 centimeter	=	0.01 meter or 0.3937 inches (about the thickness of a regular pencil)
2.54 centimeters	=	1 inch
1 kilometer	=	1000 meters or about 3281 feet (about 0.60 of one mile)

1 gram	=	0.0022 pounds (about the weight of a paper clip)
1 kilogram	=	2.2 pounds
1 liter	=	1.06 quarts
0° Celsius	=	32° Fahrenheit (temperature where water freezes)

Metric Paper Sizes (ISO Standards)

The American paper industry began to convert to metric more than five years ago, and today observant production workers note metric unit measurements on the paper cartons delivered to the plant. Also, those CAE vendors selling systems world-wide have been required to convert their original American programs in non-metric measurements to metric values, including calculations of paper and measurements in inches to millimeters, meters, and grams.

International paper sizes were first formalized in 1922 in Germany using their country's DIN standard, and continued in 1947 with the formation of the International Organization for Standards (ISO) which was created to promote the development of standards throughout the world. Under the ISO system, paper dimensions are based on the area of a square meter in the rectangular form proportion of $1:\sqrt{2}$. The result of this is if one half the longer side is halved or the shorter side is doubled, the proportion is still $1:\sqrt{2}$ and the equivalent still equals one square meter or an exact ratio.

Out of this has developed the ISO-A series of paper sizes—commonly known as the A series—which is used for business papers and other popular paper stocks. The basic size is known as A0—equal to one square meter—from which all other sizes are derived. Figure 6.7 graphically demonstrates this concept for the A series in millimeters, which includes A1 through A10 sizes (2A would be twice as big in one dimension) with exact measurements listed.

International A Sizes (in millimeters)		*Approximate Inch Dimensions*
2A	1189 × 1682 mm	46.8 × 66.2 inches
A0	841 × 1189 (basic size)	33.1 × 46.8 "
A1	594 × 841	23.4 × 33.1 "
A2	420 × 594	16.5 × 23.4 "
A3	297 × 420	11.7 × 16.5 "
A4	210 × 297	8.3 × 11.7 "
A5	148 × 210	5.8 × 8.3 "
A6	105 × 148	4.1 × 5.8 "
A7	74 × 105	2.9 × 4.1 "
A8	52 × 74	2.0 × 2.9 "
A9	37 × 52	1.5 × 2.0 "
A10	26 × 37	1.0 × 1.5 "

There are also ISO-B and ISO-C series papers. The B-series paper stocks are intended for posters and other large-size paper goods, and have been developed because the jump of A-series to these larger levels is impractical. The C-series is used for envelopes, postcards, and similar products and has been developed to correlate with insertion of A-series products when they are folded or flat. The following chart represents B-series and C-series papers with their approximate inch dimensions:

International B Sizes (in millimeters)		*Approximate Inch Dimensions*
B0	1000 × 1414 mm	39.4 × 55.7
B1	707 × 1000	27.8 × 39.3
B2	500 × 707	19.7 × 27.8
B3	353 × 500	13.9 × 19.7
B4	250 × 353	9.8 × 13.9
B5	176 × 250	6.9 × 9.8
B6	125 × 176	4.9 × 6.9
B7	88 × 125	3.5 × 4.9
(continues through B10)		

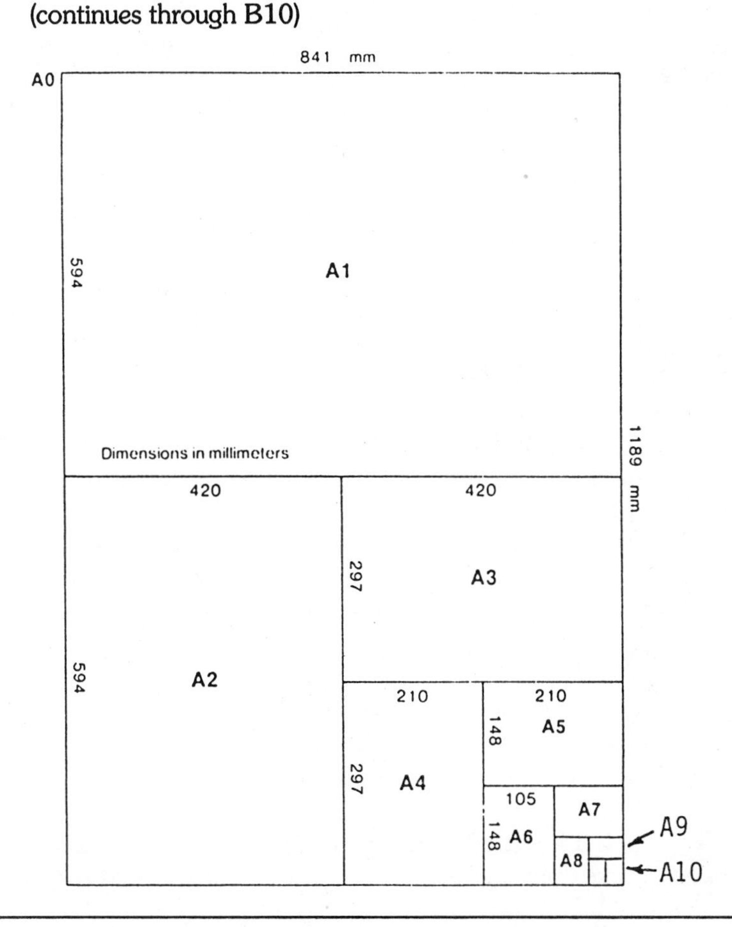

Figure 6.7

Diagram of ISO-A series with measurements in millimeters and A sizes indicated

International C Sizes (in millimeters)		*Approximate Inch Dimensions*
C4	229 × 324	9.0 × 12.8
C5	162 × 229	6.3 × 9.0
C6	114 × 162	4.5 × 6.3
DL	110 × 220	4.3 × 8.7

C4 envelopes are intended to provide A4 letters to be mailed in unfolded form, while C5 envelopes allow the same with A5 flat sheets. Because of the ratio of $1\sqrt{2}$, the C4 envelope will accommodate A3 sheets folded in half, and the C5 envelope will take an A4 sheet folded in half on long dimension. The DL category provides for two equally spaced parallel folds of an A4 sheet (regular folding of a business letter), or for the A5 sheet to be folded in half on the short dimension.

To accommodate bleeds and to add for additional paper needed for folding and trimming, British printers use a group of sizes which include paper for normal trim (R) and extra trim (SR). For example, the finished A0 sheet (see Figure 6.7) measures 841 × 1189 mm. The accepted RA0 (normal trim) dimensions enlarge the sheet to 860 × 1220 mm, and the SRA0 (extra trim) dimensions expand the sheet dimensions to 900 × 1280 mm. Suggested trims are available for most sheet sizes in the A, B, and C series.

Metric Paper Weights

The calculation of weights of paper in the metric system uses basis weight or substance as expressed in grams per square meter, g/m^2, gsm or, most commonly, "grammage." This is the weight of 500 sheets of paper of a defined type or kind whose surface area is one square meter. As was noted, the basic size of the A-series is one square meter in area.

Converting the basis weights (500 sheet weights) of the basic sizes of paperstocks from grammage to basis weights and basis weights to grammage uses the factors in the following chart:

Paper Type	Size in Inches	Factor for Grammage to Ream Weight	Factor for Ream Weight to Grammage
Writing papers	17 × 22	0.266	3.760
Cover papers	20 × 26	0.370	2.704
Newsprint	24 × 36	0.614	1.627
Book papers	25 × 38	0.675	1.480

Source: Based on TAPPI standard T410.

Figure 6.8 presents some grammages for normal basis weights of basic sizes of paper stocks using the conversion system and formulas for such conversions. For anyone unfamiliar with metric paper weights, Figure 6.8 provides a good comparison of this data.

Paper Type	Size (Inches)	Size (Millimeters)	Ream Weight (Pounds)	Grammage (gsm)
Bond	17 × 22	432 × 559	16 sub	60 gsm
			20	75
Book	25 × 38	635 × 965	40	59
			50	74
			60	89
			70	104
Cover	20 × 26	508 × 660	60	162
			70	189
			80	216
Newsprint	24 × 36	610 × 914	28	46
			30	49

To convert any substance weight to grammage or vice versa, the following formulas can be used:

$$\text{Ream weight to grammage} = \frac{\text{ream weight} \times 1.406}{\text{sheet area in square inches}}$$

$$\text{Grammage to sub weight} = \text{gsm} \times \text{sheet area (sq. in.)} \times 0.000711$$

Figure 6.8

Comparison chart of basic sizes to metric sizes and ream weights to grammage

A standard series of basis weights—R20 and R40—in grams per square meter (grammage) has been developed to be used with the metric system. R20 is based on the geometric progression based on the 20th root of 10, which equals a multiplier of 1.12 or 12% over the preceding step; R40, which is based on the 40th root of 10, provides twice the number of intermediate grammage steps. The R designates Colonel Charles Renard who devised the system. The two have been combined as the R20/R40 range and paper mills have been asked to make their stock in the following grammages:

25.0	45.0	71.0	106.0	160.0
28.0	50.0	85.0	118.0	180.0
31.5	63.0	100.0	140.0	200.0

To calculate the weight in kilograms per 1,000 sheets the following formula applies:

$$\text{M sheet kg weight} = \frac{\text{length of sheet (mm)} \times \text{width of sheet (mm)} \times \text{gsm}}{1{,}000{,}000}$$

Example 6.3 provides a problem using the above formula.

Example 6.3

We have a skid of bond paper with a grammage of 75 gsm (20 lb) and dimensions of 432 x 559 millimeters (17 × 22 inches). What would be the weight of 1,000 sheets in kilograms (kg)?

$$\text{M sheet weight} = \frac{432 \text{ mm} \times 559 \text{ mm} \times 75 \text{ gsm}}{1{,}000{,}000}$$

$$= 18.11 \text{ kg per } 1{,}000 \text{ sheets}$$

Solution Since a kilogram weighs about 2.2 pounds, multiplying 18.11 by 2.2 equals 39.84 pounds, which is approximately the nominal poundage of 1,000 sheets of basis 20 or 40M bond in the basic size.

6.5 Ordering Paper for Sheetfed Production

The Specified Form

The *specified form* provides complete information from the estimator or the purchasing agent to the paper merchant when ordering paper in sheets. All criteria for the stock to be ordered are stated clearly, explicitly, and in the following order:

Stock number, if any (as specified by the paper merchant),
Number of sheets requested,
Sheet size requested,
Basis weight and M weight (the latter in parentheses),
Color,
Specific brand name of the paper (may also be indicated as first item),

Other information, such as special finishes or grain direction,
Packing, loading, and shipping instructions.

The following examples show exactly how to write the specified form.

Example 6.4 We want to order orange Ticonderoga Text, stock number 003-221, grain long, 25 × 38, substance 60, and we need 5500 sheets.

Solution Prepare an order form that reads:

003-221, 5500 sht 25 × $\underline{38}$, sub 60 (120 M) orange Ticonderoga Text, grain long

Example 6.5 We want to order 6800 sheets, Hammermill bond, 25% rag content, stock number 001-340, cockle finish, grain short, substance 20, 22 × 34, white.

Solution Prepare an order form that reads:

001-340, 6800 sht $\underline{22}$ × 34, sub 20 (80 M) white Hammermill bond, 25% rag, cockle finish, grain short

Note in Example 6.4 that the M weight was determined by doubling the substance weight because the stock was ordered in the basic size. The second example required the use of a conversion to determine the M weight because the stock was not ordered in the basic size. (Refer to Figure 6.2 to find the basic size for bond and calculate the correct M weight as in Section 6.4.)

Specifying Paper Grain. Paper grain, the alignment of cellulose fibers in the paper, may be either long or short depending upon the way the paper was sheeted during manufacture or cut later in the printing plant. Such grain may be indicated with a straight line ether above or below the sheet dimension that applies. In Example 6.4, the grain is long and thus can be written 25 × $\underline{38}$ or 25 × $\overline{38}$. The line over or under the long dimension indicates the grain direction. In Example 6.5, grain is short; thus, $\underline{22}$ × 34 or $\overline{22}$ × 34 is the correct representation. The use of the grain line then makes optional the placement of the words "grain long" or "grain short" when writing the specified form.

Use of the grain line is important when grain is a critical factor in printing production. Further discussion of grain can be found in Section 6.7.

Paper Catalogs and Paper Prices

Most paper merchants supply the printing estimator with a large range of sample paper swatches and a paper price catalog. Together these items contain all information necessary to select appropriate paper stocks, determine prices, and aid in ordering precise kinds and sizes of paper. The catalog may be a notebook that provides for insertion of price changes in any section; pages of the catalog will generally be dated to indicate at what date such prices became effective. These catalogs are divided into sections by type of paper and may include additional sections for items such as envelopes, cut paper products, and industrial papers. See the index page from a pricing catalog in Figure 6.9 for such section divisions.

Printing papers are manufactured and costed out by weight at the mill. Most catalogs list a *price per 100 pounds*, or CWT price, that is easily converted to a price per pound. A second pricing system used by most paper merchants involves conversion of the CWT price into a *price per 1000 (M) sheets*. Because most estimators calculate paper in terms of sheets and not pounds, the M sheet price is easier to work with. It also saves computation time since poundage is not required as part of the calculation of paper cost as it is with the CWT system. A third pricing system, used for paperboards and other board stock, is a *price per 100 sheets*. The paper catalog usually lists CWT prices for most paper stocks, with accompanying M sheet or 100 sheet prices, depending upon the type of paper.

Figure 6.10 shows a page for Cascade Offset (book) from a 1989 paper pricing catalog that contains both the price per CWT and price per M sheets. In addition, prices are broken down into various carton levels. The columns are categorized across the page (from left to right) as follows: first column is a stock number, followed by the basis (substance) weight, the sheet sizes available, weight per M sheets, and the number of sheets per carton. The price per CWT and price per 1000 sheets vary depending upon the quantity being purchased. The price variations are given in the last five columns for 1 carton, 4 carton, 16 carton, 32 carton, and Zone B (shipping cost) categories.

Example 6.6

We want to order 12,500 sheets of Boise Cascade offset, white, smooth finish. Stock will be basis 60 and will be bought in the 25 × 38 size. (Stock size and quantity are determined based on production factors that have not as yet been explained; figures in this problem are for example only.) Determine the cost of this paper using both the price per 1000 (M) sheets and the price per 100 pounds (CWT).

Solution Refer to Figure 6.10

Step 1 Locate "white, smooth finish" and "basis 60" in the chart.

Step 2 Now locate the 25 × 38 size. Note that the M weight is 120 lb (which is double the basis 60, the basic size) and that each carton contains 1200 sht.

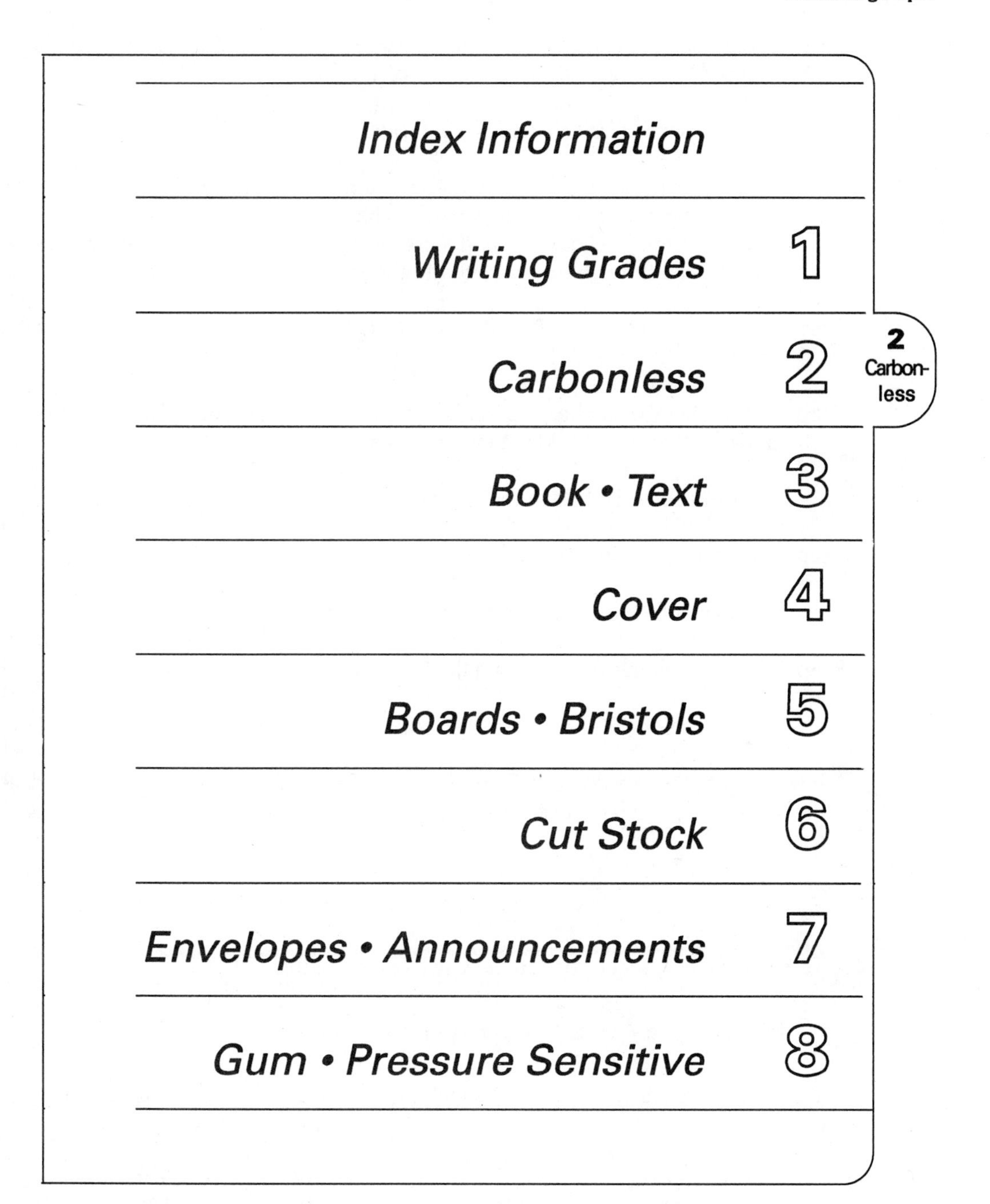

Figure 6.9

Paper catalog index. (Reproduced from a Unisource Corporation Paper Catalog, with permission)

Step 3 Determine the carton level to be purchased. Divide the total number of sheets needed by the number of sheets per carton (ctn):

12,500 sht ÷ 1200 sht/ctn = 10.416 ctn

The 4 carton price then applies because it represents a carton range between 4 and 16 cartons. Sometimes paper purchases are amalgamated, allowing for a higher carton level quantity and a greater cost savings. Amalgamation must be determined through discussions with the paper sales representative.

Step 4 Locate the 4 carton CWT and M sheet prices. The CWT price is $79.55 and the M sheet price is $95.46

Step 5 Determine the number of pounds of paper. Use the weight of 120 lb/M sht and multiply that by 12.5 M sht (number of sheets needed):

20 lb/M sht × 12.5 M sht = 1500 lb

Step 6 Find the cost using the CWT price:

1500 lb × $79.55/CWT = $1193.25

Step 7 Cross-check this answer using the M sheet price:

12.5 M sht × $95.46/M sht = $1193.25

Formulas to Determine CWT and M Sheet Prices. The CWT and M sheet prices are interchangeable. The following formula is used to change the CWT price to the M sheet price:

(CWT price ÷ 100) × poundage per M sheets = M sheet price

To change the M sheet price to the CWT price, use this formula:

(M sheet price ÷ poundage per M sheets) × 100 = CWT price

The formulas require that the 1000 sheet weight or equivalent weight be calculated accurately.

Example 6.7

At the 16 carton rate, a basis 60 sheet costs $74.35 per CWT. Determine the price per M sheets in the 25 × 38 size and then check Figure 6.10 to see whether the calculated price is correct.

CASCADE OFFSET

Approximate Calipers:
Smooth: Bs. 50 - .0038,
Bs. 60 - .0046, Bs. 70 - .0052
Vellum: Bs. 50 - .0043,
Bs. 60 - .0052, Bs. 70 - .006

PRICE PER CWT

	Bkn Ctn	1 Ctn	4 Ctns	16 Ctns	32 Ctns	Zone "B"
White, Basis 50	125.75	87.20	81.85	76.50	71.70	6.00
White, Basis 60, 70	122.20	84.75	79.55	74.35	69.70	6.00

Code	Basis	Size	M Wgt	Shts Ctn	Stk Loc	Bkn Ctn	1 Ctn	4 Ctns	16 Ctns	32 Ctns	Zone "B"
White, Smooth Finish						PRICE PER 1000 SHEETS					
05D-0196	50	17½ x 22½	41	3600	CS	51.56	35.75	33.56	31.37	29.40	2.46
05D-2482		23 x 29	70	2000	C	88.03	61.04	57.30	53.55	50.19	4.20
05D-0218		23 x 35	85	1800	C	106.89	74.12	69.57	65.03	60.95	5.10
05D-0226		25 x 38	100	1600	C	125.75	87.20	81.85	76.50	71.70	6.00
05D-0234		35 x 45	166	900	C	208.75	144.75	135.87	126.99	119.02	9.96
05D-0277	60	17½ x 22½	50	3200	CS	61.10	42.38	39.78	37.18	34.85	3.00
05D-6755		19 x 25	60	2400	CS	73.32	50.85	47.73	44.61	41.82	3.60
05D-0285		23 x 29	84	1800	CS	102.65	71.19	66.82	62.45	58.55	5.04
05D-0293		23 x 35	102	1500	CS	124.64	86.45	81.14	75.84	71.09	6.12
05D-0307		25 x 38	120	1200	C	146.64	101.70	95.46	89.22	83.64	7.20
05D-0315		35 x 45	198	800	C	241.96	167.81	157.51	147.21	138.01	11.88
05D-2490	70	17½ x 22½	58	2400	C	70.88	49.16	46.14	43.12	40.43	3.48
05D-0358		23 x 29	98	1600	C	119.76	83.06	77.96	72.86	68.31	5.88
05D-0366		23 x 35	119	1200	C	145.42	100.85	94.66	88.48	82.94	7.14
05D-0374		25 x 38	140	1000	C	171.08	118.65	111.37	104.09	97.58	8.40
05D-0382		35 x 45	232	600	C	283.50	196.62	184.56	172.49	161.70	13.92
White, Vellum Finish											
05D-0579	50	17½ x 22½	41	3600	C	51.56	35.75	33.56	31.37	29.40	2.46
05D-9908		23 x 35	85	1800	C	106.89	74.12	69.57	65.03	60.95	5.10
05D-0625		25 x 38	100	1600	C	125.75	87.20	81.85	76.50	71.70	6.00
05D-0471	60	17½ x 22½	50	3200	C	61.10	42.38	39.78	37.18	34.85	3.00
05D-0617		23 x 29	84	1800	C	102.65	71.19	66.82	62.45	58.55	5.04
05D-2504		23 x 35	102	1500	C	124.64	86.45	81.14	75.84	71.09	6.12
05D-0498		25 x 38	120	1200	C	146.64	101.70	95.46	89.22	83.64	7.20
05D-0528	70	17½ x 22½	58	2400	C	70.88	49.16	46.14	43.12	40.43	3.48
05D-0536		23 x 35	119	1200	C	145.42	100.85	94.66	88.48	82.94	7.14
05D-0641		25 x 38	140	1000	C	171.08	118.65	111.37	104.09	97.58	8.40

Bold Indicates Grain Direction

Figure 6.10

Example page from a paper pricing catalog (Reproduced with permission of Unisource Corporation)

Solution

Step 1 Use the appropriate formula and substitute:

($74.35 ÷ 100) × 120 lb/M sht = $89.22/M sht

Step 2 Checking Figure 6.8, we see that basis 60, 25 × 38 has a 16 carton price of $89.22/M sht

Example 6.8 Given a 16 carton price for basis 60 Cascade offset of $89.22 per M sheets, determine the price per CWT. Stock is 25 × 38 (see Figure 6.8).

Solution

Step 1 Use the appropriate formula and substitute:

$89.22/M sht ÷ 120 lb/M sht) × 100 = $74.35/CWT

Step 2 Checking Figure 6.10, we see that basis 60/70, 16 carton level is $74.35/CWT.

It is vital that estimators know their paper sales representatives or paper company representatives personally. These sales representatives are in a position to pass along information with regard to paper closeouts and other paper bargains. They may amalgamate certain stock purchases, providing for significant cost reductions to the printing company. Due to the high cost of paper, maintaining a good working relationship with paper sales representatives can save hundreds of dollars in paper costs yearly.

6.7 Paper Planning for Sheetfed Production

The intention of this section is to investigate and discuss various sheet sizes of paper used when completing sheetfed printing production techniques. Two types of cutting—regular and stagger—will be covered, with examples presented. Paper problems similar to those evident during printing production will be solved in step-by-step form. Calculating percentage of waste for paper stocks will also be explained.

Paper Sheet Sizes and Terms

The printing estimator must understand the various production sizes of paper and the relationships between them. The *finish size sheet* (fss) is defined as the final sheet size desired by the customer; sometimes the term *finish size* is used. The *press size sheet* (pss) is the sheet size used during actual press work for the customer's order. If a job is produced "1-up"—that is, one finish size sheet per press size sheet—the finish size and press size are often the same dimensions. (Production is not 1-up when the press sheet is slightly oversize to accommodate final trimming or images that print off the edge, called bleeds). The last important term, *parent size sheet* (pars), or *stock size sheet* (sss), is the sheet size purchased from the paper merchant and delivered to the plant. In many cases, printing production works this way: The parent (largest) sheet is cut apart to produce press sheets that are printed on the press: the press sheets are then cut apart to produce the finish size sheet for the customer. It is possible that the finish size sheet has the same dimensions as the press size sheet and the press size sheet, the same dimensions as the parent sheet. Production circumstances dictate this interrelationship of sizes.

The term *up*, which has already been introduced, is important to press size sheets. Essentially, the number-up relates to the number of finish size sheets positioned, or *imposed*, on the press sheet. Press sheets may be produced with multiple images "up," so that fewer sheets are used to print a job. The greater the number-up on the sheet—2-up, 4-up, 6-up, 8-up, and so on—the fewer the number of press sheets needed to produce a given quantity for the customer. Such production planning for the printing press saves considerable press time, allowing for more jobs to be run in any given time period. Of course, larger printing presses do require more money for purchase and have higher BHRs than do smaller presses. Basically, most printing plants acquire the largest press equipment possible and then maximize the number-up accordingly. The overall effect is a reduced cost per unit produced because the larger press is used far more efficiently than a smaller press with a smaller sheet capacity.

Regular Paper Cutting

When a press size sheet is cut to finish size sheets, if the grain of the paper goes in the same direction with all cut sheets, a regular cut has been used. Consider Examples 6.9 and 6.10, showing both the mathematics and diagrams of such regular cuts.

Example 6.9

We want to cut 8 1/2 × 11 inch letterheads from parent size $\underline{17} \times 22$ inch basic size bond. Grain can go either short or long, but the cut should be the most efficient—that is, the least wasteful possible.

Solution Cut the sheet as shown in Figure 6.11. This cut provides that 4 sht will be cut out of the $\underline{17} \times 22$ pars. The "8 1/2" will cut out of the "17" and the "11" out of the "22." Grain will run the short direction with all sheets, and there is no paper waste whatsoever.

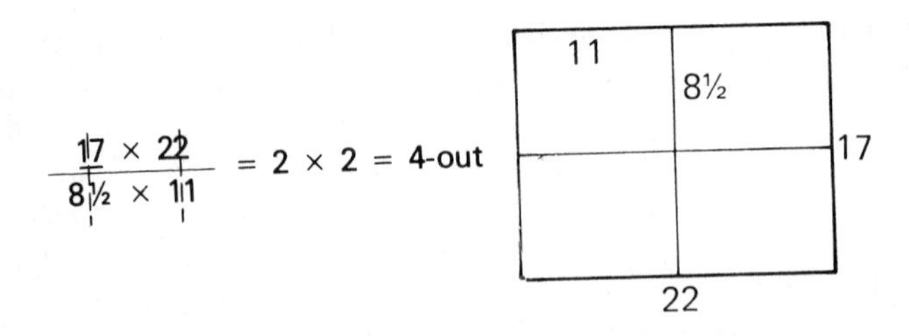

Figure 6.11

Cutting diagram for Example 6.9

Example 6.10

We want to cut 8 × 12 finish size flyers from parent <u>24</u> × 36 newsprint. Show cuts for both short and long grain directions and determine which is the less wasteful.

Solution

Step 1 Cut the sheet as shown in Figure 6.12. This cut is an even cut, with grain for each finish size piece running in the short dimension. The "8" will cut even out of "24" and the "12" evenly out of the "36" dimension. With this solution each parent sheet will yield 9 fss, all with the grain running in the short direction. There is no cutting waste with this solution.

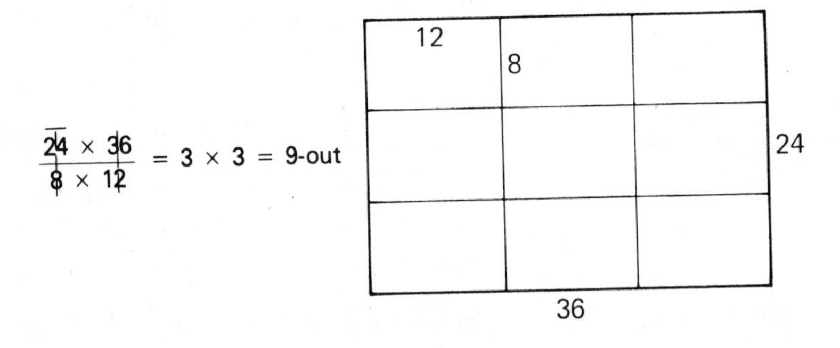

Figure 6.12

Cutting diagram for Step 1, Example 6.10

Step 2 Cut the sheet as shown in Figure 6.13. This cut provides that the grain in the finished sheet will run in the long (12 in.) direction. Note that there is some paper

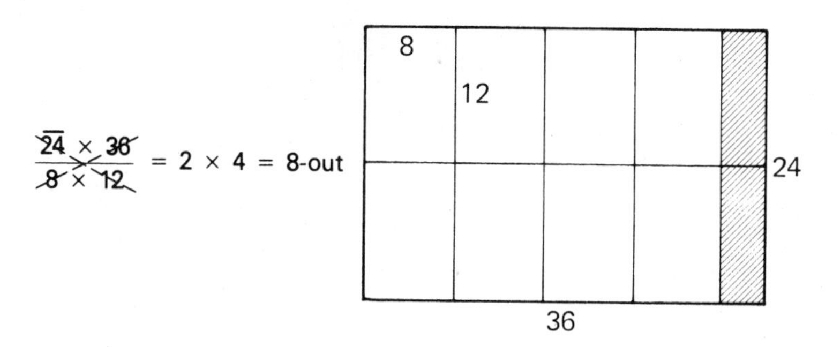

Figure 6.13

Cutting diagram for Step 1, Example 6.10

waste (see shaded area), making this cut undesirable. If grain has to be long for some reason, the best solution is to order the parent sheet with grain long.

It should be noted that the term *out* refers to the number of pieces that can be cut from a sheet of paper. The term *up*, introduced previously, relates to the number of finish size sheets imposed on a press sheet. The two terms are not interchangeable.

Basic Sheetfed Paper Problem. The plan of production is very important to paper estimating. To demonstrate the importance, Example 6.11 compares two sheetfed presses with different press size sheet capacities. Essentially, it compares a 1-up to a 4-up job. Step-by-step directions for the solution are given.

Example 6.11

A customer wants 5000 one-color letterheads printed. Finish size will be 8 1/2 × 11 inches, and substance 20 bond will be used. Plant production scheduling indicates that a Multilith Chief (maximum press size sheet 11 × 15 inches) and a Heidelberg MO (maximum press size sheet 19 × 25 1/2 inches) are both available. Determine the number and size of press sheets needed for each press in order to produce the 5000 letterheads. Press spoilage, an extra amount of paper added to compensate for press setup and makeready throwaway, will not be added in this problem.

Solution

Step 1 Determine the number-up on the press sheet for the Multi Chief and draw a diagram (Figure 6.14). Use an 8 1/2 × 11 pss for this press.

Step 2 Determine the number-up for the Heidelberg and draw a diagram (Figure 6.15). Use a 17 × 22 pss that can be printed 4-up for this press.

$$\frac{11 \times 15}{8\frac{1}{2} \times 11} = 1 \times 1 = 1\text{-up}$$

Use pss 8½ × 11.

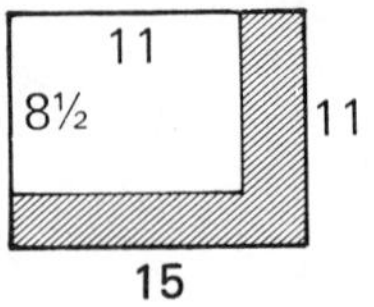

11 8½ 11 15

Figure 6.14

Cutting diagram for Step 1, Example 6.11

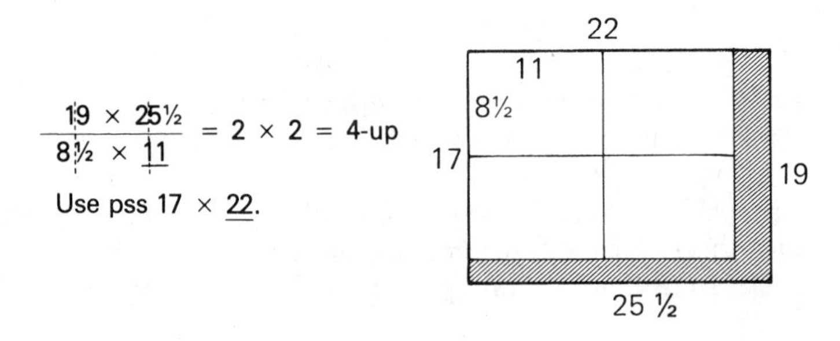

Figure 6.15

Cutting diagram for Step 2, Example 6.11

Step 3 Divide the number-up for the Multi Chief into the number of finished pieces to determine the number of press sheets needed:

5000 sht ÷ 1-up = 5000 pss

Step 4 Divide the number-up for the Heidelberg into the number of finished pieces to determine the number of press sheets needed:

5000 sht ÷ 4-up = 1250 pss

Thus far, the solution indicates that the Multilith Chief would require a 5000 pss run through the press, while the Heidelberg MO would require a run of 1250 pss. The Multilith Chief will run one image up on the 8 1/2 × 11 pss, and the Heidelberg MO will run four images up on the 17 × 22 pss. If both presses run at exactly the same production output, the Multilith Chief will require four times more press time than the Heidelberg MO. Since the BHR for each press is different—the Multilith BHR will be less than the Heidelberg's because it is smaller in size—costs for presswork will be different also. Essentially, it is up to the estimator to choose the most expeditious, least expensive production method.

To continue the problem with respect to parent sheet, assume that the Heidelberg has been selected to run the letterheads 4-up on a 17 × 22 in. pss. Checking with a paper catalog under sub 20 bond and the desired type of paper, the list of sizes for purchase includes 8 1/2 × 11, 11 × 17, 17 × 22, 17 1/2 × 22 1/2, 22 × 34, and 34 × 44 in. sizes. Since it is desirable to minimize cutting the 17 ×; 22 would be the most preferable purchase, making the press sheet and the parent sheet the same size. In such a situation, the stock would go immediately to the pressroom upon arrival at the plant, to be placed directly on the press. However, the stock requirement in the problem states that the grain of the finish sheet must go in the long direction, so the 17 × 22 short grain paper will not be suitable. The parent sheet to be purchased will be 22 × 34, which will allow the press sheet and finish sheet to meet the desired grain requirements.

To determine the number of 22 × 34 pars necessary, the following mathematics are completed.

Step 5 Divide the 17 × 22 pss (with grain long) into the 22 × 34 pars (grain short). This calculation determines the number-out of the parent sheet. Therefore:

$$\frac{22 \times 34}{17 \times 22} = \frac{34}{17} = 2\text{-out}$$

Step 6 Divide the number-out into the indicated number of required press sheets to determine the number of 22 × 34 pars necessary to complete the order. No press spoilage has been added for setup and printing of the job (it will be considered at a later point in this chapter). Therefore:

1250 pss ÷ 2-out = 625 pars

In summary, to produce the 8 1/2 × 11 letterheads for the customer, 1250 pss (17 × 22) will be cut 2-out from 625 pars (22 × 34) bond. The letterheads will print 4-up on the press sheet.

A Complete Paper Problem. Every estimator must know how to solve a complete paper problem, including the determination of paper cost and poundage. A complete paper problem with the detailed solution is presented in Example 6.12.

Example 6.12

A customer wants 40,000 handbills, 6 × 9 inch finish size, printed one color, one side, on basis 60 offset book. We have in inventory 24 × 36 inch stock for the order. The maximum press size sheet is 14 × 20 inches. Stock costs $70.00 per CWT. Determine:

The suitable press size sheet (running as many images up as possible),
The number of press size sheets needed,
The number of parent sheets needed,
The cost and poundage of paper necessary to complete the order,
The number of total impressions required on the press.

Do not add any paper spoilage for this problem.

Solution

Step 1 Divide the 6 × 9 fss into the maximum 14 × 20 pss. This calculation will provide the number-up to be printed on the sheet and the size of press sheet to be used. Therefore:

$$\frac{14 \times 20}{6 \times 9} = 2 \times 2 = 4\text{-up}$$

Use:

$$\text{pss}\ \frac{6 \times 2}{9 \times 2} = 12 \times 18$$

Note. The values 2 × 2 have been rounded to the nearest integers.

Step 2 Divide the number-up into the total number of finish size sheets (fss) desired by the customer. This calculation determines the number of 12 × 18 pss necessary to complete the order. (Press spoilage is normally added to this figure.) Calculate as follows:

40,000 fss ÷ 4-up = 10,000 pss

Step 3 Divide the press size sheet into the parent size sheet to determine the number-out that can be cut from the parent sheet (Figure 6.16).

Step 4 Divide the number cut out of the parent sheet into the number of press size sheets to determine the number of 24 × 36 pars necessary.

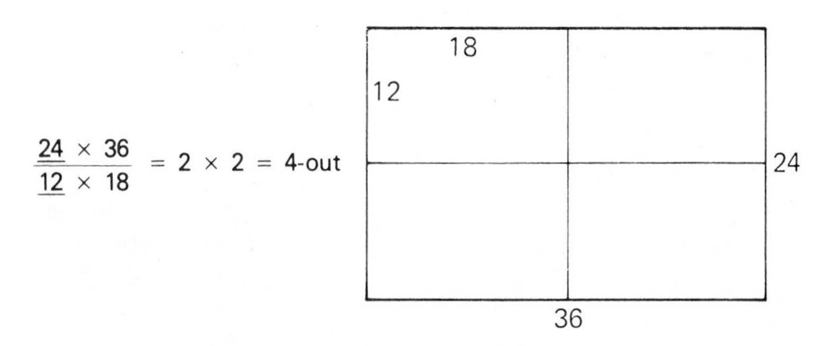

Figure 6.16

Cutting diagram for Step 3, Example 6.12

10,000 pss ÷ 4-out = 2500 pars

Step 5 Because the paper is not the basic size, complete a comparative weight conversion to determine the new M weight of the 24 × 36 book stock.

$$\frac{\underline{24} \text{ in.} \times 36 \text{ in.}}{25 \text{ in.} \times 38 \text{ in.}} \times 120 \text{ lb/M} = \frac{864}{950} \times 120 \text{ lb/M}$$

$$= 0.91 \times 120 \text{ lb/M} = 109 \text{ lb/M}$$

Step 6 Multiply the calculated new M weight by the number of parent sheets (in thousands) to determine the poundage of paper (24 × 36) to be used:

109 lb/M × 2.5 M = 272.5 lb

Step 7 Multiply the number of pounds by the cost of the paper per pound to determine the cost of the required stock. This stock at $70.00/CWT has a per-pound cost of $0.70 ($70.00/100 lb ÷ 100 lb = $0.70/lb). Therefore:

272.5 lb × $0.70/lb = $190.75

Step 8 The number of forms (fm) or colors times the number of press sheets equals the number of impressions (imp) necessary on the press. Since the order is for one color, one side, there is 1 fm needed:

1 fm × 10,000 pss = 10,000 imp

It is important that the steps indicated be followed as precisely as possible. Doing so will help to alleviate any confusion on the part of the estimator with respect to the sizes of the sheets required, will help in ascertaining the quantities printed up and cut out of the sheets used, and will aid in the addition of press spoilage (which is covered in Section 6.9).

Stagger Cutting

Regular cutting, as previously described, provides for the grain for all cut sheets to go in the same direction. *Stagger cutting,* also called *dutch cutting* or *bastard cutting*, does not give this result. While the grain will go the short way with some of the cut stock, it will go the long way with other sheets. The primary reason for stagger cutting is to maximize the number of press sheets cut from a parent sheet. It is also possible to maximize the use of a press sheet by imposing images to be printed in staggered form on the press sheet, which subsequently requires a stagger cut after printing. Consider Example 6.13.

Example 6.13

We have an order for 3000 8 1/2 × 13 inch legal size memo sheets to be printed 2-up on substance 6 bond. Our inventory shows that we already have a large amount of $17 \times \underline{22}$ bond on hand that would be suitable for this job, and we do not care to order other stock. Our customer indicates that grain will not be a problem with the final use of this product. Draw a diagram showing what this cut will look like and then determine the number of $17 \times \underline{22}$ inch parent size sheets that will be needed.

Solution

Step 1 Divide $\underline{8\ 1/2} \times 13$ in. into $17 \times \underline{22}$ to cut 2-out; then divide 8 1/2 × 13 into the remaining $\underline{9} \times 17$ in. of paper (Figure 6.17). This calculation provides for 3 pss out of the parent size sheet. Two pieces from each sheet will have grain long, and one piece will have grain short. Shaded areas in the diagram represent wasted paper that will be thrown away. Calculate as follows:

$$\frac{17 \times \underline{22}}{8\frac{1}{2} \times 13} = 2 \times 1 = 2\text{-out}$$

$$\frac{\underline{9} \times 17}{8\frac{1}{2} \times 13} = 1 \times 1 = 1\text{-out}$$

$$2 \text{ pss} + 1 \text{ pss} = 3 \text{ pss}$$

Step 2 Since we can cut 3-out of the $17 \times \underline{22}$ pars, we will be required to use 1000 pars for this job from our inventory:

$$3000 \text{ pss} \div 3\text{-out} = 1000 \text{ pars}$$

Especially with stagger cutting, the completion of an accurate diagram is necessary to be certain that the cut will work out correctly. This diagram should become a part of the job ticket to aid employees during the production of the job.

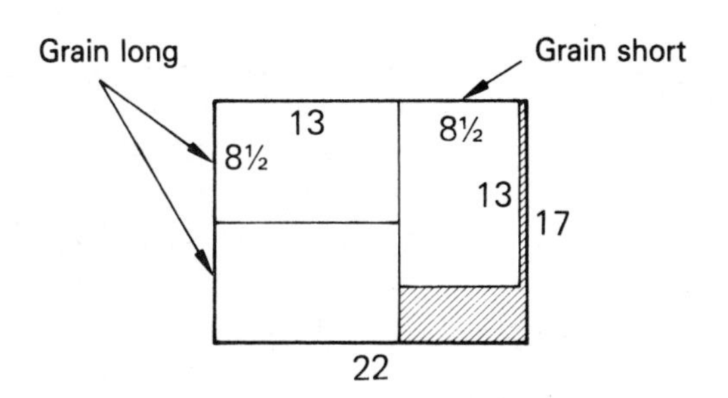

Figure 6.17

Cutting diagram for Example 6.13

There is a quick check to determine the approximate number that can be cut from a patent sheet. It involves the maximizing of the number to be cut out by using the following formula:

maximum number-out = area of the parent size sheet÷ area of the press size sheet

where area is measured in square inches. In Example 6.13, 17 × 22 inches (374 square inches) is divided by 8 1/2 × 13 inches (110.5 square inches), producing a figure of 3.38 finish size sheets (8 1/2 × 13 inches) cut from the parent sheet. The final result is 3-out, with the remaining fraction of the sheet to be considered waste.

Stagger cutting minimizes paper waste when cutting odd sheet sizes from parent sheets that are not well suited to the production of the work. Stagger cuts require more time in cutting production, with generally more cutting machine adjustments during the proccess; also, stagger cuts require that a first cut be able to be made without affecting the number out of the parent sheet. Linked with the fact that grain direction will change with a certain amount of the cut paper, many printers find stagger cuts undesirable. To avoid such problems, they order specific stock sizes for specific jobs to be produced, which also allows them to maintain lower inventories of paper stock, a situation that is usually desirable. Nonetheless, every estimator should understand how to complete a stagger cut and how to maximize the number of sheets to be cut out. Stagger cuts also require that a burst cut be able to be made without affecting the number-out of the parent sheet.

Paper Waste from Cutting

When plants purchase only certain sizes of paper stock and attempt to use these limited sizes for all jobs, the possibility of increased paper waste becomes a problem. Such waste can occur with either regular or stagger cuts but is almost a certainty with a large number of stagger cuts.

Stock waste is calculated as the ratio of the unused area of the paper sheet (square inches) to the total sheet area (square inches) and is expressed as a percentage. In the preceding stagger cutting problem, three pieces, each measuring 8 1/2 × 13 inches were cut from the 17 × 22 inch parent sheet. The three sheets total 331.5 square inches. The 7 × 22 inch sheet is 374 square inches. The difference between 374 square inches and 331.5 square inches is 42.5 square inches. By dividing 42.5 square inches by 374 square inches, a paper waste of 11.36% is determined. Generally, any waste greater than 5% of the parent sheet is considered undesirable and should be avoided. A simple formula to calculate waste percentage is:

$$\text{paper waste percentage} = 1 - (\text{area of sheet used} \div \text{total area of sheet}) \times 100$$

where area is measured in square inches.

It must be noted that percentage waste is a comparison of unused paper to the total sheet size. The term should not be confused with paper spoilage, which is the additional amount of paper added to the job prior to production to compensate for production problems, setup, and registration of the job on the press, and for throwaway stock in the bindery. Letterpress and lithographic press spoilage will be presented in the last section of this chapter.

6.8 Bookwork Impositions

The printing of books and booklets (64 pages or less) involves what are known as impositions. An *imposition* is the exact position (or lay) of book pages on a press sheet that after printing, is folded and trimmed to produce a signature. Pages on each signature are imposed in a prearranged sequence. For small books and booklets, it may be necessary to produce only one signature to complete the order; for larger books, more than one signature is usually required.

Keep in mind that these imposition techniques may be used with either sheetfed or webfed presses. However, for simplicity, this discussion will center around imposition procedures for sheetfed presses. The initial planning of such press sheets is a key ingredient when estimating books and booklets to be produced in a printing plant. Every estimator should understand basic bookwork impositions thoroughly.

Sheetwise Imposition Techniques

Sheetwise (SW) *impositions*, also called *work-and-back*, are easy to understand and produce. A press sheet with a sheetwise imposition has one-half of the pages of the signature printed on one side of the press sheet and the other half printed on the opposite side, backing up the pages that are printed on the first side. Sheetwise impositions require two different forms, one for the front and one for the back of the sheet. A *form*, then, is the imposition of pages for one side of a sheet and uses one lithographic plate. The same press *gripper edge*, or front edge of the sheet, is used during all press runs for both sides of the sheet. However, the *side guide*, used for alignment of the sheet left to right, changes from side to side as the sheet is flipped over for backup of the other form.

Consider the diagram in Figure 6.18, representing the production of a 4 page booklet using a sheetwise imposition. The imposition as shown would be termed a"1-4 SW" indicating that there is 1 signature containing 4 pages produced with a sheetwise page lay. This technique allows for the production of signatures with pages in geometric multiples—that is, the signatures will contain 2, 4, 8, 16, 32, 64, or 128 pages. It is possible to produce a 24 page booklet by combining a 16 page and an 8 page signature. Such combinations will be discussed later in this chapter.

To see how the sheetwise imposition can be applied, we will consider Examples 6.14 and 6.15. These examples involve a 32 page booklet. For this booklet, only two procedures will follow printing and will be considered here as they are related to the imposition technique: folding and stitching. *Stitching* is a binding procedure that follows folding. However, for the purposes of the Examples, stitching will be discussed first. It is important to bear in mind that for books and other types of booklets, there are procedures involved, including different binding procedures. For more discussion of these procedures, see Chapter 13.

Example 6.14

An advertising agency wants 10,000 copies of a 32 page booklet produced. Untrimmed page size is 6 1/4 × 9 1/2 inches. (Later trimming will make the finish size of the booklet 6 × 9 inches.) There will be no duplicate pages produced—that is, there will be only one page 1, one page 2, and so on. There are two presses available to produce this job: a Harris 19 × 25 (maximum press size sheet) and a Miehle 25 × 38 (maximum press size sheet). The booklet will be type matter only printed in black ink. Using the sheetwise imposition technique, we will investigate the possible production procedures. First, we will determine the number of press sheets needed for each press.

Solution

Step 1 Just as with cutting, divide the 6 1/4 × 9 1/2 into the 91 × 25 in. pss. Note that the 6 1/4 × 9 1/2 divides evenly into the respective dimensions, allowing for 8

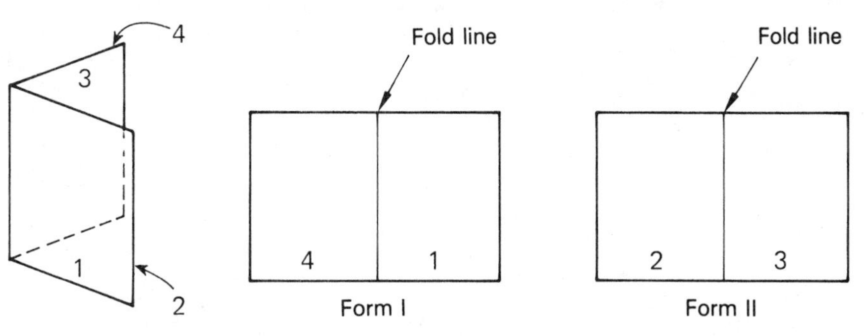

Figure 6.18

Sheetwise imposition diagram for a 4 page booklet

pp to be placed in the form (Figure 6.19). Therefore, each signature would contain 16 pp.

Step 2 Divide the 6 1/4 × 9 1/2 into the 25 × 38 in. pss. Since the sheet size is double that of the 19 × 25, the number of pages positioned on one side of the press is doubled (Figure 6.20). Thus, each signature would contain 32 pp.

Step 3 Dividing the number of pages per booklet (bkt) by the number of pages per signature (sig), we find that it will take 2 (19 × 25) sig to produce the booklet. Therefore, we must run 2-16 SW. The calculation is as follows:

32 pp/bkt ÷ 6 pp/sig = 2 sig

Therefore,

run: 2-16 SW

$$\frac{19 \times 25}{6\frac{1}{4} \times 9\frac{1}{2}} = 2 \times 4 = 8 \text{ pp/fm}$$

$$8 \text{ pp/fm} \times 2 \text{ fm} = 16 \text{ pp/sig}$$

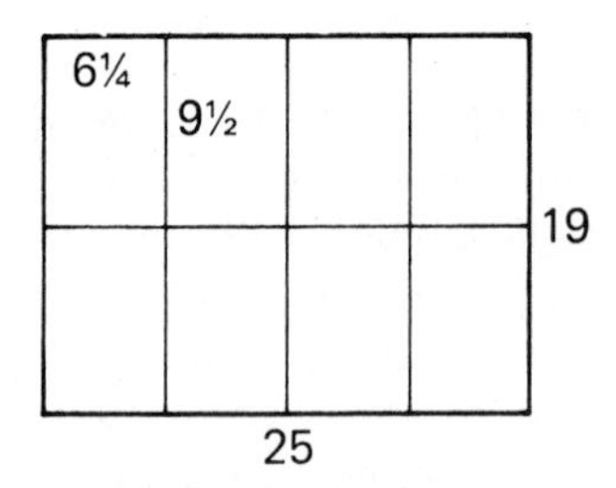

Figure 6.19

Sheetwise imposition diagram for Step 1, Example 6.14

$$\frac{25 \times 38}{6\frac{1}{4} \times 9\frac{1}{2}} = 4 \times 4 = 16 \text{ pp/fm}$$

$$16 \text{ pp/fm} \times 2 \text{ fm} = 32 \text{ pp/sig}$$

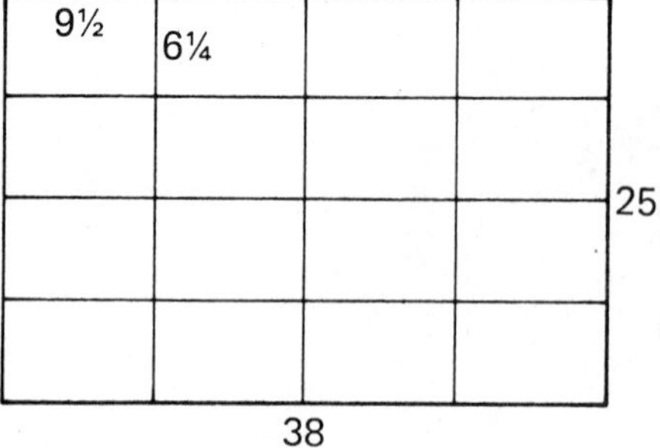

Figure 6.20

Sheetwise imposition diagram for Step 2, Example 6.14

Step 4 With the 25 × 38 pss, we note that we are able to impose 16 pp/side and are therefore able to put all 32 pp of the booklet on pss. Thus, running 1-32 SW allows the booklet to be imposed in one complete press sheet:

32 pp/bkt ÷ 32 pp/sig = 1 sig

Therefore,

run: 1-32 SW

Side and Saddle Stitching Procedures. In Example 6.14, the 1-32 SW imposition will produce one large signature on the press sheet and will require some form of binding, after folding, to hold all pages together. The 2-16 SW procedure will require gathering and stitching since two press sheets—that is, two signatures—are needed to make the 32 page booklet this way.

There are two stitching procedures that can be used in bookwork, and they apply to either a 1-32 SW or a 2-16 SW. The first is *side stitching*, which involves the placement of wire stitches through the binding edge of the signature, from top to bottom, to hold all pages securely together. The second method, *saddle stitching*, involves inserting the signatures inside each other and then placing wire stitches along the fold of the book, or saddle, as it is sometimes called. The diagrams in Figure 6.21 graphically illustrate each of these stitching techniques for the 2-16 SW.

In the case of the 1-32 SW, either binding method can be applied; since no gathering of signatures is involved, the imposition of pages is not affected by either type of stitching procedure. However, for the 2-16 SW, gathering of all signatures will be necessary and must be coordinated with the type of stitching procedure to be used. As Figure 6.21 shows, side stitching will produce one signature containing pages 1-16 and a second with pages 17-32. Saddle stitching, in contrast, will require one signature for pages 1-8 and 25-32 and a second for pages 9-24. Thus, the sequence of pages in the

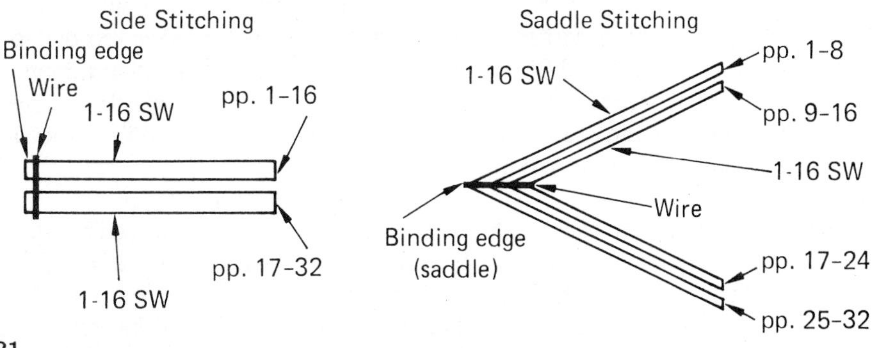

Figure 6.21

Stitching diagrams for the 2-16 SW (Example 6.14)

side stitching diagram is different from the sequence when saddle stitching is used. Different impositions will be required for the two binding procedures.

It should be understood that when small presses are used in the production of books and booklets, the number of signatures and amount of collecting and stitching in the bindery increases considerably. Such a situation may be undesirable in producing books in large quantities.

One tool that helps to determine the method of stitching and aids in the overall production planning of bookwork is the form chart. *Form charts* divide pages into outside and inside types, which by definition are to impose on opposite sides of the sheet. Form charts will indicate different page impositions depending on the type of stitching to be used. Consider the example 4 page form chart that follows the outside-inside-inside-outside pattern.

Outside		Inside
Page 1	→	Page 2
		↓
Page 4	←	Page 3

Thus, as page 1 of a booklet is an outside page, page 2, which will be printed on the back of page 1, is an inside page; it follows then that page 3, which faces page 2, will also be an inside page; and page 4, which backs up page 3, will be an outside page. This sequence is the basis for all form chart development, regardless of the number of pages in the signature.

Example 6.15

From Example 6.14, we know that the 1-32 SW requires 2 forms, each containing 16 pages, imposed on a 25 × 38 inch press sheet, while the 2-16 SW require 4 forms, each containing 8 pages and imposed on a 19 × 25 inch press size sheet. Now we will consider both side and saddle stitching for the booklet. As noted previously, the 1-32 SW will not require gathering because all pages are imposed in 1 signature (stitching will be required to hold the signature together). However, the 2-16 SW will require both gathering and stitching because 2 signatures are necessary to produce the 32 page booklet. Knowing that the customer needs 10,000 copies of this booklet allows determination of the amount of paper (without press spoilage) and the number of impressions required to complete the order. Continue to investigate the production procedures with the additional information.

Solution

Step 1 Figure 6.22 shows the form chart for the 2-16 SW with side stitching. Note that form I will contain pages 1, 4, 5, 8, 9, 12, 13, and 16—a total of 8 pp; form II will back up form I and contain pages 2, 3, 6, 7, 10, 11, 14, and 15. Forms III and IV will back up each other and contain 8 different pages, as illustrated. Because the pages are ordered sequentially, this side stitching solution is the only possible method of binding using side stitching procedures.

	Outside	Inside	
	1	2	
	4	3	
	5	6	
Form I	8	7	Form II
	9	10	
	12	11	
	13	14	
	16	15	
	17	18	
	20	19	
	21	22	
	24	23	
Form III	25	26	Form IV
	28	27	
	29	30	
	32	31	

Figure 6.22

Form chart for the 2-16 SW with side stitching (Example 6.15)

Step 2 Figure 6.23 shows the form chart for the 2-16 SW with saddle stitching. The chart indicates that form I will contain pages 1, 4, 5, 8, 25, 28, 29, and 32—a total of 8 pp; form II will back up form I and contain pages 2, 3, 6, 7, 26, 27, 30, and 31. Forms III and IV will back up each other and contain 8 different pages. Because of the way the pages are ordered, saddle stitching must be used to bind the booklet.

Step 3 Figure 6.24 correctly indicates forms and pages in those forms for the 1-32 SW. As with the other form charts, the exact lay of pages on the press sheet is not as important as the determination of the basic setup of the imposition. (Exact page lay will be covered when dummying of the signatures is discussed later in this chapter.)

Step 4 The quantity of 19×25 in. paper necessary to produce the 10,000 bkt using the 2-16 SW imposition is found by multiplying the number of signatures per copy times the number of required copies. The type of stitching used has no direct bearing on the number of sheets required. If weight and poundage data had been provided, it would be possible to determine the weight and cost of paper necessary for the order. Calculate the needed quantity of paper:

2 sig/bkt × 10,000 bkt = 20,000 sig = 20,000 pss

	Outside	Inside	
	1	2	
Form I	4	3	Form II
	5	6	
	8	7	
	9	10	
	12	11	
	13	14	
	16	15	
Form III	17	18	Form IV
	20	19	
	21	22	
	24	23	
	25	26	
Form I	28	27	Form II
	29	30	
	32	31	

Figure 6.23

Form chart for the 2-16 SW with saddle stitching (Example 6.15)

	Outside	Inside	
	1	2	
	4	3	
	5	6	
	8	7	
	9	10	
	12	11	
Form I	13	14	Form II
	16	15	
	17	18	
	20	19	
	21	22	
	24	23	
	25	26	
	28	27	
	29	30	
	32	31	

Figure 6.24

Form chart for the 1-32 SW (Example 6.15)

Step 5 The amount of 25 × 38 in. paper required to produce the 10,000 bkt is determined by multiplication of the number of signatures times the number of required copies. Even though this answer is one-half of the number required using the 19 × 25 in. pss, remember that the 25 × 38 pss is twice the size. Thus, the same amount of paper is needed with either solution, but there will be half as many sheets of 25 × 38 required:

1 sig/bkt × 10,000 bkt = 10,000 sig = 10,000 pss

Step 6 Total impressions (imp) are determined by multiplying the number of required forms (fm) times the number of copies needed. The solution here is for the 19 × 25 in. pss, which requires 40,000 imp total (without spoilage):

4 fm × 10,000 bkt = 40,000 imp

Step 7 The solution indicated here represents total impressions (without spoilage) for the 25 × 35 in. pss. Note that this solution will save 20,000 imp compared to the 91 × 25 in. pss—that is, press time will be halved. Remember, however, that the Miehle 25 × 38 press will cost significantly more per hour to operate than will the smaller press. The total number of impressions is determined as follows:

2 fm × 10,000 bkt = 20,000 imp

The Folded Dummy. The folding of a *dummy*, a model indicating how the press sheet will be folded and how the pages of the signature will be numbered, remains to be completed to finish the example. The dummy accompanies the job order throughout production and indicates specific page positions for printing and finishing operations. Any size of paper may be used for the dummy. Some estimators prefer to use the untrimmed press sheet, yet it is not necessary. For this example, begin with a 14 × 20 inch sheet.

The initial folding sequence for the sheet is shown in Figure 6.25 and is described in the following steps:

Step 1: Use an unfolded sheet measuring 14 × 20 inches.
Step 2: Fold the sheet in half with the fold to the top edge, producing a 4 page signature measuring 14 × 10 inches.
Step 3: Make a second fold by taking the left edge and aligning it with the right edge, producing an 8 page signature measuring 7 × 10 inches.
Step 4: Turn the folded edge to the top.
Step 5: Make a third fold by taking the left edge and aligning it with the right edge. producing a 16 page signature measuring 5 × 7 inches.
Step 6: Notch the top edge to ensure that each page is in the proper direction.
Step 7: Number the pages consecutively for the sheetwise signature by using the directions given in the text that follows.
Step 8: Unfold the signature to full press size and check the lay of the pages.

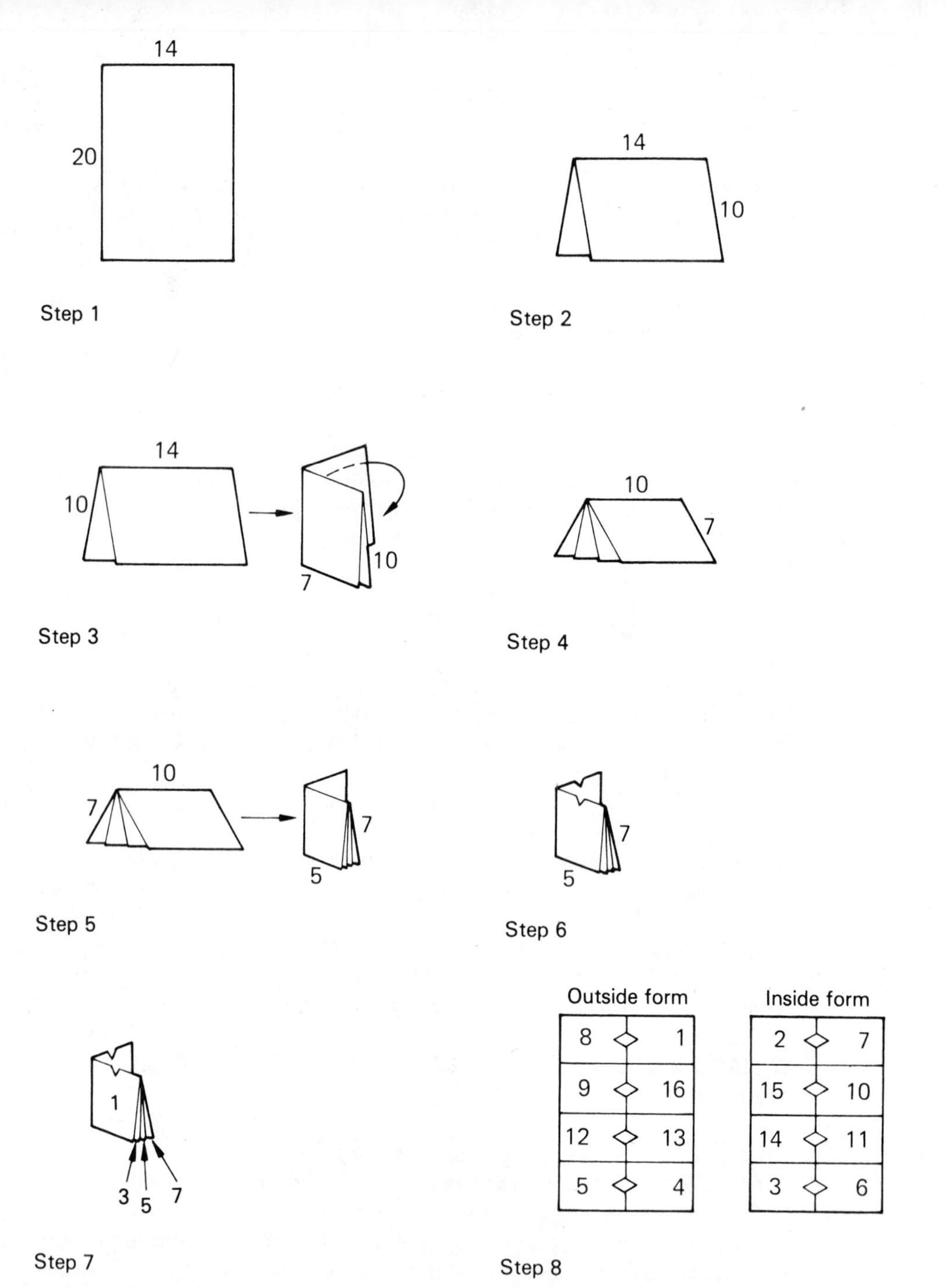

Figure 6.25

Folding a dummy for a 16 page sheetwise imposition (similar procedure used for work-and-turn impositions also)

Page numbering of the signature should be completed after it is folded. Begin by placing the binding edge to the left side, then tear open the bottom right-hand corners of the last half of the signature. Now, beginning with the front page of the signature, number each page in the bottom right- or left-hand corner, beginning with 1 and ending with 16. When numbering is completed, set aside the signature and fold another sheet of paper following exactly the same procedure. Number this second signature from 17 to 32 consecutively, thus completing the numbering of the side-stitched dummy for the 2-16 SW. Unfold each sheet and compare the pages of the sheet to the pages of the form chart. Each form should match with the pages in the form chart. Remember, the dummy provides exact page lay for the press sheet. The form chart is a useful planning tool but does not provide such exact placement information.

Follow the exact folding procedure just described with two more sheets of 14 × 20 (or any size) paper. Take the first signature and number the pages 1 through 8 consecutively; then number the remaining pages 25 through 32 consecutively. Number the second folded dummy from 9 to 24 consecutively, thus completing the numbering of the saddle-stitched dummy for the 2-16 SW. The signatures are inserted inside one another to represent the required 32 page booklet. Unfold the signatures and compare them to the appropriate form chart.

Still working with the 14 × 20 inch paper, fold it as previously described until a 16 page signature is produced with a page size of 5 × 7 inches. After the last fold has been completed, turn the signature 90° clockwise and fold it in half again, thus halving the page size (to 3 1/2 × 5 inches) and doubling the number of pages. Tear open the appropriate corners to facilitate the numbering of pages and number the signature from 1 to 32 consecutively. This dummy represents the 1-32 SW and should be compared to the appropriate form chart.

Folding Techniques. A word of caution is necessary when the folding of bookwork signatures is discussed. The preceding folding example and the example to follow utilize the *right angle folding technique*—that is, the sheet is turned 90° with each fold. It is possible to use other folding combinations, such as *parallel folds* or mixtures of *right angle and parallel folds*, when folding book signatures. Many printing and publishing operations standardize signature production with respect to size of book page and press sheet and type of folding and stitching, but there are no specific standards that apply for all companies. Thus, it is vital that the estimator plan the size, folding, and stitching for each book or booklet on an individual job basis using the appropriate company standards of production. In doing so, a dummy and form chart should be prepared.

One common technique is for the estimator to request a folded dummy from the bindery department, thus insuring imposition accuracy. Also, with some types of folding procedures, the form chart technique as described will not work precisely as indicated. The reader is referred to Chapter 13, which covers various bookbinding and finishing aspects, including illustrated diagrams for various folds common to bookwork impositions.

Work-and-Turn Imposition Techniques

Work-and-turn impositions, abbreviated W&T, are another popular imposition technique for books and booklets. *Work-and-turn impositions* have all pages of the signature to be printed on one form so that the signature is printed on one side of the press sheet instead of having half of the pages on one side and half on the other. Using work-and-turn thus reduces the number of pages that can be printed on a press sheet by one-half if the sheet size is the same as with a sheetwise imposition. However, work-and-turn requires only one form per signature instead of the two forms necessitated with sheetwise impositions. With work-and-turn, presswork involves the printing of one-half of the required number of copies initially and then backing up these printed sheets with the same form, running the sheets through the press a second time. Following this procedure, the work-and-turn press sheets, containing two identical signatures, are transported to the bindery area where they are cut in half before folding, thereby separating the two matching signatures. If production procedure permits, slitting of the work-and-turn signatures in half on the press or during folding is sometimes completed.

The production of a 4 page booklet by work-and-turn methods would be completed as indicated in Figure 6.26, which is accompanied by a form chart. Note that the form chart has no vertical center line, meaning that both outside and inside pages are imposed on the same side of the press sheet. Compare this imposition to a sheetwise imposition, where outside pages are in one form and inside pages are in a second form that backs up the first. Example 6.16 illustrates the work-and-turn imposition technique.

Example 6.16

Let us again assume that we want to produce a 32 page booklet as in Example 6.14. Recall that untrimmed page size is 6 1/4 × 9 1/2 inches, 10,000 copies are needed, the presses available are a Harris 19 × 25 and a Miehle 25 × 38, all typematter will be printed in black ink, and no duplicate pages will be produced or used. All impositions to be utilized for this problem will be work-and-turn. Describe the production procedures to be used.

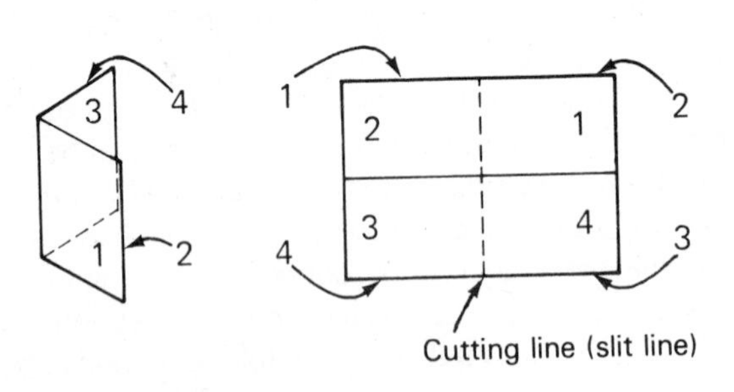

Form Chart

Outside	Inside
1	2
4	3

Figure 6.26

Work-and-turn imposition diagram and form chart for a 4 page booklet

Solution

Step 1 Begin by dividing the untrimmed page size of 6 1/4 × 9 1/2 into the 19 × 25 pss. This calculation indicates that it is possible to impose 8 pp/side.

Step 2 Since the work-and-turn press sheet is cut in half after printing, the imposition requires that each press sheet produce 2 sig that are exactly alike. Thus, the 19 × 25 pss will have 8 pp/fm: 4 outside pages and 4 inside pages imposed on the same side of the sheet (Figure 6.27). The press sheet then produces 2 (8 pp) sig with each 19 × 25 pss (one on each side).

Step 3 Dividing 8 pp/sig into the number of pages in the booklet total, it is shown that 4-8 W&T (4 sig of 8 pp each) must be produced with the 19 × 25 pss:

32 pp/bkt ÷ 8 pp/sig = 4 sig/bkt = 4-8 W&T

Step 4 Divide the 6 1/4 × 9 1/2 untrimmed page size into the 25 × 38 pss. This calculation shows that it is possible to impose 16 pp/side—double the number of pages possible with the 19 × 25 pss.

Step 5 The 25 × 38 pss will have 16 pp/fm: 8 outside pages and 8 inside pages imposed on the same side of the sheet on opposite sides of the cutting line on the press sheet (Figure 6.28). Thus, each press sheet will produce 2 (16 pp) sig exactly alike.

Step 6 Dividing 16 pp/sig into the total of 32 pp/bkt indicates that 2-16 W&T (2 sig of 16 pp each) will be necessary to complete this order using the 25 × 38 pss:

32 pp/bkt ÷ 16 pp/sig = 2 sig/bkt = 2-16 W&T

Step 7 Figure 6.29 shows the completed form charts for the 4-8 W&T (19 × 25 pss), both for side stitching and saddle stitching. Compare these charts to the form charts for the sheetwise example (see Figures 6.22 and 6.23). Note that 4 fm are needed, each containing 8 pp.

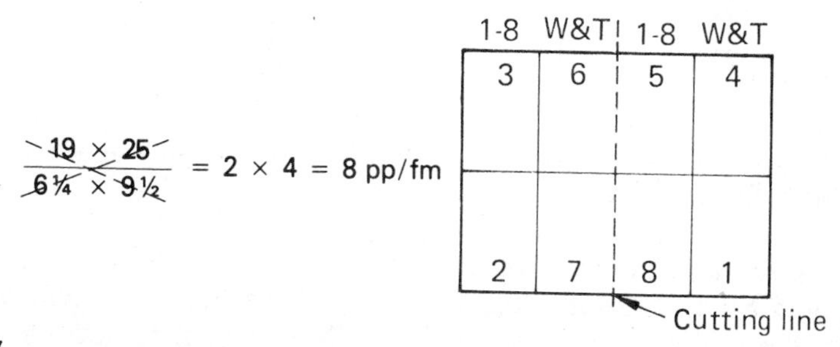

Figure 6.27

Work-and-turn imposition chart for Step 2, Example 6.16

$$\frac{25 \times 38}{6\tfrac{1}{4} \times 9\tfrac{1}{2}} = 4 \times 4 = 16 \text{ pp/fm}$$

1-16	W&T	1-16	W&T
6	3	4	5
11	14	13	12
10	15	16	9
7	2	1	8

Cutting line

Figure 6.28

Work-and-turn imposition chart for Step 5, Example 6.16

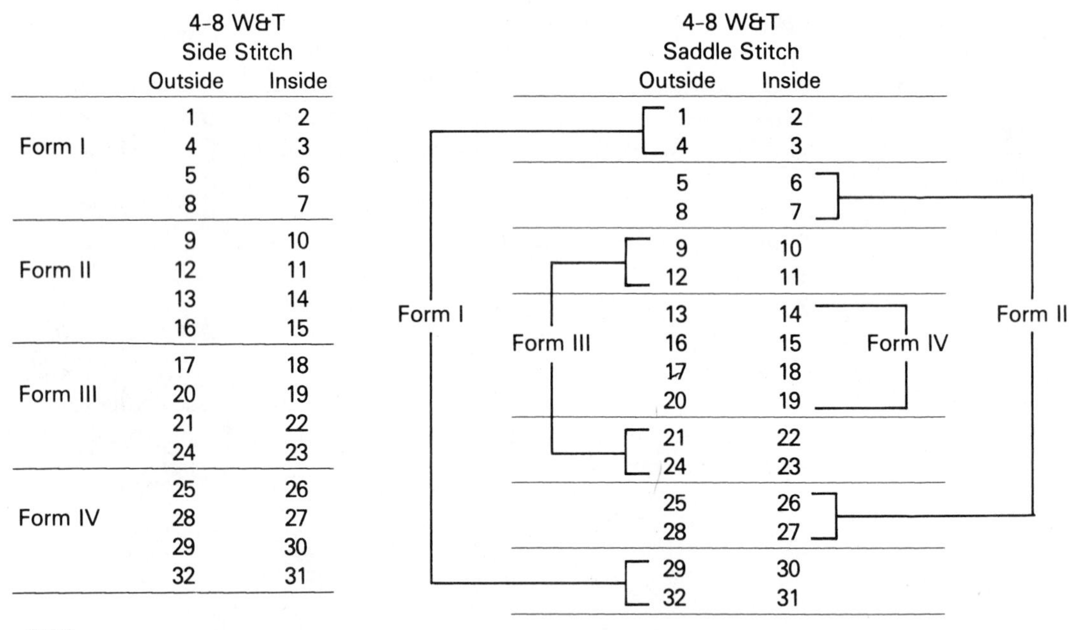

Figure 6.29

Form charts for the 4-8 W&T (Example 6.16)

Step 8 Figure 6.30 shows the completed form charts for both side stitching and saddle stitching for the 2-16 W&T (25 × 38 pss). The solutions presented cover all possibilities and no other solutions exist. Note that 2 fm are needed, each containing 16 pp. Compare these form charts to the sheetwise solutions presented in Example 6.15 (see Figure 6.24).

	2-16 W&T Side Stitch Outside	Inside
Form I	1	2
	4	3
	5	6
	8	7
	9	10
	12	11
	13	14
	16	15
Form II	17	18
	20	19
	21	22
	24	23
	25	26
	28	27
	29	30
	32	31

	2-16 W&T Saddle Stitch Outside	Inside	
Form I	1	2	
Form I	4	3	
Form I	5	6	
Form I	8	7	
	9	10	Form II
	12	11	Form II
	13	14	Form II
	16	15	Form II
	17	18	Form II
	20	19	Form II
	21	22	Form II
	24	23	Form II
Form I	25	26	
Form I	28	27	
Form I	29	30	
Form I	32	31	

Figure 6.30

Form charts for the 2-16 W&T (Example 6.16)

Step 9 The total amount of stock needed (without spoilage) is determined by multiplying the number of signatures required in that press sheet size by 1/2 (a constant representing one press sheet containing two signatures) and then by the number of copies required:

For 4-8 W&T (19 × 25 pss):
4 sig/bkt × 1/2 pss/sig × 10,000 bkt = 20,000 pss
For 2-16 W&T (25 × 38 pss):
2 sig/bkt × 1/2 pss/sig × 10,000 bkt = 10,000 pss

Step 10 The total number of impressions is determined by multiplying the number of forms required by the number of booklets needed:

For 4-8 W&T:
4 fm × 10,000 bkt = 40,000 imp
For 2-16 W&T:
2 fm × 10,000 bkt = 20,000 imp

Note that the total number of forms and impressions is the same as with sheetwise impositions when the press sheets are the same size. Thus, even though different imposition techniques are used, press time and stock used are exactly the same.

The initial folding sequence of the dummy is the same for the work-and-turn as for the sheetwise (see Figure 6.25). Since each work-and-turn press sheet is printed with the same form on both sides, each press sheet contains two of each page of the signature. Numbering of the work-and-turn signature is completed as follows: Number the first sheet of the dummy with page 1, followed with a page 2 on the reverse side; progress to the next sheet of the dummy and repeat the page 1-page 2 sequence. Now go to the third sheet of the dummy and number it page 3, with the back side page 4; then do the same for the next sheet of the dummy. This sequence—that is, 1-2, 1-2, 3-4, 3-4, 5-6, 5-6 and so on—should be followed whenever the dummy of a work-and-turn is made. After numbering, the sheet should be unfolded and compared to the form chart. Note that pages 1, 4, 5, and 8 fall on one side of the cutting line, and pages 2, 3, 6, and 7 fall on the opposite side. If the sheet is turned over, the page positions in the form compare exactly to the first side.

Comparison of the results of the 32 page booklet using sheetwise and work-and-turn imposition techniques is found in Figure 6.31. Review of Figure 6.31 shows that when the press size sheet and quantity are the same, the number of press sheets, number of forms, and total impressions required compare exactly, regardless of the type of imposition. It also indicates that the work-and-turn imposition requires extra cutting time prior to folding and requires greater gathering time in the bindery because more signatures are required. Of course, the use of slitting equipment and automated saddle binding and gathering equipment will minimize such problems in the modern printing plant. Comparisons should also be made between the various types of stitching

	2-16 SW	4-8 W&T	1-32 SW	2-16 W&T
Size of press sheet	19 × 25	19 × 25	25 × 38	25 × 38
Number of copies needed	10M	10M	10M	10M
Total pss needed	20M	20M	10M	10M
Number of forms needed	4	4	2	2
Total impressions	40M	40M	20M	20M
Number of signatures to fold	2	4	1	2
Add'l operations	None	pss cut before folding	None	pss cut before folding
Form charts stitching	Compare separately			

Figure 6.31

Comparison of sheetwise and work-and-turn imposition techniques

procedures in order to be aware of the array of solutions available in the production of such a booklet.

Combinations of sheetwise and work-and-turn impositions are used in the production of books that literally contain an even number of pages. As an example, the production of a 24 page booklet might be completed using 1-16 SW and 1-8 W&T that have the same press size sheet or may be produced using 3-8 W&T. The difference is that the former combination produces two signatures in the bindery and the latter solution produces three signatures. It is possible to produce a 56 page book run in 1-32 SW, 1-16 W&T, and 1-8 W&T or even other alternatives. It also is possible to utilize duplicate pages with the use of film contacting techniques, thus providing for consistent maximization of press sheets and lower per-page production cost. Most successful publishing houses utilize other types of imposition techniques also, such as the *work-and-tumble* (abbreviated w&t and similar to W&T, but the sheet is turned sideways for second side and cutting is done lengthwise), which allows them to produce signatures containing 6, 12, 18, 24 pages, and so on, on the press sheet. The form chart technique will work well with these sizes of signatures also; the major contingency is that folding and stitching must be coordinated to yield exact page placement and must be carefully checked before such work-and-tumble impositions are attempted.

It is important to note that the folding sequence when making the dummy should always be consistent. If not, certain pages will lie differently or shift on the press sheet from one position to another. Normally, most printing operations follow a prescribed folding sequence to which all production is geared. Of course, if folding departures are considered, they should be thoroughly investigated prior to use. If web presses are used for bookwork, such equipment has fixed cutoff and folding procedures, as well as standard webbing conditions (defined webbing patterns to produce a defined signature with appropriate page layout). When producing books with this type of equipment, any departures that vary from normal operations must also be thoroughly-considered.

The form chart is a very useful tool when planning color impositions. Normally, once the size of the book is known—that is, the total number of pages to be produced—and prior to stitching completion, pages where color will be used may be circled or otherwise marked on each page of the chart where it is to run. Stitching may then be completed with these color pages included, allowing all color to be printed in the fewest number of forms and thus minimizing production costs.

6.9 Press Spoilage for Sheetfed Production

The addition of spoilage to the number of base press sheets required for a job is necessary if the customer is to receive the number of finished pieces that have been ordered. Spoilage is the additional paper included at the beginning of the press production of a job to compensate for throwaways and unusable sheets produced during presswork, binding, and finishing operations.

The sheetfed lithographic press spoilage schedule which follows (Figure 6.32) contains example data for normal printing, binding, and finishing operations. This information is presented for example only and is not necessarily reflective of any printing company. Thus, it is desirable for such data to be developed and tailored for each printing firm. It is important to note that some printing firms develop separate spoilage data for press and finishing operations, as opposed to building a schedule which combines the two. This would be typical where the production relationship between press and finishing varies from job-to-job, that is when some jobs require significant presswork and little finishing, and other jobs require little presswork and much finishing.

The two most popular techniques for adding press spoilage are the straight percentage method and the sheet plus percentage procedure. With the straight percentage technique, the number of press sheets for the base run is multiplied by the indicated percentage, thereby calculating the number of extra sheets needed to compensate for throwaways during printing and finishing. Using this method requires that spoilage schedules be built with both makeready and running percentages summed as one percentage figure. In some plants, the straight percentage technique is used when press makeready is completed with already-printed press sheets that are used multiple times for press makeready.

The sheet plus percentage method, used in Figure 6.32, requires that the estimator begin by adding for a certain number of preparation/makeready spoilage sheets to be spoiled during that phase of press setup, then add on an additional percentage to cover running spoilage.

It is important to point out that accuracy in spoilage data is vital to reducing manufacturing costs. Progressive printing companies work diligently to insure that estimated spoilage additions reflect, as closely as possible, the actual spoilage required when producing an order. Such companies also strive to reduce spoilage additions to the lowest point, thereby reducing paper cost and making their product cost more competitive. Of course, paper spoilage data must be carefully gathered and compared, using job cost summaries, to allow these companies to maintain such precise control over paper spoilage in their plants.

Lithographic Sheetfed Press Spoilage Data

A significant amount of printing and publishing is produced using sheetfed lithographic presses. Figure 6.32 provides example spoilage figures for one-, two-, four-, five-, and six-color sheetfed presses. The decision as to the appropriate level of quality must be made: "Ordinary" covers normal plant quality, "premium" covers the types of work where some amount of critical adjustment and registration is necessary, and "process" covers signatures and press sheets that contain full color images and dot-for-dot registration.

Press preparation and make ready spoilage figures are given in the chart on the basis of number of press sheets to be added per form for flat sheet work (per form, form pair or form set) and for W&T and SW imposition spoilage additions (both sides). Flat sheet work data represents non-book impositions including advertising work, letterheads, flyers, etc.; work-and-turn and sheetwise figures are used for specific imposition

Press	Quality	Work	Press Preparation/Makeready: First color	Press Preparation/Makeready: Additional color	Running
One-color press	Ordinary	Flat sheet work (per form)	125	75	3.0%
		Work and turn	150	100	5.0
		Sheetwise	175	125	5.0
	Premium	Flat sheet work (per form)	150	100	3.5
		Work and turn	175	125	5.5
		Sheetwise	200	150	5.5
	Process	Flat sheet work (per form)	175	125	4.0
		Work and turn	200	150	6.0
		Sheetwise	225	175	6.0

Press	Quality	Work	Two colors	Running
Two-color press	Ordinary	Flat sheet work (per form pair)	175	4.0%
		Work and turn	250	6.0
		Sheetwise	300	6.0
	Premium	Flat sheet work (per form pair)	200	4.5
		Work and turn	275	6.5
		Sheetwise	325	6.5
	Process	Flat sheet work (per form pair)	225	5.0
		Work and turn	300	7.0
		Sheetwise	350	7.0

Press	Quality	Work	Four colors	Running
Four-color press	Ordinary	Flat sheet work (per form set)	250	4.5%
		Work and turn	325	6.5
		Sheetwise	425	6.5
	Premium	Flat sheet work (per form set)	275	5.0
		Work and turn	350	7.0
		Sheetwise	450	7.0
	Process	Flat sheet work (per form set)	300	5.5
		Work and turn	375	7.5
		Sheetwise	475	7.5

Press	Quality	Work	Five colors	Running
Five-color press	Ordinary	Flat sheet work (per form set)	300	5.0%
		Work and turn	400	7.0
		Sheetwise	500	7.0
	Premium	Flat sheet work (per form set)	325	5.5
		Work and turn	425	7.5
		Sheetwise	525	7.5
	Process	Flat sheet work (per form set)	375	6.0
		Work and turn	475	8.0
		Sheetwise	575	8.0

Press	Quality	Work	Six-colors	Running
Six-color press	Ordinary	Flat sheet work (per form set)	350	5.5%
		Work and turn	450	7.5
		Sheetwise	550	7.5
	Premium	Flat sheet work (per form set)	375	6.0
		Work and turn	475	8.0
		Sheetwise	575	8.0
	Process	Flat sheet work (per form set)	400	6.5
		Work and turn	500	8.5
		Sheetwise	600	8.5

Figure 6.32

Lithographic sheetfed press spoilage schedule. Data provided for example only.

methods to produce book signatures and booklets where forms are matched by the imposition process.

Press running spoilage is calculated as a percentage of the base number of total press sheets required and is not intended to be an accumulating procedure wherein makeready is included in the base figure. The added press spoilage figures will also serve as bindery spoilage to cover the setup and running of folding and stitching equipment. When estimating multicolor presswork, the figures in the chart include the amount of sheets for printing all colors in one press pass, per form group.

Example 6.17

We have a customer who has requested us to print 15, 000 letterheads, finish size 8 1/2 × 11 inches, to run three colors (red, blue, and black) in register on one the sheet. The job will print 4-up on a single-color offset press, premium quality. We will buy 22 × 34 substance 20 bond that costs $78.60 per M sheet. Determine the number of press size sheets needed, the amount and cost of stock, and the total impressions including spoilage.

Solution

Step 1 Begin by determining the size of the press sheet using the finish size and 4-up information (Figure 6.33).

Step 2 Divide the number-up into the total number of copies to be delivered to the customer to find the number of 17 × 22 pss required:

15,000 cp ÷ 4-up = 3750 pss

Step 3 Refer to Figure 6.32, one-color press , premium quality, flat sheet work; locate preparation and makeready figures for "first color," "additional colors," and "running" and apply as indicated. *Note*: The run spoilage is found by multiplying the 3.5% listed in Figure 6.32 times 3750 pss times 3 fm, one for each color. Determine spoilage figures.

(8½ × 11) × 4-up = (8½ × 11) × (2 × 2) = 17 × 22 pss

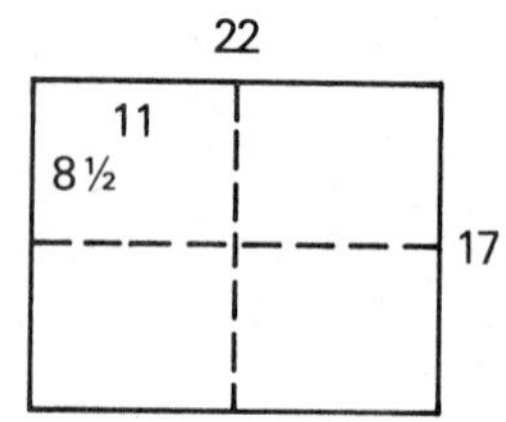

Figure 6.33

Press sheet size for Example 6.17

Total press sheets: 3750 pss
First color preparation and makeready: 150 pss
Second and third colors preparation and makeready (100 × 2): 200 pss
Run spoilage: 396 pss
Total press sheets with spoilage: 4496 pss

Step 4 Divide 17 × 22 pss into the 22 × 34 pars, which yields 2-out (Figure 6.34).

Step 5 Divide the number-out into the total number of press sheets required (found in Step 3) to determine the number of 22 × 34 pars required:

4496 pss ÷ 2-out = 2248 pars

Step 6 Determine a new M weight for the parent size 22 × 34 bond:

$$\frac{22\text{ in.} \times 34\text{ in.}}{17\text{ in.} \times 22\text{ in.}} \times 40\text{ lb/M} = 80\text{ lb/M}$$

Step 7 Determine the poundage:

2.248 M pars × 80 lb/M = 179.8 lb

Step 8 Determine the stock cost:

2.248 M pars × $78.60/M = $176.69

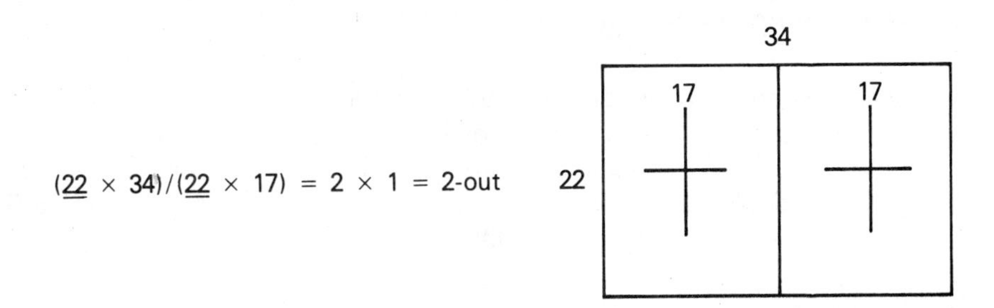

Figure 6.34

Number-out diagram for Example 6.17

Step 9 There are three different forms printing in register on one side of the press size sheet. Maximum total impressions to be used for scheduling the job during production will be 13,488, determined as follows:

3 fm × 4496 pss = 13,488 max imp

Example 6.18

In the Example 6.16 imposition problem, a customer wanted 10,000 copies of a 32 page booklet, 6 1/4 × 9 1/2 untrimmed page size. After completing and reviewing all solutions, we have decided to produce the booklets using 2-16 W&T and to saddle stitch them in the bindery. The one-color, 25 × 38 inch Miehle will be used with makeready and running spoilage at the ordinary level. We will buy 25 × 38 substance 50 book paper costing $63.50 per M sheet. All pages will print in black ink with no additional colors. Determine the amount and cost of stock and total impressions, including spoilage. A form chart showing stitching should also be provided.

Solution

Step 1 Review previous calculations with respect to size of forms, number of signatures per book, and base number of press size sheets needed:

$$\frac{25 \times 38}{6\frac{1}{4} \times 9\frac{1}{2}} = 4 \times 4 = 16 \text{ pp/fm}$$

$$\frac{32 \text{ pp/bkt}}{16 \text{ pp/sig}} = 2 \text{ sig/bkt}$$

$$2 \text{ sig/bkt} \times \tfrac{1}{2} \text{ pss/sig} \times 10{,}000 \text{ bkt} = 10{,}000 \text{ pss}$$

Step 2 Prepare a form chart (Figure 6.35) showing the imposition and saddle stitching of 2-16 W&T. Each form contains 16 pp.

Step 3 Press spoilage is based on the number of initial sheets needed—in this case, 10,000 sht of 25 × 38. Using the one-color press, ordinary level, 300 sht are added for press preparation since there are two plates and hence two preparations. An additional 5% (500 sht) is then added to cover all running spoilage. Total press sheets needed will be 10,800:

Total press sheets: 10,000 pss
Preparation and makeready: 300 pss
Run spoilage: 500 pss
Total press sheets with spoilage: 10,800 pss

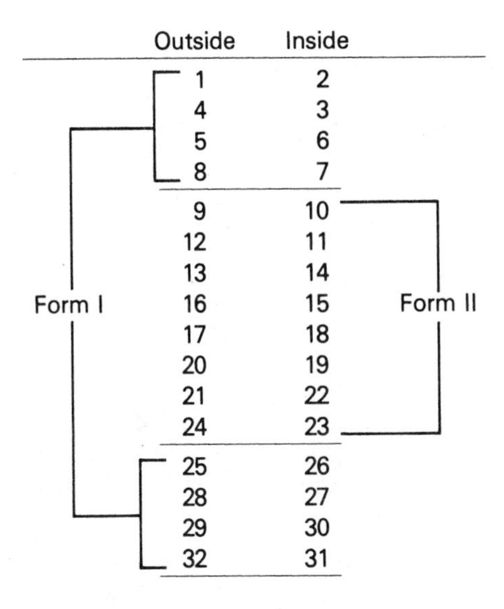

Figure 6.35

Form chart for 2-16 W&T with saddle stitching (Example 6.18)

Step 4 Calculate the poundage:

10.8 M : 100 lb/M = 1080 lb

Step 5 Determine the stock cost:

10.8 M × $63.50/M = $685.80

Step 6 Because each press sheet must be printed on both sides, total impressions may be determined by doubling the number of total press sheets including spoilage. As shown, a total of 21,600 imp will be required to complete the specified order:

10,800 pss × 2 sides/sht = 21,600 max imp

Example 6.19

We have a customer who wants us to print 45,000 four-color inserts on our Heidelberg MOVP four-color (4/C) press with a maximum press size sheet of 19 × 25 1/2 inches. The inserts will print 4/4 (four colors each side) with a finish size 9 1/2 x 12 1/2 inches, no bleeds. Stock in inventory is basic size book, grain short, sub. 60 which costs $ 75.70/M sheets. Determine the number of press size sheets needed without spoilage, the number of press size sheets with spoilage, the number of parent sheets, the cost and poundage of the parent sheets, and the number of impressions including spoilage. Process level, flat sheet

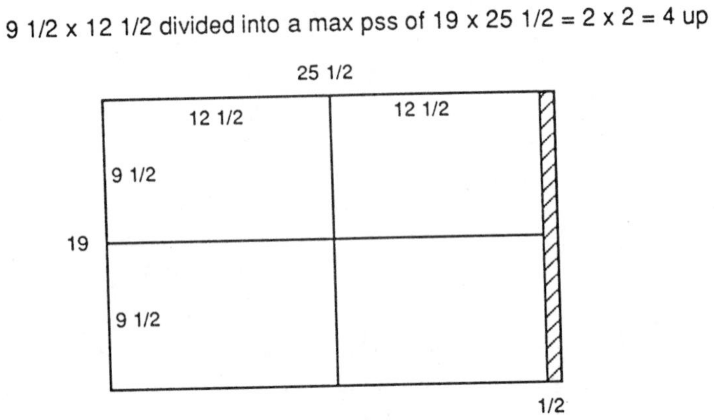

9 1/2 x 12 1/2 divided into a max pss of 19 x 25 1/2 = 2 x 2 = 4 up

25 1/2

12 1/2 12 1/2

9 1/2

19

9 1/2

1/2

Figure 6.36

Press sheet size for Example 6.19

work data since this is not a bookwork job. Use 4/C press spoilage data from Figure 6.32.

Solution

Step 1 Begin by determining the size of the press sheet using the finish size and the 4-up information (Figure 6.36).

Step 2 Divide the number-up into the total number of copies to be delivered to the customer to find the number of 19 × 25 pss required:

45,000 ÷ 4-up = 11,250 pss

Step 3 Refer to Figure 6.32, four-color press, process quality, flat sheet work, locate preparation/makeready figures for "four colors" and "running" and apply as indicated. As with previous examples, the run spoilage is found by multiplying the 5. 5% figure listed in Figure 6. 32 times the 11, 250 pss times two sets of forms. Determine spoilage figures as follows:

Total base number press sheets:	11,250 pss
First 4/C form prep:	300 pss
Second 4/C form prep:	300 pss
Run spoilage, first form set (4/C)	619 pss
Run spoilage, second form set (4/C):	619 pss
Total pss with spoilage:	13,088 pss

Step 4 Divide 19 × 25 pss into the 25 × 38 pars, which yields 2-out (Figure 6.37).

19 × 25 divided into a max pars of 25 x 38 = 2 x 1 = 2 out

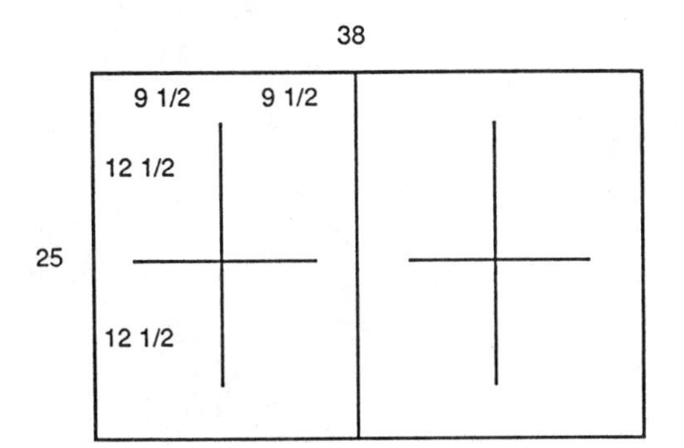

Figure 6.37

Press sheet size for Example 6.19

Step 5 Divide the number-out into the total number of press sheets required (found in Step 3) to determine the number of 25 × 38 par required:

13,088 ÷ 2-out = 6544 pars

Step 6 Determine the M weight for the 25 × 38 pars:

25 in. × 38 in. × 120 lb/M = 120 lb/M

Step 7 Determine the poundage:

6.544 pars × 120 lb/M = 785.3 lb

Step 8 Determine the stock cost:

6. 544 par × $75.70/M = $49,538

Step 9 There are two different sets of forms to print on the press sheet, one four-color set on one side and a different four-color set on the other side. Maximum total impressions to be used for scheduling the job during production will be 26,176, determined as follows:

2 passes (2 sets of 4/c forms each) × 13,088 pss = 26,176 max imp

Estimating Ink Quantity and Cost

7.1 Introduction

Printing ink is a vital ingredient in the manufacturing process, but it generally does not represent a tremendous expense compared to other job costs: usually between 2% to 5% of the average cost of a printing order. Ink estimating, however, still must be included during cost estimating for two reasons. First, because profits are sometimes low, identification of all definable ingredients for any printing job must be made, and it is reasonably easy to estimate ink costs as a portion of the total job cost. Second, the ink estimating procedure allows for the determination of approximate quantities of ink. This determination is extremely important when inks must be specially mixed or require special ordering or formulation. Should too little be ordered under such circumstances, many production problems could result.

7.2 Ink Schedules

The letterpress and lithographic ink mileage schedules, Figures 7.1 and 7.2, respectively, are provided to estimate ink quantities. Chart data represent the approximate number of thousand square inches that 1 pound of ink will cover. Thus, printing on a number one enamel using the lithographic process, 1 pound of black ink will have a mileage of approximately 425,000 square inches (see Figure 7.2).

Chart data are cross-referenced by the color of ink and the type of stock that is to be printed. Also included are tint bases, mixing white, varnish, and process color coverage information.

There are no allowances built into either schedule for leftover ink or ink loss from production and press washup. Such losses usually amount to from 2% to 10% of the total volume of ink for the lob. For example, 10% might be allowed on a 5 pound can and 5% for a 100 pound can. In addition, all chart data presented are approximations owing to the many variables occurring during production. Differences exist in what seem to be the same type and kind of paper in stock, ink absorbency factors, climatic conditions, and various production differences in the pressroom.

7.3 Variables When Estimating Ink Consumption

Type of Process

When the two ink schedules are compared, it should be noted that *the mileage factor for letterpress inks is significantly less per pound than for the lithographic inks.* The reason is that the letterpress process prints with a thicker film of ink and offset lithography prints with a far thinner ink film; thus, the mileage of lithographic inks is greater. Comparing regular letterpress black with regular lithographic black, each

Type of Stock	Color of Ink: Regular Black	Opaque Yellow	Opaque Red	Blue Lake	Green Lake	Purple Lake	Opaque Orange
No. 1 enamel	270	180	190	250	250	240	180
No. 2 enamel	230	150	160	235	220	220	160
S&SC (low finish)	180	120	130	165	125	180	125
Dull coated	150	100	110	150	100	150	110
M.F. book	135	90	90	120	135	130	95
E.F. book	120	80	85	110	120	125	80
Antique	100	65	70	70	90	90	70

Type of Stock	Color of Ink: Chrome Green	Mixing White	Tint Base	Process Colors: Cyan	Magenta	Yellow
No. 1 enamel	150	150	300	240	250	215
No. 2 enamel	130	140	220	220	220	185
S&SC (low finish)	100	105	180	200	120	145
Dull coated	90	90	160	140	140	120
M.F. book	80	75	135	120	130	110
E.F. book	80	70	120	110	120	100
Antique	70	60	95	90	90	85

Note: Numbers given indicate thousands square inches of coverage per pound of ink.

Figure 7.1

Letterpress ink mileage schedule, by color of ink (Reproduced with permission of Gans Ink and Supply Company, Los Angeles, CA)

printed on a number one enamel sheet, 1 pound of the letterpress black will cover approximately 270,000 square inches, while the same quantity of lithographic black ink will cover 425,000 square inches.

Type of Paper Stock

The smoother the surface of the paper stock or substrate, the less the amount of ink that will be required. Thus, a top-grade enamel, which is exceptionally smooth, will require less ink during printing than a rougher-coated stock. For example, the relation-

Type of Stock	Regular Black	Rubber-Base Black	Purple	Trans-parent Blue	Trans-parent Green	Trans-parent Yellow	Chrome Yellow	Persian Orange	Trans-parent Red
	Color of Ink								
No. 1 enamel	425	445	360	355	360	355	285	345	350
No. 2 enamel	400	435	355	345	355	355	270	335	345
Litho coated	380	430	350	340	350	355	260	325	345
Dull coated	375	425	320	335	340	250	250	310	340
M.F. book	400	435	350	340	350	340	250	325	340
Newsprint	290	350	250	230	250	235	165	240	240
Antique	275	335	235	220	235	220	150	225	225

Type of Stock	Opaque Red	Brown	Overprint Varnish	Opaque White	Tint Base	Process Colors Cyan	Process Colors Magenta	Process Colors Yellow
	Color of Ink							
No. 1 enamel	350	345	450	200	400	355	350	355
No. 2 enamel	345	340	435	185	390	350	347	355
Litho coated	340	335	425	175	380	340	345	355
Dull coated	325	325	415	165	375	335	340	340
M.F. book	340	335	425	175	385	340	340	340
Newsprint	190	240	—	150	265	235	240	235
Antique	175	225	—	135	250	220	225	220

Note: Numbers given indicate thousands square inches of coverage per pound of ink.

Figure 7.2

Lithographic ink mileage schedule, by color of ink (Reproduced with permission of Gans Ink and Supply Company, Los Angeles, CA)

ship between number one enamel and antique (uncalendered) paper shows that the ink mileage for antique (rough) stocks is much less than for the smoother enamel paper. It should be noted that dull-coated stocks and machine finished (MF) papers have approximately the same mileage requirements.

The absorbency of the substrate and the types of ingredients used for ink manufacture both relate to ink mileage. For this reason, ink components and paper stock absorbency factors should be professionally matched to ensure maximum mileage and optimum coverage. Orders that require significant quantities of inks should have the ink product formulated for the type of paper or substrate. This technique ensures the best combination of ink and paper, expediting production during press and bindery operations.

Color and Type of Ink

Two general rules apply to color and type of ink. First, *the darker the ink color, the greater the mileage per pound*. Thus, black inks require less ink per pound than colors, and darker colors provide greater mileage pound-for-pound than do lighter inks. The second general rule is that *the more opaque the ink, the lower the mileage, and the more transparent the ink, the greater the mileage per pound*. Referring to the lithographic schedule (Figure 7.2) as an example, 1 pound of transparent yellow printed on number one enamel has a mileage value of 355,000 square inches. On the same type of paper, chrome yellow, which is an opaque color, has a mileage factor of 285,000 square inches per pound. Process colors, because they are transparent, provide reasonably high mileage values per pound when compared to most other colors.

As noted previously, the specific ingredients in printing ink relate to mileage during press work. Chemical configurations between offset and letterpress inks differ, as do the ingredients themselves. Most lithographic inks contain various amounts of pigment, vehicle, drier, resin, and wax. Since the combination of these ingredients varies with respect to the desired type of ink product, it is difficult to provide specifics regarding mileage. Most printing companies purchase a certain type of ink product consistently and use it for most stocks they print. With a limited amount of study, it is possible for estimators to determine specific mileage based on the standard products in their plants. The data ensure estimating accuracy with these identified paper and ink combinations.

Coverage of the Form to Be Printed

Ink estimating requires knowledge of the size in square inches of the form to be printed and the approximate coverage of ink on the form. A *form*, also known as a *printer*, is one image or a collection of images to be printed in the same color. *Ink coverage*, expressed as a percentage, relates to the amount of area on the page over which the ink will be printed. Thus, the smaller the percentage, the lighter the coverage, while the larger the percentage, the heavier the coverage. Such coverage data relate directly to the quantity of ink and should be determined as accurately as possible during estimating. The type of composition matter, halftones, screen tints, type reverses, solids, and process color separations can all be evaluated at various percentages. Figure 7.3 provides some general coverage values.

To simplify the coverage estimating procedure, a page is measured in square inches; then, a percentage of coverage is determined for the aggregate of all images on that page; finally, the page size is multiplied by the percentage of coverage to determine the square inch area of the paper that will be covered by ink. The following formula is used to calculate the page size in square inches:

Type of Form	Percent of Coverage
Very light composition, no halftones	15
Normal composition, no halftones	20
Normal composition, bold paragraph heads	25
Medium composition, no halftones	35
Heavy composition, no halftones	50
Halftones	50
Screen tints*	—
Solids	100
Reverses	60–90
Process color separations:	
Magenta and cyan	40
Yellow	50
Black	20

*Approximate screen percentage

Figure 7.3

General guidelines to estimate ink coverage

page size in square inches, converted from picas = (pica page width ÷ 6 picas per inch) × (pica page depth ÷ 6 picas per inch)

For example, we have a one page flyer with a type page measuring 36 picas by 60 picas which totals 60 square inches for the entire type page:

$$(36 \text{ pi} \div 6 \text{ pi/in.}) \times (60 \text{ pi} \div 6 \text{ pi/in.}) = 6 \text{ in.} \times 10 \text{ in.} = 60 \text{ sq in.}$$

Coverage for this page, which includes both type matter and halftones together, is estimated at 35%. Therefore, 60 square inches times 35% equals 21 square inches, Which represents the amount of area covered with ink on the page when printed.

Determination of coverage is, at best, an estimate in itself. A significant majority of printed material contains combinations of type and tonal matter, as in the previous example, and the usual procedure is to arrive at an aggregate coverage percentage. Whenever possible, visual samples should be used, and every attempt should be made to be precise.

Number of Copies and Impressions

Naturally, the quantity of printed material relates to the required amount of ink for any given job. Whenever possible, ink consumption should include all press sheet setup and running spoilage additions since ink is required for these operations. Of course, the volume of ink increases or decreases proportionately to the quantity of printing ordered.

Percentage of Ink Waste

Waste of ink is one aspect of production that is difficult to pin down precisely. Normally, such waste comes from ink that has been poorly stored, residual ink difficult to remove from the container, color contamination of ink in the press fountain from a previous color (this ink must be discarded), and normal loss of ink from press washups. For the most part, such ink loss does not amount to more than 10% of the total required amount for any given job. Of course, many factors enter into how much ink is actually wasted during production, and only individual plant study can accurately determine this figure. Nonetheless, this slight additional amount should be planned and estimated in order to be certain that quantities of ink are above, and not below, minimum needed amounts to complete the printing order.

7.4 Cost of Ink

It is almost uniformly agreed that the cost of ink is secondary to its performance during production. Every production manager, at one time or another, has been faced with poor ink performance during printing or as a related follow-up problem when drying, chalking, or some other technical difficulty arises. When poor performance occurs, and it is far more frequent when lower-priced inks are consistently purchased, there are few practical or workable solutions other than to change inks. Thus, for a perhaps modest initial cost savings, the total job cost is increased tremendously because of the additional materials, time, and rescheduling necessary to redo all or part of the ruined order.

Actual ink costs, which are normally on a per-pound basis, vary from manufacturer to manufacturer. Purchasing inks in large quantities reduces the cost per pound (but, of course, ties up valuable capital that might be used in other areas more effectively). When ink is purchased in small cans, waste is usually high, thus inflating the poundage cost actually paid. Most plants prefer to buy inks in 5 pound cans or larger containers to reduce such waste and thereby cut production costs.

7.5 Ink Quantity and Cost Estimation

The preceding discussion covered all of the variables incurred when estimating ink. The following formula incorporates all such variables into a workable model used to estimate ink quantities:

number of pounds of ink = (total form area × percentage of coverage × total number of copies × anticipated percentage of ink waste) ÷ ink mileage factor

where total form area is measured in square inches, number of copies includes press spoilage, and the ink mileage factor is determined from the appropriate mileage schedule (Figure 7.1 or 7.2). Once the actual poundage of ink has been determined, that value is multiplied by the cost per pound to find the job cost for ink.

Each time any variable in the formula changes, a new ink formula must be built to reflect that change. In a typical commercial printing job, the two most common changes are in color of ink, which is reflected as a change in ink mileage, and in the coverage percentage. For example, a two-color job (red and black) prints with one half the black pages at 40% coverage, the other half of the black pages at 25%, and all red pages at 35%. Three formulas thus are needed to estimate ink for this example: one for black at 40%, one for black at 25%, and one for red at 35%. Total form area (the sum of the type page areas), the total number of copies including spoilage, and the anticipated ink waste and loss remain unchanged for all colors and all coverage percentages in this example.

The following list provides pointers for using the poundage formula:

1. The *total form area* is determined by multiplying the area (square inch value) of one type page times the number of pages to be printed in that color at the same percentage coverage. For example, a 16 page booklet with a type page measuring 40 square inches prints black ink, 30% coverage for 8 pages, and 50% coverage for the other 8 pages. Total form area would then be 8 pages × 40 square inches per page = 320 square inches to be used at each coverage percentage (30% and 50%) when the formulas are built.
2. The *coverage percentage* figure is expressed in the formula in decimal form—for example, 45% becomes 0.45 and 25% becomes 0.25.
3. The *number of copies including press spoilage* can be calculated in one step by adding 1.00 to the percentage spoilage figure, which is then multiplied by the base number of copies. For example, for a 10,000 run with 5% spoilage, the percentage figure in the formula would be (1.00 + 0.05) × 10,000 copies = 10,500 total press sheets. The same result would be achieved if 10,000 were multiplied by 5%, which equals 500 press sheets, and then the base 10,000 were added back, totaling 10,500 sheets to be printed including spoilage.
4. The *anticipated ink waste and loss percentage* is handled the same way. Use 1.00 plus the percentage of ink waste and loss. For example, if the anticipated ink waste and loss is 2%, then the formula figure would be 1.00 + 0.02 = 1.02.
5. The *ink mileage factor*, taken from Figure 7.1 or 7.2 depending on the printing process used, varies by ink color and substrate used. Schedule basis is the number of thousand square inches per pound of ink. Thus, if the schedule figure is 385, that figure translates to 385,000 square inches per pound of ink for that particular color on the substrate referenced in the schedule (in this case, from Figure 7.2, tint base on MF book).
6. Any overage of ink poundage at the thousandths is rounded up to the next

hundredth of a pound; for any fraction of a cent, rounding up is to the next penny.

Example 7.1

We have an order for 15,000 handbills to be printed 1-up, one side on our Heidelberg platen (letterpress). Finish size for the job is 8 1/2 × 11 inches, with a type page measuring 42 × 54 picas (pi). Coverage is estimated at 45% and includes a mixture of screen tints, bold display type, and a typeset block of 10 point Times Roman. We will add 5% for press spoilage and 6% for ink waste and loss during production. Ink will be regular black and stock will be MF book. Ink costs $3.55 per pound. Determine the quantity and cost of the ink.

Solution

Step 1 Find the form size in square inches:

$$\frac{42 \text{ pi}}{6 \text{ pi/in.}} \times \frac{54 \text{ pi}}{6 \text{ pi/in.}} = 7 \text{ in.} \times 9 \text{ in.} = 63 \text{ sq in./fm}$$

Step 2 Find the ink poundage using the given formula:

$$\frac{(63 \text{ sq in.} \times 1 \text{ p}) \times 0.45 \times (15{,}000 \times 1.05) \times 1.06}{135{,}000 \text{ sq in./lb}} =$$

$$\frac{63 \text{ sq in.} \times 0.45 \times 15{,}750 \times 1.06}{135{,}000 \text{ sq in./lb}} = \frac{473{,}304 \text{ sq in.}}{135{,}000 \text{ sq in./lb}} = 3.51 \text{ lb}$$

Step 3 Find the cost of the ink:

3.51 lb × $3.55/lb = $12.46

Example 7.2

We are to print 25,000 copies of a 16 page booklet on a Miehle offset. Type page size is 30 × 48 picas. All 16 pages will print with black ink at 30% coverage, 8 pages will print with a second color opaque red at 25% coverage, and 6 more pages will run in Persian orange at 20% coverage. Add 5% per color for press spoilage (a total of 15%) and an additional 5% to compensate for ink waste and loss during production. Job will be printed on dull-coated stock. Ink costs $3.80 per pound for the black, $4.70 per pound for the red, and $5.20 per pound for the orange. Determine cost and quantity of each color of ink necessary to complete this order.

Solution

Step 1 Find the page size in square inches:

$$\frac{30 \text{ pi}}{6 \text{ pi/in.}} \times \frac{48 \text{ pi}}{6 \text{ pi/in.}} = 5 \text{ in.} \times 8 \text{ in.} = 40 \text{ sq in./p}$$

Step 2 Find the number of copies including spoilage:

$$25{,}000 \text{ cp} \times 1.15 = 28{,}750 \text{ cp}$$

Step 3 Find the poundage for the black ink:

$$\frac{(40 \text{ sq in.} \times 16 \text{ pp}) \times 0.30 \times 28{,}750 \times 1.05}{375{,}000 \text{ sq in./lb}} =$$

$$\frac{5{,}796{,}000 \text{ sq in.}}{375{,}000 \text{ sq in./lb}} = 15.46 \text{ lb}$$

Step 4 Find the cost of the black ink:

$$15.46 \text{ lb} \times \$3.80/\text{lb} = \$58.75$$

Step 5 Find the poundage for the red ink:

$$\frac{(40 \text{ sq in.} \times 8 \text{ pp}) \times 0.25 \times 28{,}750 \times 1.05}{325{,}000 \text{ sq in./lb}} =$$

$$\frac{2{,}415{,}000 \text{ sq in.}}{325{,}000 \text{ sq in./lb}} = 7.44 \text{ lb}$$

Step 6 Find the cost of the red ink:

$$7.44 \text{ lb} \times \$4.70/\text{lb} = \$34.97$$

Step 7 Find the poundage for the orange ink:

$$\frac{(40 \text{ sq in.} \times 6 \text{ pp}) \times 0.20 \times 28{,}750 \times 1.05}{310{,}000 \text{ sq in./lb}} =$$

$$\frac{1{,}449{,}000 \text{ sq in.}}{310{,}000 \text{ sq in./lb}} = 4.68 \text{ lb}$$

Step 8 Find the cost of the orange ink:

$$4.68 \text{ lb} \times \$5.20/\text{lb} = \$24.34$$

Step 9 Summarize the poundage findings:

Black ink: 15.46 lb
Red ink: 7.44 lb
Orange ink: 4.68 lb

Step 10 Summarize the cost findings:

Black ink: $58.75
Red ink: $34.97
Orange ink: $24.34
Total cost: $118.06

Example 7.3

We have an order for a four-color poster measuring 18 × 24 inches finish size, with a form size of 16 × 22 inches (1 inch white margin on all sides). Based on an evaluation of the color photo to be printed, coverages are anticipated to be as follows: magenta, 45%; cyan, 65%; yellow, 40%; and black, 15%. Add 3% per form for paper spoilage and 4% for ink waste and loss during production. Stock will be litho-coated offset book, substance 70. The customer wants 7500 copies delivered. Ink costs are: magenta and cyan, $4.50 per pound; yellow, $4.15 per pound: and black, $3.80 per pound. Determine ink poundage and cost for the job.

Solution

Step 1 Determine the form size of the poster in square inches:

$$16 \text{ in.} \times 22 \text{ in.} = 352 \text{ sq in.}$$

Step 2 Determine the number of copies including spoilage. (Each color of ink requires 1 fm; therefore, 3%/fm × 4 fm = 12%.):

$$7500 \text{ cp} \times 1.12 = 8400 \text{ cp}$$

Step 3 Find the poundage for magenta ink:

$$\frac{352 \text{ sq in.} \times 0.45 \times 8400 \times 1.04}{345{,}000 \text{ sq in./lb}} = \frac{1{,}383{,}782 \text{ sq in.}}{345{,}000 \text{ sq in./lb}} = 4.02 \text{ lb}$$

Step 4 Determine the cost for magenta ink:

$$4.02 \text{ lb} \times \$4.50/\text{lb} = \$18.09$$

Step 5 Find the poundage for cyan ink:

$$\frac{352 \text{ sq in.} \times 0.65 \times 8400 \times 1.04}{340{,}000 \text{ sq in./lb}} = \frac{1{,}998{,}797 \text{ sq in.}}{340{,}000 \text{ sq in./lb}} = 5.88 \text{ lb}$$

Step 6 Find the cost for cyan ink:

5.88 lb × \$4.50/lb = \$26.46

Step 7 Determine the poundage for yellow ink:

$$\frac{352 \text{ sq in.} \times 0.40 \times 8400 \times 1.04}{355{,}000 \text{ sq in./lb}} = \frac{1{,}230{,}029 \text{ sq in.}}{355{,}000 \text{ sq in./lb}} = 3.47 \text{ lb}$$

Step 8 Find the cost for yellow ink:

3.47 lb × \$4.15/lb = \$14.41

Step 9 Determine the poundage for black ink:

$$\frac{352 \text{ sq in.} \times 0.15 \times 8400 \times 1.04}{380{,}000 \text{ sq in./lb}} = \frac{461{,}261 \text{ sq in.}}{380{,}000 \text{ sq in./lb}} = 1.22 \text{ lb}$$

Step 10 Find the cost for black ink:

1.22 lb × \$3.80/lb = \$4.64

Step 11 Summarize the poundage findings:

Magenta ink: 4.02 lb
Cyan ink: 5.88 lb
Yellow ink: 3.47 lb
Black ink: 1.22 lb

Step 12 Summarize the cost findings:

Magenta ink: \$18.09
Cyan ink: \$26.46
Yellow ink: \$14.41
Black ink: \$4.64
Total cost: \$63.60

Estimating Design, Artwork and Copy Preparation

8.1 Introduction

Certainly two of the most important job specifications to review during cost estimating are artwork and copy preparation. The preparation of artwork and other copy for printing, also termed copy preparation, can be accomplished in a number of ways. From the view of the printing estimator, perhaps the most important consideration is what the customer will supply, particularly with the advance of stand-alone design workstations and printing devices for design and copy preparation areas.

Sometimes the customer will specify that he or she will furnish the artwork camera-ready, meaning no work is necessary prior to the photographic step. In other instances, the customer may allow the printing plant to purchase artwork from outside sources or prepare it in the plant using resident staff artists and/or its own composition equipment. It must be clear from the beginning as to the type of art furnished and who will be responsible for obtaining it.

Design and preparation of artwork for printing is a vital planning element. If the work is poorly designed for a press sheet, there will undoubtedly be production problems during printing. If bleeds, folds, and trims are not considered, if colors are prepared improperly or out-of-register, if the design is creative but simply cannot be executed economically—all of these oversights mean that the job was poorly planned and coordinated during the initial copy preparation stages. The better the beginning planning and execution of artwork, the more expeditiously it will move through production.

Good graphic artists, also called commercial artists or graphic designers, understand the importance of planning during the art generation stage. The more experienced graphic designers are with respect to printing production—that is, the more they understand and can deal with the technology of printing—the better their designs will reproduce. It is difficult, if not impossible, for anyone to design graphic materials without a suitable working knowledge of printing concepts.

Today art is prepared using conventional methods, electronic procedures, or a mix of the two. Since the introduction of the Apple Macintosh in January 1984, the development of computer-based prepress and graphic design systems have had an increasing impact on the preparation of both camera-ready artwork and art. There are no longer any fixed rules as to how printers prepare artwork, design images, complete paste-ups, or image film. Some smaller printers and quick printers design, prepare art and typeset on stand-alone systems using desktop publishing software with laser output on plain paper, while others continue to prepare mechanicals manually. Larger and high volume printers have a tendency to utilize more expensive, fully-integrated electronic "front end" systems, yet some continue to use conventional design and other prepress processes.

The evolution of service bureaus and prepress trade shops for handling this complex array of electronic prepress functions provides a bridge to connect conventional technology and the electronic processes. Many of the service bureau functions formerly provided trade typesetting products to the industry.

There is no dispute that inexpensive stand-alone workstations such as the Macintosh, coupled with a LaserWriter printer and working with desktop publishing (DTP) software, has become an increasingly popular form of art production in printing.

Estimating desktop publishing—which incorporates design, copy preparation production, and typesetting elements—will be covered in Chapter 9 although it could just as easily have been included in this chapter because of its crossover applications.

Experts are unanimous in their opinion that the use of electronic prepress systems—particularly computer-aided design (CAD) front end systems—will move from infancy in the mid-1980s to adulthood in the 1990s. Costs for these systems are expected to be reduced and integration with existing electronic systems will be greatly enhanced. Methods to estimate these evolving systems will depend on the ability to define the operations each system can perform and then develop production standards representative of the operations performed by the system. Since there is no one, single recognized CAD front end system leader setting an industry standard, no attempt has been made to establish estimating procedures with this edition.

8.2 Conventional versus Electronic Prepress Systems

The flowchart in Figure 8.1, indicates the four principal sources from which much manually-generated or conventional artwork currently evolves, while Figure 8.2 is a flowchart comparing conventional art production versus electronic prepress production using a microcomputer and appropriate word processing, illustration, painting, drawing, and desktop publishing software. Figure 8.2 shows the typical application of a Macintosh or similar microcomputer system and represents fairly accurately the current mixture between conventional and electronic art production. Desktop publishing, a popular and flexible image assembling process, is covered in more detail in Chapter 9.

Conventional Artwork Production

As noted in Figure 8.1, the conventional process begins the artwork stage with previously printed images, typeset material, photographs, and original artwork. Combinations of these conventional sources in any printing job are common. Electronic copy production, such as desktop publishing, may also be used as an intermediate part of the art/design process and will be covered in detail later in this chapter.

Prior to detailing each type of artwork, which includes some basic art and copy preparation definitions and terms, it should be noted that customer input with respect to the art produced is very important. Since the customer is the purchaser and usually the ultimate consumer of the printing, design form and function are vital considerations. Sometimes customers will attempt to save money by designing their own artwork, either manually or electronically, using desktop publishing. This decision is sometimes a foolish one because many printing clients are not certain of the type of product they want, or the functions the product must fulfill. A competent printing sales representative or customer service representative, working closely with an experienced graphic artist, can aid the customer greatly in such matters. Even though the cost may

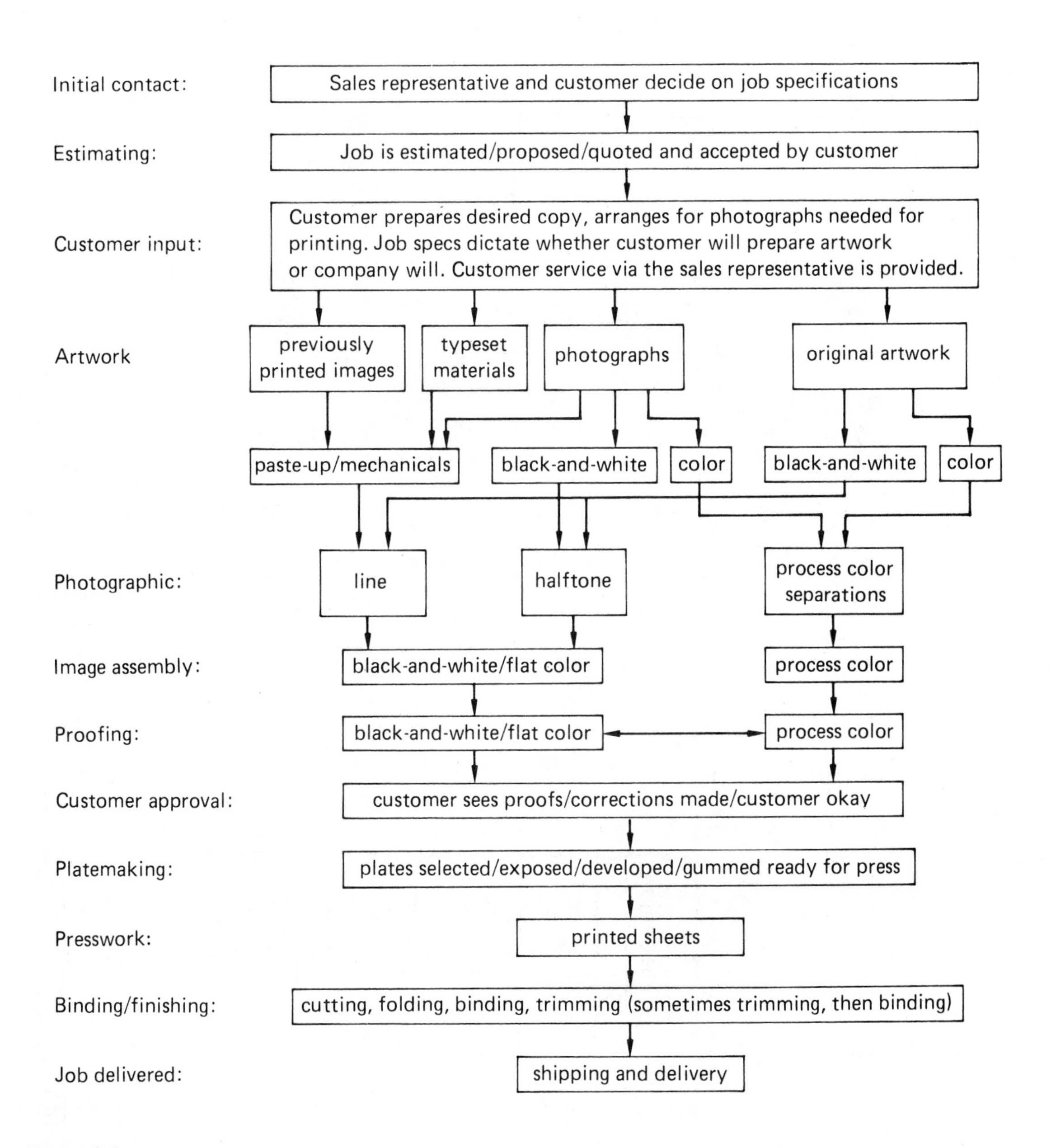

Figure 8.1

Flowchart for producing a printing order using conventional prepatory systems and the lithographic process

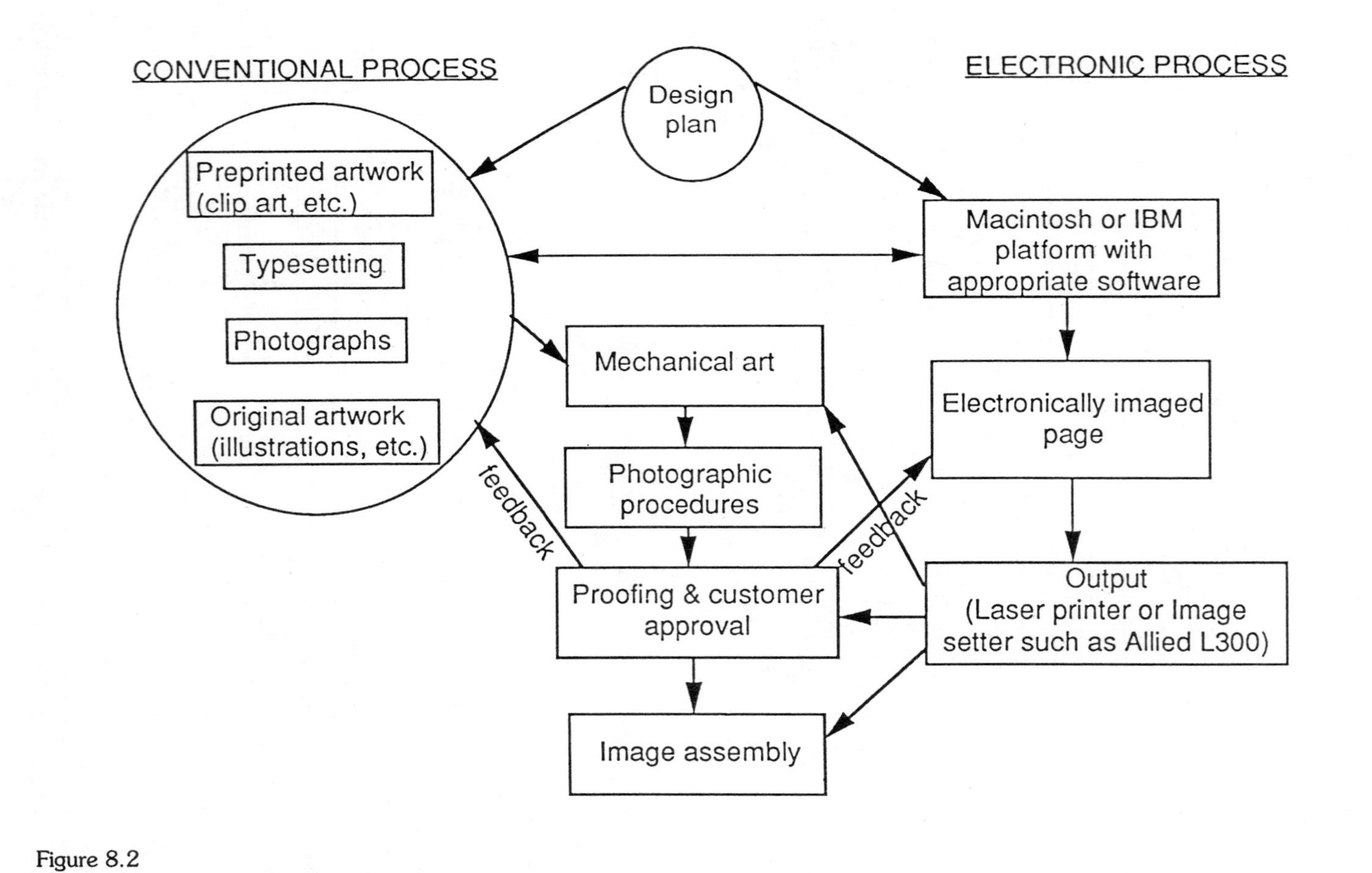

Figure 8.2

Flowchart comparing conventional art preparation with electronic art preparation common to many printing firms

be higher for such services, the print buyer will be happier with the completed design and the resultant printed product. In the intensely competitive commercial printing industry—small printer or large—satisfied customers are the single biggest source of repeat orders.

Artwork from Previously Printed Images

Artwork from previously printed images is a popular source of a large volume of artwork. Many printing companies subscribe to clip-out services that send them prepared art—that is, camera ready art—through the mail. The printer then cuts out the desired design and perhaps uses it in conjunction with typematter for the customer's order. Most preprinted artwork is intended to be used as *line art*, graphic images in line or solid block form that require no special photographic procedures. Line work is distinguished from dot pattern materials called *tints*, which break solid copy into an even pattern of small dots, effectively causing a solid to appear at a lesser printed value. Halftones, used for the reproduction of photographs in printed matter, are also a dot-patterned product and will be explained in detail later in this chapter.

In many cases, preprinted art includes borders and fancy typographic symbols; many drawings in line and cartoon images are also supplied. The preprinted material is reproduced on white uncoated book stock with a dense black image so the copy will photograph well during later process camera exposure, when line negatives are produced. Preprinted images that contain colors other than black or red cannot be considered as artwork because line films may not pick them up properly.

Note in the flowchart that previously printed images are usually prepared in paste-up or mechanical form. Essentially, a *paste-up* is the attachment of all artwork elements to white illustration board or stiff paper using rubber cement or an adhesive wax. The finished camera-ready copy is called a *mechanical* or *keyline*. Since the preprinted images are on white stock themselves, the two backgrounds will blend together; during photographic work later on, they will form a black area in the negative. All black (or red) preprinted images and typematter will be transparent in the negative. When the negative is then exposed to the plate, the light will go through the transparent image areas, producing a positive image on the plate. In addition to the graphic images and type, mechanicals also contain dimensional marks for the finish size of the sheet, as well as folding and trimming marks. Usually paste-ups are prepared in the same size as the intended final image but sometimes are slightly larger.

The paste-up is the most popular form of prepared artwork in the printing and publishing industry. Many plants employ *paste-up artists* who spend each day assembling images as camera-ready copy. When a large volume of mechanicals are produced daily, preprinted paste-up sheets are used. These sheets normally are printed on thick white cover stock with a very light blue (called nonrepro or nonphoto blue) grid system to facilitate image alignment. The light blue background is not picked up by the camera. When preparing artwork on unprinted illustration board, a nonrepro blue pencil or pen is usually used for the same purpose.

Mechanicals may be prepared in one color, as described, or in multicolor form. Normally, the first color pasted to the illustration board is considered the *key* color or image, and all additional colors are positioned—that is, registered—to it. An acetate overlay system is used for each additional color. Acetate sheets are trimmed to the overall dimensions of the white baseboard and then taped securely along one side so they will flip back and forth. Each additional color is then mounted to its own acetate overlay sheet in exact alignment with the key image. Finally, *registration marks* (which are typically two lines crossed perpendicularly and centered in a circle) are attached to each overlay and the baseboard in exact alignment, ensuring that each image may be repositioned in register later in the production sequence. This procedure is called *breaking for color*, and each line negative made during the photographic sequence will contain only the image for its respective color. If a mechanical was prepared with two overlays, the job would have three colors, and three line negatives would be imaged during the photographic procedure. Each negative represents a color.

Mechanicals are usually covered with a thin sheet of tissue when completed. This tissue sheet serves as protection for the images and as an information cover sheet for the artwork, which can be seen through it. It is most important that the artist communicate any relevant production information via this tissue overlay so that production problems will be minimized later during manufacturing. A top cover sheet of kraft paper or a thicker paper stock is then attached over the tissue sheet to protect the mechanical during production.

A word of caution is called for concerning the use of preprinted art work. Almost all printed goods are protected under the law by copyright. When art is specifically purchased for use by a printing company, the copyright for such art does not apply; however, it is illegal to pick up artwork from any copyrighted publication or other source unless consent has been given for such use. Permission should be in writing and obtained prior to the use of the image. Reproduction of previously copyrighted graphic designs and artwork without consent not only is illegal but also is considered unprofessional throughout the printing and publishing industry.

Typeset Material

Typeset material, also called *composition* or *typematter*, may be generated in many different ways. *Strike-on composition*, produced using a typewriter, is both convenient and popular. Various sophisticated typewriter models *justify* type in block form (the type is evenly aligned on both sides), but these typewriters are considered somewhat slow when compared to other typesetting methods.

The preparation of composition material using a *hot metal* Linotype or Intertype machine was the backbone of typesetting operations in past years. Here, characters are cast on a line of lead in relief form; the type is then proofed on nonglare paper with dense black ink. This method produces what is called a *reproduction proof*, or *repro*, and is usually used as copy for the mechanical. Other types of relief characters are also utilized to set type. Foundry type and the Ludlow system are both used in limited form today because of their relative slowness. The Monotype process, where characters are individually cast and assembled in one production process, is popular outside the United States and is another relief character typesetting technique.

The use of *rub-on* lettering or *transfer type* provides for the production of a limited amount of typematter. These materials can be easily transferred to artist illustration board or to the acetate overlay.

Phototypesetting and electronic typesetting represent the two methods by which the great majority of type is set today. Phototypesetting equipment generally consists of a light source, a movable film segment containing a complete assortment of characters of one kind (font) of type, and photographic paper or film upon which the image is flashed. As the film segment is positioned, character by character, in front of the photographic film or paper, the light source exposes that character to the photographic material. Phototypesetting is extremely fast and produces high-quality typeset characters in a variety of sizes. Resin-coated papers and stable-based films are both used to hold the typeset image and are machine processed with minimum production problem.

Electronic typesetting involves generating type characters from digitized, or numeric, electronic components with the final type images produced either photographically (on photographic film) or xerographically (on paper) using laser imaging equipment. The main cartridge in a laser printer or other electrostatic copying unit is known as the "engine," and has been greatly improved during the 1980s so that laser-printed images are crisp, sharp, and quickly produced.

The growth of electronic typesetting in the 1980s, particularly with stand-alone word processors and laser-driven printers, has followed closely behind the 1970s technological advances of phototypesetting units. Today many companies have both direct-input phototypesetting and stand-alone electronic typesetting units in the same typesetting area of their companies.

Photographs

Both black-and-white and color photographs are popular types of copy. When reproduced in high contrast form (black image on a white background with no intermediate gray areas), they can become a part of the paste-up. However, all photographs are initially *continuous-tone*—that is, the image on the photographic paper or film gradually changes from black to gray to white as it goes from dark to light. Because the lithographic process can reproduce only defined, high contrast images, the continuous-tone pattern—the value of shading—must be broken up into a dot pattern, or *screen* pattern. These dots are very small but are uniformly black so that the camera will pick them up. It is the number and size of the dots in a particular area of the image that determine how dark or light that area will appear to be.

A continuous-tone black-and-white photograph, when broken up into this dot pattern, is called a *halftone*; the screen used to produce the dot pattern is called a *halftone screen*. There are two major types of halftone screens used in the printing industry: the glassline screen and the contact screen. Each is used in a specific manner to divide the continuous-tone black-and-white photograph into minute dots. The *glassline screen* consists of inked fine lines at right angles into a glass body, forming a crosshatched pattern. The *contact screen* contains very fine dots through which the continuous-tone image is exposed. Various exposure sequences are used during the photographic steps of making a halftone, with the resultant image a finely dotted piece of film. Dot size is determined by the amount of exposure given to the original picture. Most printing plants use densitometers and precise electronic controls to insure production of consistent halftones and dot values within the halftones.

Color photographs, as indicated in the flowchart, do not follow the same sequence of production as their black-and-white counterparts. The reason is that color photographic images not only must be screened (they are also continuous-tone) but also must be separated into the three major colors required to reproduce them: magenta, cyan, and yellow. This process is called *color separation*. Typically, a black *printer* (or separation) will be added as a fourth color to provide for greater detail in the printed image. The technique of color separation requires considerable photographic skill, numerous controls, and special filters and films. While precise figures are unavailable, an estimated 95% or more of all color separation output in the U.S. is now produced by color scanning procedures. This subject will be covered in detail in Chapter 10.

Original Artwork

Original artwork is any graphic image designed for first-time reproduction. Original art may take many forms and may be reproduced by line, half-tone, or process color photographic procedures, For example, an original black-and-white line drawing is artwork in line form, an original mechanical for a new company letterhead with a special calligraphic type is also a line image. An artist's charcoal drawing is an example of back-and-white original artwork that must be reproduced as a halftone because it contains many shades of value and must be broken up into fine dots for accurate reproduction. Original artwork that would require color separation includes paintings, watercolors, or any other image containing a full range of color values.

Original artwork, as with electronic design and CEPS systems, can be extremely difficult to estimate precisely. For original art—such as a painting, charcoal drawing, serigraph, rendering, or any unique graphic image developed either manually or with computer aid—many factors enter into such costs: the artist's reputation, the complexity of the art, the amount of research the artist must complete, the quantity of calligraphic work needed, and so on. Later in this chapter, chart data are provided to cost out such creative art services and professional photographic services. Of course, since many artists who produce original art also produce mechanicals, cut peelcoat materials, and are sometimes involved in the design of forms design and paste-up, chart data are presented for estimating these areas also. (Peelcoats, discussed later in this chapter, are colored gelatin films carefully cut with a knife and removed from around a desired image so that the remaining gelatin provides the image.)

Electronic Art Production and Design Systems

Essentially an electronic design system (or "platform") consists of integrated hardware and software upon which images may be generated, manipulated, or modified by the designer-operator. There are two major functions of these systems: (1) to provide the designer with a flexible tool by which graphic images can be generated, massaged, and mixed with typeset material, and (2) to provide complete electronic files of graphic images as a part of the prepress manufacturing process, saving substantial time and cost to the printer in later prepress operations.

The following discussion summarizes front-end prepress systems including electronic design systems (also refer to Figure 8.3). Many of these design platforms were only recently developed, and are powerful tools for both graphic design as well as integrated prepress systems. The differences between indicated groupings is not intended to be precise. Also, it must be noted that significant changes in design workstations will continue to occur over the next ten years, linked closely with electronic advances in other prepress areas.

Group I. These are the most capital-expensive, heavy-duty production units which include Scitex, Crosfield, Hell, and DS America. Each system has the capacity to integrate all prepress functions—design, retouching, color modification, image assembly, and certain photographic processes—in one operating unit. Commonly known as Color Electronic Prepress Systems (CEPS), these systems are usually considered too expensive and powerful to be simply design units, not only because of their high cost but also because of their range of capabilities. For example, CEPS are used for various types of image manipulation, retouching, montaging, vignetting, as well as assembling color images, all completed electronically. Final output is typically to film using electrophotographic or traditional film technologies, with direct digital color proofing also available.

Each CEPS vendor is investigating or has developed lower cost design workstations which interface directly to their brand of CEPS system, thus reducing the cost and enabling designers greater access to this technology.

Group II. The intention with Group II systems is to emulate Group I without the high cost or other major disadvantages. Graphex Inc., Mosiac II (from AgfaCompugraphic), Graphics 2000 (from Linotype) and Quantel's Video Paintbox each have developed systems to fit this category. For example, the Graphics 2000 workstation from Linotype permits handling and type output of both text and graphics to the Linotronic 300 and 500 laser imagesetters. Group II systems are able to process significant quantities of images and yet may develop, over time, into Group I units as capability and system architecture are modified.

Group III. Design units at this level are aimed at completing graphic functions as a worktool, and less oriented to totally integrated production of graphic images in digital form. Crosfield Design Systems, a separate division of Crosfield, offers the

Synervision I Composition Center, developed to directly interface Crosfield's Studio 875 system. In addition, Crosfield also has entered into the stand-alone design system in conjunction with Lightspeed's Color Layout System which runs on a Macintosh II platform; the Lightspeed Design System 20, another stand-alone, runs on Sun Microsystems 3/160M hardware. Both of these have been developed specifically to meet the electronic image generation needs of the graphic designer.

Other less expensive systems—each providing different capabilities—include the Unda Color Design and Production system to run on IBM-AT or Sun workstations, Networked Picture Systems "Package Express" and "Image Express" which interface with Scitex CEPS systems, CyberPublishing with a series of units for design interface into CEPS systems, Hell's ChromaCom 1000 page assembly unit and DuPont Design Technologies Vaster Design and Imaging Systems.

Group IV. This segment—the highest of two levels of desktop publishing (DTP)—focuses on using the Macintosh II microcomputer platform, and to a lesser degree IBM-PC/XT and AT units and OS/2. A number of "off-the-shelf" design packages are currently available, with the major advantage as fairly low cost entry into the sophisticated electronic design world. Various software packages address the different designer's needs at this level. Furthermore, because the Macintosh II provides the first color unit of the Macintosh platforms, because Postscript code is the major output driver, because of the flexibility of the Macintosh system overall, and because Mac IIs are fairly low in cost compared to higher systems in the higher groups, there certainly will be significant enhancements and changes over the next few years. It is Group IV that appears to provide the most flexible platform base for future advances into CEPS for the smaller printer.

Group V. This is the lowest group and represents the most basic and inexpensive stand-alone computer platform, which includes the Macintosh 512, Mac Plus and Mac SE. When connected with laser imaging units, these machines can be used as lower-quality typesetters, with the plain-paper output used as paste-up copy for mechanical art. Figure 8.2 represents a typical Group V system, comparing the design and imaging of conventional art production versus electronic completion of this task.

The Macintosh Plus and SE stand-alone platforms—which are almost exclusively used for black-and-white imaging—also represent the most-common desktop publishing unit. Aldus' PageMaker, LetraSet's ReadySetGo, Quark XPress and others allow designers to uniquely create and integrate graphics and type at a very inexpensive level. Estimating desktop publishing is covered in Chapter 9.

For all groups, the relative fineness of image, both on the monitors used for design and customer viewing, and the output to film or paper (as a proof) represents major technical problems. At this writing the 300 dots per inch (dpi) is the accepted norm, yet most experts agree that development and use of 1,200 dpi systems will be an essential step toward increased use of design and CEPS systems in the 1990s.

GROUP I

General application and use

CEPS including image design, retouching, color modification, image enhancement, full page image assembly and high quality photographic film output

Major manufacturers of popular systems

Crosfield, Diadem Carat, DS America, Eikonix, Graphex, Hell, Scitex,

Output mode

Soft proofs, direct digital proofs and fully imaged film ready for proofing or stripping

Minimum cost (est)

$200,000+

GROUP II

General application and use

CEPS systems with less sophistication. Most systems are high quality design platforms which can be used for design, retouching, color modification, and text manipulation

Major manufacturers of popular systems

Compugraphic's Mosaic II, Quantel's Video Paintbox, Linotype's Graphics 2000

Output mode

Soft proofs, direct digital proofs, conventional proofs from film

Minimum cost (est)

$80,000

GROUP III

General application and use

Most systems are high quality design platforms which can be used for design, retouching, color modification, and text manipulation; less integrated than Level I or II.

Major manufacturers of popular systems

Crosfield's Design Systems (integrates with Crosfield's Studio 875 CEPS), Lightspeed Design 20 (on a Sun Microsystem's platform), Lightspeed Color Layout System (on a Macintosh II platform), Unda Color Design and Production System, Networked Picture Systems "Package Express and Image Express (interface with Scitex CEPS), Hell's ChromaCom 1000, DuPont's Design Technologies Vaster Design and Imaging Systems

Output mode

Varies with the system

Minimum cost (est)

$30,000

GROUP IV

General application and use

Less expensive design platforms which can be used for various design and prepress production applications. Software is "off the shelf" on Macintosh II hardware.

Major manufacturers of popular system software

Various color painting and graphics software and sophisticated integrated desktop publishing programs

Output mode

Laser-driven printers, Allied's L300 or L500 and similar systems

Minimum cost (est)

$10,000 (includes black & white laser printer)

GROUP V

General application and use

Very expensive text input and design platforms which can be used to produce desktop publishing materials in black and white form. Hardware includes Macintosh 512e, SE and Mac II.

Major manufacturers of popular systems software

Various black and white desktop publishing software and other graphics and painting programs

Output mode

Laser-driven printers, Allied's L300 or L500 and similar systems

Minimum cost (est)

$5,000 (includes black & white laser printer)

Figure 8.3

General Procedure for Estimating Electronic Prepress Systems

Estimating the production time and cost of these design and electronic systems is very difficult, since each provides different capabilities and requires varying degrees of operating skill. The following represents the typical procedure to establish such a system:

1. Once the electronic or design system is purchased and on-line, the estimator must learn the operating components of the system, either through vendor training or working directly on the unit with a skilled company employee.
2. After the system is understood, the estimator should set up operational categories which define the work performed while the unit is in production, and establish production standards in the form of hourly times, representative of this productive work.
3. A budgeted hour rate is now established using the procedure indicated in Chapter 3 of this text.
4. During the estimating process, each job is broken into blocks or components of work which fit into the operational values recognized in the system at Step 2. This represents a detailed breakdown of job duties to be performed by the operator and will likely require intimate knowledge of the system on the part of the estimator. In many plants having such systems currently on-line, the estimator takes training in the system offered either by the vendor or a company employee who is competent in system use.
5. The estimating process is completed by multiplying the operational times by the BHR, then adding material costs (including appropriate markups) which apply.

8.3 Sources of Artwork

Artwork and prepared copy can come into the printing plant from a number of different sources and arrive in different formats. At this writing the most common for many printers remains conventional mechanical art, used as reflection copy with a process camera, and color separations which are reproduced by photographic scanner. Once rendered in photographic film form, these images are then manually assembled during the stripping process. Estimating mechanical art production, scanner production and manual (or "table") stripping will be detailed later in this book in the appropriate chapters.

For those printers who have purchased electronic platforms (Group I) of different types and varieties, their prepress production flow has been significantly changed, from artwork through proofing to lithographic plate for the press. While some of these firms still do varying amounts of conventional prepress as described in the preceding paragraph, they also produce and transmit artwork electronically via satellite, overnight disk transfer, and telecommunication exchange. Furthermore, because the initial cost of such electronic equipment in the printing plant is very high and skilled operators are expensive to train and have available, printers with such systems must maintain high productivity in order to keep costs low. Also, Group I CEPS systems provide extremely flexible imaging procedures, rendering images that might normally be extremely expensive or impossible to produce manually as art. This represents a major advantage and a primary customer service requirement, since complex color images can be easily modified overnight or within hours of receipt of an electronic file. There is almost uniform agreement among experts that the growth of CEPS in both systems and volume of work, which included front-end design platforms that feed CEPS equipment, will occupy significant market share in the 1990s for the printing and publishing industry.

Customer Prepared Artwork

Sometimes when discussing the job specifications with the sales representative or customer service representative, customers indicate they will supply the printer with all camera-ready art or will provide photographic film ready for stripping. When the customer provides "film," the printing production sequence bypasses the art and photographic sequences completely; the film arrives, is stripped, plated and the job is printed, finished, and delivered. For customers who elect to provide their own artwork—which is becoming increasingly common today—the art may be generated either conventionally or electronically.

Electronic artwork would be produced directly on a design workstation as previously described. Group V systems using desktop publishing software are popular. To do this, two essential ingredients must be input into the electronic design workstation: typeset material and graphic images, normally in the form of color separations. Typeset material can be formatted for electronic use on most computers or word processors using the most reputable word processing programs. Graphic images—a color photograph, original color painting, charcoal drawing, or any other color or black-and-white image—requires digital input through an electronic scanner. There are two essential categories of electronic scanners: microcomputer scanners and color scanners. Microcomputer scanners are generally low in cost (up to $3,000) and have been developed to electronically scan images for desktop publishing use.

Digital color scanners, typically with some form of laser photographic imaging system, represent a major expense (up to $200,000). These complex electronic units render color images in digital form to be stored electronically (for later use with a CEPS Group I platform) or imaged directly to photographic film as a production unit in prepress manufacture of a job. To complete this operation, the color photograph is mounted on the scanner drum, which rotates while a light optically reads the image, linked to digital software of the scanner. The color photograph is broken into magenta, cyan, yellow, and black separations (or "printers") and reduced to fully electronic form; it will be either stored in massive computer memory or imaged directly to photographic film using a laser light beam, or both. Highly skilled personnel and sophisticated electronic support are required. Estimating digital color scanning will be detailed in Chapter 10.

Once input, both digitized typeset material and digitized graphic images can then be transferred or input into an electronic design workstation whereby the artist can manipulate or massage both type and graphic elements to produce final art in digital form. When this is done, the final electronic material is transmitted back to a CEPS unit which produces the final art directly to photographic film, for image assembly, platemaking, and subsequent printing. Proofing of the job, so the customer can visually approve it, can be done either on the color monitor (called a soft proof), direct to a digital proofing unit, or manually after the film is generated from the CEPS system.

For customers who are trying to save money on the job and who may not be familiar with the printing processes involved, customer-supplied art may be the most undesirable. (This type of situation applies essentially to commercial printing; in book manufacturing, the customer usually supplies camera-ready art unless the printer also supplies the typesetting.) It is not unusual for some printers to recommend that the customer furnish all camera-ready artwork and suggest competent artists who have the knowledge and ability to produce this material. Experienced print buyers with full knowledge of the printing process sometimes prefer to supply the printer with "film only," thus bypassing the art production stage since imaged film incorporates camera-ready art.

In-Plant Artist Prepared Artwork

Depending on the particular type of customer served by the printer, many printing establishments provide design and art services using their own skilled personnel. Having a resident artist is particularly desirable when the majority of products the company prints require special graphic design skills. Such product lines might include posters, packaging items, and annual reports. Most printing plants that produce a significant amount of business forms have in-plant artists for form design and paste-up.

An in-plant artist offers a number of advantages which include control of both the quality of design and mechanical art thereby streamlining production of the job after art, control of the timeliness of production of art, so rush or special jobs can be more easily accommodated. This improves company profitability since art is produced in-house and thus is a chargeable profit center and increased security to the printer since, under the Printing Trade Customs, the artwork is owned by the printer when produced as part of the customer's order.

In-plant artists produce mechanical art using either manual paste-up techniques or computer design platforms such as the Macintosh linked with a laser-driven or digital typesetting output unit. While manual paste-up of mechanicals is still popular in the industry, the rapid development and use of computer-based design and production units have begun to have significant impact on the art/design end of the printing industry. Estimating mechanical/paste-up production is discussed later in this chapter while estimating desktop publishing is covered in Chapter 9 which follows.

Artwork Purchased from a Trade Composition House

Many trade typesetting or trade composition houses provide graphic design and mechanical art production, generally in conjunction with the collateral typesetting services they offer. Such art production is actually a reasonable extension of typesetting since typematter can be a major portion of most prepared artwork, especially with products such as books, booklets, pamphlets, and advertising pieces. Also, with the blending of typesetting and graphic image production into inexpensive stand-alone microcomputers in the form of desktop publishing, trade composition houses provide an excellent interface between artist, typesetter, and author.

The purchase of artwork from a trade composition house normally insures that the prepared material will be of a quality suitable for graphic reproduction. The reason is that trade composition plants work almost exclusively with printing companies and build their reputation on the reproducibility of their type and graphic products. Also, many printing plants work closely with only a few trade typesetting houses and get to know thoroughly the skill level of personnel and quality of services they provide.

Artwork Purchased from a Freelance Graphic Artist

Freelance graphic artists and designers are an excellent source of artwork for the printer provided that the artist is aware of printing processes and technical printing requirements in depth. Many freelance artists charge a flat rate for their services, which may include creative idea development, research, thumbnail sketches (small rough sketches that convey ideas for the final artwork), and perhaps the completion of mechanicals. Fees vary according to the reputation of the artist, special skills such as calligraphic work, the type of work under consideration, the size of the job, the general availability of the graphic artist at the time, and the relative amount of electronic typesetting and design capacity the artist can provide.

Freelance graphic artists may be approached by individuals securing artwork on their own for later printing or by a printing company that has an identified customer who needs camera-ready artwork developed. It is possible that a freelance artist will be working for more than one customer at the same time, so deadlines can be a problem. For this reason, it is advisable that each printing company have a list of freelance graphic designers to contact. Being able to call on a variety of artists provides a versatility of design ideas for the company since each graphic artist produces a different range of visual images.

Further, as the skill and reputation of the freelance artist grows, his or her services may be harder to arrange and become higher in price. It is therefore necessary that an estimate for the freelancer's services be secured on a job-by-job basis or that the printing estimator be provided with some method to determine the cost of artwork. Of course, the availability of the artist also must be determined. Charts for estimating the relative costs of certain creative art services provided by the freelance artist are presented later in this chapter.

Artwork Purchased from a Studio

The art studio is essentially a group of artists working together and sharing common facilities. Sometimes each member of a studio represents a different discipline and is called upon when his or her particular services are needed. The art studio is like any other business: It is concerned about sales, production, billing, and internal business matters.

The person in charge of studio projects is called an art director and is versed in both the business and practical aspects of art and design. Most art directors determine those projects that will be accepted, assign persons to work on certain jobs, and generally serve as coordinator for such efforts. The art director may work closely with the printing company in determining specifics as to the design for a customer's order.

Art studios, because they have more personnel than the independent freelance artist, have the capacity to undertake reasonably large projects. They have the ability to produce many different types of artwork mechanicals and develop creative graphic ideas. They also usually have the skills to coordinate all elements of the art, including typesetting, copyfitting, copy markup and composition, color separations, proofreading, and photographic studio services. The art studio can offer the printing plant an entire array of art and copy preparation services so that all art elements of the customer's order can be produced for efficient printing production.

Artwork Supplied from an Ad Agency or Public Relations Firm

Some printers provide printing services and goods to advertising and public relations agencies, with the agency sometimes supplying camera-ready artwork or film. While agency operations vary, both agency-supplied artwork or imaged film can come from a variety of sources available to the agency, including the agency's own internal art staff, freelancers hired by the agency, or typesetting or trade houses engaged by the agency. Larger agencies will typically make extensive use of electronic imaging systems and may, in fact, have even purchased electronic prepress platforms that integrate into the printer's CEPS system.

The relationship between advertising agency or public relations firm and the printer runs the gauntlet from cooperative and open to quite strained. Typically, agencies recognize their position of control over the printer, particularly since the volume of printing they purchase may be in the millions of dollars, and may be very profitable to the printing company. Because some agencies tend to control tightly, printers are sometimes asked to compromise production schedules to accommodate what they may consider an unreasonable job change or modification. Also, experienced printing personnel may sometimes be asked to cut production corners so that budgets can be met, while the agency is unwilling to bend on their own design concepts, such as the size or style of product.

One of the areas where the conflict between agency and printer becomes most noticeable is in the quality and timeliness of artwork for the job. Sometimes agencies provide substandard artwork to the printer, yet expect a high quality final product. Sometimes the art is the wrong size, perhaps poorly prepared for photographic production, or comes in piecemeal instead of as a unified whole. In sum, while it is fair to say that advertising agencies and public relations firms do represent customer demands that strain the printer's goodwill, it is also understood that the volume of work these agencies purchase is so high as to encourage significant printer interest.

8.4 Estimation of Artwork, Photographic Services, and Copy Preparation

Creative Art Services

Figure 8.4 may be used to estimate the creative art services for a resident in-plant artist. It will also serve as a guideline to compare with freelance or studio prices. Since so many different elements go into the creative processes by which artwork is produced, the prices are guidelines only. They must be adjusted for the relative skill and reputation of the artist or studio, the geographic region, and other less tangible factors.

Example 8.1

A customer desires a new logo for company letterheads, business cards, and other business communications materials. An estimate must be made for the preparation of this artwork, to be completed through the in-plant artist. The customer wants a three-color image and has indicated a preference for calligraphic display if possible. The artist will prepare all art from roughs through comprehensives and will produce the calligraphic logo in camera-ready form.

Approximately 1 hour will be spent talking with the customer, reviewing various magazines and completing other research. Artist consultation will require 1 hour. Preparation of thumbnails and roughs will take 2 hours. Comprehensives for customer approval will take 2 hours to prepare. Finally, calligraphic art for this job can be prepared in 1 hour.

Solution Refer to the schedule in Figure 8.4 for hourly rates.

Step 1 Calculate the research and idea development fee:

1 hr × $40.00/hr = $40.00

Service	Hourly Rate
Research and idea development	$40
Artist consultation	65
Artist preparation of thumbnails and/or rough drawings	55
Artist preparation of comprehensive drawings as finished art	60
Calligraphic work	70

Note: Dollar figures in this schedule presented for example only. Deduct 20% for full day services by the artist or studio.

Figure 8.4

Creative in-plant artwork services schedule

Step 2 Determine the consultation fee (includes travel time but no expenses):

1 hr × $65.00/hr = $65.00

Step 3 Determine the cost of preparing thumbnails and roughs:

2 hr × $55.00/hr = $110.00

Step 4 Calculate the cost of preparing comprehensives:

2 hr × $60.00/hr = $120.00

Step 5 Calculate the cost to prepare calligraphic art (camera-ready):

1 hr × $70.00/hr = $70.00

Step 6 Determine the total estimated cost (without expenses):

$40.00 + $65.00 + $110.00 + $120.00 + $70.00 = $405.00

Professional Photographic Services

As with creative services, many elements and factors must be taken into consideration when estimating professional photographic work. Rates are contingent upon the skill of the photographer, his or her reputation, the quality of work done, level of quality desired, and geographic region. Some art studios have resident photographers; others subcontract this service. Freelance designers sometimes provide such services; others contract them out to a professional photographic freelancer, depending upon the difficulty of the work and other factors. Figure 8.5 provides information for estimating photographic services. Dollar amounts are for example only.

Example 8.2

An in-plant photographic department will shoot a series of photographs for a 16 page fashion brochure. Based on past experience, it will take one-half day of studio time, with 4 models each hired for the 4 hour period. Color 4 × 5 inch transparencies will be produced using outside processing, with a total film and processing cost estimated at $320.00. In addition, some black-and-white photos will be taken and posterized using the plant's darkroom facilities (to be printed in color in the brochure). Development and printing will take about 1 hour for the black-and-white photos, with an estimated $45.00 in material costs. Approximately 2 hours will be required for producing the posterizations, with an estimated cost of materials of $42.00. Determine the total estimated cost for these professional photographic services.

Solution Refer to the schedule in Figure 8.5 for hourly rates.

Step 1 Calculate the studio session fee:

4 hr × $125.00/hr = $500.00

Step 2 Determine the model fee for 4 models:

4 models × 4 hr/model × $90.00/hr = $1440.00

Step 3 Determine the black-and-white photo development and printing fee including $45.00 for materials:

1 hr × $42.00/hr = $42.00
$42.00 + $45.00 = $87.00

Service	Hourly Rate
Sessions: studio	$125
on location	175
Model fee (each)	90
Development and printing (black-and-white)	42 + materials
Development and printing (color)	56 + materials
Special darkroom services	70 + materials
Film correction/print correction retouching	38

Note: Dollar figures in this schedule presented for example only. Rates include all necessary equipment unless specifically ordered. Materials charged as specified.

Figure 8.5

Professional photographic services schedule

Step 4 Determine the estimated color film cost including processing (given): $320.00

Step 5 Calculate the fee for special darkroom services including $42.00 for materials:

2 hr × $70.00/hr = $140.00
$140 + $42.00 = $182.00

Step 6 Determine the total estimated cost:

$500.00 + $1440.00 + $87.00 + $320.00 + $182.00 = $2529.00

Paste-Up and Mechanical Production

Figure 8.6 can be used to determine the time necessary for the makeup of mechanicals as camera-ready art. The size of the illustration board (format site) in square inches must be known in order to locate the correct time in the schedule. Chart times should be adjusted according to the artist's skills. It is assumed that all copy ready for paste-up will be clean, crisp, and sharp. Initial layout time provides for gathering all necessary artwork together; laying out illustration board; indicating sheet size, trim, and fold marks; and adding registration marks if needed. Additional time must be added for each piece to be pasted and for overlay acetate sheets for extra colors. Tissue and kraft paper overlays are included with basic layout time.

The following formulas are developed step-by-step in Example 8.3 and illustrated further in Example 8.4:

total time for mechanical production = (board layout time × number of boards) + (number of pieces to be pasted × production time per piece) + (overlay attachment production time × number of boards) + (number of pieces to be pasted per overlay × production time per piece)
total production cost for mechanical art = (mechanical production time × artist BHR) + cost of materials used

Example 8.3

A customer requests an estimate for a 4 page brochure, untrimmed page size 8 3/4 × 11 1/4 inches, to be printed in black with a second spot color throughout. Each paste-up will be completed at 100% (same size) on artist illustration board. An overlay will be used for the second spot color. Our artist has indicated that there will be 6 pieces (pc) to paste for each black page and 4 pieces for each overlay. BHR for the artist is $37.90. Determine the time and cost to produce the 4 pages of mechanicals for camera.

Solution Refer to the schedule in Figure 8.6.

Preparation	Format Size: Under 100 sq in.	101-200 sq in.	201-300 sq in.	301-400 sq in.	Over 400 sq in.
Layout time	0.15 hr	0.20 hr	0.25 hr	0.30 hr	0.40 hr
Hour per piece	0.03 hr per piece pasted				
Add'l time per overlay	0.08 hr	0.12 hr	0.16 hr	0.20 hr	0.25 hr
Hour per piece	0.03 hr per piece pasted				

Note: Paste or waxer makeup is determined by adding layout time and total paste-up time together. Use appropriate square inch value for format size of the job. Determine the number of pieces pasted. Additional colors/overlays are extra and apply as indicated. Use appropriate BHR. Times are given in decimal hours.

Figure 8.6

Mechanical/paste-up schedule

Step 1 Calculate the layout time:
8 3/4 in. × 11 1/4 in. = 98 sq in.
0.15 hr/p × 4 pp = 0.60 hr

Step 2 Determine the number of black pieces pasted:

6 pc/p × 4 pp = 24 pc

Step 3 Determine the black paste-up time:

24 pc × 0.03 hr/pc = 0.72 hr

Step 4 Determine the time required for overlay attachments:

1 overlay/mechanical × 4 mechanicals = 4 overlays
4 overlays × 0.08 hr/overlay = 0. 72 hr

Step 5 Determine the overlay paste-up:

4 pc/overlay × 4 overlays = 16 pc
16 pc × 0.03 hr/pc = 0.48 hr

Step 6 Determine the total time for paste-up:

0.60 hr + 0.72 hr + 0.32 hr + 0.48 hr = 2.12 hr

Step 7 Calculate the total cost (without materials):

2.12 hr × $37.90/hr = $80.35

Developing Page Rates for Estimating Art Production

In some cases estimators prefer to estimate artwork using "per page" time and cost rates. This can be done quite simply by manually cost estimating the time and cost of art production for one or a cross-section of jobs, then dividing both time and cost figures by the number of pages included in the estimate.

For example, for the 4 page brochure estimated in Example 8.3, the total cost is $80.35 and will take 2.12 hours to produce. Developing a per page rate for this job would then be accomplished by dividing $80.35 by 4 pages which equals $20.09 per page, and dividing 2.12 hours by 4 pages to yield an estimated production time of 0.53 hours per page. These page rates, when averaged with other similar 2/C jobs, could be used to estimate any type of two color work (black type with second PMS or flat color), assuming the number of pieces to paste totals about ten pieces for both colors.

When developing any type of cost or hourly standard for a finished product—such as a piece of mechanical art in the above example—it is important that the product be clearly identified and that the figures be used only for similar types of work. Also, since the original per page rates were built on an identified cost estimating process, revision of the per page standards should be done using the same cost estimating model. Averaging the data by cost estimating a number of jobs which share similar production requirements, e.g., they are all two color jobs with a black keyline and one overlay PMS color and all about the same size, provides a result that is generally more accurate than simply using one job the estimator believes to be representative of the finished product.

Establishing hourly mechanical art page production time and cost:

hourly page production time = total production time for preparing similar mechanical art ÷ number of pages of finished mechanical art

hourly page production cost = (total production time for preparing similar mechanical art × employee BHR) + cost of materials used ÷ number of pages of finished mechanical art

Example 8.4

A customer wants a printer to prepare the mechanicals for a theater poster. Finish size (paste-up size) is 19 × 25 inches (100%). Type and line work will run in black ink with additional flat color overlays for red and blue. The in-plant artist has indicated that the black printer (form) will have 10 pieces to paste and the red and blue printers will have 6 pieces each. Artist BHR is $39.50. Determine production time and cost for mechanical preparation.

Solution Refer to Figure 8.6.

Step 1 Determine the layout time:

19 in. × 25 in. = 475 sq in.
0.40 hr/p × 1 p = 0.40 hr

Step 2 Determine the paste-up time for black pieces:

10 pc × 0.03 hr/pc = 0.30 hr

Step 3 Determine the time required for overlay attachments:

0.25 hr/overlay × 2 overlays = 0.50 hr

Step 4 Determine the overlay paste-up time:

6 pc/overlay × 2 overlays × 0.03 hr/pc = 0.36 hr

Step 5 Determine the total paste-up time:

0.40 hr + 0.30 hr + 0.50 hr + 0.36 hr = 1.56 hr

Step 6 Calculate the total cost (without materials):

1.56 hr × $39.50/hr = $61.62

Peelcoat Production

Peelcoat materials consist of a lacquer-based coating on acetate, such as Zipatone and Ulano Rubylith or Amberlith, and are used by graphic designers and artists to take away or add an image mechanically. For example, a client wants a second color to be printed inside the line drawing of a balloon, which is pasted up on white illustration board. To do so, the artist attaches a sheet of peelcoat masking material just as if it were an acetate overlay and tapes it to one edge of the artwork. Since the gelatin side must be cut and peeled away, it is placed facing away from the artwork, toward the artist. As the peelcoat material lies flat and secure over the balloon image, the artist carefully cuts lightly through the gelatin coat, making certain the acetate base is not punctured. Once the entire balloon area has been cut, the artist peels away all the surrounding red (or orange) gelatin and leaves only that which remains in the center of the balloon. Register marks are then added in exact alignment between the two images. Later, two film negatives will be made: one, a negative containing the line work of the balloon and the other, a negative containing a clear area where the second color for the balloon will be.

Figure 8.7 is provided for estimating peelcoat production time and material cost.

The values given in the peelcoat schedule should be substituted into the appropriate formula:

total time for peelcoat production = (total number of linear inches to be cut ÷ 100) × production time per 100 linear inches

total cost for peelcoat production = (peelcoat production time × employee BHR) + cost of materials used

Example 8.5

A customer wants 10 halftones in silhouette form in a small booklet. Thus, the background of the image must be dropped out so that only the image itself will be seen. In this case, the images are all 5 × 7 inch halftones of the customer's employees. They will be peelcoated following halftoning, with cutting and peeling done on the film negatives. Estimate the maximum area to be cut as represented by the perimeter total of each picture (5 × 7 inches) because outline cuts are ragged. BHR for the artist is $39.50. Determine time and cost for peelcoat production including material cost.

Solution Refer to the schedule in Figure 8.7.

Step 1 Determine the number of linear inches to be cut and peeled:

perimeter area = 2(5 in.) + 2(7 in.) = 24 in./image

24 in./image 10 images = 240 in.

Step 2 Determine the total time needed to cut and peel where 100 in. is the base amount:

(240 in. ÷ 100 in.) × 0.30 hr/100 in. = 0.72 hr

Step 3 Calculate the total cost:

0.72 hr × $39.50/hr = $28.44

	Type of Cut		
	Squared Cut	Outline Cut	Special Cut
Time required to cut 100 linear inches of peelcoat material	0.20 hr	0.30 hr	0.40 hr

Note: Peeling of cut film included. Cost of materials is $0.010 per square inch. Type of cutting must be accurately determined. Times are given in decimal hours.

Figure 8.7

Peelcoat schedule

Step 4 Calculate the total material cost:

(5 in. × 7 in.)/pc × 10 pc × $0.010/sq in. = 35 sq in./pc × 10 pc × $0.010 sq in. = 350 sq in. × $0.010/sq in. = $3.50

Step 5 Calculate the estimated total cost:

$28.44 + $3.50 = $31.94

Form Ruling and Paste-Up Production

Many printing operations, at least from time to time, produce various types of forms (such as business forms) for their customers. Figure 8.8 provides basic times for ink pen ruling and paste-up of type in the production of such forms. The area of the form in square inches must be determined in order to find the correct values in the schedule.Those values are then substituted as necessary into the following formulas:

total time for form ruling and paste-up production = (production time per form for ruling × number of forms to be produced) + (number of pieces to be pasted + production time per piece)
total production cost for form ruling and paste-up = (form ruling and paste-up production time × employee BHR) + cost of materials used

It must be understood that the actual details of form design have been accurately determined prior to execution of ruling and paste-up. As with other types of artwork, times are relative and must be adjusted for the artist's skill and experience in this area.

Example 8.6

A customer wants a printer to prepare the artwork and then print an order form for a mail-order business. It will be produced on 8 1/2 × 13 inch bond, with a different form printed on each side of the sheet. Preliminary layout indicates form ruling at the average level for each of the 2 forms. In addition, the customer has indicated that there will be an estimated 25 pieces to paste on the front side and 20 pieces on the back (such as type over column heads and form data).Cost of the artist's time is $41.00 per hour. Determine the time and cost for this job.

Solution Refer to the schedule in Figure 8.8.

Step 1 Determine the square inch value (area) per form:

8 1/2 in. × 13 in. = 110.5 sq in.

Step 2 In Figure 8.8, the time needed to rule each form (101-200 sq in.) is given as 0.40 hr.

Step 3 Determine the total ruling time:

0.40 hr/fm × 2 fm = 0.80 hr

Step 4 Determine the time needed to paste the pieces (there are 25 pc for side one and 20 pc for side two):

(25 pc + 20 pc) × 0.03 hr/pc = 45 pc × 0.03 hr/pc = 1.35 hr

Step 5 Determine the total estimated production time:

0.80 hr + 1.35 hr = 2.15 hr

Step 6 Determine the estimated total cost for the artwork for the forms:

2.15 hr × $41.00/hr = $88.15

Form Layout and Time Required Ruling	Form Size				
	Under 100 sq in.	101-200 sq in.	201-300 sq in.	301-400 sq in.	Over 400 sq in.
Simple	0.15 hr	0.25 hr	0.40 hr	0.60 hr	0.75 hr
Average	0.25 hr	0.40 hr	0.60 hr	0.75 hr	1.00 hr
Difficult	0.40 hr	0.60 hr	0.75 hr	1.00 hr	1.25 hr
Hour per piece	0.03 hr per piece pasted				

Note: Time for initial ruling must be evaluated at simple, average, or difficult based on the number of lines and difficulty of the form, as well as the size of the form in square inches. Add for all pieces pasted. Times are given in decimal hours. Use the appropriate BHR.

Figure 8.8

Form ruling and paste-up schedule

Estimating Copyfitting, Typesetting and Desktop Publishing

9.1 Introduction

There is no doubt that typesetting, as with the rest of the prepress segment of the printing industry, is in a state of continuous and significant change. As previously discussed in both Chapter 5 and Chapter 8, this has been largely driven by the advent of electronic systems that integrate typesetting, art and design, copy preparation, and graphic image production. Such technological advances will almost certainly continue at an ever-accelerating pace through the 1990s and beyond.

While electronic imaging systems can affect production operations of any printer, many smaller printers, quick printers, and some commercial printers continue to complete the typesetting and related art production areas using conventional technology including direct-input typesetters, existing stand-alone phototypesetting units, and conventional paste-up methods. Yet the time of the dedicated typesetting system is passing, with the advent of various types of microcomputer and other platforms that provide fast processing of typeset material, high memory capacity, and direct links to high-speed output units which include laser-driven engines.

As indicated in the preceding chapter, conventional preparatory processes interrelates copyfitting and typesetting production with the preparation of artwork and copy for printing. Usually the graphic designer or artist will coordinate the selection and use of type with the design of the product, so the final graphic image—ultimately reproduced in quantity in the printing plant—will meet the customer's needs, in addition to having a coordinated appeal.

Most graphic designers learn to work with type effectively even though there are many schools and philosophies regarding type selection and use. The purpose of this chapter is not to discuss the design aspects of typography but to acquaint the reader with the important areas of copyfitting and typesetting. Copyfitting (also known as casting off) is a mathematical technique to determine the amount of space a specified portion of written or typewritten manuscript will occupy, given certain type specifications. Typesetting is the generation of characters in a selected typeface and size of type, within certain given dimensions, from the manuscript copy. When copyfitting is completed, it usually precedes the typesetting function.

The purpose of this six-part chapter is to provide a basis from which typeset products can be estimated. Section 9.2 provides information on basic typographic procedures and terms. Section 9.3 deals with copyfitting definitions with which the estimator must be familiar and introduces copyfitting methods and procedures to enable the establishment of an accurate process for evaluating space and other conditions between typewritten or word-processed manuscript and typeset material. Section 9.4 provides estimating data and procedures for slug machine composition and selected photographic and electronic composition systems. Example problems will be presented and solved in a step-by-step format. Section 9.5 provides data and information on estimating third generation phototypesetting systems. The last section—new to this edition—provides procedures by which desktop publishing can be estimated, including text input, graphic image input, image manipulation, and printing on an imagesetting unit.

9.2 Important Typographic Procedures and Terms

The following seven procedures—linecasting, phototypesetting, strike-on, foundry, press-on, calligraphic, and electronic composition—summarize typesetting methods found throughout the printing and publishing industry.

Typesetting Methods

Linecasting (Hot Metal). Hot metal typesetting utilizes a system of individual brass matrices, each containing a type character in recessed form. The operator of a linecasting machine, such as Linotype or Intertype, uses a keyboard to assemble these matrices in the desired word pattern. This line of matrices is then cast into a solid line of lead material (with relief characters) commonly termed a *slug*. The slugs are then assembled one after another to produce a column, or *galley*, of typeset material. The galleys are proofed and read for accuracy and then corrections are made line by line. Reproduction, or repro, proofs may be made for paste-up use; the type may also be used for direct impression on paper utilizing the letterpress printing process.

Sometimes the term *slug machine composition* is used to mean linecasting methods as described. While phototypesetting systems have essentially taken over the major typesetting duties in the printing and publishing industry, slug machine typesetting still is used in small printing plants and some newspapers. For this reason, Section 9.4 will show how these linecasting systems can be estimated. In addition to slug machine operations using a Linotype or Intertype, two other hot metal procedures are used for typesetting: The Ludlow process is used primarily for setting large type sizes and the Monotype for casting individual characters that are then assembled as a complete line.

Phototypesetting and Electronic Composition. The process of setting type using photographic methods requires two sequential operations: keyboarding or inputting the manuscript using word processing or a computer system, and the subsequent setting of the input material on a phototypesetting unit. Keyboarding involves the transforming of manuscript written by the author, copywriter, or secretary into a form that it can be then processed to produce type.

There are essentially three categories of keyboarding units—the direct-input typesetter, the dedicated word processor, and the microcomputer. The direct-input typesetter functions when an operator keyboards manuscript into the machine, with directly related typesetting production as output, in one continuous operation. Direct-input systems were popular in the 1970s and early 1980s, yet provide inherent character storage limitations and fairly slow production since the typesetting machine performs at roughly the same speed as the operator.

Dedicated word processors, also popular in the 1970s and 1980s, were developed specifically to input text and generally had application for larger offices and in situations where many people wished to be networked together using a single word processing sequence. Each brand of dedicated word processor was made with its own unique operating system and ran individualized word processing software, making data transmission between units difficult to do.

Since the late 1970s, both direct input typesetting systems and dedicated word processors have been replaced with microcomputer systems—commonly known as workstations or front-end "platforms"—on which a variety of typesetting, design and other graphic preparatory functions can be accomplished interactively. In addition to the flexibility of such platforms, these micro systems offer the company/user fast processing, large RAM memory and storage capacity via inexpensive hard disks, ability to use application or "canned" software which is fairly inexpensive, increasingly good CRT displays and convenient connection to laser printers and typesetting units. In fact, the evolution of printers or "image setters"—began with the Apple LaserWriter which was introduced in 1985, followed with Allied Linotype's Linotronic 300 and 500 units in 1986, and coupled with the acceptance of Postscript as the standard page description printer code.

Another significant advance in the mid-1980s was the introduction of Aldus' PageMaker, which allowed for the interactive use of text and graphics, further enhancing the flexibility of the stand-alone microcomputer and threatening direct-input and dedicated composition systems. In addition, the development and sale of numerous, inexpensive word processing programs, compatible with PageMaker, Letraset's ReadySetGo, Quark XPress, and other desktop publishing programs, plus the ability of these programs to save keyboarded material in a generic form transportable to numerous other formats, known as ASCII files, moved the microcomputer ahead of either direct-input or dedicated systems.

Once keyboarding is completed, text files can be stored in various forms on hard or floppy disk format for retrieval, modification, or printing. Copies are very easily made so transmitting captured data by sending it via overnight delivery, or using telecommunication or satellite methods, greatly enhance the flexibility between manuscript generation and finished type or image.

Keyboard input can be completed at literally any stage of the text drafting process—for example, by the author writing the original draft on the computer, by an editor completing editorial modification on the computer, or perhaps by a secretary working directly from the author's handwritten manuscript. Such ease in capturing and modifying text data offers significant typesetting cost benefit and provides extreme flexibility throughout the text writing and editing sequence. Changes in the text file are possible any time after the initial input is completed, since all work is saved and stored in magnetic form on hard or floppy disks, or magnetic tape.

Printing keyboarded material can be completed using dedicated phototypesetting units or electronically on a laser system, on plain paper. Because many word processing programs provide spelling check software, spelling errors are generally reduced; however, proofreading is still needed to correct grammatical and syntax problems. Corrections can be made on the proofs and final printing can either be directly to film, or to photographic paper and plain paper for paste-up into mechanical art.

One of the most significant changes to the typesetting process has been the development of desktop publishing software which allows the final image—both text and graphics—to be fully completed on the computer as one file, and printed as a fully prepared image on one page. It is not unusual for DTP systems to be used to produce camera-ready art, reducing impact of paste-up as a manufacturing step in the printing process. Figure 8.2 is a flowchart comparing conventional art preparation with desktop art production.

Estimating stand-alone phototypesetting systems are addressed in Section 9.3 and estimating desktop publishing is detailed in Section 9.4. Keyboarding input is covered as a part of each of these schedules.

Strike-On Typesetting. *Strike-on typesetting* involves the physical impact of characters onto paper or similar types of image carriers. The most common strike-on unit is the typewriter, but most conventional typewriters do not justify type. (Justification is the setting of type in block form, with lines flush left and flush right against each margin.)

One popular strike-on unit that does justify type is the IBM Selectric Composer, even though two typings of the material are required. Characters produced with the Selectric system are generally sharp and crisp, and interchangeable typefaces using a ball font system are another advantage. (The term font is defined as a collection of characters of one size and style of type.)

Foundry Type. *Foundry type* consists of individual characters cast in metal in relief form that are stored as complete fonts in type drawers. In foundry typesetting, characters are manually selected and assembled individually in a composing stick. Setting foundry type is a slow, tedious process that makes it unpopular for volume production. In addition, small typefaces are difficult to handle and work with.

Press-On Type. *Press-on,* or *transfer*, lettering is popular when a limited amount of type must be set. Carbon letters, carried on a plastic sheet, are individually aligned and rubbed onto the artwork, creating the desired word or words. As with foundry type, the procedure is essentially manual. Therefore, high-output applications are limited.

Calligraphy. *Calligraphy* is a manual process, completed by an artist or calligrapher, wherein individual characters or words are handwritten in a distinctive manner. Special pens and other tools are required by the calligrapher to produce unique letter designs and patterns. Since special skills are necessary and the procedure is completely manual, calligraphy is used for a very small amount of type production.

Typeface Information

The printers *point system* is used as the basis for measurements in typesetting. The following equivalents show the relationship between points, picas, and inches:

1 point = 1/72 inch = 0.013837 inch
12 points = 1 pica
6 picas = 1 inch
72 points = 1 inch

Note: The inches are an approximate value.

Typefaces up to 72 points high are measured by the point size of the type; type larger than 72 points is measured in inches. Text typefaces—that is, those used for the composition of most reading material—are commonly offered in heights of 5, 6, 7, 8, 9, 10, 11, and 12 points. Material set in block form—that is, as justified material—without typographic variation is called *straight matter*.

While the height of the letter is important, the body size of the line on which the letter is cast or set is also important. For example, 8 point type may be set on an 8 point body or perhaps on a 9 or 10 point body. Any extra amount of space added to the point size of type is called *leading* and is the white space between each set line. Thus, for example, 8 point type on an 8 point body would be *set solid,* 8 point on a 9 point body would be *leaded 1 point,* and 8 on 10 would be *leaded 2 points.* Normally, text typefaces are leaded to a maximum of 2 points, although this decision a typographic one.

Typefaces that measure over 12 points in height are classified as *display typefaces*. Such larger faces are used, for example, for headlines in newspaper composition, for paragraph heads, and for chapter titles. Copyfitting normally relates to typeset material in text form. Display typefaces may be copyfitted, but as characters grow in size, copyfitting becomes a much less precise procedure.

Both point and body size refer to the vertical measure of type. The horizontal measure, called the *set size* or *set width*, is the linear distance that the character occupies on the page. Typewritten characters each take up the same amount of space, whereas typeset characters each take up a different linear measure. Thus, while the point size will be the same for all characters in the same font, set width varies from letter to letter. For example, the letter *i* takes up less amount of horizontal distance than the letter *n*, and the *n* takes up a different linear amount than the *m*. Capital letters vary just as lowercase letters in terms of set size. The variable set width necessitates accurate copyfitting procedures.

Some typefaces are designed in an *expanded* or *condensed* form, meaning that the typeface characters are spread out or made narrower depending upon the original design. Each form has different uses. For example, condensed faces save paper by allowing more characters in a given area. Both designs tend to diminish readability to some extent. In addition to expanded and condensed faces, many typefaces are produced with matching bold or italic counterparts used for emphasis of words and phrases.

The term *em* is defined as the square area of the point size of the type and is referred to in typesetting as an *em quad*—that is, a piece of spacing material that equals the square area of the typesize. For example, an em in 9 point type would be 9 points wide and 9 points high. Sometimes, the em quad is called a *mutt quad*. The *en quad*, or *nut quad*, is a measure of space one-half the set width of the em of the same size body. For example, in 12 point type, the en quad would measure 6 points wide by 12 points high. Normally, em quads are used when a defined spacing unit is needed during typesetting (such as the centering of lines in slug machine linecasting). En quads are commonly used as spacing between sentences. Number characters are usually placed on a set width the size of the en quad so that they will align with each other when used in tabular form.

Some estimators use the em measure as a method of estimating composition output. While this procedure can be used for any type of composition, it is primarily aimed toward estimating straight text matter up to 12 points. When using this system, the machine operator will be required to set a certain number of ems per hour depending on typeface size—for example, 4000 ems per hour of 6 point type, 2500 ems per hour of 11 point, and so forth, on a sliding scale of fewer ems as the typeface increases in size. Schedule data will be presented for this estimating technique using slug machine equipment in Section 9.4 of this chapter.

One term used during copyfitting procedures is the *square pica*, which is defined as a square measuring 12 points on all four sides. While the point is the basis for measuring type, the pica and square pica are common measures in the copyfitting process. Essentially, typeset material is divided into squares, each measuring 12 points on a side, much like a carpenter would use square feet as a standard measurement.

9.3 Copyfitting Procedures

Copyfitting Defined

Copyfitting is a procedure used to determine the amount of space manuscript copy will occupy in typeset format and can be completed using either mathematical or electronic procedures. The following describes the three ways copyfitting is typically completed.

Manual Copyfitting from Typewritten Manuscript. This process utilizes manual copyfitting counting and calculation procedures. An author writes an article which is ultimately produced in typewritten form, perhaps as double-spaced manuscript pages, on a pica or elite pitch typewriter. The number of typewritten characters is then determined for the entire article, either by counting characters, counting words or by using an average number of characters per page multiplied by the number of pages of manuscript. Once total characters are known, this figure can be mathematically translated into typeset copy, given the customer's chosen size and style of type, line length, leading requirements, and other graphic spacing requirements.

Because the manual copyfitting process from typewritten manuscript is the most complex, and because the second procedure (which follows) utilizes the copyfitting portion of this process, manual copyfitting from typewritten manuscript is addressed fully later in this chapter, using described procedures and Figure 9.2.

Manual Copyfitting from Electronic Manuscript. Another author writes an article on his computer using one of many different types of word processing programs, electronically capturing the characters. When the draft article is finished, the size of the article in characters or words can be determined from the electronic file, which can be done in various ways depending on the word processing program. Once total words or characters are known, the article can be mathematically copyfit using the selected type size and style of type, line length, and leading as chosen by the author or client. As with the previous procedure, Figure 9.2 or a similar copyfitting schedule would be needed to complete this process.

Electronic Copyfitting from Electronic Manuscript. The same article, existing in electronic media such as a hard disk is transmitted via overnight carrier (or directly from the computer by telecommunication modem) to a typesetting service bureau. Here the file is carefully coded with the customer's choice of type face and type size and other requirements such as runarounds for photographs, and the electronic file through a typesetter, actually producing the job. Since the typesetter can perform at very high speeds, the typesetting will be done in a short time, perhaps a few minutes. Typical output is to photographic paper, although film could be used if desired.

The typeset job is now reviewed for space requirements, visual appearance, and other reader-oriented issues. Decisions are made to modify the size or style of type, change leading between lines, change the line length or make other modifications, thus changing the typeset length of the article, its visual appearance, etc. Should such changes be desired, the inbedded code in the electronic text file is modified to make the needed changes, and the job typeset a second time. The process, as described, continues until the final result is approved by the customer or client. With experienced typesetting service bureau personnel, rerunning the job more than a few times is not typical.

Basic Copyfitting Information

The copyfitting procedure can be used two ways. Either the typesetter can determine the amount of manuscript copy that should be supplied in order to fit within a specified number of pages, typeset area, and specified typesize and style or, given a manuscript and type specifications, the typesetter can tell the customer how many lines, pages, or other appropriate measure the supplied manuscript will make. For example, a customer wants an 8 page booklet produced with a page size of 28 x 42 picas. This information allows for identification of a maximum amount of space available for typeset material. Through copyfitting, the typesetter will be able to indicate to the customer, with reasonable accuracy, how many words, characters, or manuscript pages the customer must supply to fill the defined space area. Of course, the addition of photographs, line drawings, and original artwork must be planned into the space requirements also and will subtract from the amount of characters that should be written. Copyfitting is sometimes done using a percentage evaluation of typeset area to total area available. Because typesetting should not be a hit-or-miss production procedure, which can be expensive, it is essential that the graphic artist or designer be able to copyfit typematter effectively.

Copyfitting, however, is not a totally precise process. Copyfitting data are generally based on averages of characters. For example, manuscript pages are prepared with a ragged right-hand margin. While the left-hand margin is *flush* with each line beginning at the same point, the right-hand margin zigzags with each line ending at a different point (see sample manuscript page, Figure 9.1). Thus, when characters are not counted one by one, exact character determination is impossible. Usually, to speed up the process for many manuscript pages, an average word count or average page measure procedure is used. (These procedures will be explained later.) Characters on the typed page do not vary in width; the lowercase *t* and the capital *M* take up the same amount of space. Constant width is not a characteristic of typefaces used in composition. Almost all typefaces are designed with a variable set width, as previously mentioned. Essentially, copyfitting is an averaging procedure; however, every attempt should be made to be as precise as possible so that manuscript and set typematter fit together. Otherwise, composition costs can be very high, with associated frustrations on the part of the typographer and designer.

Determining the Number of Manuscript Characters

There are three procedures used to determine the number of characters on a manuscript (or typewritten) page:

Actual character count,
Word count,
Average page measure.

Actual character count involves the individual counting of characters including spaces and punctuation. This procedure is very accurate but is also time-consuming, especially when a large number of manuscript pages must be evaluated. The second method, counting words, is a popular procedure in some segments of the publishing industry since manuscripts written by authors are usually contracted on a per-word basis. The problem with this system is twofold: It is difficult to determine the exact number of words per typed page unless they are each counted individually (which is time-consuming), and the number of characters per word varies. Even when using the word count procedure, experts disagree as to the number of average characters per word: Some use 6, while others use 5.6, 5.2, or perhaps 5 characters per word. In

now a fast, convenient operation -- a true departure from the many processing steps needed to produce a deep-etch or albumin plate in the 1950's. The practical development of the Cameron belt press could have significant impact with respect to existing publishing methods -- it is a versatile process and provides great output with minimum production difficulty.

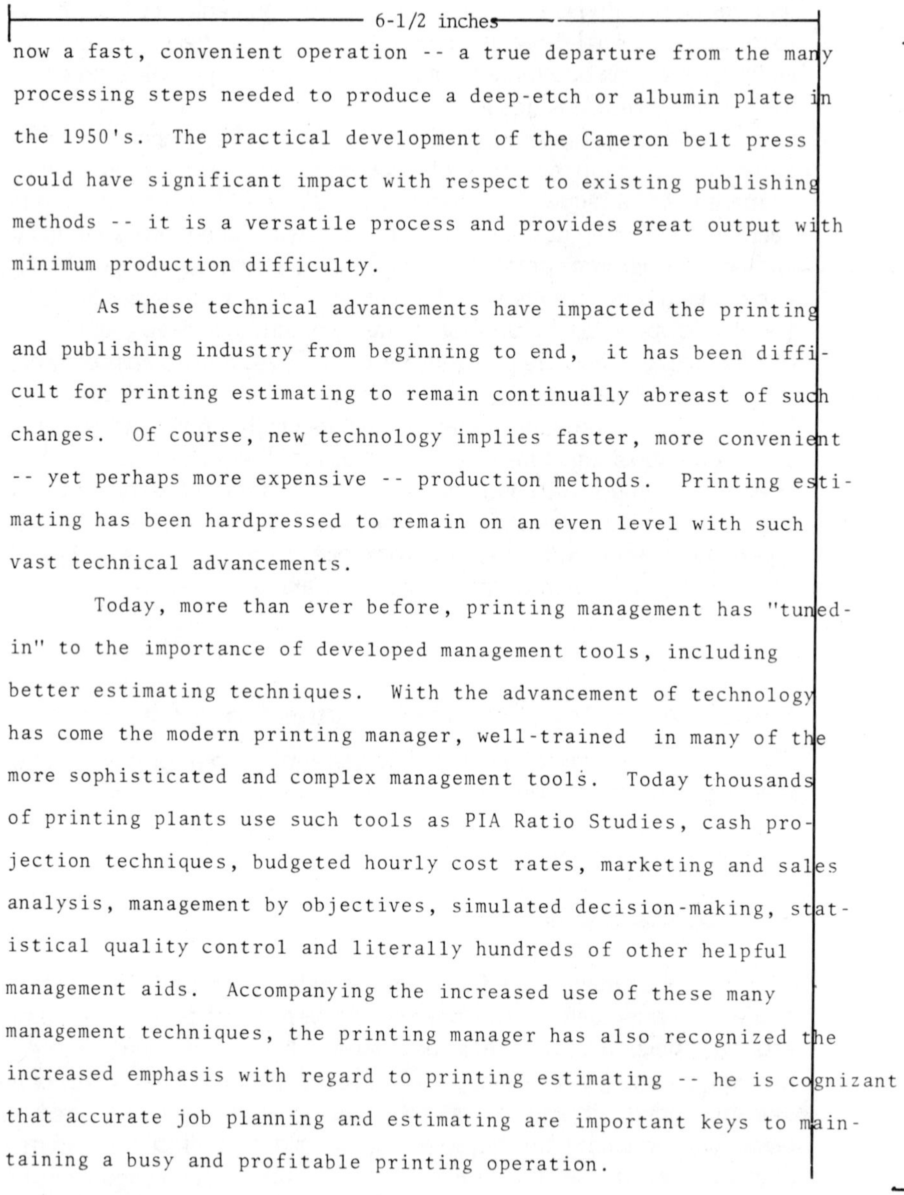

As these technical advancements have impacted the printing and publishing industry from beginning to end, it has been difficult for printing estimating to remain continually abreast of such changes. Of course, new technology implies faster, more convenient -- yet perhaps more expensive -- production methods. Printing estimating has been hardpressed to remain on an even level with such vast technical advancements.

Today, more than ever before, printing management has "tuned-in" to the importance of developed management tools, including better estimating techniques. With the advancement of technology has come the modern printing manager, well-trained in many of the more sophisticated and complex management tools. Today thousands of printing plants use such tools as PIA Ratio Studies, cash projection techniques, budgeted hourly cost rates, marketing and sales analysis, management by objectives, simulated decision-making, statistical quality control and literally hundreds of other helpful management aids. Accompanying the increased use of these many management techniques, the printing manager has also recognized the increased emphasis with regard to printing estimating -- he is cognizant that accurate job planning and estimating are important keys to maintaining a busy and profitable printing operation.

Figure 9.1

Sample page of manuscript with an average size of 6½ × 9 inches of typewritten material (shown reduced)

essence, the word counting technique is not accurate because there are too many variations that cannot be easily adjusted.

Using the average page measure system is the fastest and most accurate for the time required. Since manuscript material, when typed, has a consistently flush left, ragged right configuration, and since all typewriters preferred for such manuscripts have equivalent spacing per character, the average page measure method is both easy and accurate. Note that on the page of manuscript in Figure 9.1, which demonstrates this procedure, a line is drawn down the ragged-right margin of the typed material in such a position that shorter lines and longer lines balance each other. Thus, the line represents an average between long and short lines of the manuscript. Spacing top to bottom is determined with respect to the type of spacing between lines: single-, double-, or triple-spacing. Double-spacing is normally recommended for most manuscripts, but single-spacing and triple-spacing are also used. A manuscript submitted for typesetting in handwritten form is usually penalized with a higher cost by the typesetter. Because of variations in handwriting that make it hard to read, copyfitting is much more difficult, and the set material is more prone to mistakes.

When the average page measure is used, both horizontal and vertical measurements of the manuscript pages are identified in inches. For example, Figure 9.1 (shown reduced) has dimensions of 6 1/2 inches across and 9 inches deep, or an average page measure of 6 1/2 × 9 inches. (Throughout this book the first figure given is the horizontal measure, and the latter dimension is the vertical measure.) In order to translate such page dimensions into numbers of characters, the fixed typewriter dimensions must be known. For typewriters recommended in the production of manuscript for typesetting, the following measurements apply:

Horizontal Spacing, or Character Pitch (across)

10 characters per inch (pica typewriter)
12 characters per inch (elite typewriter)

Vertical Spacing (up and down)

6 lines per inch (single-spaced)
3 lines per inch (double-spaced)
2 lines per inch (triple-spaced)

The following formulas are used to determine the number of manuscript characters:

number of characters per typewritten line = typewritten page width in inches × character pitch

number of lines per typewritten page = typewritten page depth in inches × spacing lines per inch

number of characters per typewritten page = number of characters per typewritten line × number of lines per typewritten page
total number of manuscript characters = number of characters per typewritten page × number of manuscript pages

In Figure 9.1, with a horizontal measure of 6 1/2 inches and a pica typewriter (10 characters per inch), there are 65 characters per average line (some lines are longer and some are shorter). With a depth of 9 inches and double-spaced material, there are a total of 27 lines on this page. Total characters are then 65 characters per line times 27 lines per page, which equals 1755 characters. When manuscript pages contain a significant number of short lines, total page characters can be adjusted accordingly.

Some general comments on the procedure are in order. First, no typewriters or word processors should be used that set variable width manuscript; all characters should be either pica or elite pitch at 10 or 12 characters per inch. Second, it must be remembered that the average page measure procedure is still an averaging process so that total characters may be slightly different when counted individually. Third, in order for this procedure to work well, there should be a consistency between typewritten pages from one to the next: The spacing should be the same, indentations and margins the same, and so forth. This consistency insures accuracy, especially with large manuscripts.

Example 9.1

A typesetter is given 27 full pages of manuscript and has determined each page has an average measure of 7 1/2 × 9 inches. An elite typewriter was used, and all material was double-spaced. Determine the number of total characters (char) that have been written.

Solution Refer to the measurements for horizontal and vertical spacing as already given.

Step 1 Determine horizontal spacing for elite type:

7 1/2 in./line × 12 char/in. = 90 char/line

Step 2 Determine vertical spacing when double-spaced:

9 in./p × 3 lines/in. = 27 lines/p

Step 3 Determine characters per page:

90 char/line × 27 lines/p = 2430 char/p

Step 4 Determine total manuscript characters:

2430 char/p × 27 pp = 65,610 char

Once total characters written have been calculated, the second portion of copyfitting—translating these characters into typeset material—must be completed. This procedure requires the use of a copyfitting schedule for the specific typefaces and sizes of type in any given typesetting operation.

Developing a Copyfitting Schedule

The chart in Figure 9.2 is generally applicable to all typefaces with the exception of those that are extremely condensed or expanded. Common copyfitting procedure is based on the length of the lowercase alphabet measured in points (a point equals approximately 1/72 inch). Thus, it is possible to develop a copyfitting schedule similar the one provided for the range of typefaces used in a particular plant. Keep in mind that the schedule is based on the aggregate length of 26 lowercase alphabet characters, each with a different set width. Literally all type manufacturers specify the length of the lowercase alphabets for each typeface they produce, but if this information is not available, the lowercase alphabet can be set and measured carefully in points. In order to explain the development of the copyfitting schedule in Figure 9.2, the examples that follow have been completed for 10 point type with an alphabet length of 132 points.

Adjusted Characters per Linear Pica. Because type is kerned (fit together) and to adjust for the larger number of thinner characters in the alphabet, the characters per linear pica figure is normally adjusted, meaning that the actual measure of alphabet length is not used directly for such calculations. Adjustments are on a sliding scale, with 5–6 point type at 100%, 8 point at 95%, 10 point at 93%, and 12 point at 90%. The following formulas are used to find the *adjusted characters per linear pica,* the number of characters that will fit into one 12 point (1 pica) linear measure:

alphabet length in points = alphabet length in picas × percentage adjustment for point size
adjusted picas for lowercase alphabet = alphabet length in points ÷ 12 points per pica
adjusted characters per linear pica = 26 alphabet characters ÷ adjusted picas per alphabet

For the example chart data in Figure 9.2, the adjusted characters per linear pica is determined as follows:

Step 1: Determine the adjustment when 132 points (pts) is the alphabet length for 10 point type:

132 pts × 0.93 = 122.76 pts

Point Size	Alphabet Length (points)	Adjusted Characters per Linear Pica	Characters per Square Pica: Set Solid	Characters per Square Pica: Leading 1 Point	Characters per Square Pica: Leading 2 Points	Square Picas per 1000 Characters: Set Solid	Square Picas per 1000 Characters: Leading 1 Point	Square Picas per 1000 Characters: Leading 2 Points
5	60	5.08	12.19	10.16	8.71	82.0	98.4	114.8
	62	4.92	11.81	9.84	8.43	84.7	101.6	118.6
	64	4.81	11.54	9.62	8.25	86.7	103.9	121.2
6	66	4.72	9.44	8.09	7.08	105.9	123.6	141.2
	68	4.62	9.24	7.92	6.93	108.2	126.3	144.3
	70	4.51	9.02	7.73	6.77	110.9	129.4	147.7
	72	4.40	8.80	7.54	6.60	113.6	132.6	151.5
	74	4.30	8.60	7.37	6.45	116.3	135.7	155.0
	76	4.21	8.42	7.22	6.32	118.8	138.5	158.2
	78	4.11	8.22	7.05	6.17	121.7	141.8	162.0
	80	4.00	8.00	6.86	6.00	125.0	145.8	166.6
	82	3.92	7.84	6.72	5.88	127.6	148.8	170.0
	84	3.81	7.62	6.53	5.72	131.2	153.1	174.8
	86	3.74	7.48	6.41	5.61	133.7	156.0	178.3
7	88	3.66	6.27	5.49	4.88	159.5	182.1	204.9
	90	3.60	6.17	5.40	4.80	162.0	185.2	208.3
	92	3.52	6.03	5.28	4.69	165.8	189.4	213.2
8	94	3.46	5.19	4.61	4.15	192.7	216.9	241.0
	96	3.40	5.10	4.53	4.08	196.1	220.8	245.1
	98	3.35	5.03	4.47	4.02	198.8	223.7	248.8
	100	3.31	4.97	4.41	3.97	201.2	226.8	251.9
	102	3.26	4.89	4.35	3.91	204.5	229.9	255.8
	104	3.20	4.80	4.27	3.84	208.3	234.2	260.4
	106	3.14	4.71	4.19	3.77	212.3	238.7	265.3
	108	3.11	4.67	4.15	3.73	214.1	241.0	268.1
	110	3.04	4.56	4.05	3.65	219.3	246.9	274.0
	112	2.99	4.49	3.99	3.59	222.7	250.6	278.6
9	114	2.94	3.92	3.53	3.21	255.1	283.3	311.5
	116	2.90	3.87	3.48	3.16	258.4	287.4	316.5
	118	2.85	3.80	3.42	3.11	263.2	292.4	321.5
	120	2.81	3.75	3.37	3.07	266.7	296.7	325.7
	122	2.76	3.68	3.31	3.01	271.7	302.1	332.2
10	124	2.69	3.23	2.93	2.69	309.6	341.3	371.7
	126	2.66	3.19	2.90	2.66	313.5	344.8	375.9
	128	2.63	3.16	2.87	2.63	316.5	348.4	380.2
	130	2.59	3.11	2.83	2.59	321.5	353.4	386.1
	132	2.55	3.06	2.78	2.55	326.8	359.7	392.2
	134	2.50	3.00	2.73	2.50	333.3	366.3	400.0
	136	2.46	2.95	2.68	2.46	339.0	373.1	406.5
	138	2.43	2.92	2.65	2.43	342.5	378.4	411.5
	140	2.40	2.88	2.62	2.40	347.2	381.7	416.7
	142	2.38	2.86	2.60	2.38	349.7	384.6	420.2

Figure 9.2

Copyfitting schedule

Point Size	Alphabet Length (points)	Adjusted Characters per Linear Pica	Characters per Square Pica			Square Picas per 1000 Characters		
			Set Solid	Leading 1 Point	Leading 2 Points	Set Solid	Leading 1 Point	Leading 2 Points
11	144	2.36	2.57	2.36	2.18	389.1	423.7	458.7
	146	2.34	2.55	2.34	2.16	392.2	427.4	463.0
	148	2.32	2.53	2.32	2.14	395.3	431.0	467.3
12	150	2.30	2.30	2.12	1.97	434.8	471.7	507.6
	154	2.25	2.25	2.08	1.93	444.4	480.8	518.1
	158	2.21	2.21	2.04	1.89	452.5	490.2	529.1
	162	2.15	2.15	1.98	1.84	465.1	505.1	543.5
	166	2.11	2.11	1.95	1.81	473.9	512.8	552.5
	170	2.04	2.04	1.88	1.75	490.2	531.9	571.4
	175	2.00	2.00		1.71	500.2		584.8
	180	1.96	1.96		1.68	510.2		595.2
14	185	1.91	1.64		1.43	609.8		699.3
	190	1.84	1.58		1.38	632.9		724.6
	200	1.75	1.50		1.31	666.7		763.4
16	206	1.70	1.28		1.13	781.3		885.0
	212	1.63	1.22		1.09	819.7		917.4
18	218	1.60	1.07		0.96	934.6		1041.7
	225	1.54	1.03		0.92	970.9		1087.0
	233	1.49	0.99		0.89	1010.1		1123.6
	240	1.44	0.96		0.86	1041.7		1162.8
	250	1.40	0.93		0.84	1075.3		1190.5
	260	1.34	0.89		0.80	1123.6		1250.0
20	270	1.30	0.78		0.71	1282.1		1408.5
	280	1.26	0.76		0.69	1315.8		1449.3
24	290	1.23	0.62		0.57	1612.9		1754.4
	300	1.18	0.59		0.54	1694.9		1851.9
	350	0.91	0.46		0.42	2173.9		2381.0
	400	0.70	0.35		0.32	2857.1		3125.0

Figure 9.2

Continued

Step 2: Determine the length in picas of the lowercase alphabet:

122.76 pts ÷ 12 pts/pi = 10.23 pi

Step 3: Determine the adjusted characters per pica:

26 char ÷ 10.23 pi = 2.55 char/pi

Characters per Square Pica. This value, as will be shown in the following section, is desirable when determining a suitable typeface and size of type. To accurately determine the characters per square pica, the point size of type, adjusted characters per linear pica, and the amount of leading (space between lines of type) must be known. For text matter set in up to 12 point type, normal leading is solid, 1 point, or 2 points. The greater the amount of leading, the more white space between lines of type that are set. The formula for calculating characters per square pica is:

characters per square pica = (12 points per pica ÷ point value of typeface including leading) × characters per linear pica

Using the example information provided, the following may then be derived:

Set solid:

(12 pts/pi ÷ 10 pt) × 2.55 char/pi = 3.06 char/sq pi

1 point leading:

(12 pts/pi ÷ 11 pt) × 2.55 char/pi = 2.78 char/sq pi

2 point leading:

(12 pts/pi ÷ 12 pt) × 2.55 char/pi = 2.55 char/sq pi

In the 2 point leading situation for 10 point type, the characters per linear pica and characters per square pica are identical. The reason is that 12 points is the equivalent of 1 pica, and with 2 point leading, we have 10 point type on a 12 point body.

Square Picas per 1000 Characters. The value of square picas per 1000 (M) characters is used to determine column depth of set material, which is then used to determine the number of pages in bookwork calculations. Such data are easily derived from the character per square pica figure as follows:

square picas per M characters = M characters ÷ characters per square pica

Set solid:

1000 char ÷ 3.06 char/sq pi = 326.8 sq pi/M char

1 point leading:

1000 char ÷ 2.78 char/sq pi = 359.7 sq pi/M char

2 point leading:

1000 char ÷ 2.55 char/sq pi = 392.2 sq pi/M char

A chart developed from the preceding formulas provides for an accurate and versatile copyfitting system. Keep in mind that each typeface will have a different alphabet length. For example, a 6 point expanded face might have an alphabet length of 92 points, while a 6 point condensed face might have an alphabet length of 50 points. Even though Figure 9.2 does provide generalized copyfitting data, it is best for each typesetting establishment to develop its own figures.

Solving Three Major Types of Copyfitting Problems

The calculations of both number of manuscript characters and copyfitting data have been presented. What remains is to demonstrate how these two procedures can be used together. There are three major types of copyfitting problems:

To determine column depth (space to be filled) in picas when type size (alphabet length) and number of characters written are known,
To determine type size (alphabet length) when column depth (space to be filled) and number of characters written are known,
To determine number of characters written when column depth (space to be filled) and type size (alphabet length) are known.

Example problems and step-by-step solutions are provided to demonstrate each type of copyfitting situation. Figure 9.2 is used to solve the problems.

Determining Column Depth in Picas. This type of problem is actually a calculation of space that will be filled given a specific typeface and number of characters written. It can be solved using two different chart data figures, and the solutions should correlate closely because all data were derived from the adjusted alphabet length base.

Example 9.2

We have been given 35 full manuscript pages (msp) typed with an elite typewriter, average page size 6 1/2 × 9 inches, double-spaced. It will be set in 8 point Souvenir on a 10 point body (alphabet length 106 points). Typeset pages will measure 28 × 44 picas. Determine the column depth of the typeset material in picas and the number of booklet pages that this typeset material will occupy. Note that there are two methods for solving this problem: Solution A utilizes the characters per linear pica figure from Figure 9.2; Solution B uses the square picas per 1000 (M) characters data.

Solution A Use characters per linear pica chart data from Figure 9.2.

Step 1 Determine total manuscript characters with elite type:

Characters per line:
6 1/2 in./line × 12 char/in. = 78 char/line
Lines per manuscript page:
9 in./msp × 3 lines/in. = 27 lines/msp
Characters per manuscript page:
78 char/line × 27 lines/msp = 2106 char/msp
Total characters for entire manuscript:
2106 char/msp × 35 msp = 73,710 char

Step 2 On the copyfitting chart, locate 8 pt type, alphabet length 106 pts. The adjusted characters per linear pica value is 3.14 char.

Step 3 Use the following formula to determine the number of characters per typeset line that is 28 pi long:

number of characters per pica × line length in picas = number of characters per typeset line
3.14 char/pi × 28 pi = 88 char/line

Step 4 Use the following formula to determine the number of typeset lines:

total number of characters per manuscript ÷ number of characters per line = number of typeset lines
73,710 char ÷ 88 char/line = 838 lines

Step 5 Use the following formula to determine the depth of total typeset material (column depth):

$$\frac{\text{number of total typeset lines} \times \text{body size of type}}{\text{12 points per pica (constant)}} = \text{column depth in picas}$$

$$\frac{\text{838 lines} \times \text{10 pts/line}}{\text{12 pts/pi}} = \text{699 pi}$$

Step 6 Determine the number of booklet pages with a 699 pi depth:

699 pi ÷ 44 pi/p = 15.88 pp/bkt

Step 7 This booklet, set in 8 pt Souvenir with 2 pt leading, will have approximately 73,710 total characters and can be produced in a 16 pp signature format. Each page will measure 28 × 44 pi. No allowance has been made for halftones or other images. If other material is desired, then the number of characters would need reduction, the page size would have to be increased, or the type would have to be set with less leading or in a smaller face.

Solution B Use the square picas per M characters chart data from Figure 9.2.

Step 1 From the previous solution, the total manuscript characters are determined to be 73,710 char.

Step 2 Note in Figure 9.2, that the square picas per M characters for 8 pt type, alphabet length 106 pts, with 2 pt leading is 265.3 sq pi/M char.

Step 3 Use the following copyfitting formula to find the depth for this example:

(square picas per M characters × total characters × 0.001)
÷line length in picas = depth for all typeset materials in picas

$$\frac{265.3 \text{ sq pi/M char} \times 73{,}710 \text{ char} \times 0.001}{28 \text{ pi}} =$$

$$\frac{265.3 \text{ sq pi/M char} \times 73.71}{28 \text{ pi}} = 699 \text{ pi}$$

Step 4 Use the following formula to find the number of booklet pages:

total depth in picas ÷ depth of page in picas = number of typeset pages per booklet
699 pi ÷ 44 pi/p = 15.88 pp

Step 5 The same conclusion is reached as for the first solution.

Comparison of the two solutions will show that the same answers have been derived because the same basic data were used to compute both chart figures. Sometimes sight differences in schedule data and rounding will result in slightly different final answers.

Determining a Suitable Size of Type (Alphabet Length). Many times, when composition is estimated, the total area (column depth) is known, as well as the number of characters written for that space. What must be calculated are the type face, identified by alphabet length, and appropriate leading so that the space will be filled with the total written characters. The following example demonstrates this type of copyfitting problem.

Example 9.3

We have been given 35 full manuscript pages, average page size 6 1/2 × 9 inches, typed with an elite typewriter, double-spaced. It will be set for a 16 page booklet containing only typematter, no halftones or other images. Each typeset page will measure 28 × 44 pi. Select two typefaces, alphabet lengths, and leading that will provide for all 16 pages to be filled with type, cover to cover.

Solution

Step 1 Determine total manuscript characters:

Characters per line:
6 1/2 in./line × 12 char/in. = 78 char/line
Lines per manuscript page:
9 in./msp × 3 lines/in. = 27 lines/msp
Characters per manuscript page:
78 char/line × 27 lines/msp = 2106 char/msp
Total manuscript characters:
2106 char/msp × 35 msp = 73,710 char

Step 2 Use the following formulas to determine total square pica area to be filled:

pica width of page × pica depth of page = total square picas per page
28 pi × 44 pi = 1232 sq pi/p
square picas per page × number of pages per booklet = total square picas per booklet
1232 sq pi/p × 16 pp = 19,712 sq pi

Step 3 Use the following formula to calculate necessary characters per square pica:

total manuscript characters ÷ total square picas per booklet = characters per square pica
73,710 char ÷ 19,712 sq pi = 3.74 char/sq pi

Step 4 Search Figure 9.2 in all columns under "characters per square pica" for any typeface at 3.74 or slightly larger. Upon investigation, the following typefaces (alphabet lengths) can be used for this job:

9 pt on a 9 pt body (set solid), alphabet length 114–120:
8 pt on a 10 pt body (leaded 2 pts), alphabet length 100–106.

Step 5 It should be noted that this problem correlates with Example 9.2, resolving column depth. The typeface might well be Souvenir set 8 on 10, which has an alphabet length of 106. Other faces within these alphabet ranges are also possibilities.

It is advisable to select a typeface and alphabet length within a characters per square pica range of 0.20 but always higher than the calculated figure. The reason that the selected number must be higher in value, and not lower, is that the typeface becomes smaller as the number value increases. Choosing a smaller size insures that all the type will fit into the available space, perhaps with some space remaining. If a lower numerical figure is used, the typeface is larger and requires more space, and all the manuscript copy might not fit.

Determining the Number of Characters to Be Written (Total Manuscript Pages). The third copyfitting procedure is most useful when the size of the book or pamphlet has been determined (column depth) and the typeface and size of type have been selected. Initial calculations provide for the determination of the number of characters that must be written to fill the space. This information is then translated into manuscript pages so the author will know how many pages to produce. This problem and the solution correlate with the preceding two examples.

Example 9.4 We have selected Souvenir, 8 on 10 (alphabet length 106 points), to set our 16 page booklet, filled cover to cover with type. All manuscript pages will be typed with an elite typewriter, double-spaced, with an average page size of 6 1/2 × 9 inches. Typeset pages will measure 28 × 44 picas. Determine the number of total characters to be written and the number of manuscript pages that the customer must prepare using the specified margins and spacing.

Solution

Step 1 Determine the total square pica area to be filled:

Square picas per page:
  28 pi × 44 pi = 1232 sq pi
Total square picas:
  1232 sq pi/p × 16 pp/bkt = 19,712 sq pi/bkt

Step 2 Locate on the copyfitting chart (Figure 9.2) the characters per square pica of 8 pt type on a 10 pt body (leaded 2 pts), alphabet length 106 pts: 3.77 char/sq pi.

Step 3 Use the following formula to calculate total characters to be written:

total square picas × characters per square pica = total characters to be written
19,712 sq pi × 3.77 char/sq pi = 74,314 char

Step 4 Determine the number of characters per manuscript page:

Characters per line:
6 1/2 in. /line × 12 char/in. = 78 char/line
Lines per manuscript page:
9 in./msp × 3 lines/in. = 27 lines/msp
Characters per manuscript page:
78 char/line × 27 lines/msp = 2106 char/msp

Step 5 Use the following formula to determine number of manuscript pages to be typed:

total characters to be written ÷ number of characters per manuscript page
= number of manuscript pages to be typed
74,314 char ÷ 2106 char/msp = 35.29 msp

Step 6 Refer to the two solutions to Example 9.2 (which correlate with this problem) and note that 35 full manuscript pages fill only 15.88 typeset pages. It follows that 35.29 msp (as calculated) will then fill the 16 pp booklet from cover to cover with type set 8 on 10 (alphabet length 106 pts).

9.4 Slug Machine Composition Procedures

Estimating composition can be accomplished in almost as many ways as there are to set type. The two most popular are the *em method* and the *keystroke procedure*; others include estimating type by the galley, by the linear inch, by the square inch, by the set line, and by the typeset page. The following estimating data utilizes the em method for slug machine composition and the keystroke procedure for phototypesetting. It is possible to interchange these systems—that is, to estimate phototypeset material using the em procedure—but chart data are not presented here for estimating in this manner.

Slug Machine Composition: Em Method

When using ems to estimate composition, the process essentially involves dividing the amount of composition to be set into squares of the size of the type to be set. Smaller typefaces, which set quickly, will have a larger number of ems to be set in any given time period; larger typefaces have a smaller number of "set ems per hour" because they set slowly. To solve for ems, use the following formulas:

total ems per page = number of ems per line × number of ems per page depth

$$\text{ems per line} = \frac{\text{pica page width} \times 12 \text{ points per pica (constant)}}{\text{point size of type}}$$

$$\text{ems per page depth} = \frac{\text{pica page depth} \times 12 \text{ points per pica (constant)}}{\text{body size of type}}$$

Example 9.5

We have an 8 page booklet with a type page measuring 26 × 44 picas. It has been set in Bodoni Book, 9 point type leaded 1 point with a 26 pica line length. The booklet is 60% typematter, with the remainder halftones and line charts. Calculate total booklet ems.

Solution

Step 1 Substitute into the formula for total ems per page:

$$\text{ems/p} = \frac{26 \text{ pi} \times 12 \text{ pts/pi}}{9 \text{ pt}} \times \frac{44 \text{ pi} \times 12 \text{ pts/pi}}{10 \text{ pt}}$$

Step 2 Solve for ems per line and ems per page depth (dp):

$$\frac{26 \text{ pi} \times 12 \text{ pts/pi}}{9 \text{ pt}} \times \frac{44 \text{ pi} \times 12 \text{ pts/pi}}{10 \text{ pt}} = 34.67 \text{ ems/line} \times 52.8 \text{ ems/dp}$$

Step 3 Now calculate ems per page:

34.67 ems/line × 52.8 ems/dp = 1831 ems/p

Step 4 Calculate the total booklet ems, 10 pt type, with 60% typematter:

1831 ems/p × 8 pp × 0.60 = 8789 ems

Using the em measure does not adjust for typefaces that are expanded or condensed. It is an averaging procedure for most typefaces that have been developed, but extraordinary set width modifications render the system inaccurate.

Em calculations will be provided with the two following slug machine composition problems as part of the problem solution. Figure 9.3 is a slug machine composition schedule. The data assume that all material is submitted in suitable typewritten form and that it has been properly marked up for typesetting (typemarked). This table includes (in the second part) adjustments and machine time setup factors. Proofreading time is included in the chart times. The data in Figure 9.3 are used in the following examples.

Example 9.6

We have an 8 page booklet to set on our Intertype, page size 24 x 36 picas, typeface is Garamond 8 on 9 (alphabet length 110 points). We estimate that the book will be 75% typematter, with the remainder halftones, line drawings, and white space. There will be a mix of Garamond light and italic throughout the job. Determine the number of total characters to be set, total ems, time, and cost for all slug machine composition. The BHR for our machine is $28.70.

Solution

Step 1 Determine total characters (see Figure 9.2):

Square pica area per page:
  24 pi × 36 pi = 864 sq pi/p
Unadjusted book square pica area:
  864 sq pi/p × 8 pp = 6912 sq pi
Adjustment for artwork and white space:
  6912 sq pi × 0.75 = 5184 sq pi
Total characters:
  5184 sq pi × 4.05 char sq pi = 20,996 char

Step 2 Calculate total ems per page:

$$\text{ems/p} = \frac{24\text{ pi} \times 12\text{ pts/pi}}{8\text{ pt}} \times \frac{36\text{ pi} \times 12\text{ pts/pi}}{9\text{ pt}} = 36 \times 48 = 1728\text{ ems/p}$$

Step 3 Calculate total ems to be set per booklet with a 75% adjustment:

1728 ems/p × 8 pp × 0.75 = 10,368 ems

Step 4 Use the given formula to calculate the slug machine total composition time (see Figure 9.3):

total composition time = (basic composition time × adjustment factor) + additional machine time

First, solve for basic composition time:

Point Size	Typeset Ems per Hour	Typeset Hours per M Ems
6	4200	0.2381
7	4000	0.2500
8	3500	0.2857
9	3000	0.3333
10	2750	0.3636
11	2500	0.4000
12	2300	0.4348
14	2100	0.4762

Machine Time Setup Factors	Additions in Hours
Preparation	0.30
Each complete machine change	0.20
Change of line width only (each)	0.05
Change of magazine only (each)	0.05

Schedule Adjustments	Percentage Increase
Mix of regular face with boldface and/or italics	25
Butted slugs (over 30 picas)	50
Foreign languages set in English	50
Centering by hand (up to 30 picas)	50
Runarounds and counted lines around material	50
Running heads and folios	50
Lines with all capital letters	50
All small capitals or caps and small caps	100
Tabular work	100
Using vertical aligning leaders	100
Using period alignment in charts	100
Single line composition	100

Figure 9.3

Slug machine composition schedule

(total ems ÷ 1000 ems) × typeset hours per M ems = basic composition time

(10.368 ems ÷ 1000 ems) × 0.2857 hr/M ems = 2.97 hr

Second, determine the adjustment factor by applying a 25% increase for light and italic mix:

$1.00 + 0.25 = 1.25$

Third, establish additional machine time as 0.30 hr for preparation of machine. Now, substitute all values into the formula for total composition time:

total composition time = (2.97 hr × 1.25) + 0.30 hr = 4.02 hr

Step 5 Use the following formula to determine the cost:

total composition time × BHR = composition cost
4.02 hr × $28.70/hr = $115.38

Example 9.7

We have an order for a display poster to be printed for a government agency. It will be predominantly typematter covering working rules and conditions and will be posted in industrial areas. Finish size of the poster is 17 × 22 inches. All type will be set in Baskerville 10 on 12 (alphabet length 136 points), set in two columns (col), each 40 picas wide by 110 picas deep (butt slug the 40 pica lines into two 20 pica lines). Allow a 15% adjustment for paragraph heads (hand set in foundry type) and for white space. BHR for the Linotype is $27.50. Determine the total number of set ems and the time and cost for slug machine composition.

Solution

Step 1 Determine total characters (see Figure 9.2):

Unadjusted total square pica area per poster:
 40 pi/col × 110 pi/col × 2 col = 8800 sq pi
Adjustment for white space (100% – 15% = 85%):
 8800 sq pi × 0.85 = 7480 sq pi
Total characters per poster:
 7480 sq pi × 2.46 char/sq pi = 18,400 char

Step 2 Calculate total ems to be set per poster:

$$\text{ems/poster} = \frac{40 \text{ pi} \times 12 \text{ pts/pi}}{10 \text{ pt}} \times \frac{220 \text{ pi} \times 12 \text{ pts/pi}}{12 \text{ pt}} = 48 \times 220$$
$$= 10{,}560 \text{ ems/poster}$$

Note: 220 pi allows for depth of two 110 pi columns per poster.

Step 3 Solve with 85% allowed for adjustment:

10,560 ems × 0.85 = 8976 ems

Step 4 Calculate slug machine composition time (see Figure 9.3)

Basic time:
(8976 ems ÷ 1000 ems) × 0.3636 hr/M ems = 3.27 hr
Adjustment factor (50% for butt slugging):
1.00 + 0.50 = 1.50
Additional machine time:
0.30 hr for preparation

Now find total time:

(3.27 hr × 1.50) + 0.30 hr = 5.21 hr

Step 5 Calculate total cost:

5.21 hr × $27.50/hr = $143.28

9.5 Phototypesetting and Electronic Composition

Figure 9.4 is provided for estimating phototypesetting and electronic composition production. While this schedule covers four kinds of phototypesetting equipment, there are many different keyboard and typesetting units in production today. For example, a smaller printer in an established business might have a second generation phototypesetter that does excellent work and would be much too costly to replace given its limited use. Another smaller printer might have a more advanced third generation direct input typesetter, yet desire to move to a fourth generation system such as the Macintosh with a Linotronic 300 output unit. Second generation equipment—direct photographic typesetting and state of the art 15 or so years ago—has largely been replaced with third generation phototypesetting systems which are mostly digital in format and provide much faster typesetting output.

Fourth generation systems such as Scitex and Macintosh—which integrate graphic images and type as described in the preceding chapter—are now gaining a foothold in the industry largely because they are imagesetters and not just typesetters. Imagesetters have the production capacity to produce integrated pages—full pages—containing text and graphic images together, as opposed to typesetters which only set type that must then be manually pasted with graphic images. Section 9.6 of this chapter

Type of Manuscript	Keyboard Production (hr/M char)	Phototypesetting Unit Production—Multiplier of Keyboard Production										Proofreading Production (hr/M char)
		Mergenthaler VIP		APS Micro 5			Mergenthaler 202			Compugraphic 8400		
		5–18 pt	Over 18 pt	5–30 pt	31–60 pt	61–120 pt	6–12 pt	13–72 pt	73–120 pt	6–24 pt	25–72 pt	
Straight text matter with mix of italics and boldface	0.0952	0.12	0.18	0.01	0.01	0.01	0.01	0.01	0.02	0.01	0.01	0.0222
Capital and small capital letters with text	0.1190	0.14	0.21	0.01	0.01	0.01	0.01	0.01	0.02	0.01	0.01	0.0222
Tabular forms	0.1818	0.11	0.17	0.01	0.02	0.02	0.01	0.02	0.03	0.01	0.02	0.0333
Foreign text matter	0.1250	0.13	0.20	0.01	0.02	0.02	0.01	0.02	0.03	0.01	0.02	0.0222
Technical material with mixed text and symbols	0.2000	0.11	0.17	0.01	0.02	0.02	0.01	0.02	0.03	0.01	0.02	0.0444

Figure 9.4

Production schedule for estimating phototypesetting composition

details the estimating of desktop publishing, a popular and inexpensive fourth generation graphic image system. Chapter 10 addresses CEPS such as Scitex, Crosfield, and Hell which are extremely sophisticated and provide integrated output for multicolor images. CEP systems are typically found in sophisticated service bureaus, larger printing firms and prepress trade plants which produce finished film, proofs, or perhaps imaged plates for a printing company completing the final printing and finishing of the job.

The data presented in Figure 9.4 for estimating keyboard production and proofreading are based on hours per 1000 characters. This technique is also termed the *keystroke measure* of production. Estimating the production of the phototypesetting output is based on a percentage of keyboarding time since the greatest amount of time in production is spent keyboarding. The advantage of the keystroke procedure is that total manuscript characters quickly translate into typesetting production times, making the calculation of ems or any other base unnecessary. Of course, it is possible to estimate slug machine composition by the keystroke measure; a common industry standard in such a case is 0.10 hour per each 1000 characters set (10 point type).

Figure 9.4 divides the type of manuscript copy into five categories: straight text matter, capital and small capital letters mixed with text, tabular forms, foreign text matter, and technical material with symbols and mixed text. These classifications are the most common, but others can be developed to fit individual manuscript and typesetting circumstances.

The estimating procedure—using Figure 9.4—involves determining hourly keyboard production using the number of characters (in thousands) and then taking that hourly production figure and applying the multiplier to determine the phototypesetting unit production time. Proofreading is determined on the basis of keystrokes or characters, just as is keyboard production. The following formulas are used to estimate phototypesetting time and cost:

keyboard production time = number of characters ÷ 1000 × keyboard hours per M characters
keyboard production cost = keyboard production time × keyboarding BHR
phototypesetting unit production time = keyboard production time × phototypesetting unit multiplier
phototypesetting unit production cost = phototypesetting unit production time × phototypesetting BHR
proofreading production time = number of characters ÷ 1000 × proofreading hours per M characters
proofreading production cost = proofreading production time × proofreading BHR
total production time = keyboard production time + phototypesetting production time + proofreading production time

where the appropriate values are found in the phototypesetting composition schedule.

In the typical estimating situation with second generation typesetting equipment, such as the Mergenthaler VIP, keyboard production represents about 70% of total production time, the phototypesetting unit about 10%, and proofreading about 20%. With third generation typesetters—which are approximately 10 times faster than second generation in production output on the average—keyboard production increases to approximately 80% of the total production time, typesetting drops to about 1%, and proofreading to about 19%. Some printing plants and trade composition houses include the cost of proofreading in the BHR and do not estimate this segment of production, while others have a BHR for keyboard and phototypesetting unit production and a second lesser BHR separately identified for proofreading. Regardless of the system, proofreading must be accounted for.

Phototypeset material is produced easily in either film or paper form, both have excellent, high contrast images. Resin-coated photographic papers have become very popular because of their excellent dimensional stability and their cost savings over film. The paper-set material is used in the preparation of mechanical art (see Chapter 8). If photographic film positives are produced from the phototypesetter, they can move directly to platemaking with positive-acting plates or can be converted to film negatives using common film contacting procedures (see Chapter 10).

Example 9.8

A customer has provided us with 140 full pages of typewritten manuscript, average page size 6 1/2 × 8 inches, pica typewriter, double-spaced. The material will be typeset on a Mergenthaler 202, 9 point on 10, in Times Roman (alphabet length 118 points). The material will be set in straight text matter with italics. on a line length of 28 pics. Keyboard and phototypesetting BHR is \$56.70; proofreading BHR is \$22.40. Determine the total characters to be set, column depth of the set material, and production time and cost.

Solution Refer to Figure 9.2 and Figure 9.4 as necessary.

Step 1 Determine total characters from the manuscript:

Characters per line:
6 1/2 in./line × 10 char/in. = 65 char/line
Lines per page:
8 in./p × 3 lines/in. = 24 lines/p
Characters per page:
65 char/line × 24 lines/p = 1560 char/p
Total characters:
1560 char/p × 140 pp = 218,400 char

Step 2 To determine the column depth of set material, use the following formula for column depth:

$$\text{column depth} = \frac{\text{square picas per M characters} \times \text{M characters}}{\text{line length in picas}}$$

Step 3 Substitute the proper values in the formula (see Figure 9.2) to solve for column depth:

$$\text{col dp} = \frac{292.4 \text{ sq pi/M char} \times 218.4 \text{ M char}}{28 \text{ pi}} = \frac{68{,}860 \text{ sq pi}}{28 \text{ pi}} = 2281 \text{ pi}$$

Step 4 Calculate production time and cost (see Figure 9.4):

Keyboard production time:
218.4 M char × 0.0952 hr/M char = 20.79 hr
Phototypesetting unit production time:
20.79 hr × 0.01 = 0.21 hr
Phototypesetting production time total:
20.79 hr + 0.21 hr = 21.0 hr
Phototypesetting production cost:
21.0 hr × $56.70/hr = $1190.70
Proofreading production time:
218.4 M char × 0.0222 hr/M char = 4.85 hr
Proofreading production cost:
4.85 hr × $22.40/hr = $108.64

Step 5 Summarize your findings:

Total characters: 218,400 char
Total column depth: 2281 pi
Total production time required: 21.0 hr + 4.85 hr = 25.85 hr
Total production cost: $1190.70 + $108.64 = $1299.34

Example 9.9

A request has been received for a phototypesetting estimate of a chemistry book to be set with many technical symbols and formulas. The customer wants 10 point on 12 point Century Schoolbook (alphabet length 130 points), with a page size of 34 × 56 picas. The material will be set on a Compugraphic 8400. The book will signature into 320 pages, but the calculation for typesetting will be for 300 typeset pages. In addition, the total characters will be adjusted to 80% to allow for halftones, line drawings, charts, and white space. Keyboard and phototypesetting unit BHR is $49.70; proofreading BHR is $24.20. Determine total characters to be set, total column depth in picas, and time and cost for typesetting.

Solution Refer to Figures 9.2 and 9.4 as necessary.

Step 1 Determine total characters to be set:

Square picas per page:
34 pi × 56 pi = 1904 sq pi/p
Total square picas:
1904 sq pi/p × 300 pp = 571,200 sq pi
80% adjustment:
571,200 sq pi × 0.80 = 456,960 sq pi
Total characters:
456,960 sq pi × 2.59 char/sq pi = 1,183,526 char

Step 2 Determine the total column depth using the given formula:

total column depth = page depth × number of pages
56 pi/p × 300 pp = 16,800 pi

Calculate an 80% adjustment:

16,800 × 0.80 = 13,440 pi

Cross-check:

$$\frac{386.1 \text{ sq pi/M char} \times 1183.5 \text{ M char}}{34 \text{ pi}} = 13{,}440 \text{ pi}$$

Step 3 Calculate the production time and cost:

Keyboard time:
1183.5 M char × 0.2000 hr/M char = 236.7 hr
Phototypesetting unit production time:
236.7 hr × 0.01 = 2.37 hr
Phototypesetting production time total:
236.7 hr + 2.37 hr = 239.07 hr
Phototypesetting production cost:
239.07 hr × $49.70/hr = $11,881.78
Proofreading production time:
0.0444 hr/M char × 1183.5 M char = 52.5 hr
Proofreading production cost:
52.5 hr × $24.20/hr = $1270.50

Step 4 Summarize your findings:

Total characters: 1,183,526 char
Total column depth: 13,440 pi
Total time required: 239.07 hr + 52.5 hr = 291.57 hr
Total cost: $11,881.78 + $1270.50 = $13,152.28

9.6 Desktop Publishing, Desktop Typesetting and Imagesetters

As would be expected, the rapid evolution of fourth generation imagesetters—which have the capability to generate integrated text and graphics output in one unit—offer significant advantages over preceding generations of typesetting systems. Figure 9.5 summarizes the five currently identified groups of these units (also see Chapter 8) and the following discussion covers some of the more important advantages.

1. Perhaps the most significant advantage of electronic prepress systems is the complete working flexibility of stored digital data. For example, a digitized process color image from a color scanner can be manipulated and massaged on a Group I CEPS system, mixed with digitized type, again modified, color corrected, proofed for customer okay, changed a third time with the addition of revised text, reproofed for final customer okay, then output directly to film for image assembly. Interchangeability of digital data between systems and levels is also possible, although standards have yet to be fully developed here which would ensure full compatibility.

2. All five groups of electronic prepress systems provide a company with potentially greater throughput of jobs, an increasingly improved level of quality for more complex work, and streamlined production. For firms which have installed sophisticated CEPS systems—Groups I and II—employees will shift from a "craftsmen" orientation to workers who are much more "technology-driven." As a disadvantage, these "high-tech" firms will likely face manpower supply problems and training issues as they shift from conventional prepress to electronic methods.

3. By the year 2000, enhancements in imagesetting output devices—300 dot per inch (dpi) laser printers and higher resolution (2500 dpi) photographic imagesetters—are expected to improve greatly in both quality and speed. Group I and Group II systems will be able to directly image lithographic plates or produce precisely registered film for expedited stripping. Imagesetters for Groups III, IV, and V will improve in quality and production output as well. Currently, lower-end (Group IV and V) final images are printed on a laser printer to inexpensive paper, either as completed pages for paste-up into mechanical art or for direct photographic imaging on a process camera. If desired, a quality imagesetter—Allied Linotype's Linotronic 300 and 500 and Compugraphic's 9600 are examples of popular units—will provide output in either

GROUP I

General application and use

CEPS including image design, retouching, color modification, image enhancement, full page image assembly and high quality photographic film output

Major manufacturers of popular systems

Crosfield, Diadem Carat, DS America, Eikonix, Graphex, Hell, Scitex

Output mode

Soft proofs, direct digital proofs and fully imaged film ready for proofing or stripping

Minimum cost (est)

$200,000+

GROUP II

General application and use

CEPS systems with less sophistication. Most systems are high quality design platforms which can be used for design, retouching, color modification, and text manipulation

Major manufacturers of popular systems

Compugraphic's Mosaic II, Quantel's Video Paintbox, Linotype's Graphics 2000

Output mode

Soft proofs, direct digital proofs, conventional proofs from film

Minimum cost (est)

$80,000

GROUP III

General application and use

Most systems are high quality design platforms which can be used for design, retouching, color modification, and text manipulation; less integrated than Level I or II.

Major manufacturers of popular systems

Crosfield's Design Systems (integrates with Crosfield's Studio 875 CEPS), Lightspeed Design 20 (on a Sun Microsystem's platform), Lightspeed Color Layout System (on a Macintosh II platform), Unda Color Design and Production System, Networked Picture Systems "Package Express and Image Express (interface with Scitex CEPS), Hell's ChromaCom 1000, DuPont's Design Technologies Vaster Design and Imaging Systems

Output mode

Varies with the system

Minimum cost (est)

$30,000

GROUP IV

General application and use

Less expensive design platforms which can be used for various design and prepress production applications. Software is "off the shelf" on Macintosh II hardware.

Major manufacturers of popular system software

Various color painting and graphics software and sophisticated integrated desktop publishing programs

Output mode

Laser-driven printers, Allied's L300 or L500 and similar systems

Minimum cost (est)

$10,000 (includes black & white laser printer)

GROUP V

General application and use

Very expensive text input and design platforms which can be used to produce desktop publishing materials in black and white form. Hardware includes Macintosh 512e, SE and Mac II.

Major manufacturers of popular systems software

Various black and white desktop publishing software and other graphics and painting programs

Output mode

Laser-driven printers, Allied's L300 or L500 and similar systems

Minimum cost (est)

$5,000 (includes black & white laser printer)

Figure 9.5

Recognized groups of integrated and stand alone prepress systems

film form ready for immediate film assembly or to high resolution photographic paper for pasteup into mechanical art.

4. The flexibility of Group I and II CEPS systems, driven by customer demands and improved technology, will greatly streamline complex prepress color imaging. Enhancements in color monitors for soft proofs and direct digital proofing are certain to occur. Also, direct electronic links between the Group I and Group II systems and inexpensive, stand-alone computer-aided design (CAD) systems will make the entire prepress operation faster, less expensive, and more sensitive to customer-driven imaging issues.

5. The cost of Group IV and V units is low enough, and the platforms flexible enough, that literally every printing company will be able to offer desktop publishing and typesetting services. Software improvements will continue as well, making more sophisticated image development a routine process.

6. Postscript code has become recognized as the currently accepted page description standard, making communication among all five levels extremely easy and flexible. Postscript is also easy to learn to program, thus making personalized, complex designs available for anyone who learns this skill.

7. At all levels, the cost of reliable and inexpensive mass storage devices is a vital issue. Extremely large hard disk storage is now within affordable ranges, and floppy disk, smaller hard disks and magnetic tape storage are all available. The future will bring the development of even larger and faster storage devices, including optical laser disks as the first step.

8. Data transmission systems, particularly satellite transmission, will improve. Data compression procedures will certainly improve, allowing for faster, more accurate transmissions via telecommunication. Telecommunications via modems is expected to increase in speed from the current 1200 and 2400 baud rate to 9600 baud and higher.

Estimating Desktop Publishing. Desktop publishing (DTP) is a process whereby text and graphic images are input into a microcomputer, then manipulated and massaged to produce an integrated, final graphic product. The typical hardware requirements of a DTP system include a microcomputer system such as a Macintosh or an IBM-PC with a hard disk and one megabyte RAM storage, and a laser printer or Allied Linotronic phototypesetter. DTP software for the Mac includes a general word processing package such as Microsoft Word or MacWrite, various computer graphics programs such as Adobe Illustrator, MacDraw, MacPaint, or SuperPaint and an image manipulation program such as Aldus' PageMaker, Letraset's Ready-Set-Go, Quark's XPress, or Xerox's Ventura Publisher (for DOS systems). Peripheral hardware might include a scanner (for inputting photographs and predrawn images) and a graphics tablet, while there are numerous software programs for developing, modifying, or manipulating images.

Desktop publishing technology, while providing typesetting and graphic images at a low to medium quality level, continues to improve while hardware costs decrease and users are provided more flexibility, largely through software enhancements. Still, critics of DTP state that the quality of the images is not consistently good and that DTP systems have lead to printed and published materials that look amateurish. Nevertheless, the future of DTP using a Group IV or V platform is bright since general developments in the field will focus on improvements in quality and final image.

One area of growth almost directly related to the DTP surge is the development of hundreds of service bureaus across the country, in many cases operations that were previously typesetting companies. Such service bureaus provide DTP services through the interface of various typesetting and inputting functions, meeting the fast turnaround now taken for granted by customers. Also the general acceptance of Postscript as the standard electronic output code to drive laser or phototypesetting units has enhanced the phenomenal acceptance of imagesetting systems.

Figure 9.6 is provided for estimating desktop publishing. It is divided into five sequential steps: setup of the files and system, typesetting and spellcheck/proofreading, graphics input, image manipulation/modification, and printing. As will be noted, breakdown of certain groups into relative levels of difficulty is necessary. Production times are presented here for example only and will require adjustment based on the type of system software and hardware as well as operator skill. In fact, the relative skill of the operator in the graphics input and image manipulation steps will be essential to ensuring accuracy, so care should be taken here to review and job cost this area frequently. Example 9.10 is provided to demonstrate the application of Figure 9.6.

Budgeted hour rates for desktop publishing include recovering for the cost of computer hardware and appropriate software. The cost of copy paper (and toner) for the laser printer or photographic paper (and processing chemicals) for the imagesetting unit has been included as a departmental direct supply. Should a job require a specific software package that must be purchased for that job which will have a one-time use, or should there be additional DTP material costs directly related to a specific job or customer, these dollar amounts would be added on with an appropriate markup.

Desktop Publishing (DTP) Production Time and Cost:

total time for DTP production = preparation/setup + typesetting & proofreading + graphics input + image manipulation + printing

total cost for DTP production = total time for DTP production × BHR cost

Example 9.10

We have been asked to estimate the time required to produce an 8 page brochure using our Macintosh SE and our LaserWriter Plus. The following provides details of this job:

Typesetting: There are an estimated 13,200 manuscript characters to be set as straight text matter. Proofreading and spellchecking must be completed on the text.

Graphics: MacDraw will be used to generate 6 simple images and Adobe Illustrator will be used to produce another 4 complex (difficult) images. There will also be four continuous tone photos to be input from our Microtech 300 GS scanner at average level.

1. Preparation/setup	Fresh startup for new job: 0.20 hrs Modify existing files: 0.10 hrs		
		Typesetting input	*Spellcheck and Proofreading*
2. Typesetting & proofreading spellcheck	Straight text matter:	0.1000 hrs/M characters	0.0222 hrs/M characters
	Complex text matter:	0.1667 hrs/M characters	0.0350 hrs/M characters
	Technical text including equations and symbols:	0.2000 hrs/M characters	0.0444 hrs/M characters
	Foreign text:	0.1250 hrs/M characters	0.0222 hrs/M characters
	Formatting ASCII text:	0.03 hrs/M characters	

		Simple	*Average*	*Difficult*
3. Graphics input	Drawing programs:	0.10 hr/image	0.15 hr/image	0.30 hr/image
	B&W painting programs:	0.15 hr/image	0.20 hr/image	0.35 hr/image
	Color painting programs:	0.20 hr/image	0.25 hr/image	0.40 hr/image
	Illustration programs:	0.30 hr/image	0.75 hr/image	1.00 hr/image
	Scanner input:	0.20 hr/image	0.30 hr/image	0.45 hr/image
4. Image manipulation/ graphic image modification	Simple: 0.20 hrs per page Moderate: 0.30 hrs per page Complex: 0.40 hours per page			
		Simple	*Average*	*Difficult*
5. Printing	Laser Printer (paper)	0.03 hrs/page	0.06 hrs/page	0.15 hrs/page
	Phototypesetter (paper)	0.05 hrs/page	0.10 hrs/page	0.30 hrs/page

Figure 9.6

Production schedule for estimating desktop publishing.

Manipulation: PageMaker will be used as the image coordinating program, which we estimate will be done at moderate difficulty for four finished pages, complex difficulty for two pages and simple level for two finished pages.

Printing: Average level on our Apple LaserPlus printer.

Add for one fresh startups and three modify startups since the job will be done over a period of two days. BHR for our Macintosh SE and dedicated LaserWriter printer is $24.75.

Solution Refer to the schedule in Figure 9.6 for hourly production standards.

Step 1 Calculate the setup time:

1 fresh startup × 0.20 hr = 0.20 hr
3 modify startups × 0.10 hr = 0.30 hr
Total setup time: 0.20 + 0.30 = 0.50 hr

Step 2 Calculate typesetting and proofreading:

Typesetting:13.2 M char × 0.1000 hr/M char = 1.32 hr
Proofreading: 13.2 M char × 0.0222 hr/M char = 0.29 hr
Total typesetting and proofreading: 1.32 hr + 0.29 hr = 1.61 hr

Step 3 Calculate graphic input time:

Drawing: 6 simple images × 0.10 hr ea = 0.60 hr
Illustration: 4 complete images × 1.00 hr ea = 4.00 hr
Scans: 4 photos (average) × 0.30 hr ea = 1.20 hr
Total graphic input time: 0.60 hr + 4.00 hr + 1.20 hr = 5.80 hr

Step 4 Calculate manipulation time:

Moderate level: 4 pg × 0.30 hr = 1.20 hr
Complete × level: 4 pg × 0.40 hr = 1.60 hr
Total manipulation time:1.20 hr + 1.60 hr = 2.80 hr

Step 5 Calculate printing time:

Laser printer (average): 8 pg × 0.10 hr = 0.80 hr
Total printing time: 0.80 hr

Step 6 Calculate total estimated time for this job:

0.50 hr + 1.61 hr + 5.80 hr + 2.80 hr + 0.80 hr = 11.51 hr

Step 7 Calculate total hourly cost for this job:

11.51 hr × $24.75/hr = $284.87

Estimating Photographic and Electronic Prepress Systems

10.1 Introduction

This chapter addresses the photographic procedures used in the printing, publishing, and allied industries. It is safe to say that literally any printing company today—from quick printer or small commercial printer to very large printing firm—can accurately and reliably produce high quality photographic film products with minimum technical difficulty. Industry suppliers and equipment manufacturers of graphic arts photographic products—as with the other supply segments of the industry—have technologically advanced their products and services, effectively providing any printer with high quality, troublefree materials.

It is important to note that the photographic area—as with all other prepress manufacturing areas—has been broken into conventional and electronic segments. The movement from conventional to electronic imaging is occurring at an accelerated pace. Nevertheless, conventional photographic processes are expected to remain a strong force in many printing plants in the 1990s. As electronic imaging becomes less costly, more user-friendly and generally more accepted by designers and print buyers, conventional photographic techniques will be slowly abandoned.

Production of most printed goods no longer follows the standard manufacturing sequence used just a few years ago. Ten years ago, a printing job in a typical printing company began with mechanical art, which was photographically imaged to film. The job was then stripped, proofed for customer approval, corrected, and plated. Presswork and finishing operations concluded the process. Today, with varying degrees of electronic imaging in place—such as desktop publishing and CEPS—prepress operations are no longer clear-cut. For example, certain sophisticated printing companies utilize electronics throughout prepress manufacturing and use little actual film in producing the job. Other firms—less "high tech"—use electronics to perform only selected prepress, typesetting and design operations, and perform most prepress manufacturing using conventional technology. Still others use few electronic systems, preferring to produce jobs in the conventional manner.

10.2 The Changing Face of Graphic Arts Photography

As with all other areas of printing manufacturing, the photographic segment of the industry has been affected by numerous changes in technology. The following addresses the major technological impacts which have occurred during the 1980s.

Movement Away from the Process Camera. The process camera, considered the backbone of graphic arts photography for many years, has become only one of a number of the ways images are now generated in film form. Two major trends are apparent here. First is the movement to electronic imaging of film which includes color electronic prepress systems (CEPS), various electronic design and Macintosh platforms, color scanners and flat-field scanners. Second is the movement which incorporates conventional photographic and film technology and electronics together, and includes the Opti-copy system, and systems such as the Gerber AutoPrep 5000 which is an automated mask-generating tool. While technical improvements have been made with respect to process cameras—precise density control and imaging control devices as well as sensitometric units—the process camera appears to be slowly losing ground as the primary imaging method in the industry.

Rapid-Access Processing and Improved Film Products. No longer do most printing firms process films using tray development, nor use complex first-generation processors for film processing. The 1970s saw the evolution of first-generation automatic film processors, moving production from inconsistent tray processing to machine procedures. These early film processors had various chemical and technical limitations. In the 1980s, as technical chemical advances and modifications in photographic developers and emulsions were made, second-generation, rapid-access (RA) processing evolved. RA processing offered enhanced exposure control while stabilizing both physical and chemical processing variables. As the 1980s progressed, RA processing continued to improve, increasing film productivity since processing time was shortened and film remakes reduced. Development of an enhanced second generation of rapid-access films—RAL films—continues to improve this area of photographic production. These film products have improved density values, greater exposure latitude, finer grain, and enhanced stability and exposure consistency.

Greatly Improved Color Scanner Technology. In the past ten years there has been a rapid movement to color scanners for producing color separations, eliminating both enlargers and process cameras, the previous workhorses for this process. Graphic Arts Technical Foundation estimates that in 1980 there were approximately 800 color scanners in North America, while in 1988 there were about 2,400 units. The movement has been enhanced by the advent of the fully digital scanner, away from the analog-digital machines of the early 1980s. These modern digital scanners—using laser and electronic imaging procedures—can perform gray component replacement (GCR), undercolor removal (UCR) and undercolor addition (UCA) and serve as a vital electronic input tool for CEPS systems. In addition, scanner productivity has been greatly improved through reduced setup times, color preview systems, and various devices that provide image analysis before scanning. These result in faster throughput of jobs as well as fewer rescans, representing a cost savings to both printer and customer.

Growth of CEPS. Linked with the acceptance and technological improvements of digital color scanners has been the leap into imaging systems that have the potential to replace literally all manual operations in the prepress area over time. Color Electronic Prepress Systems—high-tech electronic systems which offer sophisticated image manipulation once an image is rendered in digital form from a scanner, typesetter, or computer—are fairly common in many printing plants and service bureaus today. CEPS are both a design and production tool. Output is typically finished film, ready for stripping. Adoption of digital data exchange standards, development and use of optical disk storage, improved data compression and satellite transmission, improved monitor resolution, integration with lower-level and less expensive platforms such as the Macintosh II, direct digital proofing and workstations for graphic design production represent factors that will ultimately force CEPS systems into many printing plants. Already we are seeing mechanical art production replaced by electronic imaging, a reduction in the use of PMT and other diffusion-transfer materials, less interest in conventional photographic procedures including process cameras and contacting procedures.

Movement of an Increasing Number of Photographic Procedures to Image Assembly, Where Jobs are "Built" During Production. Ten years ago, photographic departments in most printing plants did all photographic imaging—camerawork, contacting and duplicating, image manipulation and special photographic effects. In the early 1980s, customers began to demand five- and six-color work which required sophisticated photographic procedures such as traps, film composites, and precise fake color combinations. To accomplish these requirements, jobs began to be "tailor-made" in the image assembly area, which meant that the contacting and duplicating, normally done in the photographic department, moved into the stripping area. Today, printers producing sophisticated color work make use of either electronic CEPS systems or conventional manual/photographic systems. Those using conventional systems tend to rapidly complete the process camera portion of the job from mechanical art, then "tailor" the job to the customer's needs in image assembly. In sum, the stripping area has increased dramatically in its use of photography—particularly daylight contacting—while the photographic area has lost ground.

Diminishment of Enlargers as a Graphic Arts Photographic Tool. With the rapid movement to color scanning—and the numerous technological improvements in that area—photographic enlargers for color separation are almost never utilized. Furthermore, with the integration of many varieties of digital formats into graphic arts production, it appears likely that the enlarger will not find much further production use in the industry.

10.3 Overview of Conventional Photographic Procedures and Equipment

As shown in Figure 8.1 and explained in Chapter 8, there are four major types of artwork that enter the conventional graphic arts darkroom: preprinted images, typeset material, photographs, and original artwork. Each type of artwork may receive somewhat different treatment to provide a usable image for the follow-up prepress operations of image assembly, proofing, and platemaking. The discussion that follows provides an overview of terms and procedures for conventional graphic arts photographic operations.

Conventional Preprinted Images and Mechanical Art

Preprinted images, usually pasted up as mechanicals, come to the darkroom as reflection copy, with solid black images such as typematter secured to a white background. When the image is placed in the process camera copyboard and exposed to a sheet of line film at the desired size (reduced or enlarged), the white areas of the copy reflect light and become black in the negative (after development), while the black image areas absorb the largest portion of the light and remain clear in the negative.

The majority of paste-up artwork is opaque and allows no light to pass through, and the resultant film negative produced on the process camera is a high contrast transmission image—light will pass through in selected areas that are clear and will be held black in the dense black areas. Lithographic platemaking requires this high contrast film image.

Typeset Materials. This area was addressed in detail in Chapter 9 and need not be covered in detail here. In sum, typesetting can be completed either using conventional typesetting systems (with the output coordinated as a part of the completed pasteup art), electronically (with the output as complete pages which stand alone), or any mix of the two. Desktop publishing (DTP) provides for typeset material to be electronically integrated alongside graphic images, with final image output to a laser-driven printer or a more costly and higher-quality imagesetter. This can serve as finished art, pasteup art, or as an electronic file transportable to a higher level electronic imaging system.

Photographs. Photographs, the third category of artwork to enter the conventional graphic arts darkroom, may begin either as black-and-white, continuous-tone pictures that must be screened or halftoned or as full-color photographs that require color separation prior to printing. It is possible for photographic materials to be incorporated into mechanical art using PMTs.

When halftones are being produced using conventional photographic techniques, various exposure procedures are used to control "end point" dot structure—that is, the dots in the darkest areas (shadows) and those in the lightest portions (highlights), as well as mid-tone values. Densitometers and precise exposure controls are used, coupled with automatic film processing to ensure consistent image development.

Reproducing color images, using conventional direct or indirect photographic techniques, require that the color image be broken into four distinctly different "separations" using filters. The red filter produces the cyan (blue-green) printer, the blue filter produces the yellow (red-green) image, and the green filter produces the magenta (red-blue) separation. The fourth printer—black—is separated using a yellow filter and enhances the detail of the final image. Direct separations are normally completed on a process camera and provide the final four printers without intermediate film steps, while indirect separation requires the production of intermediate film images for color correction purposes. Since approximately 95% of all color separation production is done electronically by color scanner—from color transparencies or reflective color prints—the details of both direct and indirect separation need not be covered here in detail. Suffice it to say that neither direct or indirect separation is completed at any significant volume at this time.

It is important to note that automatic film processing—either conventional "lith" processing or rapid-access processing—is essential to accurate color separations. This is true regardless of whether the photographic technique is conventional or electronic. Automatic film processing controls both physical and chemical development variables, which inherently affect the results if not monitored and maintained carefully.

Original Art. Original artwork, the fourth category of conventional art preparation, can be divided into black-and-white and color images, requiring reproduction in line or tonal form. Line work might include an inked line drawing with no shading or a drawing with line shading in the form of crosshatching. Frequently, high contrast stats or photomechanical transfers (PMTs) of line drawings are used as mechanical art. These stats may be made to a certain size through enlargement or reduction from the original, then incorporated into the mechanical art with type and other graphic images

Tonal artwork images could include black-and-white charcoal drawings or a black-and-white artist's wash. Both necessitate halftoning procedures to translate the shades of value into high contrast dots. Here again, an intermediate step of providing either a high contrast stat or a proof of the haltone for use in a line work mechanical is likely to be necessary. Full-color paintings, watercolors, or other naturally colored images require color separation using either the direct or indirect method. Whatever the form of the original artwork, it is vital that the resultant photographic product of any graphic arts darkroom procedure be a high contrast image because the lithographic plate does not carry gray.

Much original art—particularly color images—are electronically separated using a color scanner. The most common procedure to do this is to photograph the original art image using a 35 mm (or larger format) high-quality camera, yielding a color transparency of the original image. This transparency—accurate in color translation from the original—is then used as the image from which all electronic scanning is completed. Control of color between the original image, the transparency, and the resultant separations becomes a critical issue, even though digital color scanners have the capacity to color correct for some differences. Color correction is also used to adjust for ink pigment variations and color or reflection variations in paper stock.

Electronic color scanners are standard in many medium and larger printing plants. For printers who are unable or unwilling to invest in this technology, color separation trade houses provide this service, usually along with many other prepress services such as stripping, proofing and platemaking. Quick printers and smaller commercial printers usually purchase color separation images from such trade shops, while medium size and larger printing firms typically own their color scanning equipment and related electronic units.

Conventional Photographic Imaging Equipment

Process Cameras. Process cameras are used to photographically render mechanical art to the exact size desired for final production. Such cameras photograph two-dimensional images and reduce or enlarge by movement of their copy and lens planes, while the *focal plane* (also called a *vacuum back* where vacuum suction holds film securely during exposure) remains stationary. Many can reduce the copyboard image to as small as one-fourth size (25%) and enlarge up to three times the original size (300%). Same-size images are exposed at 100%. Small plants generally use a gallery camera, such as the Robertson Meteorite or Kenro Vertical, which is totally enclosed in the darkroom and limited with respect to copyboard and film size. Large plants may use either floor or overhead process cameras that have the focal plane (vacuum back) built into the darkroom, with the remainder of the camera in an adjoining area. These cameras are classified as darkroom cameras and generally have large copyboard and film size capability.

Process cameras may be purchased with various types of light sources including pulsed xenon, quartz, and incandescent. Artwork may be either reflection mechanicals or transmission (film) copy when cameras are able to handle both types; usually reflection copy is used. Process cameras are excellent for the production of both line and halftone images; halftones are commonly screened with contact screens which contain many continuous tone dots into which the original photograph is broken. Process cameras may be used in the production of color separations using either direct or indirect separation techniques and color filters, but this is not common today.

Photographic Enlargers. Graphic arts photographic enlargers were at one time very popular for completing color separation production. Each separation exposure required careful control and considerable operator skill. In the past ten years, as color scanners have evolved into fast, flexible, color separation units, enlargers have played less and less of a role in the industry.

Enlargers can still be found when production of black-and-white halftones are needed, since these can be exposed directly from original continuous tone film images in one step. Enlargers are also still used as a controllable light source for exposing orthochromatic film materials during the contacting process.

Opti-Copy. Opti-copy is a photographic projection technique by which an image can be reduced or enlarged and then exposed directly to film in the desired size. Opti-copy systems are commonly found in printing firms that photocompose multiple images on one sheet of film, such as labels to run 24-up or coupons to run 16-up. To do this, one image, or an accurately related group of images (4-up), are used to project multiples directly to film. The major advantages of an Opti-copy system are its accuracy of image and high production output when compared to conventional step-and-repeat systems found in platemaking.

Film Contacting and Duplicating. *Film contacting* is the production of a film image that has opposite tonality from the original and is completed using negative-acting materials. Thus, a film negative is contacted to produce a positive, or a positive is contacted to negative form. *Film duplicating* utilizes positive-acting photographic materials in the production of images that have the same tonality. For example, a film negative will be used to make other negatives or a positive to make other positives.

Film contacting procedures are found in both photographic and film assembly areas of printing companies. Because many commercial printers specialize in multi-color production with jobs tailor-made to customers, completing contacting and duplicating procedures in film assembly has become common.

For printing companies doing significant amounts of complex color work, fitting images together precisely becomes a critical matter. To accomplish this, photographic spreads and chokes are completed by contacting, so that images fit—or trap—properly. Contacting is also used to produce film composites, whereby many film images are photo-composed onto one piece of film, primarily to save time during the stripping and platemaking processes which follow. This is sometimes termed "composed film " and is referenced in the industry on live jobs to "composed film furnished." This means that the final film given to the printer will be the fewest number of pieces for film assembly, thus saving time both in film assembly and platemaking and proofing.

Equipment for accurate contacting consists of a vacuum frame, suitable light source, and an accurate timing mechanism. The vacuum frame ensures positive, direct contact between previously-imaged film and raw (unimaged) film. Generally a large vacuum frame setup can be obtained fairly inexpensively compared to other imaging equipment such as a process camera or electronic color scanner.

Both darkroom and daylight film products are available. Darkroom films are orthochromatic (red-blind) and require that the contact setup be placed in a safelighted, controlled environment. Daylight film products, popular in stripping departments, allow the contact frame and associated equipment to be situated in a fairly bright—yet spectrally-controlled—environment. Both daylight and darkroom films are available as contact films (which reverse polarity, going from negative to positive to negative, etc. with each photographic step) and duplicating films (which keep the same polarity, negative to negative to negative, or positive to positive to positive, through each photographic step). Film processing with both contact and duplicating films can be completed manually or automatically, with automatic rapid-access processing systems fast, reliable, and popular today.

In sum, film duplicating and contacting procedures can be used to do the following:

1. Make duplicates of existing film images
2. Make reverses of existing film images
3. Produce spreads (slightly enlarged) from existing negative film images
4. Produce chokes (slightly reduced in size) from existing positive film images
5. Make composite film images, collecting many same-color images on one piece of film

The usual procedure for film duplicating (or contacting) begins after the pasteup or half tone has been reduced or enlarged to the desired size and reproduced in film form. Since the procedure requires contact between films, sizing of the image must be completed prior to the contacting procedure. Working at the vacuum frame and under the appropriate safelight for the film, the unimaged film is placed on the open frame, with the imaged original over it; the frame is then closed and the vacuum drawdown applied. When the vacuum pressure reaches the desired level, thus insuring intimate contact, a timed exposure is given. When completed, the exposed film sheet is processed and the resultant duplicate or contact negative should be identical to the original negative.

As a technical note, it is possible to contact films either *emulsion-to-emulsion* (E-E) or *emulsion-to-base* (E-B). Photographic films consist of a clear base, typically from 0.004 to 0.007 inch thick, with a light-sensitive emulsion coated on one side. This emulsion is essentially a photographic gelatin holding light-sensitive silver salts in suspension. The problem when contacting is that the thickness of the base can cause changes in the size of very fine dots and lines or may cause image loss altogether.

When emulsion-to-emulsion contacting procedures are used, which require an intermediate film image (when right-reading, the intermediate is emulsion-up and not strippable), the film emulsion is always placed in direct contact with the next piece of film. Thus, the original negative is contacted to a piece of film E-E, producing an intermediate negative, and this intermediate is then contacted to another piece of film (again, E-E), producing the desired duplicate negative. A major disadvantage of E-E contacting is that the intermediate ultimately is discarded and plays no part in future production operations of the job, which is costly.

Emulsion-to-base contacting is perhaps more popular because no intermediate is required. However, E-B contacts usually require longer exposures and may slightly change fine halftone dots and line images since exposure is completed through the film base. When estimating contacting and duplicating procedures, plant production generally dictates the technical specifics that apply.

Film contacting has important applications in printing production. For example, if a job is to be imaged on a lithographic plate 16-up, film duplicating could be used to take one negative and make the remaining fifteen. This might provide some economic benefit to the customer as opposed to using step-and-repeat techniques, whereby one film image would be moved about during platemaking to produce the 16-up result. Or if a line of type was to reverse in a black background and then print in a PMS color—which is called a trap—a spread of the overprint type would be made, allowing the PMS color to slightly overlap the reversed area and provide a quality result. If a job has many stripped flats to burn on one plate, using contacting to photo-compose these many images into one piece of film—called compositing—truly streamlines production in some cases.

Because of the profound economic effect that film contacting has on job quality and printing production, it is vital that the estimator has a clear understanding of this process. Of all the conventional photographic procedures available, contacting is the most versatile and dynamic, and therefore somewhat difficult to estimate.

Conventional Proofing Systems and Materials

Photographic proofing serves one or more of the following purposes. Proofs are used:

As an intermediate visual look at a job-in-process (or a segment of a job-in-process),

As a portion of a job requiring photographic work as a segment of art preparation where the photographic work being proofed will then become a part of the mechanical paste-up,

As a final visual look at a job for internal plant approval or for customer review and approval.

Monochromatic Proofs. The most popular black-and-white—or monochromatic—proofing material is Dylux, which is a paper substrate manufactured with an encapsulated imaging material, broken by exposure to actinic light. Since no contact with any liquid is necessary for processing, Dylux proofs, also known as bluelines or blueline proofs, are extremely dimensionally-stable. Exposure requires an intense ultraviolet light source, typical of contact and platemaking units found in most printing companies. Exposing the Dylux proof to a clearing filter will render the image permanent.

Other monochromatic proofs include brownline or silverprint materials, rapid-access silver halide products such as DuPont's Bright Light Papers (BLP) and 3M's LOP papers, various azo products, and stabilization proofs such as Ektamatic and Fotorite materials. Certain monochromatic proofing materials are available in single-faced (coated one side, or C1S) or double-faced (coated both sides, or C2S) for proofing two sided jobs and bookwork impositions to check backup positions.

Color Proofing Materials. Color proofing materials are generally categorized into two types: press proofs, and off-press proofs. Off-press proofs are then further subdivided into overlay proofs and single sheet proof classifications.

Press Proofs. Color press proofs are still used extensively when a larger quantity of proofs is desired, particularly for advertising and packaging purposes. Such proofs require the manual stripping of separated materials, then platemaking and printing on a press. When done by a color separation trade house, the press sheets are normally ganged with many images, usually many different jobs proofed on one sheet. After presswork is complete, the flats are torn apart and the separations sent to the customer. Press proofs are generally expensive, since all the work done is for review only and largely nonproductive in terms of delivering finished copies to the customer. The primary advantage of press proofs—and the main reason for their popularity in light of their high cost—is that they most closely approach the final printed version of the color image.

Overlay Proofs. These proofs consist of single acetate sheets of color pigment which, when exposed, processed, and superimposed over one another provide a composite image of the final print. Overlay proofs are relatively low in cost and can be produced quickly and easily in most printing plants. Color specification standards such as SWOP (Standard Web Offset Publications) can be matched. Opinion on the quality of the proofs varies among industry users, particularly related to the fact that pigment lots vary from box to box and that the acetate sheets impart an undesirable grayness to the final image, thus rendering the accuracy of the proof questionable to some customers.

3M's Color Key and Enco's NAPS (Negative-Acting Proofing System) and PAPS (Positive-Acting Proofing System) are standard products in this group. These materials, which are pigmented diazo products, are exposed to the appropriate film image using ultraviolet light and then developed manually or automatically using the required processing chemicals. Color Key is available in standard process colors plus 26 Pantone Matching System (PMS) color choices, while NAPS/PAPS is available in process colors plus a variety of standard flat colors. There are differences between Color Key and Enco's products, but, for all intents and purposes, the final physical proofing results appear much the same.

DuPont's Chromacheck is an overlay negative-acting proofing system which produces images using a post-exposure peel-apart system. No processing is required and the process is completely dry. The colored pigment is an adhesive photopolymer material; exposure to ultraviolet light renders the image areas adhered to the base, and the proof image is produced when the two films are peeled apart using a vacuum easel or an autopeeler system which can be purchased from DuPont.

Single-Sheet Proofs. While this category could be subdivided by type of proofing system, the following represent the more popular single-sheet proofs currently used in the printing industry.

Cromalin (DuPont). This is a popular, yet somewhat expensive, proof using a photopolymer adhesive and dry toners. The procedure requires a heated laminator to apply a thin acetate film—carrying a photo-sensitive polymer coating—to an enameled substrate such as Kromecoat cover. Film separations with the desired image are then positioned over this paper/acetate laminate, and exposure to ultraviolet light is administered. The thin acetate carrier sheet is then peeled away, leaving photopolyer areas which are sticky from the exposure. Immediate development of the sheet is completed, where dry color toners are carefully applied and adhere to the sticky areas, rendering a final image. Processing can be done either automatically using an automatic toning console (ATM) or manually. The procedure is then repeated—laminating, exposing to the next film image, peeling, and processing of the next color pigment—with the final result a trapping of colors on top of each other. Cromalin is available in both positive- and negative-acting types and toner can be matched to SWOP or PMS standards.

Matchprint II (3M). Originally known as Transfer-Key, Matchprint II has evolved into a popular single-sheet proofing product. The materials required are a base sheet, pigmented diazo photosensitive film, a heated laminator and an automatic Matchprint processor (which can also be used for processing 3M ColorKey). The proofing process involves laminating a pigmented photosensitive film material to a base sheet, peeling away the thin acetate carrier (leaving only the pigment), then exposing the silver film image to the proof material with ultraviolet light. Automatic processing follows, and then a repeat of the sequence for each color. Various bases are available such as Publication Base (groundwood papers such as newsprint) and Commercial Base (coated offset book paper). Matchprint II—introduced in 1985 —is available in both positive- and negative-acting formats and can be matched to SWOP, SNAP, or PMS standards.

10.4 Overview of Electronic Imaging Systems

As discussed in Chapters 8 and 9, electronic imaging systems integrate prepress production so that design, typesetting, imaging, stripping, and proofing can be done either on stand-alone platforms or with fully integrated equipment. Figure 10.1 is reproduced here (from Chapter 9) to segment the many electronic imaging systems into five groups. The following provides information relative to the photographic and imaging portions of these electronic systems.

Photographic Scanners

Photographic scanners are a marriage of electronic and photographic technology, and improvements in color scanners have paralleled improvements in both areas over the past twenty years. In general, the process of color scanning involves placement of a color print or color transparency on the copy drum of the scanner unit. As the drum rotates, the print or transparency is scanned from end to end with a controlled light. Next to the copy drum is a film drum, or film cassette, which holds film that receives the scanned image. This cassette is light-tight, allowing scanner operation to be completed in normal room lighting.

As the light source moves slowly across the copy drum, it "reads" reflected light, sending signals through a color filter and then into the scanner electronics. A second light source located in the film cassette receives these signals and exposes the film accordingly. As needed, the electronics of the scanner can be adjusted to modify the electronic signal sent to the film cassette, thereby "color correcting" the final image on the scanned film.

There are essentially two classifications of color scanners: analog-digital and digital. Analog-digital (or analog) scanners represent fairly old technology and were

GROUP I

General application and use

CEPS including image design, retouching, color modification, image enhancement, full page image assembly and high quality photographic film output

Major manufacturers of popular systems

Crosfield, Diadem Carat, DS America, Eikonix, Graphex, Hell, Scitex

Output mode

Soft proofs, direct digital proofs and fully imaged film ready for proofing or stripping

Minimum cost (est)

$200,000+

GROUP II

General application and use

CEPS systems with less sophistication. Most systems are high quality design platforms which can be used for design, retouching, color modification, and text manipulation

Major manufacturers of popular systems

Compugraphic's Mosaic II, Quantel's Video Paintbox, Linotype's Graphics 2000

Output mode

Soft proofs, direct digital proofs, conventional proofs from film

Minimum cost (est)

$80,0C0

GROUP III

General application and use

Most systems are high quality design platforms which can be used for design, retouching, color modification, and text manipulation; less integrated than Level I or II.

Major manufacturers of popular systems

Crosfield's Design Systems (integrates with Crosfield's Studio 875 CEPS), Lightspeed Design 20 (on a Sun Microsystem's platform), Lightspeed Color Layout System (on a Macintosh II platform), Unda Color Design and Production System, Networked Picture Systems "Package Express and Image Express (interface with Scitex CEPS), Hell's ChromaCom 1000, DuPont's Design Technolcgies Vaster Design and Imaging Systems

Output mode

Varies with the system

Minimum cost (est)

$30,000

GROUP IV

General application and use

Less expensive design platforms which can be used for various design and prepress production applications. Software is "off the shelf" on Macintosh II hardware.

Major manufacturers of popular system software

Various color painting and graphics software and sophisticated integrated desktop publishing programs

Output mode

Laser-driven printers, Allied's L300 or L500 and similar systems

Minimum cost (est)

$10,000 (includes black & white laser printer)

GROUP V

General application and use

Very expensive text input and design platforms which can be used to produce desktop publishing materials in black and white form. Hardware includes Macintosh 512e, SE and Mac II.

Major manufacturers of popular systems software

Various black and white desktop publishing software and other graphics and painting programs

Output mode

Laser-driven printers, Allied's L300 or L500 and similar systems

Minimum cost (est)

$5,000 (includes black & white laser printer)

Figure 10.1

Recognized groups of integrated and stand alone prepress systems

used as a direct production scanner. Because memory capacity was limited, analog scanner output was directly to film, which was then used in production of the customer's job. Also, analog units typically did not use laser light when producing the scanned separations, with the final images lacking extremely fine dot structure.

Digital scanners evolved as technology improved the analog-digital units. By the mid-1980s literally all new scanners were digital, typically with prices of $150,000 or more. Digital units can either be used as direct production machines or scan images into electronic memory for later use on CEPS systems. Also, modern digital scanners allow the operator to perform many image modification procedures including gray component replacement (GCR), undercolor removal (UCR), and undercolor addition (UCA). To improve productivity, the operator can work on setting up one job while the scanner is outputting another. Also, setup times and rescanning times have been reduced using various tools such as prescan analysis and color monitors, which allow the operator to view a fairly accurate image before it is separated. Today's digital scanners also have improved film loading and processing mechanisms.

Perhaps one of the most significant future trends is the development of color scanning procedures tied to low-cost microcomputer platforms (see Group IV in Figure 10.1). With the development of stand-alone color copying and the rapid advance of low-end scanners for black-and-white DTP, it seems fairly safe to predict that the world of color scanning will move toward less costly, lower-end units over the next decade.

Digitizing Platforms

As noted in Figure 10.1, various levels of electronic imaging are available to any printing company. Each has been discussed in Chapter 8 as design and image development tools, and Chapter 9 relative to desktop publishing and typesetting. The lines between the groups are not clear-cut, with future changes and technological advances likely to reduce the number of groups to less than five. Generally, Group I is limited to fairly exclusive use by sophisticated printing, publishing and trade shops providing full-service color imaging, while any individual can purchase or rent Group V equipment for producing black-and-white DTP images. Groups II, III, and IV are intermediate electronic design and imaging units and can be found throughout the industry.

Color Electronic Prepress Systems

As indicated on Figure 10.1, CEPS systems (Group I) include integrated electronic design, retouching, color modification, image enhancement, full page image assembly, and high quality photographic film output. While it is difficult to precisely identify all the reasons CEPS have become the focused imaging process of the late 1980s, the following represents what are generally recognized as the more important goals:

1. To produce an exceptionally high quality image, superior to conventional prepress imaging procedures;

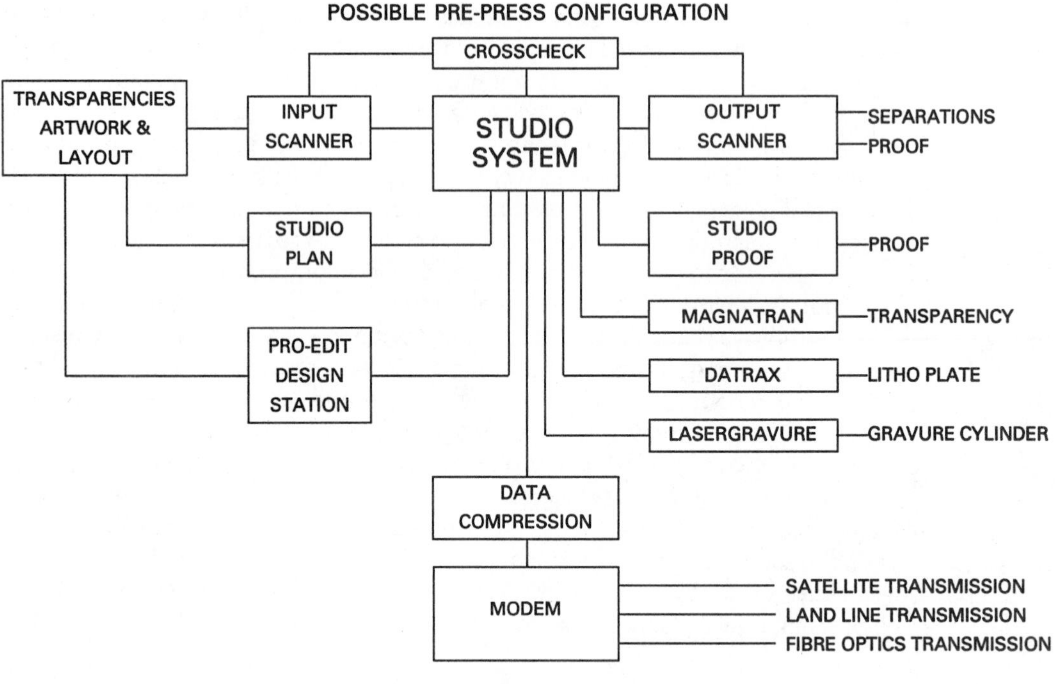

Figure 10.2

Possible pre-press configuration block diagram using the Crosfield 870/880 Studio System. (Reproduced with prmission of Crosfild Electronics Inc., U.S.A. Headquarters.)

2. To reduce production time and cost, particularly when sophisticated, high quality image production is desired;
3. To be able to design and produce complex color images accurately and quickly, and in a customer-responsive manner;
4. To provide the customer with faster and more accurate approvals, reducing major (and expensive) on-press job changes;
5. To increase the efficiency and productivity of prepress employees, allowing for increased volume of sophisticated color images;
6. To provide flexible tools by which graphic images can be enhanced and uniquely designed, at an economical cost to the customer.

In sum, CEPS systems have evolved through technological advances, driven by increasingly competitive customer demand for complex color on a fast-turnaround basis. CEPS serve to diminish production bottlenecks evident throughout the prepress manufacturing area, thus streamlining production; they also improve the quality of graphic images demanded by the customer, and interface design and production together as one manufacturing process. Figure 10.2 provides a block diagram of the

Crosfield 870/880 CEPS system to indicate the general links and interfaces of one manufacturer of such equipment. Other CEPS system configurations—Scitex, Hell, DS America, and others—vary somewhat from Figure 10.2, since each manufacturer takes a different approach to image processing.

Direct Digital Proofing and Color Monitor Enhancements

One of the most significant limitations of CEPS has been that no effective proofing system has been available until late 1988. Prior to that time proofs were generated using conventional film technology, which required that the image be finished and output in film form, then proofed using conventional Cromalin or other proofing operations. This was time-consuming and fairly expensive. At this writing direct digital proofing—directly from the CEPS system, without photographic film—is on the verge of become commercially available. Crosfield/3M, Kodak, DuPont, Coulter, DX Imaging, and Polaroid all have systems in differing stages of development, including beta test sites. Various proofing technologies are being tested including polymers, laser-exposed coatings, and electrostatic methods.

The development of significantly improved color monitors (producing what is termed a "soft proof") for CEPS is also receiving major industry interest. There are numerous issues here, yet two of the most important are fineness of image and relative control of color, so that the monitor image resembles final printed image as closely as possible.

10.5. Photographic Production Planning

As with most other prepress areas, estimating both conventional and electronic prepress operations requires detailed production planning. It is absolutely essential that the estimator have a complete knowledge and understanding of the processes used in his plant to photographically produce the images required, whether they are conventional, electronic, or a mix of the two. It is also important that the estimator understand and interrelate the type of artwork provided to the photographic area, all job requirements indicated important to the customer, and any specific production details critical to the job. Clearly the estimator must also evaluate each production plan in light of the cost to both the customer and the company.

Copy and Artwork Specifications

In efficient printing companies, an estimator, a customer service representative, or a production planner reviews all artwork as it enters production. It is at this point the job is broken up into various segments and potential problems initially addressed. In general the better the artwork and copy at this stage, the more smoothly the job will be able to be produced. If any abnormalities are apparent, the customer service representative or estimator should be informed and assess these relative to the cost impact on the customer.

Specific Job Requirements

Sometime during the review process each job should be broken up into component parts, with the production planner developing a list or log of anticipated line, halftone, color separation, contact, and other prepress work to be completed. The job should be laid out and dummied, with the estimator or production planner carefully analyzing how the print order will move through the manufacturing sequence. Dozens of factors are important here, such as camera copyboard size, contact frame sizes and production output, availability of equipment, anticipated quality level of product, and trade vendors meeting their turnaround requirements.

Plant Production Details

Manufacturing and production requirements vary from job to job. It has been said that no job done in a printing company is produced exactly like any other.

The estimator must be thoroughly familiar with the specific production opportunities available using current equipment and personnel and should work toward the most economical manufacturing sequence. For example, it may be desirable to produce the job 1-up on the press sheet, or perhaps 4-up or 8-up. The latter choices would necessitate multiple image production during design and art production, or perhaps additional film exposures in the photographic area or step-and-repeat techniques when the job moves into image assembly. It is essential that the estimator consider all production possibilities, factoring in both conventional and electronic prepress operations, presswork configurations, and finishing procedures.

Cost to Customer and Company

Certainly one of the most important jobs of the estimator is to evaluate each manufacturing step in light of the cost variations between production plans. Such cost determination is affected by the budgeted hour rate of the plant center where the work will be done, the productivity of the employees doing the work, the complexity of the job, and the relative time which will be required to complete the job. Such cost evaluations should include a fair assessment of all materials required, including anticipated additional charges to be paid by the customer for author's alterations (AA's) and customer-requested job changes. Of course any errors made by company employees during the production of the job must be borne by the company and cannot be fairly charged to the customer.

10.6 Estimating Line, Halftone, and Diffusion-Transfer Production Costs

Figure 10.3 is provided for estimating line, halftone, and diffusion-transfer production. Chart times are presented in decimal hours per exposure for both manual and automatic film processing. To the right side of the schedule is the standard production time for the exposure and processing of diffusion-transfer materials, such as PMTs. These products also require automatic processing, which is completed using a special tabletop unit, not the processor unit used for film products. PMTs have become a popular intermediate photographic step in the preparation of paste-ups and mechanicals.

Chart times include placement of the copy in the copyboard, adjustment of the camera for reduction or enlargement, normal exposure, and processing with the applicable tray or automatic unit. Schedule data assume that the artwork to be used is of good quality. Adjustments in production times are to be made for poor-quality artwork and filter exposure operations. Time and cost for remakes of film images that are not usable during job production have not been included in the schedule data.

Film and chemistry costs have been summed together and calculated on a basis of cost per square inch of processed film. Costing data for plastic-base line film, polyester-base "lith" halftoning film, and PMT products are provided. Such costs should be adjusted for each individual plant, as should the indicated chart times.

	Production Time in Hours		
	Line per Piece	Halftone (contact screen) per Piece	Diffusion-Transfer Materials
Manual processing	0.15	0.20	n/a
Automatic processing	0.12	0.15	0.07

Note: Time and cost for remakes not included. Processing is for all film sizes.

Adjustments: For poor quality artwork, increase times 25%. For filter exposures, increase times 20%. Consider duotones as two halftones.

Material and processing costs (based on square inches of required materials): Plastic-base line film: $0.016/sq in.; polyester-base (halftoning) film: $0.021/sq in.; PMT negative-receiver paper: $0.012/sq in. (both pieces included)

Figure 10.3

Schedule for estimating line and halftone production costs

Example 10.1

A customer requests an estimate for a 16 page booklet with cover, untrimmed page size 6 1/4 × 9 1/2 inches. Job specifications indicate that all 16 pages will be paste-up typematter to be shot same-size. They will be gang-shot 4-up on the copyboard and exposed to standard 16 × 20 inch plastic-base line film. In addition, there will be 10 halftones to fit with the line work, and they will be shot individually on 5 × 7 halftoning film. The 2 outside cover pages will print in black and red, in register, and will be prepared as mechanicals with overlays; they will be exposed separately on 8 × 10 inch polyester film. The inside front and back covers will be blank. All film processing will be manual. The BHR for the process camera including film processor is $64.50. Estimate production time and cost for the camera including material costs.

Solution Refer to Figure 10.3.

Step 1 Use the following formula to determine the camera production time required to shoot the booklet line work when 16 pp (pieces, pc) are shot 4-up on 16 × 20 in. line film and the manual processing time for each exposure is 0.15 hr:

(number of pieces of art to be photographed ÷ number-up on copyboard) × production time per piece = production time
(16 pc ÷ 4-up) × 0.15 hr/pc = 4 pc × 0.15 hr/pc = 0.60 hr

Step 2 Determine the camera production time required to shoot 10 halftones separately (1-up) on 5 × 7 in. film at 0.20 hr each:

10 pc × 0.20 hr/pc = 2 hr

Step 3 Determine the camera production time required to shoot the cover line work. (The 4 line images—2 black and 2 red—will be shot separately on 8 × 10 in. polyester halftoning film and will require 0.15 hr each to shoot.):

4 pc × 0.15 hr/pc = 0.60 hr

Step 4 Determine total camera production time:

0.60 hr + 2 hr + 0.60 hr = 3.20 hr

Step 5 Use the following formula to determine total camera production time cost (not including materials):

production time × BHR = production time cost
3.20 hr × $64.50/hr = $206.40

Step 6 Use the following formula to determine material and film cost:

number of pieces of material and film × number of square inches per piece × material and film cost per square inch = material and film cost

For line film:
4 pc × (16 in. × 20 in.)/pc × \$0.016/sq in. = \$20.48
For halftones:
10 pc × (5 in. × 7 in.)/pc × \$0.021/sq in. = \$7.35
For the cover:
4 pc × (8 in. × 10 in.)/pc × \$0.021/sq in. = \$6.72
Total material and film cost:
\$20.48 + \$7.35 + \$6.72 = \$34.55

Step 7 Use the following formula to determine the total cost for process camera production:

production time cost + material cost = total production cost
\$206.40 + \$34.55 = \$240.95

Step 8 Summarize your findings:

Total production time: 3.20 hr
Total production time cost: \$206.40
Total material cost: \$34.55
Total production cost: \$ 240.95

Example 10.2

We have a job that consists of 32 line shots, 18 halftones, and 14 PMTs. Line art will be shot on 11 × 14 inch plastic-base line film, halftones on 8 × 10 inch halftoning polyester film, and PMTs on 8 1/2 × 11 inch PMT paper. All exposures will be l-up, and automatic processing will be used. The BHR for our camera is \$58.80. Estimate production time and cost for camera work for this job including material costs.

Solution Use Figure 10.3 and the appropriate formulas.

Step 1 Determine time required to shoot line work:

32 pc × 0.12 hr/pc = 3.84 hr

Step 2 Determine time required to shoot halftones:

18 pc × 0.15 hr/pc = 2.70 hr

Step 3 Determine time required to shoot PMTs:

14 pc × 0.07 hr = 0.98 hr

Step 4 Determine total camera production time:

3.84 hr + 2.70 hr + 0.98 hr = 7.52 hr

Step 5 Determine total camera production time cost (not including materials):

7.52 hr × $58.80/hr = $442.18

Step 6 Determine material and film cost:

For line film:
32 pc × (11 in. × 14 in.)/pc × $0.016/sq in.= $78.85
For halftones:
18 pc × (8 in. × 10 in.)/pc × $0.021/sq in.= $30.24
For PMTs:
14 pc × (8 1/2 in. × 11 in.)/pc × $0.012/sq in.= $15.71
Total material and film cost:
$78.85 + $30.24 + $15.71 = $124.80

Step 7 Determine the total cost for process camera production:

$442.18 + $124.80 = $566.98

Step 8 Summarize your findings:

Total production time: 7.52 hr
Total production time cost: $442.18
Total material cost: $124.80
Total production cost: $566.98

10.7 Estimating Film Contacting and Duplicating Production Costs

Figure 10.4 is provided for estimating film contacting and duplicating production time and cost. To do this, the estimator should identify the following:

1. The size of the vacuum frame to be used
2. The type of processing to be used

3. The number of film images to be contacted at one time (the number up)
4. Whether the contacting process will be emulsion-to-base (E-B) or emulsion to emulsion (E-E)
5. Whether film compositing or spreads or chokes will be required
6. The type of imaging material to be used

The data in Figure 10.4 assumes that the original film materials are dry and suitable for contact exposure. The schedule is organized in a step-by-step fashion, beginning with preparation and setup, moving to contacting production time, and then making adjustments as they apply. Materials costs are added based on square inches of required products. Remakes are not included. The appropriate BHR for the contacting center will be used.

Just as when determining the number of line and tonal exposures using the process camera, estimating contacting and duplicating procedures require that the estimator or production planner develop a system for breaking each job into basic production operations. This is typically done by identifying the number of contacts or duplicates that are required, as well as the number of composites, spreads, or chokes needed. If an estimator is unfamiliar with film contacting/duplicating techniques and materials, he or she should spend time learning the essentials by observing production.

The following formulas should be used to estimate film contacting and duplicating time and cost:

production time = preparation and setup + contacting production time + adjustments (as required)
production cost = production time × BHR
material and film cost = number of pieces of film or paper X number of square inches per piece X film cost per square inch
total production cost = production cost + material cost

Example 10.3

Our job log for an important customer indicates that we have to produce the following materials using contacting procedures: 16 -10 × 12 duplicate negatives (E-B) to be completed 4-up, 8 duplicate halftones (E-E, 2-up) on 8 × 10 inch dupe film, and 4 composite negatives (E-B, 1-up) each to receive 6 exposures (from a total of 24 negatives) on 11 × 14 dupe film. Vacuum frame measures 22 × 34 inches with a BHR of $47.70. Add for one setup since all this work will be done at one time. Automatic film processing.

Solution Use Figure 10.4, the film costs given and the appropriate formulas.

Step 1 Determine the preparation and setup time:

1 setup (large frame) × 0.10 hr = 0.10 hr

1. Preparation and setup

Note: Covers all preparatory time including contact frame and film processing setup.

Small frame (20 × 30 inches or less): 0.05 hrs

Large frame (over 20 × 30 inches): 0.10 hrs

2. Contacting film production

Note: Schedule times below provide for direct contact and processing of imaged film to raw film, such as E-B contacting. Double times for E-E contacting to provide for production of both intermediate and final film images.

	1-up*	2-up	3-up	4-up	Each add'l over 4-up
Manual Processing					
Small frame (20 × 30 inches or less):	0.14	0.16	0.18	0.20	0.04
Large frame (over 20 × 30 inches):	0.16	0.18	0.20	0.22	0.04
Automatic Processing					
Small frame (20 × 30 inches or less):	0.10	0.12	0.14	0.16	0.02
Large frame (over 20 × 30 inches):	0.12	0.14	0.16	0.18	0.02

*Number up refers to the number of pieces of film contact exposed at one time and not the number of images contained on one piece of film.

3. Adjustments

Compositing: Count exposures and add 0.02 hrs for each additional exposure of working films after the first exposure, which is provided in the above schedule.

Step-and-repeat: Count exposures and add 0.02 hrs for each exposure of working films after the first exposure, which is provided in the above schedule.

Spreads or chokes: For each spread or choke, add 0.08 hours to the above schedule times

Material Costs: (includes all processing costs)
- Contact film: $0.015 per square inch
- Duplicating film: $0.019 per square inch
- Dylux paper: $0.0020 per square inch (C1S)
  $0.0030 per square inch (C2S)
- Photographic papers: $0.011 per square inch

Note: Time and cost for remakes not included.

Figure 10.4

Production schedule for estimating film contacting and duplicating

Step 2 Determine the production time required to dupe the 16 negatives, 4-up on 10 × 12 dupe film using E-B procedures:

(16 pcs ÷ 4-up) × 0.18 hr ea = 0.64 hr

Step 3 Determine the production time required to produce 8 halftones using E-E procedures, 2 up on 8 × 10 dupe film:

(8 pc ÷ 2-up) × 0.14 = 0.56 × 2.0 (E-E times are doubled) = 1.12 hr

Step 4 Determine the production time required to dupe-composite 24 individual negatives to 4 final composites, E-B procedures on 11 × 14 dupe film:

(4 pc ÷ 1-up) × 0.12 = 0.48 + (.02 × 20 additional exposures) = 0.88 hr

Step 5 Determine the total production time:

0.10 + 0.64 + 1.12 + 0.88 = 2.74 hr

Step 6 Determine the production cost (not including materials):

2.74 hr × $47.70 = $130.70

Step 7 Determine the total material cost for duplicating film:

16 pc × (10 in. × 12 in.)/pc. × $0.019 = $36.48
8 pc × (8 in. × 10 in.)/pc. × $0.019 = $12.16 × 2 = $24.32
4 pc × (11 in. × 14 in.)/pc. × $0.019 = $2.93
Total material cost:
$36.48 + $24.32 + $2.93 = $63.73

Step 8 Determine the total production cost:

$130.70 + $63.73 = $194.43

Step 9 Summarize your findings:

Total production time: 2.74 hr
Total production time cost: $130.70
Total material cost: $63.73
Total production cost: $194.43

10.8 Estimating Photographic Proofing Costs

Figure 10.5 is provided for estimating graphic arts photographic proofing procedures. Proofing techniques in Figure 10.5 are divided into five major categories. The first two techniques utilize safelighted darkroom materials with appropriate contact frames, process cameras or enlargers, while the latter three involve the use of ultraviolet light sources typically used for lithographic platemaking and daylight contacting. Proofing done by using this schedule goes to a maximum of 24 x 30 inches. Larger proofs, covered by Types III, IV, and V of the schedule, utilize large vacuum contact frame and

	Exposure and processing time (apply adjustments as needed) Up to 24 × 30 inch proofs	Material and processing costs (per square inch)
Type 1: Darkroom proofing papers Includes all types of orthochromatic, lith, velox and polycontrast photographic papers. Time for exposure and processing under normal conditions.	0.15 hr	$0.012
Type 2: Stabilization papers All types of stabilization papers including Kodak Ektamatic and Fotorite products. Time for normal exposure and machine processing	0.10 hr	$0.10
Type 3: Dylux papers Exposure and related production for Dylux materials	0.15 (first exposure) 0.04 (each add'l exposure)	C1S: $0.002 C2S: $0.003
Type 4: Overlay proofing Exposure and processing of 3M Color Key and Enco NAPS/PAPS. Includes mounting time on selected substrate.	Manual processing (per color): 0.20 (first exposure) 0.04 (each add'l exposure) Automatic processing (per color): 0.15 (first exposure) 0.04 (each add'l exposure)	Color Key Process colors:$0.014 PMS colors: $0.016 NAPS/PAPS Process colors:$0.013 Flat colors: $0.015
Type 5: Single-sheet proofing Exposure and processing of DuPont Cromalin and 3M Matchprint II materials. Times include laminating film to substrate, exposure and processing for one layer; add for each proof layer as needed. Material costs include base substrate.	Manual processing (per color): 0.20 (each exposure) Automatic processing (per color): 0.15 (each exposure)	Matchprint II* Process colors: $0.017 PMS colors: $0.019 Cromalin* $0.020 (film and toner) *per laminated and fully-toned layer including all materials.

Note: Time and cost for remakes not included.

Figure 10.5

Schedule for estimating proofing production.

are sometimes completed in conjunction with platemaking, as covered in Figure 11.2 of the next chapter.

It is important to note that exposure data and material and processing costs are presented here for example only and are best developed carefully for each individual printing plant. All costs are on a dollar-per-square-inch basis. Schedule times include all operations necessary to setup, position, expose, and process the applicable proofing material. Time and costs for remakes are not included and must be added separately if needed. Example 10.4 utilizes Figure 10.5 data and is presented here to demonstrate how the schedule might be used.

Example 10.4

We have a job that requires estimating the following segments based on our production analysis: 30 velox (halftone) proofs, 8 × 10 inches each; 15 continuous-tone prints on Ektamatic SC paper, 9 prints on 8 × 10 inch paper and 6 prints on 16 × 20 inch paper; and 6 four-color process separations on 10 × 12 inch Enco NAPS. BHR for the veloxes and continuous tone proofs is $38.70. The contact frame BHR for the NAPS overlay proofs is $49.75. Determine production time and cost for this job, including material cost.

Solution Use Figure 10.5.

Step 1 Determine production time required for the 30 veloxes (Type I):

30 pc × 0.15 hr/pc = 4.50 hr

Step 2 Determine production time required for the 15 Ektamatic proofs (Type II):

15 pc × 0.10 hr/pc = 1.50 hr

Step 3 Determine production time required for process color separation proofs (6 sets with 4 pc/set results in a total of 24 pc) (Type IV):

24 pc × 0.25 hr/pc = 6 hr

Step 4 Determine total production time:

4.50 hr + 1.50 hr + 6 hr = 12 hr

Step 5 Determine total production time cost (not including materials):

For veloxes and Ektamatic proofs:
4.50 hr + 1.50 hr = 6 hr
6 hr × $38.70/hr = $232.20
For color separations:
6 hr × $49.75/hr = $298.50

Total production time cost:
$232.20 + $298.50 = $530.70

Step 6 Determine total material cost:

For veloxes:
30 pc × (8 in. × 10 in.)/pc × $0.012/sq in. = $28.80
For Ektamatics:
9 pc × (8 in. × 10 in.)/pc × $0.010/sq in. = $7.20
6 pc × (16 in. × 20 in.)/pc × $0.010/sq in. = $19.20
For NAPS:
24 pc × (10 in. × 12 in.)/pc × $0.020/sq in. = $57.60
Total material cost
$28.80 + $7.20 + $19.20 + $57.60 = $112.80

Step 7 Determine total production cost:

$530.70 + $112.80 = $643.50

Step 8 Summarize your findings:

Total production time: 12 hr
Total production time cost: $530.70
Total material cost: $112.80
Total production cost: $643.50

10.9 Estimating Color Separation Production Costs

Figures 10.6 and 10.7 provide data for estimating color separation times and costs. Figure 10.6 represents data for completing color separation procedures using a process camera, enlarger, or contacting equipment, cross-referenced for both indirect and direct masking procedures. As a general rule, direct separation masking requires two color correction masks (ccm), and indirect procedures require an individual ccm for each printer in the separation. Material costs, including contact film applicable to indirect separation procedures, are provided. Example 10.5 demonstrates the application of schedule data in Figure 10.6.

Figure 10.7 provides data for estimating color separation production using color scanning equipment. Both older analog-digital scanners and newer, electronically advanced digital scanning equipment is represented. As will be noted, schedule use involves a step-by-step procedure beginning with image review, then image mounting, followed by scanner setup and prescan adjustments, then scanning time where the separation is actually exposed or digitized, and automatic film processing. Since digital

Equipment or Procedure	Direct Separation (no masking) First Exposure	Direct Separation (no masking) Additional Exposures (each)	Direct Separation (with ccms) Separation Exposure (each)	Direct Separation (with ccms) ccm (each)	Indirect Separation (with ccms) Separation Exposure (each)	Indirect Separation (with ccms) ccm (each)	Indirect Separation (with ccms) Contacts (each)
	Production Time in Hours per Separation						
Process camera or enlarger	0.5	0.4	0.4	0.5	0.35	0.5	0.15
Contacting	0.2	0.15	0.15	0.2	0.15	0.2	0.15
Color scanner	0.4	0.3	n/a	n/a	n/a	n/a	n/a

Note: Time and cost for remakes not included.
Material and processing costs: color correcting masking film: $0.20/sq in.; separation film: $0.019/ sq in.; contacting/duplicating film: $0.019/ sq in.

Figure 10.6

Schedule for estimating color separation production costs

		Digital Scanner	*Analog-digital Scanner*
1. Image review and check (each image)		**0.10**	**0.10**
2. Image mounting (each image)		**0.05**	**0.05**
3. Scanner setup and prescan adjustment (each image)		**0.10**	**0.20**
4. Scanning time (per 4/C set)	**Small format:**	**0.20**	**0.40**
	Medium format:	**0.30**	
	Large format:	**0.40**	
5. Auto processing (per 4/C set)		**0.10**	**0.10**

Material costs: (includes all processing costs)
Halftone scanner film (used with contact screen): $0.018 per square inch
Continuous tone film: $0.015 per square inch
Laser film (Helium Neon): $0.014 per square inch
Laser film (Argon Ion): $0.011 per square inch

Note: Time and cost for rescans or remakes not included.

Figure 10.7

Schedule for estimating color scanning production

scanners are an input source for CEPS systems, digital scanning equipment is sometimes used without direct film output, in which case the scanned image is magnetically stored for later CEPS use. In such instances, there would be no actual film output at the conclusion of scanning, and automatic processing (Step 5) would not be included in the estimate. Application of Figure 10.7 data is demonstrated in Example 10.6.

Schedule data for both Figure 10.6 and 10.7 assumes high-quality original transparencies or reflection art. Using either production process, times and costs for remakes or rescans is not included. Chart times and costs are presented here for example only.

Digital scanner production time and cost:

digital scanner production time = image review (per image) + image mounting (per image) + scanner setup and prescan adjustment (per image) + scanning time (per scanned set) + automatic processing (per scanned set)

digital scanner production cost = digital scanner production time × digital scanner BHR

material and film cost = number of pieces of film or paper × number of square inches per piece × film cost per square inch

total digital scanner production cost = digital scanner production cost + film and material cost

Example 10.5

Using our process camera and direct masking procedures, we want to estimate the time and cost for completing 16 four-color separations (sep) from high-quality reflection prints. All separations will be done 1-up on 8 × 10 inch pan separation film. There will be 2 ccms per separation set of 4 negatives. We need to allow for remakes of 6 negatives, which are expected and must be estimated. Proof will be Color-Key, 8 × 10 inch size (Type IV proofing in Figure 10.5) with no remakes. Camera BHR is $52.70; contact frame BHR for proofing is $39.80. Determine total production time and cost for this job including material cost.

Use Figure 10.6.

Solution

Determine camera production time and cost and material cost:

Step 1

Time per set:
(2 ccm/set × 0.5 hr/ccm) + (4 neg/set × 0.4 hr/neg) = 2.6 hr/set
Total time for 16 separations:
16 sep × 2.6 hr/set = 41.6 hr
Remake time:
6 neg × 0.4 hr/neg = 2.4 hr
Total camera time:
41.6 hr + 2.4 hr = 44 hr

Total camera time cost:
44 hr × $52.70 = $2318.80
Total number of separation film pieces:
(16 sep × 4 neg/set) + 6 neg = 64 pc + 6 pc = 70 pc
Total separation film cost:
70 pc × (8 in.× 10 in.)/pc × $0.019/sq in.= $106.40
Total number of ccm film pieces:
16 sep × 2 ccm/set = 32 pc
Total ccm film cost:
32 pc (8 in. × 10 in.)/pc × $0.02/sq in.= $51.12
Total film material cost:
$106.40 + $51.12 = $157.52

Step 2 Determine proofing production time and cost and material cost:

Total number of pieces:
16 sep × 4 neg/set = 64 pc
Total proofing time:
64 pc × 0.25 hr/pc = 16 hr
Total proofing time cost:
16 hr × $39.80/hr = $636.80
Total proofing material cost:
64 pc × (8 in.× 10 in./pc × $0.02/sq in.= $102.40

Step 3 Summarize your findings:

Total production time: 44 hr + 16 hr = 60 hr
Total production time cost: $2318.80 + $636.80 = $2955.60
Total material cost: $157.52 + $102.40 = $259.92
Total production cost: $2955.60 + $259.92 = $3215.52

Example 10.6

We have a job which consists of 44 transparencies to be color separated using our Crosfield Magnascan 645 digital scanner. Based on our production review we have the following breakdown by scan group: 6 scans with 6-up mounted on the scanner copy drum (total of 36 separations), 2 scans with 3-up (total of 6 separations) and 2 scans with 1-up (total of 2 separations). Argon Ion laser film will be used in the 20 × 24 size for all output. Automatic film processing. Scanning time at a medium format. Add 10% additional time and materials for anticipated rescans. Scanner BHR is $153.50. Also estimate the cost to proof using Cromalin materials and our automatic toning machine (ATM). Proofs will be ganged onto five 22 × 26 inch Cromalin proof sheets. Use Figure 10.5 and a contact frame/ATM combined hourly rate of $47.90.

Solution Use Figure 10.7 (scanner production) and Figure 10.5 (proofing)

Step 1 Determine scanner production time:

Image review: 44 images × 0.10 hr ea = 4.4 hr
Image mounting: 44 images × 0.05 hr ea = 2.2 hr
Scanner setup and prescan adjustment: 44 images × 0.10 = 4.4 hr
Scanning time: 11 scan groups × 0.30 hr ea = 3.3 hr
Automatic processing: 11 scan groups × 0.10 hr ea = 1.1 hr

Step 2 Total scanner production time:

4.4 hr + 2.2 hr + 4.4 hr + 3.3 hr + 1.1 hr = 15.4 hr

Step 3 10% additional production time:

15.4 hr × 1.10 = 16.94 hr

Step 4 Total scanner time cost:

16.94 hr × $153.50/hr = $2600.29

Step 5 Scanner film cost:

10 pc (20 in × 24 in) × $0.011/sq in = $52.80

Step 6 10% additional materials:

$52.80 × 1.10 = $58.08

Step 7 Proofing time (see Figure 10.5, Type 5 Cromalin proofing):

4 colors × 5 Cromalin proofs = 20 exposures × 0.15 hr = 3.00 hr

Step 8 Total proof production cost:

3.00 hr × $47.90 = $143.70

Step 9 Cromalin material cost (see Figure 10.5, Type 5 Cromain proofing)

20 laminations (22 in × 26 in) × $0.020/sq in = $228.80

Step 10 Summarize your findings:

Total production time: 15.73 hr + 3.0 hr = 18.73 hr
Total production time cost: $2600.29 + $143.70 = $2743.99
Total material cost: $58.08 + $228.80 = $286.88
Total production cost: $2743.99 + $286.88 = $3030.87

10.10 Estimating Color Electronic Prepress Systems (CEPS)

As changes have rapidly taken place in the conversion from conventional to electronic prepress, methods to estimate both cost and production have become more complex. Of all the areas difficult to estimate, CEPS systems (Group 1, Figure 10.1) represent the hardest to define and complete for a number of reasons:

1. Each system—Crosfield, Scitex, Hell, DS America, and others—operates using different hardware and software. Thus, there are different input procedures, massaging and image modification techniques, output methods, and equipment operating speeds. Suggesting any type of standard estimating system which covers all CEPS, since there are few operating similarities between equipment types and software, is impossible. This is complicated by the fact that CEPS is in a constant state of improvement through software revisions and hardware upgrades.

2. Operator productivity, which is critical to meeting tight schedules and using equipment efficiently, varies considerably. This makes accurate cost estimating very difficult. Some factors affecting this productivity are the operator's level of training, environment where the system is located, ergonomics issues, and the operator's inherent knowledge, learning speed, and manual dexterity when working on the system.

3. Accurate estimating is complicated since some of the operations are electronically completed, making specific times difficult to predict. Also, many of the CEPS systems are multi-tasking, meaning that more than one job can be in production at any one time.

4. The estimator must have a thorough understanding of system details, which is difficult to learn given the many other estimating duties required. While there are variations, some companies have the system operator complete CEPS estimating—somewhat like a consultant—taking productive time away from completing work on the system. In other firms, an estimator who has a fairly good knowledge of the system, who has worked as a backup operator or trainee on the system, completes the estimating.

5. During production, some customers request graphic changes in jobs which require costly CEPS production. While customers anticipate paying for these changes, the exact cost may be difficult to predict. Add to this the fact that the original estimate may be somewhat inaccurate or a "best guess." This situation is further complicated by the fact that most customers are working within certain budget constraints. All of these factors contribute to the dynamic and difficult costing and pricing problems faced by CEPS trade shops and printers who own CEPS production equipment.

The following is intended to provide initial guidelines for estimating CEPS production. There are many intangibles and hard-to-define production conditions when estimating CEPS. In sum, CEPS estimating is best done using accurate production standards and costs tailored to the company owning the CEPS units.

Current Estimating Methods

In general, there are two approaches for costing CEPS systems: Real-time plus materials and component estimating.

Real-Time Plus Materials. Essentially this is a time and materials approach based on actual time used and materials consumed. It is typically completed after-the-fact, but most print buyers typically request a "ballpark" estimated figure before production begins on the job. The company must be very careful any time a "ballpark" dollar amount is discussed, since customers tend to consider such an approximation with more seriousness than just a "ballpark" figure.

This real-time and materials process typically works as follows: the company receives color separations for scanning, mechanical art, electronic files of type or hard-copy manuscript, and other graphic components in either electronic or conventional form. The job is initially reviewed by a skilled production planner in conjunction with the CEPS operator, prepress supervisor, or knowledgeable estimator for apparent problems or difficulties. If all the preliminary materials appear suitable, the job begins production and work commences. Sometimes the customer asks for a "ballpark" estimate at this point.

Throughout the production cycle the customer or print buyer and company discuss various job requirements, particularly those which have changed during production. Color correction on separations, separation dot etching, image modifications, and any other desired changes are tailored to the customer's needs. Both CEPS and manual procedures may be utilized. With the flexibility available via CEPS, it is safe to say that many jobs are modified more extensively than originally designed or estimated. Customers may request a "proof to satisfaction" condition, whereby the printer is required to provide proofs of preliminary work to a point of customer approval; this can be both expensive and time consuming, and may be a frustration to the printer.

Once the job is approved, the time and materials components are tracked and the customer is invoiced for actual use. There should be an accurate BHR for the CEPS system, including peripheral devices. Printers should charge for any material item or time consumed that is directly chargeable to the customer. Profit should be added relative to market conditions, services provided for the job, and customer-desired needs.

Component Estimating. This estimating process requires that a production planner or estimator carefully work out a production plan representative of the production elements of the customer's jobs. To do this accurately, it is essential that the customer identify all materials and component parts to be supplied.

Component estimating is defined as breaking the job into specific operations—or components—necessary to produce the job. Details, which might otherwise be overlooked, are important. Since the BHR for many CEPS systems is in the range of $300 to $500 per hour, and since some of the manufacturing work can be done less expensively using conventional imaging procedures, an accurate production plan is critical to the component estimating process.

Figure 10.8 is a Scitex Operations Worksheet, reproduced here with permission of Walker Graphics/Techtron of San Francisco, CA. This worksheet provides a fairly complete view of the component estimating process for a popular CEPS system. Use of the form would be as follows: first the particulars related to the date, customer and other identifying material are completed. Then the specific CEPS operations are identified (left column). Operation codes are then applied, production hours calculated, then prices using the firm's BHR rate for the CEPS, and then the final for each component element and the total. The worksheet resembles a computer screen since most CEPS estimating systems are computer-assisted.

Generally the specific hourly production standards vary from company to company and system to system. Thus, Scitex production methods are different from Crosfield's, which are different from Hell's or DS America's. Because of the intense competition in the CEPS area, somewhat driven by the extensive capital investment required for a CEPS system, most companies consider their BHR and hourly production standards proprietary.

Step-by-Step Development of a Component Estimating System for Electronic Imaging Platforms Including CEPS

Estimating the cost of any electronic imaging system—design, CEPS, or any other electronic platform—is very is difficult. The best approach to developing such a component estimating system is presented in the following step-by-step sequence:

1. Once the CEPS, design system, or other electronic platform is purchased and on-line, the estimator should identify the operating components of the system. Vendor training or working directly on the unit with a skilled company employee are the two-most common ways this is done.
2. Once the system is understood, the estimator should set up operational categories which define the work performed while the unit is in production, and then establish production standards in the form of hourly times, representative of this productive work.

SCITEX OPERATIONS WORKSHEET

B:SCITEX.REV

DATE:

AGENCY:

DOCKET:

SUBJECT:

OPERATION	CODE	HOURS	PRICE	AMOUNT
PRE-PLAN	800	0.00	0.00	
SCANNING TRANSP.	821	0.00	0.00	
REFLECTIVE	822	0.00	0.00	
CONTONES	823	0.00	0.00	
LINE WORK	824	0.00	0.00	
MULTICUT	825	0.00	0.00	
STRIPPING				
Silo	808	0.00	0.00	
Scale	810	0.00	0.00	
Rotate	811	0.00	0.00	
Ghost	812	0.00	0.00	
L Wrk Col	813	0.00	0.00	
L Wrk Insert	814	0.00	0.00	
L Wrk Create	815	0.00	0.00	
CT Create	817	0.00	0.00	
CT Place	818	0.00	0.00	
CT Blur	819	0.00	0.00	
Fast Execute	840	0.00	0.00	
Assign	841	0.00	0.00	
Line Art	842	0.00	0.00	
Clr Brk Ty Ly	843	0.00	0.00	
Clr Brk Ty Im	801	0.00	0.00	
Editor	802	0.00	0.00	
Strip in Image	803	0.00	0.00	
RETOUCHING				
Incl. Etch.	804	0.00	0.00	
Reg. Etch.	805	0.00	0.00	
Spec.Effects	807	0.00	0.00	
ERAY				
Scan	831	0.00	0.00	
Final	832	0.00	0.00	
RES. & SCALE	844	0.00	0.00	
PU & Resize	845	0.00	0.00	
Read/Write	847	0.00	0.00	
FILE STORAGE FEE		0.00		0.00
TOTALS		0.00 Hrs		0.00

Figure 10.8

Scitex Operations Worksheet for estimating Scitex 350 Response CEPS sytem. (Reproduced with permission of Walker Graphics/Techtron.)

3. Next, a budgeted hour rate is established using the procedure indicated in Chapter 3 of this text.
4. During the estimating process, each job is broken into blocks or components of work which fit into the operational values recognized in the system at Step 1. This represents a detailed breakdown of job components to be performed by the operator. Figure 10.8 is an example worksheet for estimating production using a Scitex Response CEPS system.
5. The estimating process is completed by multiplying the operational times by the BHR, then adding material costs (including appropriate markups) which apply. Profit markup and pricing decisions are then made by company management.

It must be emphasized that for a CEPS component estimating system to be accurate and reliable, the estimator must have detailed knowledge of the system. In many plants having such systems currently on-line, the CEPS estimator is trained in the system operation along with the company's CEPS operators. Training is typically given either by the vendor or the company employee considered the "key operator" of the system. Most electronic platforms, particularly those in Group I (Figure 10.1) are frequently changed through both hardware and software revisions. If component estimating is to remain accurate, CEPS estimator training process should be an ongoing process.

11

Estimating Film Assembly, Platemaking and Proofing

11.1 Introduction

Film assembly, proofing, and platemaking are vital prepress functions in the lithographic sequence. Film assembly, also commonly termed stripping, is a procedure by which images on film are assembled in an ordered fashion to produce the desired results. Stripping typically precedes the lithographic platemaking function and is checked for accuracy, image placement, and quality before platemaking, using common proofing methods as discussed in Chapter 10.

In this chapter, film assembly procedures are divided into two major types: conventional (or manual) film assembly, also commonly termed table stripping, and electronic image assembly. As with all other areas of prepress production, film assembly is undergoing major technological change. In the mid-1980s electronic imaging and stripping devices began to appear. Some were linked to CEPS units, which produced imaged film ready for proofing and platemaking, while others were stand-alone stripping platforms such as the Gerber AutoPrep. The stripping department in today's typical multicolor commercial printing company requires a mix of both electronic equipment and manual skills. Some jobs are completely assembled using table stripping procedures, while others are fully assembled using CEPS or electronic stripping equipment. Of course, some jobs are manufactured using a mix of the two.

Lithographic proofing and platemaking has also been affected by recent technological advances. The most significant development at this time is the shift of plate coatings from solvent-based emulsions to aqueous-based materials. Various factors have caused this movement including environmental issues, unpleasant odors during plate processing, and potential health issues which might affect platemaking and proofing employees. Even with this shift to improved emulsions, lithographic platemaking and related proofing production has become a fairly controlled, precise manufacturing operation. Automatic plate and proof processing, improved consistency of platemaking and proofing emulsions providing greater exposure latitude, and adoption of fully anodized plate bases and dimensionally-stable proofing materials have made the platemaking and proofing area largely a trouble-free, expeditious production area.

11.2 Conventional Film Assembly

Film assembly resembles copy preparation to a considerable degree. Both deal with exact measurements, preciseness of layout, defined image registration, and a resultant visual product. However, copy preparation is a pre-photographic procedure normally completed by a pasteup artist, and film assembly deals almost exclusively with line and halftone film images in either positive or negative form. Since the majority of art for conventional prepress is prepared in positive form as reflection copy, assembly of negative film materials represents the usual practice in most manual stripping departments in the U.S.; European printers tend to work more with positive film images.

General Table Stripping Techniques

Once the camera negative has been exposed, processed, and dried, it is transported into the image assembly area. Initially, the *stripper,* the individual who assembles these film pieces together, visually evaluates the images for quality, density, sizing, and other production requirements. In conjunction with this initial review, the stripper analyzes the artwork from which the negatives were made and reads all pertinent instructions regarding the positioning of images on the plate (and ultimately the press sheet). He or she checks to be certain that all film elements required for the job are provided. The stripper also evaluates the job relative to special film requirements such as spreads, chokes, compositing, special proofing needs, or any other extraordinary consideration.

Once the materials have been verified as accurate and complete, the stripper selects the proper masking base to be used for the job. There are two types of base materials commonly used for table stripping: goldenrod masking base and clear acetate, better known as clearbase. Goldenrod masking materials are sheets of orange, yellow, or amber paper or vinyl, normally cut to the size of the press plate. (The goldenrod color allows the light from the light table to be seen but protects covered areas of the lithographic plate from being exposed during platemaking. Most plants stock precut masking base—both goldenrod and clearbase—to the plate sizes which fit company presses. Thus, it is important that the stripper have some idea as to the press selected for the job before stripping begins.

Assume we want to strip a one-color (1C) job which has one film negative. The following procedures describe the two ways this probably would be done in a typical commercial printing plant.

Stripping a One-color (1/C) Job to Goldenrod Masking Material. Working at a light table with a T-square and triangle or at a line-up table, the stripper—drawing on the goldenrod sheet with pencil or pen—accurately locates the position of the image. It is important that this work be done precisely since little movement is possible once the job has been plated and goes to press.

The stripper, still at the light table, slides the film negative under the goldenrod sheet, and positions the negative image relative to the marks drawn in pen or pencil. Holding the film securely, two or three small goldenrod areas are cut away where the negative is located and tape is used to secure the negative to the goldenrod sheet. The flat—an assemblage of film and masking base—is now turned over on the back (the image is now wrong-reading) and the negative taped securely to the back of the masking base. The final step is to cut out the goldenrod masking material from the image areas of the negative, thus allowing white light to pass through during platemaking and proofing.

Stripping a 1/C Job to Clearbase Masking Material. Working at a light-table with a T-square and triangle or at a line-up table, the stripper—drawing on the clearbase sheet with an acetate pen—accurately locates the position of the image. Again, it is important that this work be done precisely since little movement is possible once the job has been plated and goes to press.

The stripper, still working at the light table, turns the clearbase upside-down and positions the film negative—wrong-reading—over the marks made in pen. Taping of the film image to the clearbase material (the negative is still wrong-reading) is now completed. At this point the stripper secures a sheet of goldenrod masking material in the same size as the clearbase, lines the sheets of material up, then—at a light table with the film image right-reading—cuts away the goldenrod where the image appears, thus allowing white light to pass through during platemaking and proofing. When plates and proofs are made, the clearbase flat and goldenrod are superimposed together, blocking light in all areas but those where the windows appear in the goldenrod mask.

Pin Registration Procedures

Many jobs require more than one color as in the preceding example. To effectively and economically strip multicolor work using conventional methods, pin registration is used.

Pin registration is a process by which masking base materials are punched or positioned before stripping so that the relationship between flats, and the film images positioned on the flats, will remain fixed. The pins are usually made of metal (plastic is also used) and come in two styles: 1/4-inch round hole and slotted pins (a rectangular slot approximately 1/4-inch by 3/8-inch long, with two parallel flat sides).

There are two common ways pin registration is accomplished during image assembly: stripping tabs and commercial pin registration punches and matching pin bars.

Pin Registration Tabs (Stripping Tabs). Stripping tabs are nylon or acetate strips approximately one inch by three inches in dimension with a round hole or slot hole punched in one end. For use, masking sheets which are to be registered to each other—normally done before stripping—are carefully aligned on the light table. Two tabs are then taped along one side of the masking sheet, roughly equidistant from the center and at least one-half the distance to the corner of the masking material. The tabs hang out from the edge of the sheet. If a three-color required three flats, there would be three masking sheets prepared in this manner, and each would be stripped to hold the appropriate images for the color to be represented. Stripping would be done on the table using pins and exposures to plates, and proofs would also use pins suitable for insertion into the platemaking or proofing vacuum frame.

Commercial Pin Registration Systems. These device are available for purchase and use in table stripping, and have two essential parts: a punching unit and associated pin bars used on the table, and in the vacuum frame for platemaking and proofing. The punching unit, usually a table-top device, allows for insertion of one side of a sheet of unpunched masking base into it (about two inches), with manual depression of a bar or handle that punches a fixed group of round and slot holes along the sheet edge. Since the same punch is used over and over, the system becomes standardized throughout the stripping area, ensuring a consistency from job to job and day to day. Exactly matching the pin positions in the punching unit are pin bars, normally made of metal for heavy-duty, continuous use. These are placed both on the table and in the various vacuum frames for platemaking and proofing.

Commercial pin registration systems, because of their consistency, reliability, speed of use, and accuracy, are very popular today. Most of the systems are not considered expensive compared to the hours saved if required to restrip a poorly registered job. Manufacturers of commercial pin registration equipment provide a standard pin placement arrangement, instead of designing different pin positions for each printer. Tailor-made systems are also available. The holes are a combination of round hole and slot types along the masking sheet edge to allow for slight dimensional changes which inevitably occur in masking materials.

Table Stripping of Process Color (Multicolor Stripping Procedures)

While single-color stripping is fairly easy to understand and do, the majority of work done in the commercial printing industry is multicolor, or many colors printed in precise registration to each other. Multicolor production is divided into two categories: flat color work and process color work. Flat color production is simply different pigment colors of ink—commonly using Pantone Matching System (PMS) colors—prepared and printed in relation to one another. Process color work involves printing four colors—magenta, cyan, yellow, and black—in precise registration to each other, thus reproducing a full-color photograph. Flat color images—line work such as type or line drawings—are typically a product of the process camera, while process color work is produced using color scanners and CEPS systems. Chapter 10 describes the production and estimating of these products in film form.

Multicolor work, if completed using conventional table stripping, begins with the mechanical artwork which has been preseparated into color layers, called overlays; each overlay represents a different ink color of the final image. Photographic imaging using a process camera is completed, translating each overlay into negative film form, with the film negatives sent to the stripping department for placement on masking sheets so that proofing and platemaking can be completed. Assume, for example, that we want to strip a four-color (4/C) process poster, measuring 16 x 20 inches, with black type. The type image would be completed from reflection mechanical art and photographed to one negative—called a black type printer—while the four color separation printers (magenta, cyan, yellow, and black) would be produced from a color scanner. Typical abbreviations for specifying process colors and marking flats are: K=black, M=Magenta, C=cyan, and Y=yellow.

Working on the light table, and using a commercial pin registration system the stripper would do the following to strip the job:

1. Check the line negative and four separations film negatives for quality. This is a visual check to be certain the materials are properly sized and suitable for stripping.

2. Get four sheets of clearbase and two sheets of goldenrod masking material and punch each using the commercial pin punch.

3. Using the black type mechanical art—sometimes called a keyline—and working with one of the goldenrod sheets on pins at the light table, the stripper would accurately mark the location in pencil or pen for the black type printer. The stripper would also mark this masking sheet with required production measurements such as the press centerline, press sheet gripper, and plate gripper. This goldenrod sheet is now a key (or master) flat to which all other images will ultimately register.

4. After all measurements and positioning marks have been carefully determined, the black type negative would be stripped to the goldenrod masking base. This is done by sliding the negative type image under the goldenrod base (which is on pins), lining it up to the hand-drawn imaging position marks, then taping the negative to the goldenrod sheet securely. Once this is done, a window is opened up in the goldenrod material, so everything on the masking sheet is covered with goldenrod but the type to print. The process described is called "laying and cutting" since the negative has been laid-in (or positioned) followed with the opening (or cutting) away of goldenrod.

5. With the black type negative on pins and wrong-reading (flipped-over), the stripper now takes a clearbase masking sheet and positions it over the stripped material. The cyan (blue-green) separation—or printer—is now turned wrong-reading (emulsion-up) and positioned carefully to marks on the black keyline, thus registering the separated image to the black type. When located exactly, the cyan printer is taped to the clearbase material and will serve as a key to which all the remaining process color separations are stripped. Because this was registered to the black keyline (master) flat, it is typically considered a complimentary flat to the black flat.

6. The black type flat is removed from pins and placed aside, and the stripped cyan printer is repositioned on pins, wrong-reading. Over this the second sheet of clearbase is positioned on pins, and the magenta separation accurately registered and taped to the cyan image. This requires careful alignment using a magnifying device. It is important to note that this alignment process can be slow and tedious work. Most separations are imaged with register marks so that the stripper can "lay-in" the separations fairly quickly, speeding up production. The magenta image stripped to clearbase is a complimentary flat.

7. The process of laying-in the yellow and black separations is repeated, each with a separate sheet of clearbase material. Each flat is a complimentary flat to the black type master. All separation stripping thus described is done to the cyan flat as the master, stripping wrong-reading. At this point the black type master flat on goldenrod is stripped and all four color separations have been laid-in to four clearbase supports.

8. The stripper now puts the black type flat on pins, right-reading, and positions a new goldenrod sheet over it, on pins. Using the keyline black type flat, the stripper cuts open a window representative of the precise area to be covered by the color image, thus creating the edge of the image area. When done, this goldenrod sheet has one area—or window—cut in it, and is called a "common-window" flat. It will be used with each clearbase separation during proofing and platemaking to provide a precise border and thus define the process color image size. Some companies use peelcoat materials such as Ulano's Rubylith or Amberlith for this common-window flat.

9. In order to control quality on press, many printing companies strip additional flats which include color bars and marks for positioning, cutting, and folding. Color bars, stripped to each clearbase separation with an accompanying window in the "common-window" flat, allow for accurate ink density control during printing. The "marks flat," which can be incorporated into the black type flat (with an appropriate mask to block the type images) or stripped as a separate flat to double-expose to all four plates with other images, has become fairly common for faster makeready on press, and when follow-up bindery operations are needed.

10. The final group of flats—called a flat file—represents the complete collection of images for this poster job. The following describes each:

- Black (K) type (keyline) flat on goldenrod which is the master flat
- Cyan (C) separation flat on clearbase, which is a complimentary flat, with cyan color bars along the tail edge of the flat
- Magenta (M) separation flat on clearbase, which is a complimentary flat, with magenta color bars along the tail edge of the flat
- Yellow (Y) separation flat on clearbase, which is a complimentary flat, with yellow color bars along the tail edge of the flat
- Black (K) separation flat on clearbase, which is a complimentary flat, with black color bars along the tail edge of the flat
- Common-window flat on goldenrod, which is a complimentary flat
- Marks flat (if desired)

The process just detailed represents the most common table stripping procedure for color separation in the printing industry. It should be noted that there are other methods that can be employed, such as stripping the separations to blueline (Dylux) materials.

Complex Color Image Assembly Procedures

It is important to note that many color jobs done today are five-, six-, or more colors. While four color process requires only the printing of cyan, magenta, yellow, and black (as described) customers sometimes wish to varnish—add gloss—or print a special PMS color to the conventional process color images. To do this, printers have begun to purchase six-, seven-, and eight-color presses, allowing such printing to be done in one pass, which is most productive. Stripping a six-color job thus involves the film assembly of four color process as described above, with the addition of varnishing flats, and PMS flats.

Screen Tints. Screen tints provide for even toning of a normally solid area. Sometimes called plate screens, they are pieces of film that contain an overall dot pattern. Screen tints are purchased by the number of lines and percentage of printable dot. For example, a 150-30 indicates that the screen tint is a 150 line screen with a

30% printable dot. The commercially-available tints usually range from 10% to 90% in 10% increments, and in values from 85 line to 300 line, at increments standard in the industry.

The general procedure to tint an image is as follows: identify the negative or area of a negative to be tinted. The flat is then turned over so that the image is wrong-reading (emulsion-up). The tint screen is then cut slightly larger that the image area and taped emulsion-up over it so that the emulsion is up for both the image and the tint. When the flat is turned over, the emulsion of the screen tint will directly contact the plate or proof, as will the rest of the image. This technique minimizes the spread of light, producing dot gain.

Flat color, as previously discussed, is a special pigment mix or black. Thus, adding a screen tint to an image printing in black or a PMS color simply tones that color down to a lesser value. Yet since process colors can combine to produce a fairly accurate spectrum or rainbow of all colors in reproducing color separations, the use of screen tints and process colors can produce color matches of hundreds of colors. This is known as fake color. The advantage of fake color is that when magenta, cyan, yellow, and black are being printed on the press sheet, literally hundreds of other colors can be manufactured using screen tints, to print using the same four process colors.

For example, combining a 30% cyan tint and a 40% magenta tint in one image produces a purple color, somewhat toned down from solid purple since tints are used and the whiteness of the stock blends into the colors. Color selecting—tint matching—is commonly selected from standard tint mixing charts either developed by the printing firm for internal use, or from various sources such as ink suppliers. One popular color tool is Pantone's Color Simulator which allows the printer to reproduce a selected PMS color using fake color or reproduce a fake color using PMS color.

It is important to note that fake color stripping requires that tints be laid into flats so that they do not cause a moire' pattern, which is an undesirable optical effect caused by two or more screen tints angled such that they interfere with each other. Different devices, such as GATF's Screen Angle Guide, have been developed to aid the stripper in angling such tints. Some strippers, to save time, simply lay in matching tints for fake color work by eye, rotating the tints until no undesirable pattern is visually evident.

Image Fitting Techniques. As discussed in Chapter 10, one area common to both the photographic and film assembly in a typical commercial printing plant doing complex color work is image fitting. The very small size change allows for images to overprint or underprint in a given area, effectively producing what is known as a trap. Image fits are commonly done though conventional contracting of imaged negatives and positives using clear acetate spacers and diffusion sheets, producing a spread or choke; orbital rotation equipment, such as the Byers MicroModifier, is also common. Daylight contact and duplicating films and rapid-access processing are typical. Both E-E and E-B orientations are used. Chapter 10 covers this area in some detail.

Producing Spreads and Chokes. A spread begins with an imaged film negative, which can be contacted to either duplicating or contacting film. Clear acetate spacers are used between the imaged film negative and raw (unimaged) film to cause the image to become slightly larger, with greater amounts of spread related to increasing the quantity of spacers used. If contact film is used the resultant spread film image is positive, and if duplicating film is used the resultant fattened is negative. Diffusion sheets are also used to scatter light during the exposure process.

A choke requires an imaged film positive, which can be contacted to either duplicating or contacting film. Again, clear acetate spacers are used between the imaged positive (on top) and the raw film, causing the choked image to be reduced in size; the greater the number of spacers, the greater the resultant image shrink. Diffusion sheets are used to scatter light during the exposure process. If contact film is used the resultant choked image is negative, and if duplicating film is used the resultant choked image is positive.

Orbital (Micromodifier) Techniques. The purchase and use of the orbital equipment for fitting images is done to improve the quality of final images over manual contacting procedures. The primary problem with spreads and chokes done using acetate spacers—as previously described—is that fine type, sharp image edges and corners and other fine image details tend to quickly close up or plug. Increasing the amount of spacers increases this tendency.

Orbital technology essentially does away with spacers and diffusion sheets. Basically the system works as follows: A negative (for a spread) or positive (for a choke) is mounted on a stable (unmoving) clear plastic platform, and a piece of raw (unimaged) contact or duplicating film secured to an orbital bed directly below the imaged negative or positive. Rotation of the orbital plate during exposure provides even imaging to the raw film, with greater rotation yielding a fatter spread or thinner choke. Very fine image detail is maintained with the rotation very precisely controlled. This allows for traps to be precisely and accurately generated quickly, instead of a more "hit and miss" approach with spacers. The orbit is precise and adjustable to control the trap thickness accurately.

The most popular orbital machine is the Byers MicroModifier. As with conventional spread and choke procedures, daylight contacting or duplicating film and rapid access processing are most typically used.

Figure 10.4 provides estimating data for spreads and chokes using conventional spacers and diffusion exposure processes.

Film Compositing Procedures. Also as discussed in Chapter 10, film compositing procedures may be completed either in the photographic or image assembly departments during production of the complex color job. Sometimes this process is called photo-composition.

In general, producing composite film materials is done for the following reasons:

1. To reduce or minimize plate and proof exposures by collecting all the same images to go onto one plate onto film, saving production time since film exposures are significantly less in time than individual exposures of images to plates. This has a side benefit in that a plate remake for a job "on press"

is substantially less using composed film.

2. To gather together images that are easiest or best produced in film form, thus making follow-up stripping production easier.
3. As a check for image quality, image fit or image combinations.

Photo-composing of negative film materials to duplicating film, thus maintaining negative image polarity, is the most common composition process; however, producing composite negatives can also be done from film positives using contact film. Raw materials needed to composite images include contact or duplicating daylight film, rapid-access silver halide paper or Dylux. Pin registration procedures should be common to both the imaged film materials and the raw film or paper which is receiving the compositing exposures.

Consider the following example: We have a 4/C poster to strip, with the black printer containing four different black separations and three black lines of type, each on a separate piece of film. We want to produce a composite black printer containing all these images on one piece of negative film, for one exposure to a lithographic plate.

To do this, we would begin with a sheet of daylight duplicating film. Rapid-accesss processing will be used. Emulsion-to-emulsion (E-E) contacting will also be used to minimize dot spread, which means that all compositing exposures will be made to an intermediate piece of raw dupe film, which will be processed, then contacted E-E to a second sheet of raw dupe film to produce a final strippable duplicate negative. (Careful planning is usually required to ensure that the final working film composites to be used for stripping have emulsions on the desired side.) The following steps would be taken to do this:

1. Begin by punching two pieces of raw film with the pin registration punch used for registration of the negatives to be composited.
2. Place one piece of raw film emulsion up (light-side up) in the vacuum frame and place the first black separation in direct emulsion contact with the raw film. Close the frame, engage the vacuum and allow suitable drawdown time, and complete the exposure.
3. Remove the first separation, place the second separation over the raw film and complete the exposure for the second image.
4. Remove the second separation, place the third separation over the raw images and complete the exposure for the third image.
5. Remove the third separation, place the fourth separation over the raw film and complete the exposure for the fourth image.
6. Remove the fourth separation, place the first type image over the raw film and complete the exposure.
7. Remove the first type image, place the second type image over the raw film and complete the exposure.
8. Remove the second type image, place the third type image over the raw film and complete the exposure.
9. At this point the film sheet has received seven exposures, each containing different negative image. Processing of this fully exposed, (raw film) interme-

diate negative is now completed.

10. When processing is done, the intermediate negative—which contains four separations and three lines of black type—is contacted E-E to the second sheet of raw duplicating film, using the standard pin registration system, vacuum frame, and exposure. Once processed, the result is a duplicate film negative ready for stripping.

Figure 10.4 provides estimating data for film compositing production.

Manual Peelcoats and Photographic Etch & Peel Film Materials. Peelcoat materials are a masking product that is used extensively both for table stripping and in mechanical art production. As indicated in Chapter 8, peelcoat material is a thin, transparent lacquer-based film, coated on an acetate carrier. For use, the artist or stripper places it lacquer side up over the image and gently and carefully cuts and peels the desired areas away, letting white light from the light table show through. These white areas represent the image areas on the plate or proof. Popularly known as "Rubylith" or "Amberlith," peelcoats are used to produce "knockouts" whereby the background area of a photograph or other image is dropped out or masked away. They are also used in place of conventional goldenrod masking base when producing common-window flats for process color stripping. Peelcoat materials can be punched or have tabs attached if used in conjunction with pin registration.

Photographic peelcoat material is also a popular product for certain types of manual stripping. Commonly known as "etch and peel" products, their application is valuable when traps meet at extremely fine points or when stripping very fine line work that requires knockouts or image masking of any type. Essentially photographic peelcoat materials are exposed to the film image, which may be a choke or spread providing the necessary trap, using an intense ultraviolet light source. Immersion in a special developing bath for approximately three minutes follows, then a careful washed out in a water bath. Once the peelcoat material is fully dry, it is then workable for peeling.

Step-and-Repeat Techniques

Step-and-repeat techniques allow for the maximizing of the same film image on a lithographic plate. Thus, one camera negative or a group of negatives can be exposed to a plate numerous times, each exposure increasing the number of images on the plate and thereby reducing the number of required press sheets during printing. While step-and-repeat techniques reduce photographic and image assembly production times, they can increase platemaking exposure times extensively.

The step-and-repeat sequence can be executed in a number of different ways. There are manual techniques that utilize wedges and butterflies but that are not particularly accurate with large plates and many exposures. The use of pin registration for step-and-repeat is popular in most printing plants, and many different individual systems are available. Accuracy is extremely good when using pin registration for step-and-repeat.

The pin registration step-and-repeat technique begins with a row or column of images that are to be repeated. For example, four of the same image are received by the stripper from the photographic area. He or she then assembles the images across the top of the masking sheet, positioning them as a row, four across. Then a pin registration system, along one or both sides of this flat, is prepared. During plate exposure, this row of images is exposed along the top of the plate, providing four images across. Then that row is moved down the distance of the pins, and a second exposure is given for the next row. (Extra masking paper is used to cover adjacent plate areas.) This sequence is repeated until the four images have been moved the length of the printing plate. If four exposures were made of the row of four images, the plate would print 16-up, with all images positioned in precise and exact relation to one another.

Large-volume printing plants have generally investigated, if not purchased, a modern step-and-repeat machine. While configurations and models vary, as does the cost of such equipment, many models provide for mounting of the plate in a vertical position on the plate bed. The film negative, carefully imaged and prepared for the technique, is placed in a film carrier, behind which an ultraviolet light source is provided. Carefully machined gear tracks along all four sides of the perimeter of the unit then allow the film carrier to be positioned exactly where desired on the printing plate, normally within 0.001 inch.

Although step-and-repeat machines represent the ultimate in equipment cost and multiple exposures are required for each plate, the cost savings in press time because of maximized images on every printing plate can be significant. In addition, because only a limited number of film images are necessary and very little image assembly is required, time and cost savings can be high in camera work and image assembly operations. For large presses and high-volume printing operations, the purchase and use of step-and-repeat equipment can be readily justified.

Opti-copy represents another way to step-and-repeat images. In sum, Opti-copy is an extremely accurate photographic projection system used to project one or more images directly to film or a lithographic plate. Because Opti-copy units are fairly expensive—and because other more conventional step-and-repeat units are already installed and operational—their use in the printing industry is limited to companies that have significant production capacity for Opti-copy work.

Film Correction Techniques

Opaquing of line negatives is perhaps the most common film correction technique. This technique involves the application of a clay material over pinholes and other film areas that are transparent but should be opaque. While a small artist's brush and a jar of opaque material are used by some, other strippers prefer an opaquing pen similar to a felt-tip marker. It is also possible to image the film with opaque, especially when working with film positives. However, actual image generation by the stripper is considered an awkward procedure since final artwork is best prepared by artists during the copy preparation stage. Any film area that lacks density—that is, does not have a suitable blackness—requires opaquing prior to the platemaking operation.

Tonal correction on halftones is not commonly completed in the stripping department. Should tonal modifications be desired, chemicals like Farmer's Reducer can be used to modify dot size and tonal density. An acid resist is also used if the desired tonal modification is local rather than over the entire sheet of film. In many cases, when pronounced tonal modification is required, it may be more expeditious to rephotograph the image, changing exposures to yield a film image with more suitable tonal value. Opaquing on halftones is not a usual practice since it is very difficult to correct individual dots.

The *cutting-in* of line and tonal film images into other pieces of film is a reasonably simple technique. Basically, the procedure involves the cutting out and removal of a film image that is not desired and the simultaneous insertion of the correction or halftone. Lith tape, which is red and does not allow the plate to be exposed along the cut line, is then used to secure the cut-in into the fitted area.

Film scribing is another common correction technique, especially with line work. Basically, the stripper is able to create a line image, or a single line, through the removal of the film emulsion. Scribing is a normal correction when fine lines have filled in from poor copy preparation or photographic development problems. The technique is usually completed using special scribing tools available to the stripper.

11.3 Electronic Image Assembly

At this time the movement to electronic image assembly systems has taken two general directions: fully integrated systems in the form of CEPS, and systems that take over some of the stripping functions, particularly those that are hard to execute or that can be streamlined into a computer-aided manufacturing (CAM) mode. The most common and popular device here is the Gerber AutoPrep 5000.

CEPS

Color Electronic Prepress Systems, summarized as Group I in Figure 10.1 and discussed in various detail in Chapters 8, 9, and 10, represent a tool of great capacity for all prepress areas. This is particularly true in film assembly. Since the output of CEPS is fully-imaged, registered, ready-to-strip film, the process of assembling images is almost completely electronic. Thus, with a fully functioning CEPS system, the stripper's primary job is to lay out masking sheets, lay in preregistered film, and perform related duties that support the CEPS.

Gerber AutoPrep 5000

This is a unique and most helpful stripping device, developed by Gerber Scientific

Instrument Company to improve the accuracy of the stripping process and remove much of the drudgery of stripping. The AutoPrep best fits into companies that do an extensive amount of complex color stripping, particularly table stripping, since it converts many difficult stripping jobs to a computer-aided manufacturing (CAM) mode which is accurate, precise, and productive.

Essentially the most sophisticated model—the AutoPrep 5000 Model 3—has three working components: a DataPrep Station (or inputting terminal) where the job is digitized into the system, a Pen Plotter for paper plotting of impositions, layout, and for sizing images, and an AutoPlot Station which is a sophisticated film plotter that produces photographic masks using digitized DataPrep station data. Software for the inputting station is held in internal memory, with floppy disk transmission of jobs between the DataPrep station and the AutoPlot unit.

Basically the operation of the AutoPrep follows these steps:

1. The operator begins with a properly-sized film image; artwork can be used if the size considerations are determined beforehand. The artwork or film image is taped or secured to the flat working table of the DataPrep terminal, under which are located hundreds of electronic sensors. A color monitor is positioned directly in front of this working area and is essential to working on the system.

2. The operator, working with a digitizing puck, selects commands from an arrangement of options built to the left or right of the the working table (the system can accommodate both left- and right-handed operators). The hand-held puck also has a number of buttons controlling operations and functions, making the inputting operation very fast once the commands have been learned and understood.

3. Watching the color monitor and working with the puck and table commands, the operator determines how the job will be broken out for stripping. It is important that the operator has a thorough knowledge of complex table stripping since this unit is an extension of that process.

4. The operator follows the desired stripping plan and digitizes the images from the artwork or negative into the system. The color monitor shows graphically what is being done and provides written command instructions.

5. When done, the operator removes the floppy disk containing the digitized program for the job, and moves to the AutoPlot station to make the photographic masks for the job. The pen plotter could be used to view the job for size and other considerations at this point, if desired.

6. The operator inserts the floppy disk into the AutoPlot system and follows a monochromatic screen menu to complete the film plotting process. Special plotting film is punched using the company's pin registration system and mounted around a large drum, emulsion-up. A safelight front cover is closed and the plotting begins.

7. Based on menu instructions given by the operator, the plotter produces each photographic mask. Rapid-access processing is completed for each piece of exposed film. The plotter will produce one mask, stop for a new piece of film, produce the second mask, stop for film change, and so on. The plotter head is driven by the electronics of the system following the digitized program from the DataPrep Station; the plotter head has a light source and various apertures which are selected to expose the film areas. The plotter is fairly noisy, yet fast, in the production of photographic masks.

8. When all plotting is completed, the masks produced are used in conjunction with other table-produced materials, color separations, etc., to produce the job.

It is important to note that the Gerber AutoPrep system is an intermediate solution between fully electronic stripping and manual table stripping. While the AutoPrep 5000 Model 3 is fairly expensive, smaller models are available at less cost.

The AutoPrep is fast and accurate, particularly for difficult, complex stripping work. Extremely fine traps and lines can be produced extremely quickly, and the system can strip at any angle or in any size in either inch or metric measurements. Operator speed increases dramatically with experience, and skilled operators can digitize complex work at exceptional speeds, saving significant time over table stripping methods. Because traps and other fine-fit work can be done on the unit, time is also saved in contacting and other manual photographic procedures.

11.4 Lithographic Platemaking and Proofing

Major Categories of Lithographic Plates

Lithographic plates may be divided into two major categories: surface and etched. *Surface plates*, which have an ink-receptive coating adhered to the surface of the plate, are further divided into presensitized and wipe-on groups. *Etched plates* are categorized into deep-etch and multimetal divisions.

Presensitized plates, which are purchased ready for exposure to a flat, represent the largest category of all lithographic plates in terms of total sales dollars. They are popular because of ease of use during production, excellent image quality, and good consistency of image from plate to plate. Even though they are more expensive than the wipe-on plate (described next), they do not require preliminary coating during production, which saves time. Presensitized plates are in use in thousands of commercial printing and publishing operations throughout the world. There are many manufacturers of presensitized plates, including Enco (Hoechst Celanese), the 3M Corporation, Polychrome Corporation and Western.

The second type of surface lithographic plate is the *wipe-on*. These plates require that the platemaker wipe on a light-sensitive coating on the aluminum base and then dry it thoroughly. Exposure and attendant processing are then completed. The most significant advantage of wipe-on plates is that they are far less expensive than the presensitized type because coating is done in the plant prior to use. In addition, wipe-on plates are generally a good image carrier on the press and can be readily obtained in large quantities. Disadvantages include the additional time necessary during production for the coating and drying steps and the fact that the wipe-ons have a tendency to be inconsistent from time to time, especially when defined production techniques for coating and development are not followed carefully. Wipe-on plates are used in large quantities in the newspaper industry and in many web printing and publishing operations.

Of the two major categories of etched plates, the *deep-etch* type has excellent image-reproducing capability and good mileage during printing. The major disadvantage of the deep-etch type is that it requires many intricate steps in production, making it both costly and sometimes inconsistent. The acceptance of presensitized plates, which are easy to obtain and provide for fast exposures and simple processing with high-mileage capability, has eroded the deep-etch plate market. Very few printing companies make or use deep-etch plates today.

Multimetal plates represent the longest-running plate for the printing and publishing industry. Sometimes called trimetals, these image carriers have a base of aluminum or steel to which copper (an ink-receptive metal) and chromium or stainless steel (water-receptive) are electroplated. Etching is also part of the manufacturing procedure. Multimetals generally run two million impressions or more when press conditions are carefully controlled. Special equipment is required for their production, making them initially expensive. They are popular in the packaging industry and for long-run magazines and other publications where circulation is in the millions.

Overview of Bases and Coatings

A surface lithographic plate is a combination of a metal base and a light-sensitive coating or emulsion. Various metals, such as zinc, have been used in past years, but today aluminum serves as the base metal for literally all lithographic plates. The thickness of the aluminum sheet may very depending upon the size of press, and many plates are grained because the roughened surface provides better coating adhesion and improved water-carrying capability. In addition to graining, aluminum plates are also passivated—silicated or anodyzed—to reduce the chemical reaction that sometimes occurs between the base and certain coatings. Passivation also helps to improve plate performance on the press.

Once the base has been grained and passivated, it is ready to be coated. There are three major types of coating products used for presensitized plates: diazo, photopolymer, and silver halide compounds. *Diazo coatings* are by far the most used because the chemical product is readily obtained, inexpensive, and easy to store and use. Various additional chemical ingredients are used with the diazo sensitizer to allow for different mileage results.

Photopolymer coatings are popular today because of the increased run lengths they offer over diazo products, easy exposure and processing considerations, and faster roll-up on the press. Polymer coatings are more expensive than diazos because they are synthetic and must be carefully manufactured. However, increased plate mileage and excellent press performance make them very popular.

Silver halide coatings are used for short-run lithographic surface plates when limited quantities are desired. The coating resembles a photographic emulsion but is applied to a metal or plastic base instead of paper or polyester. Kodak PMT (diffusion-transfer) and Itek plates are both examples of silver halide image carriers. Another short-run lithographic plate is the electrostatic master. However, the ink-receptive carbon toner used for this plate is not a photographic coating but a product fused to the base material during the making of the plate.

Working Parameters of Lithographic Plates

All surface lithographic plates may be categorized with respect to exposure and processing orientation. Those lithographic plates that expose from negatives to produce a positive image, called *negative-acting* or *negative-working*, reverse image tonality. Plates that produce the same tonality are called *positive-acting* or *positive-working*. Because negative film images are the usual output from the photographic area in most printing plants, negative-acting plates, which reproduce a positive image, represent the majority of image carriers used in the printing industry today.

The processing of surface plates is either additive or subtractive. *Additive surface plates* require the addition of a developing lacquer to harden the plate image and make it visible. *Subtractive surface plates* are purchased in prelacquered form and require removal of lacquer from the nonimage areas during processing. Subtractives outsell additives because they are easier to develop and seem more consistent from plate to plate.

The processing of lithographic plates is in a state of transition. Prior to 1985 literally all lithographic plates were processed using solvent-based chemicals, which included alcohol and other somewhat toxic ingredients. Due to pressures from various sources, lithographic plate manufacturers introduced aqueous-based chemicals for processing plates, as a direct substitute for the solvent-based products. While chemical changes were necessary in the photographic plate coatings, the shift to aqueous-based products is currently in transition. At this writing both solvent-based and aqueous-based products are available and purchased by most lithographic plate manufacturers.

There is little noticeable difference between solvent-based and aqueous-based lithographic plates in terms of physical appearance. There is little difference in exposure or processing times or productivity issues during platemaking. Aqueous plate performance on press has received some criticism, which is typical of most newly-introduced products, particularly those that are direct links to high-cost production centers.

Chemical products for processing plates are not interchangeable between brands of plates, even though coating ingredients may be similar. An automatic plate processor cannot process both additive and subtractive varieties, again because of the differences in processing chemistry and development techniques.

During production, the lithographic plate initially receives all exposures required from either negative or positive film images. Manual or automatic film processing then follows, using the appropriate developing agents and rub-up procedures. With negative-acting plates, the clear film areas allow the intense ultraviolet light to pass through, rendering these areas insoluble to developing agents and producing the plate image. Areas covered by black film or masking material stop the ultraviolet light and, thus, are soluble to developing chemistry and are removed during plate processing to create the nonimage areas of the plate. Following development, gum arabic is applied to protect the plate surface and enhance performance during the press run. Normally, the platemaking employee completes both the exposure and the processing operations.

Automatic Plate Processing

Automatic plate processing is commonly found in most printing companies today, largely for the same reasons as automatic film processing: it is productive, fast, provides assured consistency of developed image, controls processing variables, reduces manpower, and saves space. Processors running solvent-based chemistry cannot be used with aqueous-based plates and vice versa. Changeover from solvent-based to aqueous-based chemistry with automatic film processing is a fairly simple procedure and affects primarily the development area of the processor; gumming and finishing processor units remain largely unchanged.

Plate processors vary in cost depending on the size of the unit and the type of plate to be processed. Additive processors are used for wipe-on plates and are common in newspaper and publishing plants. Subtractive processors are more popular in the commercial printing segment because they can develop two-sided plates simultaneously or one-sided plates placed back to back after exposure. In addition, subtractive processors utilize a recirculating bath system, allowing for the reuse of developer chemicals, which saves on developer material cost. Both additive and subtractive processor models provide for plate gumming as the final processing step.

Establishment of a Platemaking System

One of the important keys to profitability in printing and publishing is standardization of manufacturing processes. Many printing managers recognize this fact and have locked their production in the plateroom to a series of plates that utilize one common processing unit and the same exposure development chemistry and gumming products and procedures.

For example, most plate manufacturers provide a standardized line of subtractive plates with the variable for selecting as the mileage component of the plate coating. To illustrate, Enco (Hoechst Celanese) provides the solvent-based Enco "N" series. The sequence begins with the N-25 (negative-acting, 25,000 impressions per side) as the shortest run plate of the series, followed with the N-50 (negative-acting, 50,000 impressions per side) and then the Enco N-100 and N-200. Thus, with the purchase of one 42 inch Encomatic plate processor, any printing plant is able to select the Enco plate suitable for the job, by size (up to 42 inches) and length of run. All platemaking is done using common exposure and processing chemistry. Enco offers a similar aqueous-based line as well, beginning with the A-30 (negative-acting, 30,000 impressions per side) then the A-60, and so on. Most other major lithographic plate manufacturers provide similar processing equipment and lines of plates.

Proofing Products

In many printing plants, the equipment used for platemaking is also used for exposure of proofing materials. Proofing, as discussed in Chapter 10, is divided into monochromatic (single-color) and multicolor proofing categories. Primary reasons for proofing are for checking the job during production as an internal quality control mechanism, for customer or client approval, or as a portion of production where the proof is a part of image development or incorporated into the artwork.

Of all monochromatic proofs, by far the most popular is the negative-acting Dylux or blueline proof. Exposure to an intense ultraviolet light breaks open an encapsulated imaging chemical which immediately begins to turn blue in the image areas. Because no contact with water or other processing chemistry is needed, Dylux proofs are very dimensionally stable. Dylux is available in single-faced (C1S) for proofing single-sided material, and double-faced (C2S) for proofing two-sided materials such as book work impositions. Various exposure levels provide different levels of blue density, thus Dylux can be used to check multicolor registration and alignment.

There are essentially three types of color proofing procedures in current use in the industry: press proofing, overlay proofing, and single-sheet proofing. Press proofs require that color images be stripped, plated, and produced using conventional lithographic presswork. This is expensive and time-consuming but provides extremely accurate proof results.

Overlay proofs—Enco's NAPS/PAPS and 3M's Color Key are popular—are acetate carrier sheets coated with a pigmented diazo emulsion. When a negative is exposed with an ultraviolet light, the coating becomes insoluble to the solvent-based developer, and after development the image remains on the acetate carrier. DuPont's Chromacheck is another form of overlay proof which utilizes dry materials and a peel-apart system.

Single-sheet proofs—DuPont's Cromalin and 3M's Matchprint II—require lamination of carrier sheets to a substrate, with proof colors "building" on the substrate in a superimposed fashion. While Cromalin and Matchprint II use different techniques, they are the most popular proofing products when proofing out complex color images.

It should be noted that other proofing products are available. Many of these products are available in both negative- and positive-acting types and most may be either manually or automatically processed.

For further information see Chapter 10.3 Conventional Proofing Systems and Materials.

11.5 Estimating Conventional Film Assembly Times and Costs

Estimating table stripping is more difficult than most other areas of printing production due to the following reasons:

1. Stripping techniques and procedures vary from one employee to another. There are many ways that a given job might be stripped, even with image assembly employees working next to each other in the same company.
2. There are many small operations required in completing table stripping, particularly when completing complex color work. Some jobs may require spreads and chokes, others knockouts or composites, still others special effects such as drop shadow boxes. Also, how these are produced can vary, since more than one procedure can be used to yield the same result.
3. Because table stripping is largely a manual skill, establishment of accurate production standards is difficult to determine.
4. Many estimators don't know how even a simple job would be stripped, having never worked in the film assembly area before. It is difficult to accurately estimate any area of printing production with which one is unfamiliar.
5. Estimating many jobs is done without looking at the artwork, requiring the estimator to take a "best guess." The image assembly area is where most of the repair and correction procedures occur for improperly prepared art.
6. Jobs moving through film assembly are subject to a significant level of author's alterations, perhaps more than any other production area in printing manufacturing. Some jobs may be stripped and proofed, then changed to meet shifting customer demands and restripped and reproofed, then changed again to meet customer needs, restripped and reproofed a third time. Some customers require a "proof to satisfaction" condition when the job is begun, requiring the printer to produce as many proofs—with changes—as will be required for customer approval.

While most of these changes are chargeable to the customer, the intensely competitive reality of the business may not allow for full, complete charges in every case. Also, keeping accurate track of these customer-requested changes can be difficult and time-consuming.

Figure 11.1 is provided for estimating table stripping. It has been divided into three major sections: time for flat preparation, production time for laying/cutting, and additional time for special operations. Each production time segment should be determined and then the times added together, with adjustments as they apply. BHRs are applied as determined, based on estimated hours of production, with the addition of material costs that apply.

It is absolutely essential that a production plan or "flat configuration plan" be developed prior to applying the schedule data in Figure 11.1. Normally this plan, completed by the estimator, breaks the job into the required component parts as it will likely be stripped during production. Such a flat configuration plan should succinctly detail the number of master and complimentary flats to be stripped, the number of pieces of film to be laid-in and windows to be cut, and any special or additional operations that will be required. The more knowledge the estimator has about table stripping, the more quickly and accurately the flat configuration plan can be developed.

1. Flat Preparation and Marks Flat Production

Approx press size in inches (sq. in.)	Flat preparation time in hours per flat: Key/master flat	Complimentary flat (each add'l per same flat set)	Marks flat (hours each)
10 x 15 (150)	0.10	0.04	0.06
14 x 20 (280)	0.15	0.05	0.08
17 x 22 (374)	0.18	0.08	0.10
19 x 25 (475)	0.20	0.10	0.12
23 x 29 (667)	0.25	0.12	0.14
24 x 36 (864)	0.30	0.14	0.18
25 x 38 (950)	0.35	0.16	0.20
38 x 50 (1900)	0.40	0.20	0.24
44 x 60 (2640)	0.45	0.22	0.28
52 x 77 (4004)	0.50	0.25	0.32

2. Laying and Cutting Time (in hours per piece stripped)

Film size for stripping in sq.inches (typical film size)	SIMPLE			AVERAGE			DIFFICULT		
	Lay Only	Cut Only	Lay&Cut	Lay Only	Cut Only	Lay&Cut	Lay Only	Cut Only	Lay&Cut
Up to 35 sq . in. (5 x 7 or less)	0.05	0.02	0.07	0.07	0.03	0.10	0.08	0.04	0.12
36 to 80 sq. in. (8x10)	0.07	0.02	0.09	0.10	0.03	0.13	0.10	0.04	0.14
81 to 120 sq. in. (10 x 12)	0.08	0.03	0.11	0.13	0.04	0.17	0.13	0.05	0.18
121 to 200 sq. in (11 x 17)	0.10.	0.03	0.13	0.15	0.04	0.19	0.17.	0.05	0.22
201 to 300 sq. in. (15 x 20)	0.11	0.04	0.15	0.18	0.05	0.23	0.21	0.06	0.27
301 to 400 sq. in (17 x 22)	0.13	0.04	0.17	0.22	0.05	0.27	0.24	0.06	0.30
Over 400 sq. in. (20 x 24)	0.16	0.04	0.20	0.25	0.05	0.30	0.29	0.06	0.35

3. Additions

Chokes/spreads	**See Figure 10.4 to estimate**
Composites	**See Figure 10.4 to estimate**
Dupes/contacts	**See Figure 10.4 to estimate**
Tints	**0.03 hr per each black-and-white tint laid (any size)** **0.05 hr per each matched process color laid (any size)**
Color bars	**0.08 hr per each color bar stripped to clearbase** **(incl. cutting window on goldenrod)**
Knockouts	**0.15 hr (simple cut) 0.20 (average cut) 0.30 (difficult cut)** **(peelcoat materials extra based on $0.004 per sq. inch.)**
Cut-ins	**0.10 hr per each cut-in completed (includes taping)**
Scribing	**0.15 hr per each 100 linear in. of scribed line on film emulsion**

Note: Time and cost for restripping or reworking jobs not included.

Figure 11.1

Film assembly production schedule

Once the flat configuration plan is completed, schedule data (Figure 11.1) is applied to determine production times. The following provides information related to applying that data in a step-by-step format:

Step 1. Flat preparation time includes time for the employee to read all work order instructions, review film materials, plan the stripping procedure, organize the work, punch flats for pin registration, and lay out the flats including making image position marks. The key/master flat time represents the production time needed for each master flat to be stripped; a master flat is one to which all other flats in the flat file register. Complimentary flats are all flats that register to a given master flat and have reduced production times since the master flat covers most of the job complexity.

Certain jobs require that a flat be stripped to detail the folding, cutting, and other "marks" that will be needed for press, bindery (or both) manufacturing operations of the job. The "Marks flat" column is to provide times for the addition of such a flat. Times cover all flat preparation and stripping production. The typical printing order in production requires one marks flat per job.

Step 2. Laying and cutting time in hours per piece is referenced by the size of the film and type of stripping required as simple, average, or difficult. Each category is divided into three divisions: laying only, cutting only, and laying and cutting. The following describes the application of this part of the schedule:

- **Goldenrod stripping:** Conventional goldenrod stripping of a line or halftone negative requires that the film image be laid-in and taped to the goldenrod masking support, then a window cut in that same goldenrod support for that negative. These are typically completed by the stripper as sequential production operations and would utilize the "Lay & Cut" portion of the schedule since both laying in of the film image and cutting the window are done as sequential production operations.
- **Clearbase stripping:** Clearbase stripping of film images such as color separations requires that the separations be carefully "laid-in" (registered to another separation) and taped to the clearbase support, with follow-up "cutting only" of common-windows in a separate sheet of peelcoat or goldenrod masking material. Table stripping of a typical four-color separation requires careful registration—"laying only"—of four printers, with "cutting only" of one window per color separation. When stripping color separations to clearbase, the stripper normally completes all "laying-in" operations for all film separations of all printers before cutting the common-window flat.

Some strippers—instead of stripping line or halftone negatives directly to goldenrod as described under the "goldenrod stripping" procedure above—prefer to strip these film materials to clearbase and cut windows in separate peelcoat or goldenrod base. If this were the common production practice in a company, the "Lay only" and "Cut only" data in Figure 11.1 would be used instead of "Lay & Cut" data for stripping line and halftone work as well as color separation production.

Step 3. The additions segment of the schedule is to add for any special operations or other production activities completed on the job. These may include scribing, cut-ins of film materials for correction purposes, the addition of color tints, the addition of color bars to color separation clearbase flats, and knockout production using Rubylith or Amberlith peelcoat materials. If traps are required using conventional choke and spread contacting procedures, the film contacting and duplication schedule in Chapter 10 would be used.

The following formulas summarize how manual image assembly production time and cost are estimated, using Figure 11.1:

Total table stripping production time = flat preparation time (key flats + master flats + marks flats) + laying and cutting time (see following formulas) + additions (as applicable).

Laying and cutting time for goldenrod stripping = number of pieces of film to be laid-in and cut × production time per piece

Laying and cutting time for clearbase stripping = number of separations film images to be laid-in × production time per piece + number of windows to be cut (in common-window flat) × the production time per window

Total table stripping production cost = total table stripping production time × table stripping BHR

It is important to note that material costs have not been included in the image assembly estimating schedule (Figure 11.1). In general, the cost of masking materials and related production tools are minimal in cost. Because of this, these costs are recovered through the BHR process (Line 8 of the BHR sheet as a department direct supply). Of course, if film contacting is needed, or if material costs or production tools are expected to be excessively high on any job, the addition of these material costs (plus appropriate markup) should be made during estimating.

Example 11.1

We want to estimate the table stripping production time and cost to produce 7 1/4 × 8 1/2 inch announcements, to run 6-up, and print black and red, two sides (2/2) on a Heidelberg MOZP (maximum press size sheet 19 × 25 1/4). The BHR for our table stripper who will do this job is $57.90.

Film supplied: Six duplicate line negatives each of four images (12 black type and 12 red line drawings) so that stripping can be completed 6-up. There will also be one black and red duotone per finished image, or a total of 12 duotones (12 black and 12 red film images). The duotones will strip to clearbase 6-up, with separate common-windows on peelcoat; each plate will double-burn the duotone images.

Stripping details: Stoesser pin registration system. Vinyl goldenrod for line work, clearbase and peelcoat for duotones. Windows 8 × 10 inches for line work and

5 × 7 for duotones, all at simple level. Use 19 × 25 flat data. There will be 12 black-and-white tints for the black type flats (40%) and 12-20% tints for the red line flats. The first black flat will be a master flat, all others will register to that flat as compliments. Add for one marks flat for registration and as a trim-out guide, to burn to each black plate.

Flat Configuration Plan

Side 1 (2 plates: 1 black 6-up plate and 1 red 6-up plate)

Flat I Black type (master) with 6 pcs film, goldenrod
Flat II Black duotone (compliment) with 6 pcs film, clearbase
Flat III Red line work (compliment) with 6 pcs film, goldenrod
Flat IV Red duotone (compliment) with 6 pcs film, clearbase
Flat V Common-window (compliment) for duotones, cut 6 windows, peelcoat

Side 2 (2 plates: 1 black 6-up plate and 1 red 6-up plate)

Flat VI Black type (compliment) with 6 pcs film, goldenrod
Flat VII Black duotone (compliment) with 6 pcs film, clearbase
Flat VIII Red line work (compliment) with 6 pcs film, goldenrod
Flat IX Red duotone (compliment) with 6 pcs film, clearbase
Flat X Common-window (compliment) for duotones, cut 6 windows, peelcoat
Flat XI Marks flat (compliment) to burn with one black plate, for trim-out infformation

Solution Use Figure 11.1.

Step 1 Determine flat preparation and marks flat production time:

1 master flat (19 × 25 inches) × 0.20 hr ea = 0.20 hr
9 compliments (19 × 25 inches) × 0.10 hr ea = 0.90 hr
1 marks flat (19 × 25 inches) × 0.12 hr ea = 0.12 hr
Total preparation/startup production time: 0.20 + 0.90 + 0.12 = 1.32 hr

Step 2 Determine laying and cutting time:

24 pcs (8 × 10, lay & cut to goldenrod, simple) × 0.09 hr ea = 2.16 hr
24 pcs (5 × 7, lay to clearbase, simple) × 0.05 hr ea = 1.20 hr
12 windows (5 × 7, cut to peelcoat, simple) × 0.02 hr ea = 0.24 hr
Total laying and cutting time: 2.16 + 1.20 + 0.24 = 3.60 hr

Step 3 Determine additional time:

24 tints (black-and-white) × 0.03 hr ea = 0.72 hr

Step 4 Total image assembly time:

1.32 hr × 3.60 hr = 0.72 hr = 5.64 hr

Step 5 Total image assembly production cost:

5.64 hr × $57.90/hr = $326.56

Step 6 Image assembly material cost:

None (recovered through BHR)

Step 7 Summarize your findings:

Total production time: 5.64 hr
Total production time cost: $326.56
Total material cost: $0
Total production cost: $326.56

Example 11.2

We want to estimate the image assembly time and cost to table strip a four-color process art flyer for a local college. It will run 4/4 on a Komori Lithrone 426 (maximum press size sheet 20 × 26) on an 18 × 24 pss.The front side of the flyer will have 8 separations to strip and the back side will have 6 more. Each side will also have a black type printer (1 negative) to be stripped as a keyline master and double-burned to the black process color plate. BHR for our table stripper who will do this job is $61.75.

Stripping details: Stoesser pin registration system.Vinyl goldenrod for black line work, clearbase and peelcoats for separations. Windows 17 × 22 inches for black line and 5 × 7 for laying in all separations, at average level. Cutting for common-window flat windows at simple, 5 × 7. Use 17 × 22 flat data. Add for one color bar per each stripped separation, or a total of 8 for the job. Each black type flat will be a master flat for that side, and all others flats—including process color flats—will register to that flat as compliments. Add for one marks flat for registration and as a trim-out guide, to burn to each black plate.

Flat Configuration Plan

Side 1 (4 plates: 1 black,1 magenta,1 cyan,1 yellow)

Flat I Black type (master) with 1 film negative, goldenrod
Flat II Black separation flat (compliment) with 8 film negatives, clearbase
Flat III Magenta separation flat (compliment) with 8 film negatives, clearbase
Flat IV Cyan separation flat (compliment) with 8 film negatives, clearbase
Flat V Yellow separation flat (compliment) with 8 film negatives, clearbase
Flat VI Common-window (compliment) for separations, cut 8 windows, peel-coat

Side 2 (4 plates: 1 black,1 magenta,1 cyan,1 yellow)

Flat VII Black type (compliment) with 1 film negative, goldenrod
Flat VIII Black separation flat (compliment) with 6 film negatives, clearbase
Flat IX Magenta separation flat (compliment) with 6 film negatives, clearbase
Flat X Cyan separation flat (compliment) with 6 film negatives, clearbase
Flat XI Yellow separation flat (compliment) with 6 film negatives, clearbase
Flat XII Common-window (compliment) for separations, cut 6 windows, peel coat
Flat XIII Marks flat (compliment) to burn with one black plate, for trim-out information

Solution Use Figure 11.1.

Step 1 Determine flat preparation and marks flat production time:

2 master flat (17 × 22 inches) × 0.18 hr ea = 0.36 hr
11 compliments (17 × 22 inches) × 0.08 hr ea = 0.88 hr
1 marks flat (17 × 22 inches) × 0.120 hr ea = 0.10 hr
Total preparation/startup production time: 0.36 + 0.88 + 0.10 = 1.34 hr

Step 2 Determine laying and cutting time:

2 pcs (17 × 22, lay & cut to goldenrod, average) × 0.27 hr ea = 0.54 hr
56 pcs (5 × 7, lay to clearbase, average) × 0.07 hr ea = 3.92 hr
14 windows (5 × 7, cut to peelcoat, simple) × 0.02 hr ea = 0.28 hr
Total laying and cutting time: 0.54 + 3.92 + 0.28 = 4.74 hr

Step 3 Determine additional time:

8 color bars × 0.08 hr ea = 0.64 hr

Step 4 Total image assembly time:

1.34 hr + 4.74 hr + 0.64 hr = 6.72 hr

Step 5 Total image assembly production cost:

6.72 hr × $61.75/hr= $414.96

Step 6 Image assembly material cost:

None (recovered through BHR)

Step 7 Summarize your findings:

Total production time: 6.72 hr
Total production time cost: $414.96
Total material cost: $0
Total production cost: $414.96

11.6 Estimating Electronic Image Assembly Times and Costs

Figure 11.2 has been provided to estimate production using a Gerber AutoPrep Model III consisting of a DataPrep input terminal, a pen plotter and an AutoPlot film plotter. This computer-aided manufacturing (CAM) device is growing in popularity in the printing industry due to its extreme accuracy, and the relative ease to complete even the most complex stripping job. Example 11.3 is provided to demonstrate the application of estimating using the Gerber AutoPrep.

The following formulas are used to estimate image assembly production time and cost using Figure 11.2:

Total Gerber AutoPrep production time = Preparation/setup time per job + (Stripping input time × the number of film flats to be produced) + Film/pen plotting time.

Total Gerber AutoPrep production cost = Total Gerber AutoPrep production time × Gerber AutoPrep BHR

AutoPrep film cost = number of pieces of film used × size of film in square inches × cost per square inch

Example 11.3

We want to produce a part of a complex job using our Gerber AutoPrep system. Working from a keyline negative, we must create six film masks. Based on our assessment of the job, two will be at simple, one at moderate, and three at complex. It is a new job and add three additional reopen setups since not all the work will be done at at one time. Our AutoPlot film plotter will produce the final film masks in it. 20 × 26 inch size with three plots at simple, one at average, and two at difficult. Gerber BHR is $122.60 per hour.

Solution Use Figure 11.2.

Step 1 Determine preparation/setup time:

1 (fresh startup) × 0.20 hr = 0.20 hr
3 (reopen startups) × 0.10 hr ea = 0.30 hr
Total preparation/startup production time: 0.20 + 0.30 = 0.50 hr

Step 2 Determine stripping/digitizing time:

2 masks (simple) × 0.15 hr ea = 0.30 hr
1 mask (moderate) × 0.25 hr ea = 0.25 hr
3 masks (complex) × 0.40 hr ea = 1.20 hr
Total stripping/digitizing production time: 0.30 + 0.25 + 1.20 = 1.75 hr

Step 3 Determine plotter time for AutoPlot system:

3 plots (simple) × 0.20 hr = 0.60 hr
1 plot (average) × 0.30 hr = 0.30 hr
2 plots (difficult) × 0.40 hr = 0.80 hr
Total AutoPlot production time: 0.60 + 0.30 + 0.80 = 1.70 hr

Step 4 Total AutoPrep production time:

0.50 hr + 1.75 hr + 1.70 hr = 3.95 hr

Step 5 Total AutoPrep production cost:

3.95 hr × $122.60/hr = $484.27

Step 6 AutoPrep film cost:

6 pc (20 in. × 26 in.) × $0.019/sq in. = $59.28

Step 7 Summarize your findings:

Total production time: 3.95 hr
Total production time cost: $484.27
Total material cost: $59.28
Total production cost: $484.27 + $59.28 = $543.55

1. Preparation/setup	Fresh startup to begin new job: 0.20 hrs Reopen existing job and rework: 0.10 hrs
2. Stripping (digitizing) input* (*Evaluate job and determine the number of final film flats to be produced. Level of difficulty can vary for different film layers in any job.)	Simple: 0.15 hrs per layer of one job Moderate: 0.25 hrs per layer of one job Complex: 0.40 hours per layer of one job

3. Plotter	*Simple*	*Average*	*Difficult*
Pen Plotter	0.05 hrs/page	0.10 hrs/page	0.30 hrs/page
AutoPlot (film plotter) (includes processing)	0.20 hrs/page	0.30 hrs/page	0.40 hrs/page

Note: Time and cost for remakes or replots of film materials not included.
Film cost: AutoPlot film $0.019/sq in.

Figure 11.2

Schedule for estimating Gerber AutoPrep 5000 Model 3 production

11.7 Estimating Platemaking and Proofing Times and Costs

Figure 11.3 contains schedules for estimating lithographic platemaking and proofing production time and cost of materials. Chart data is divided into manual plate and proof processing and automatic plate and proof processing, which are further divided into manual and automatic (step-and-repeat) exposures. First the plate size or maximum press size for the press should be located. Use a higher square inch value for plates and sheets that do not fit exactly into a listed size. The cost of plate and proofing materials is a separate part of the schedule, oriented around the square inch value of the proof or size of the plate. Processing chemicals have been included in the plate and proofing costs indicated in the schedule. Example 11.4 and 11.5 which follow, demonstrate how to apply Figure 11.3 data.

As with estimating image assembly, a "plate and proofing configuration plan" should be developed to frame out the production that will be done during this manufacturing sequence. There are interconnected links between image assembly and platemaking/proofing which necessitates that planning be an early consideration when estimating both areas. Use the following formulas in conjunction with the data in Figure 11.3 to estimate platemaking and proofing time and cost.

platemaking and proofing production time = (number of first exposure flats × first exposure production time) + (number of additional exposure flats × additional exposure production time)

Manual Plate and Proof Processing Time in Hours					Automatic Plate and Proof Processing Time in Hours			
Manual Exposure		Automatic Exposure*		Maximum Press Size Sheet or Plate Size in inches (sq in.)	Manual Exposure		Automatic Exposure*	
First	Each Additional	First	Each Additional		First	Each Additional	First	Each Additional
0.11	0.03	0.06	0.02	10 × 15 (150)	0.08	0.03	0.05	0.02
0.16	0.05	0.08	0.03	14 × 20 (280)	0.12	0.05	0.06	0.03
0.19	0.06	0.11	0.04	17 × 22 (374)	0.14	0.06	0.08	0.04
0.24	0.07	0.14	0.04	19 × 25 (475)	0.18	0.07	0.11	0.04
0.27	0.08	0.15	0.04	23 × 29 (667)	0.20	0.08	0.12	0.04
0.32	0.10	0.16	0.04	24 × 36 (864)	0.22	0.10	0.13	0.04
0.35	0.10	0.18	0.04	25 × 38 (950)	0.26	0.10	0.14	0.04
0.37	0.10	0.20	0.04	38 × 50 (1900)	0.30	0.10	0.15	0.04
0.41	0.10	0.21	0.04	44 × 60 (2640)	0.35	0.10	0.16	0.04
0.45	0.10	0.22	0.04	52 × 77 (4004)	0.37	0.10	0.17	0.04

*Step-and-repeat machines.

Plate Size in Inches (sq in.)	Enco N-25	Enco N-50	Kodak SX	3M Viking G-1	3M Viking G-2	Western Duratec SR	Western Duratec MR	Enco A-30	Enco A-60
10 × 15 (150)	$ 2.65	$ 3.02	$ 2.84	$ 2.86	$ 2.94	$ 2.56	$ 2.94	$ 2.52	$ 2.84
14 × 20 (280)	4.90	5.64	5.29	5.33	5.49	4.87	5.49	4.70	5.29
17 × 22 (374)	6.55	7.54	7.07	7.12	7.33	6.39	7.33	6.28	7.07
19 × 25 (475)	8.31	9.58	8.98	9.04	9.31	8.11	9.31	7.98	8.98
23 × 29 (667)	10.74	12.51	12.23	12.42	12.89	10.55	12.13	10.27	11.67
24 × 36 (864)	13.91	16.21	15.85	16.09	16.70	13.67	15.72	13.30	15.12
25 × 38 (950)	15.30	17.82	17.42	17.69	18.35	15.03	17.29	14.63	16.63
38 × 50 (1900)	14.76	31.92	31.38	31.92	34.05	29.26	31.92	26.60	29.52
44 × 60 (2640)	41.02	44.35	43.61	44.35	47.30	40.66	44.35	36.96	41.03
52 × 77 (4004)	62.22	67.27	66.15	67.27	71.75	61.66	67.27	56.06	62.22

Costs for plates indicated above include development and finishing chemistry.

Proofing costs including processing:

OVERLAY PROOFS
Color Key
Process colors: $0.14/sq. in.
PMS colors: $0.016/sq. in.

SINGLE SHEET PROOFS
*Matchprint II:***
Process colors: $0.017/sq. in.
PMS colors: $0.019/sq. in.

DYLUX
C1S: $0.002
C2S: $0.003

NAPS/PAPS
Process colors: $0.013/sq. in.
Flat colors: $0.015/sq. in.

*Cromalin:***
$0.020/sq. in. (film + toner)

**per laminated and fully-toned layer

For all proofing, use platemaking exposure times indicated above.

Note: Time and cost for plate or proof remakes not included.

Figure 11.3

Lithographic platemaking and proofing schedule including plate and proofing material costs

platemaking and proofing production cost = platemaking and proofing production time × platemaking and proofing BHR

plate cost = number of plates × cost per plate

proof cost = number of proofs × size of proof in square inches × cost per square inch

total cost for platemaking and proofing = production cost + platemaking and proofing material cost

Example 11.4

We want to estimate the proofing and platemaking time for Example 11.1 previously presented. To review, the product was 7 1/4 × 8 1/2 inch announcements, stripped to run 6-up and to be printed black and red, both sides (2/2) on a Heidelberg MOZP. There were duotones to print in red and black, as well as separate red and black line images. The stripped job requires four plates, 2 black and 2 red, with each plate receiving exposures from two different flats.

Platemaking and proofing details: Stoesser pin registration system. 19 × 25 Enco A-60 aqueous plates will be made. Manual exposure and automatic plate processing. Color Key proofs, manually-exposed and processed, will also be made in the plate size to check registration and will be imaged with the marks flat (on one black proof) to check trim-out and other production marks. PMS colors will be used. BHR for platemaking/proofing is $54.90.

Platemaking and Proofing Configuration Plan

Side 1 (2 plates: 1 black 6-up plate and 1 red 6-up plate)

Plate I Black type + black duotone (with common-window flat) + marks flat
Plate II Red line work + red duotone (with common-window flat)

Side 2 (2 plates: 1 black 6-up plate and 1 red 6-up plate)

Plate III Black type + black duotone (with common-window flat)
Plate IV Red line work + red duotone (with common-window flat)

Solution Use Figure 11.3.

Step 1 Determine platemaking production time:

Plate I (black printer, 19 × 25 inches, auto processing, manual exposure):
1 first exposure (black type) × 0.18 hr ea = 0.18 hr

2 additional exposures (black duotones and marks flat) × 0.07 hr ea = 0.14
Production time for Plate I: 0.32 hr
Plate II (red printer, 19 × 25 inches, auto processing, manual exposure):
1 first exposure (red line work) × 0.18 hr ea = 0.18 hr
1 additional exposure (red duotones) × 0.07 hr ea = 0.07
Production time for Plate II: 0.25 hr
Plate III (black printer, 19 × 25 inches, auto processing, manual exposure):
1 first exposure (black type) × 0.18 hr ea = 0.18 hr
1 additional exposure (black duotones) × 0.07 hr ea = 0.07
Production time for Plate III: 0.25 hr
Plate IV (red printer, 19 × 25 inches, auto processing, manual exposure):
1 first exposure (red line work) × 0.18 hr ea = 0.18 hr
1 additional exposure (red duotones) × 0.07 hr ea = 0.07
Production time for Plate IV: 0.25 hr
Total platemaking production time: 0.32 + 0.25 + 0.25 + 0.25 = 1.07 hr

Step 2 Determine proofing production time:

Proof I (black printer, 19 × 25 inches, manual processing, manual exposure):
1 first exposure (black type) × 0.24 hr ea = 0.24 hr
2 additional exposures (black duotones and marks flat) × 0.07 hr ea = 0.14
Production time for Plate I: 0.38 hr
Proof II (red printer, 19 × 25 inches, manual processing, manual exposure):
1 first exposure (red line work) × 0.24 hr ea = 0.24 hr
1 additional exposure (red duotones) × 0.07 hr ea = 0.07
Production time for Plate II: 0.31 hr
Proof III (black printer, 19 × 25 inches, manual processing, manual exposure):
1 first exposure (black type) × 0.24 hr ea = 0.24 hr
1 additional exposure (black duotones) × 0.07 hr ea = 0.31
Production time for Plate III 0.31 hr
Proof IV (red printer, 19 × 25 inches, manual processing, manual exposure):
1 first exposure (red line work) × 0.24 hr ea = 0.24 hr
1 additional exposure (red duotones) × 0.07 hr ea = 0.07
Production time for Plate IV: 0.31 hr
Total proofing production time: 0.38 + 0.31 + 0.31 + 0.31 = 1.31 hr

Step 3 Determine total platemaking and proofing production time:

1.07 hr + 1.31 hr = 2.38 hr

Step 4 Determine total platemaking and proofing production cost:

2.38 hr × $54.90/hr = $130.66

Step 5 Platemaking and proofing material cost:

4 plates (Enco A-60, 19 × 25) × $8.98 ea = $35.92
4 proofs × 19 × 25 × 0.016/sq.in = $30.40
Total plate and proof materials: $35.92 + $30.40 = $66.32

Step 6 Summarize your findings:

Total production time: 2.38 hr
Total production time cost: $130.66
Total material cost: $66.32
Total production cost: $196.98

Example 11.5

We want to estimate the proofing and platemaking time for Example 11.2 previously presented. To review, the product was a 4/4 color art flyer for local college, stripped for a Komori Lithrone 426 (maximum press size sheet 20 × 26). There were a total of 14 separations, 8 on the front side and 6 on the back. A total of eight plates are required.

Platemaking and proofing details: Stoesser pin registration system. Use 19 × 25 plate data for 3M Viking G-2 plates. Manual exposure and manual processing. Proofing will be done with Matchprint II material, manually-exposed and automatically processed for customer approval and internal registration check. The marks flat will be proofed (on one black proof) to check trim-out and other production marks. Matchprint process colors will be used. BHR for platemaking/proofing is $57.40.

Platemaking and Proofing Configuration Plan

Side 1 (4 plates: one black, one magenta, one cyan, and one yellow)

Plate I — Black type + black separations (with common-window flat) + marks flat
Plate II — Magenta separations (with common-window)
Plate III — Cyan separations (with common-window)
Plate IV — Yellow separations (with common-window)

Side 2 (4 plates: one black, one magenta, one cyan, and one yellow)

Plate V — Black type + black separations (with common-window flat)
Plate VI — Magenta separations (with common-window)
Plate VII — Cyan separations (with common-window)
Plate VIII — Yellow separations (with common-window)

Solution Use Figure 11.3

Step 1 Determine platemaking production time:

Plate I (black printer, 19 × 25 inches, manual processing, manual exposure):
1 first exposure (black type) × 0.24 hr ea = 0.24 hr
2 additional exposures (black separations and marks flat) × 0.07 hr ea = 0.14
Production time for Plate I: 0.38 hr
Plate II (magenta printer, 19 × 25 inches, manual processing, manual exposure):
1 first exposure (magenta separations) × 0.24 hr ea = 0.24 hr
Production time for Plate II: 0.24 hr
Plate III (cyan printer, 19 × 25 inches, manual processing, manual exposure):
1 first exposure (cyan separations) × 0.24 hr ea = 0.24 hr
Production time for Plate III: 0.24 hr
Plate IV (yellow printer, 19 × 25 inches, manual processing, manual exposure):
1 first exposure (yellow separations) × 0.24 hr ea = 0.24 hr
Production time for Plate IV: 0.24 hr
Plate V (black printer, 19 × 25 inches, manual processing, manual exposure):
1 first exposure (black type) × 0.24 hr ea = 0.24 hr
1 additional exposure (black separations and marks flat) × 0.07 hr ea = 0.07 hr
Production time for Plate V: 0.31 hr
Plate VI (magenta printer, 19 × 25 inches, manual processing, manual exposure):
1 first exposure (magenta separations) × 0.24 hr ea = 0.24 hr
Production time for Plate VI: 0.24 hr
Plate VII (cyan printer, 19 × 25 inches, manual processing, manual exposure):
1 first exposure (cyan separations) × 0.24 hr ea = 0.24 hr
Production time for Plate VII: 0.24 hr
Plate VIII (yellow printer, 19 × 25 inches, manual processing, manual exposure):
1 first exposure (yellow separations) × 0.24 hr ea = 0.24 hr
Production time for Plate VIII: 0.24 hr
Total platemaking production time: 0.38 hr + 0.24 hr + 0.24 hr + 0.24hr + 0.31 hr + 0.24 hr + 0.24 hr + 0.24 hr = 2.13 hr

Step 2 Determine proofing production time:

Side One
Proof Layer I (black printer, 19 × 25 inches, auto processing, manual exposure):
1 first exposure (black type) × 0.18 hr ea = 0.18 hr
2 additional exposures (black separations and marks flat) × 0.07 hr ea = 0.14
Production time for Proof Layer I: 0.32 hr
Proof Layer II (magenta printer, 19 × 25 inches, manual processing, manual exposure):
1 first exposure (magenta separations) × 0.18 hr ea = 0.18 hr
Production time for Proof Layer II: 0.18 hr
Proof Layer III (cyan printer, 19 × 25 inches, manual processing, manual exposure):

1 first exposure (cyan separations) × 0.18 hr ea = 0.18 hr
Production time for Proof Layer III 0.18 hr
Proof Layer IV (yellow printer, 19 × 25 inches, manual processing, manual exposure):
1 first exposure (yellow separations) × 0.18 hr ea = 0.18 hr
Production time for Proof Layer IV: 0.18 hr
Side Two
Proof Layer I (black printer, 19 × 25 inches, auto processing, manual exposure):
1 first exposure (black type) × 0.18 hr ea = 0.18 hr
1 additional exposure (black separations) × 0.07 hr ea = 0.07
Production time for Proof Layer I: 0.25 hr
Proof Layer II (magenta printer, 19 × 25 inches, manual processing, manual exposure):
1 first exposure (magenta separations) × 0.18 hr ea = 0.18 hr
Production time for Proof Layer II: 0.18 hr
Proof Layer III (cyan printer, 19 × 25 inches, manual processing, manual exposure):
1 first exposure (cyan separations) × 0.18 hr ea = 0.18 hr
Production time for Proof Layer III: 0.18 hr
Proof Layer IV (yellow printer, 19 × 25 inches, manual processing, manual exposure):
1 first exposure (yellow separations) × 0.18 hr ea = 0.18 hr
Production time for Proof Layer IV: 0.18 hr
Total Matchprint proofing production time: 0.32 hr + 0.18 hr + 0.18 hr + 0.18 hr + 0.25 hr + 0.18 hr + 0.18 hr + 0.18 hr = 1.65 hr

Step 3 Determine total platemaking and proofing production time:

2.13 hr + 1.65 hr = 3.78 hr

Step 4 Determine total platemaking and proofing production cost:

3.78 hr × $57.40/hr = $216.97

Step 5 Platemaking and proofing material cost:

8 plates (Viking G-2, 19 × 25) × $9.31 ea = $74.48
8 proof layers × 19 × 25 × 0.017/sq in = $64.60
Total plate and proof materials: $74.48 + $64.60 = $139.08

Step 6 Summarize your findings:

Total production time: 3.78 hr
Total production time cost: $216.97I
Total material cost: $139.08
Total production cost: $356.05

Estimating Sheetfed Presswork

12.1 Introduction

Following this introduction, this chapter is divided into five parts. Section 12.2 discusses the essential changes occurring in the sheetfed press area, both philosophically and technologically. This is followed by Section 12.3 which details methods by which sheetfed press productivity can be enhanced or improved. Section 12.4 covers changeover points, a valuable mathematical tool for selecting the most economical unit when more than one press could be used for a given job. Estimating letterpress production is detailed in Section 12.5, while Section 12.6 provides discussion and examples with respect to accurately estimating lithographic sheetfed presswork. The chapter concludes with a list of tips which summarize sheetfed press estimating factors of importance.

12.2 The Changing Environment of the Sheetfed Pressroom

Sheetfed press production, as with numerous other production centers in the technologically-advancing printing plant, is undergoing structural changes that point to significant long-term effects for the industry.

Up through the early 1980s, many printing managers emphasized the importance of the pressroom over other areas of printing production, and focused much of their attention on pressroom performance related to job profitability. After all, the pressroom was both labor-intensive and capital-intensive, and provided the company with a focused printed product as output, not parts or segments of products as did prepress and finishing production. Maintaining pressroom efficiency was considered by some as *the* critical, controlling factor relative to job and company profitability, and many managers believed the pressroom was the area where the company "made or lost money."

Yet changes are occurring which require the press manufacturing area of a printing plant to be viewed differently than in the past. These changes are briefly summarized as follows:

1. In many printing firms, there has been a noticeable philosophical shift in management thinking, away from a "production" focus and toward an emphasis on "marketing" and customer service. Progressive printing firms have moved away from past decades of being production-driven, and forced—largely by competitive pressures—to become both customer-sensitive and market-driven. The 1980s have seen an increasing number of company owners and managers realize that the true key to profitability in printing is not through precise control of production costs—even though cost control is important—but in serving customers and focusing on markets and clients that fit into the printers product/process orientation. The pressroom has become recognized as one key—but not the only key—to achieve true customer satisfaction and generate profits to continue in business.

It should be noted that many quick printers and smaller commercial printers have never focused as intensely on the pressroom and press-related issues compared to their larger printing company counterparts. One reason for this is that the competitive environment for a typical quick printer or smaller commercial printer requires a market-driven focus, using "rush" work and direct counter sales as major tools for customer service, customer loyalty, and repeat business. Because of this quick and smaller commercial printers have evolved with a philosophy of the "press (or copier) as a vehicle" concept, making the "ink on paper" manufacturing process only one aspect of the business. The typical quick printer concentrates on services that go along with printed materials, not just on the printing. While this concept may seem simplistic, it is clear that implementation of the "press as a vehicle" philosophy by larger printers is an important concept in an industry so intensively competitive as commercial printing.

2. Along with the philosophical movement from production to marketing and customer service, many printing managers have become aware of the true meaning of the concept of value added, whereby all productive work done on a given job should directly increase the value of that product to the customer. Thus, the actual presswork of the job is but one of many job ingredients that add value to the customer's final product. Today, prepress and finishing operations have become recognized as important contributors to adding value to a customer's product. For example, selling creative design skills for a job, doing the color separations required, completing sophisticated typesetting, foil-stamping, die-cutting, and special bindery operations are critical to giving the final printed product more value to the customer, and thus contribute more profit to the printer. In sum, the pressroom is recognized as an important component of the value added process, but certainly not the only one.

3. The pressroom is no longer the production area which always has the greatest capital- or labor-intensity. The 1980s brought continuous and unabated change in prepress, particularly expensive electronic equipment. Bindery equipment became more specialized, and to offer such services for value added purposes, printers realized they needed to install such equipment. Also, to compete and survive—to produce quality printing faster in response to customer demand—managers have been forced into installing this sophisticated prepress and bindery equipment in their own manufacturing plants. Not only is much of this electronic equipment expensive, but also requires considerable manpower and training efforts, increasing the need for skilled, trained, knowledgeable employees. In effect, both labor intensiveness and equipment costs have required printing management to take a broader, more encompassing view of the numerous profit centers in any printing plant.

4. Automation in the pressroom has also reduced the labor intensiveness over the long-term. Linked to the manpower demands brought on by prepress and bindery, press equipment manufacturers have worked to streamline production activities on the printing press. For example, electronic plate scanning systems are commonly used to determine ink densities before the plates are mounted on the press, meaning substantially reduced makeready times. Feeders and delivery units have been redesigned to reduce paper waste and setup times, further reducing press preparatory costs. Electronic master control consoles for larger sheetfed and web presses are common, reducing significantly the number of skilled press operators necessary to produce an extremely high quality product at higher press speeds than even ten years ago. In sum, automation and electronic controls have helped to shift the previous high number of pressroom employees downward, leaving the pressroom with fewer yet more highly skilled craftsmen/technicians.

5. This newer press equipment—automated and electronically-controlled—reduces makeready costs, meaning faster production and less customer cost. Knowledgeable printing managers have known for years that the faster the makeready for a particular job can be completed, prior to printing the first saleable sheet, the more competitive and profitable the printing company will be. In the 1980s equipment manufacturers, to sell more equipment, have incorporated numerous technological advances specifically to address this issue.

6. During the 1980s there has been an increased focus on improving the productivity of the high volume sheetfed pressroom, not only through technical press improvements but through the increased installation of 5- and 6-unit multicolor printing presses. These large-unit multicolor presses improve productivity significantly because they require less press time by reducing the number of passes for the complex color work now demanded by customers. In some printing companies, the strategy of multicolor press purchases has streamlined the production management process as well. As an example, purchasing all sheetfed presses in the same size, with variable numbers of printing units on each press, provides great scheduling flexibility. A six-color, one side order could be run on a six-color press in one pass or, if that press is occupied, on 4-color and 2-color presses in two passes. Stripping is standardized to one size of press equipment and finishing is unchanged regardless of press choice.

Such improvements—scheduling efficiency, reduced prices through improved productivity, and controlled job delivery so fast-turnaround work can be more easily accommodated—are all essential keys to satisfied customers. They also ensure long-term profitability in the pressroom and, more globally, to the printing company as we enter the next decade.

An issue of major importance related to both sheetfed and web press areas is the matter of pressroom overcapacity. Some experts argue that the significant increase of multicolor sheetfed presses—coupled with faster and wider web presses—has set up a situation whereby there is too much available press capacity for existing printing volume from customers. Other experts say that there is no overcapacity since the sale of presses by equipment manufacturers has not yet been reduced, the key measure of overcapacity in their opinion.

Whether or not there is an inherent overcapacity in the printing industry is an issue that remains unclear. However, there is no question that profitable management of the pressroom, in light of the increasing number of multicolor presses going on line and the higher costs of this sophisticated equipment, is a vital issue for printing managers everywhere.

The above points are not intended to reduce the importance of the press area. The importance of excellent presswork—producing the needed quantity and quality of product for the customer—cannot be overemphasized. The pressroom yields a critical graphic product that the customer will see and use, even though prepress and finishing operations contribute to the product's value added as well. In sum, there is no doubt that the pressroom represents a vital, critical area of printing production, and is an important control point upon which the company may build its reputation.

12.3 Increasing Pressroom Productivity

As previously mentioned, productivity in the pressroom is an essential consideration, particularly in these times of intense competition and movement by customers to increasingly complex multicolor work, with compressed delivery dates. The printing estimator or production planner should consider the recommendations discussed in the following sections to improve pressroom productivity.

Maximizing the Use of Press Sheets

No pressroom can be considered productive and efficient unless equipment use is maximized. Every revolution of every press provides the capacity to print a maximum surface area. Truly efficient plants work diligently to maximize their press size sheets using the following general approaches:

1. Images are maximized on the press sheet using film duplication/stripping techniques or step-and-repeat procedures.
2. Press sheets are imposed with ganged images of different customer's jobs. Here quality level, quantity, ink color, type of paper and other job particulars must be matched.
3. When completing bookwork production, press sheets are imposed with the maximum number of pages per sheet or signatures are ganged on a larger press sheet. Also, customers are informed when ordering bookwork products that smaller page sizes will require less paper saving them money, which may also allow more pages per press sheet and less signatures per book.

Scheduling Production Efficiently

Efficient production scheduling is a critical key to profitability in the pressroom. Turnkey computer systems are available here which can be of major help to both production planning, estimating, and scheduling personnel. The following scheduling points also deserve consideration.

1. Consistently identify and correct pressroom bottlenecks. This may require an extreme solution such as purchase of a new press, or simply a new imposition method or approach. Certain jobs that consistently represent a bottleneck in the pressroom might be better purchased from a printer more suited to the work.

2. Carefully select the press during estimating and attempt to stay with that press during actual production. This is not always possible, but closely aligning estimating and production scheduling provides greater control of production costs and can be a key issue for improved company profitability.

3. Schedule multicolor jobs in the most effective pattern. One way to ensure this is to buy all presses in the same size, varying the number of printing units among the array of presses. For example, a company would purchase all 40-inch Komori presses and buy two two-color presses, two four-color presses, and two six-color presses. Given this situation, stripping procedures could be standardized to the 40-inch size, and a four-color job could be scheduled to run in one pass on one of the four-color presses, or if either press was unavailable, in two passes on the two-color press. Such standardization of production allows greatly enhanced flexibility to meet delivery dates and freedom to schedule multicolor work in the most effective, cost-saving manner.

Increasing Press Efficiency

Press output can be related to the speed at which the press runs: the faster the press operates, the greater the output. However, press speeds are actually a result of two component forces: the mechanical/electronic efficiency of the equipment and the skill level of the press operator or press crew.

Many engineering advancements have been made on sheetfed press equipment in recent years. Improvements in makeready systems, plate registration and hanging procedures, feeding and delivery operations, sheet detectors, dampening and paper registration devices, image transfer, and inking systems represent major improvement areas. Modern metals, plastics and rubberized parts allow for higher operating speeds. Sophisticated electronic systems—including plate scanners, automated ink controls during printing, and electronic registration devices—are now commonly found in many printing plants. Even though these engineering modifications represent major production efficiencies in reducing makeready times and costs, and increasing running speeds, they are not without increased cost. It is safe to say that the pressroom still represents significant capital expenditures to a typical printing company.

Coupled with these engineering advances is an increased emphasis in training programs and education for the already skilled press operator. Seminars, on-the-job training, and instruction by press manufacturers are making the operator more aware of both the theoretical and practical aspects of press operation. Press operators are enthusiastically taking advantage of the enlightened educational attitude evident in the industry.

When these two factors are combined—improved press engineering and a more experienced operator—press preparation, makeready, running, and washup are completed more efficiently. Greater efficiency means increased productivity and improved output from the pressroom area.

Modifying Job Specifications

A number of changes can be suggested to the customer that will reduce job cost while increasing pressroom output. None of these changes should be implemented without first notifying the customer and obtaining their approval. The following list summarizes some alterations which might be offered:

1. Reduce the finish size (trimmed size) of the job, thereby allowing for an increase in the number of images imposed on the press sheet. Also consider reducing the amount of trim margin built into the press sheet.
2. Standardize paper stock so that only stocks that provide for efficient press output are used. Avoid paper that is too thick or thin and stocks that are not suited to the selected printing process.
3. Make it plant policy to put all color to be printed in the fewest possible forms, but maximize color in the forms that are selected.
4. Modify the quality level of the job, perhaps from excellent to good, which may allow for increased press speeds and reduced makeready time. This consideration is especially important for customers who consistently purchase a higher quality product than their use requirements dictate.
5. Produce work at quantity levels that are most suited to the individual production flow and press equipment of a given plant. Such work may involve short-run work for Plant A and long-run jobs for Plant B. Because of the high speeds using web offset and gravure, consider sending longer run work out or shifting it from sheetfed to in-plant webs after reviewing all related production factors.

12.4 Changeover Points for Estimating Presswork

Changeover points (COP), sometimes termed *break-even points* (BEP) allow the estimator or production planner to compare different press outputs, thereby facilitating selection of the least costly press for the job under consideration. To accurately establish such a system, the estimator must identify prepress, makeready, and washup production costs (fixed costs), the BHR and production output rate for the presses to be compared, and a *basic unit* (BU) of output represented by a product common to all presses.

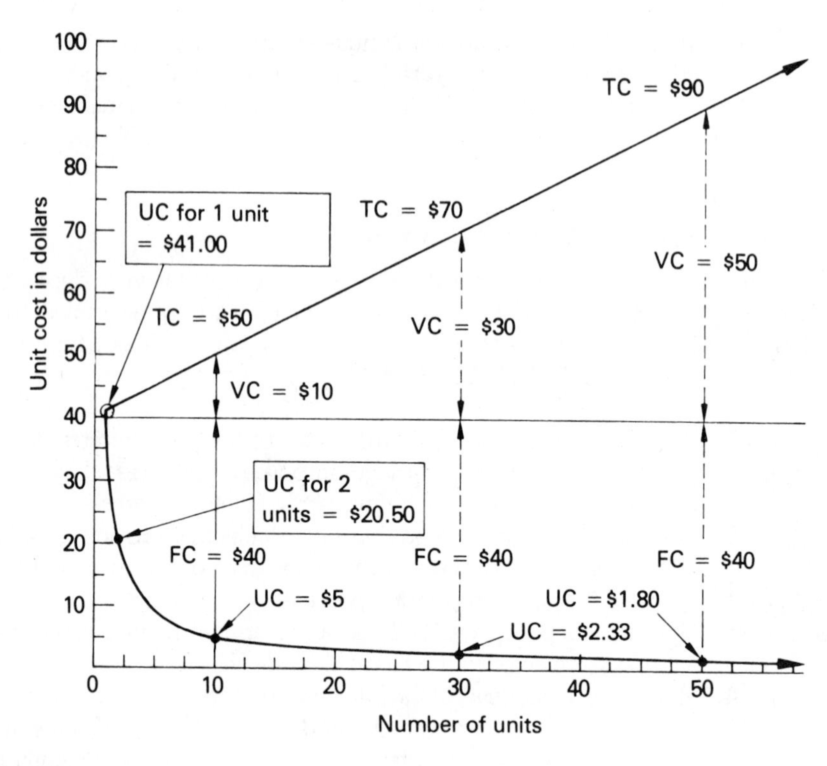

Figure 12.1

Relationship of fixed, variable, total, and unit costs for printing manufacturing

Figure 12.1 demonstrates diminishing unit cost with volume increases, a situation that is typical of printing manufacturing. *Fixed cost* (FC), also called *setup cost*, includes those costs incurred during printing production that do not change, such as taxes, depreciation, insurance, and preparatory costs prior to presswork. *Variable cost* (VC) includes costs that increase as production continues, such as the cost of paper and presswork. *Total cost* (TC) represents the sum of FC and VC at a given quantity, and *unit cost* (UC) is determined by dividing TC by the number of units produced.

The following formulas and number substitutions clearly illustrate the relationship between FC, VC, TC, and UC. In Figure 12.1, the FC is $40.00 and the VC is $1.00 per unit; these dollar figures are for illustrative purposes only.

total cost = fixed cost + (variable cost per unit × number of units produced)
unit cost = total cost ÷ number of units produced

Thus, at 1 unit:

TC = \$40.00 + (\$1.00/unit × 1 unit) = 41.00
UC = \$41.00 ÷ 1 unit = \$41.00

At 10 units:

TC = \$40.00 + (\$1.00/unit × 10 units) = \$50.00
UC = \$50.00 ÷ 10 units = \$5.00

At 30 units:

TC = \$40.00 + (\$1.00/unit × 30 units) = \$70.00
UC = \$70.00 ÷ 30 units = \$2.33

As can be seen in Figure 12.1 and the number substitutions, the UC diminishes rapidly as the number of units increases. The reason is that while VC increases, the constant FC is spread out over a larger number of units produced, thus diminishing the cost per unit. The conclusion is that the greater the quantity produced and sold for any given printing order, the less the UC, even though the TC increases as the quantity increases. Knowledge of this concept of diminishing UC and the definitions of FC and VC is necessary to better understand COP development.

As previously indicated, COP analysis requires identification of specific data: FC for the presses to be compared, the applicable BHR, and appropriate production output in impressions per hour (imp/hr). The BU of production, a specific size or type of product common to all presses under consideration, must be carefully identified.

Examples 12.1 and 12.2 illustrate how to calculate COPs. The following formulas should be used for your calculations:

$$\text{dollars per M basic units} = \frac{\text{BHR} \div \text{impressions per hour (imp/hr)}}{\text{number-up on press sheet}} \times 1000 \text{ units}$$

$$^{*}\text{changeover point} = \frac{\text{fixed cost}_1 - \text{fixed cost}_2}{\text{dollars per M basic units}_2 - \text{dollars per M basic units}_1}$$

$$\text{total cost} = \text{fixed cost} + (\text{dollars per M basic units} \times \text{number of units})$$

Example 12.1

We want to determine the COP to compare a Multilith 1250 and a Miehle 125, two presses in our plant. The BU for this comparison will be 8 1/2 × 11 inch one-color letterheads. The Multilith can produce these letterheads 1-up on an 8 1/2 × 11 press size sheet, while the Miehle can print them 4-up on a 17 × 22 press size sheet. The accompanying table represents current FC and VC data for each press. (The dollar figures presented here are for example only: a company developing COP should use its most current FC and VC.) Determine the COP both mathematically and graphically for these two presses.

	Multilith 1250	Miehle 125
Prepress Production	FC	
Preparatory	$15.60	$23.75
Makeready	4.85	15.30
Washup	3.85	12.50
Total setup cost	24.30	51.55
Presswork	VC	
BHR	$19.25	$35.90
Production output	5500 imp/hr	4125 imp/hr

*The COP formula is derived algebraically from $y = mx + b$, used for linear graphing.

Solution Use the formulas that precede the example text as necessary.

Step 1 Calculate the dollars per 1000 basic units ($/MBU) for each press using the appropriate formula:

For the Multilith 1250:
[($19.25 ÷ 5500 imp/hr) ÷ 1-up] × 1000 units = $3.50/MBU
For the Miehle 125:
[($35.90 ÷ 4125 imp/hr) ÷ 4-up] × 1000 units = $2.175/MBU

Step 2 Determine the TC for each press at 30 M pieces using the appropriate formula:

For the Multilith 1250:
$24.30 + ($3.50/MBU × 30 M) = $129.30
For the Miehle 125:
$51.55 + ($2.175/MBU × 30 M) = $116.80

Step 3 Mathematically calculate the COP using the appropriate formula:

$$\text{COP} = \frac{\$51.55 - \$24.30}{\$3.50 - \$2.175} = 20.6 \text{ MBU (approximately)}$$

The COP of 20.6 MBU means that the Multilith 1250 is the cost-effective press up to 20,600 BU (8 1/2 × 11 letterheads running l-up), after which the Miehle 125 (running 8 1/2 × 11 letterheads 4-up) provides a better cost advantage. Thus, at 20,600 letterheads, the costs are about the same with either press. The greater the number away from 20,600 BU—either more than or less than 20,600 BU—the greater the cost savings for the respective printing press.

Step 3 Using graph paper and labeling the vertical axis as "UC in dollars" and the horizontal axis as "BU in thousands," plot the Multilith and Miehle FCs along the cost axis (which is FC only, at zero quantity) and then plot the TC for each press at the 30,000 quantity level. (See Step 2 for TC calculations for each press.) Once these pairs of points are located accurately, use a ruler to connect the Multilith ponts and then the Miehle points. They should cross at approximately 20,600 BU, as noted in Figure 12.2. They do cross at about 21,000 in the figure, which is approximately equal to 20,600.

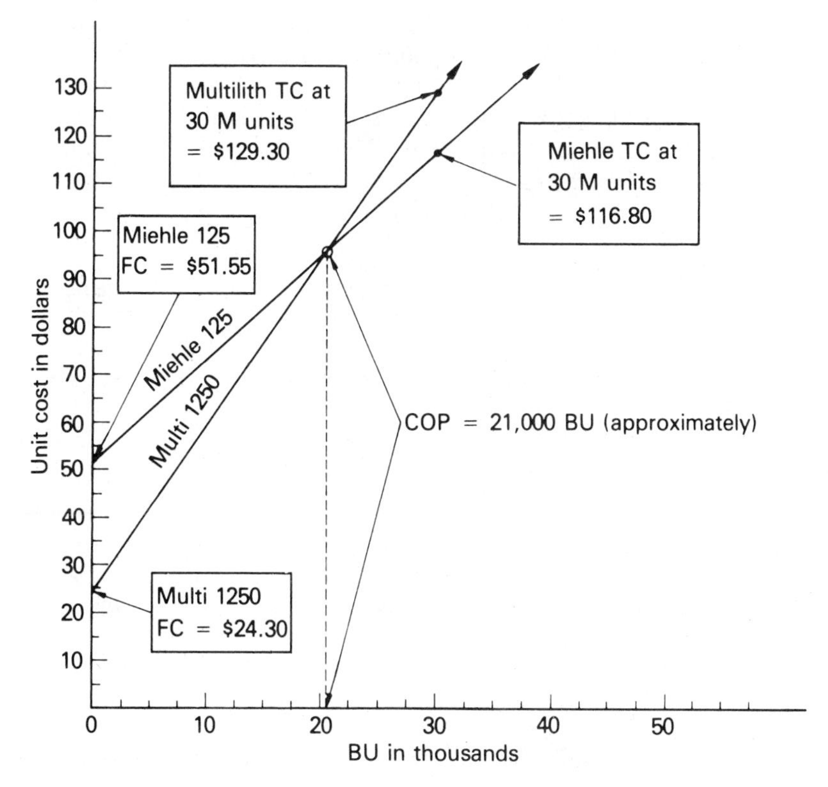

Figure 12.2

Changeover (break-even) point for a Multilith 1250 versus a Miehle 125

It is most important to note that COPs are affected by even slight changes in FC or VC factors, by modification of equipment output, or by a change in the BU. The system can be used to compare sheetfed to sheetfed presses, sheetfed to web, or web to web. As long as the BU is the same, any number of presses can be plotted on the same graph to compare multiple press outputs.

Many estimators use the COP system to enable them to make quick production and press cost comparisons for standard company product lines. Of course, all input data should be accurate and updated to reflect the most current costs incurred during production. A second COP example problem comparing the output of book signatures between a Heidelberg KORD and a Miehle 36 is provided next.

Example 12.2

We have a Heidelberg KORD and a Miehle 36, each used almost exclusively for bookwork signature production. We will use 8 3/4 × 11 1/2 inch untrimmed signatures, common to plant production with these presses. The KORD has an 18 × 25 1/2 maximum press size sheet and can run 8 pages per signature on a 17 1/2 × 23 press size sheet. The Miehle 36 has a 25 × 36 maximum press size sheet that can impose 16 pages per signature on a 23 × 35 press size sheet. Thus, our BU will be 16 page signatures with an 8 3/4 × 11 1/2 untrimmed size. The Heidelberg will run 2 signatures of 8 pages each to total the Miehle's 1 signature of 16 pages. Using the information just given and the data in the accompanying chart, determine the COP for these presses by both mathematical and graphical methods.

	Heidelberg KORD	Miehle 36
Prepress Production	FC	
Preparatory	$57.80	$105.30
Makeready	27.20	49.24
Washup	23.60	43.10
Total setup cost	108.60	197.64
Presswork	VC	
BHR	$55.60	$69.70
Production output	3420 imp/hr	5700 imp/hr

Solution

Step 1 Calculate the $/MBU for each press using the appropriate formula:

For the Heidelberg KORD

[($55.60 ÷ 3420 imp/hr) ÷ 1 sig] × 1000 units × 2 sig = $32.51/MBU

For the Miehle 36:

[(\$69.70 ÷ 5700 imp/hr) ÷ 1 sig] × 1000 units × 1 sig = \$12.23/MBU

Step 2 Determine the TC for each press at 8 M pieces using the appropriate formula:

For the Heidelberg:

\$108.60 + (\$32.51/MBU × 8 M) = \$368.68

For the Miehle:

\$197.64 + (\$12.23/MBU × 8 M) = \$295.48

Step 3 Mathematically calculate the COP using the appropriate formula:

$$\text{COP} = \frac{\$197.64 - \$108.60}{\$32.51 - \$12.23} = 4.4 \text{ MBU (approximately)}$$

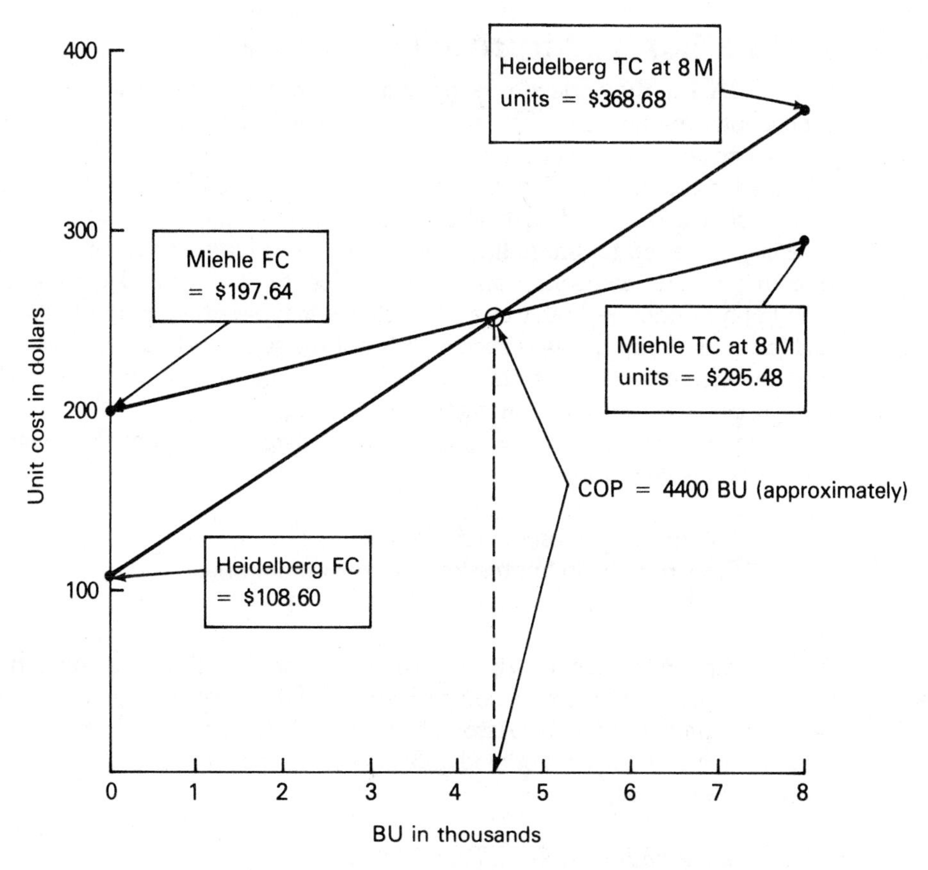

Figure 12.3

Solution to Example 12.2

The COP of 4.4 MBU means that the Heidelberg KORD is cost effective up to approximately 4400 BU (8 3/4 × 11 1/2 in.; two 8 pp sig), after which the Miehle 36 (8 3/4 × 11 1/2 in.; one 16 pp sig) provides a better cost advantage. Thus, at approximately 4400 sig, the costs are about the same with either press. The greater the number away from 4400—either more than or less than 4400—the greater the cost advantage for the respective printing press.

Step 4 Using graph paper and labeling the vertical axis as "UC in dollars" and the horizontal axis as "BU in thousands," plot the Heidelberg and Miehle FCs along the cost axis (which is FC only, at zero quantity) and then plot the TC for each press at the 8000 quantity level. (See Step 2 for TC calculations for each press.) Once these pairs of points have been accurately located, use a ruler to connect them. They should cross at approximately 4400 BU, as noted in Figure 12.3.

12.5 Estimating Letterpress Presswork

Figure 12A.1 in the appendix to this chapter is to be used for estimating letterpress production time and cost. The schedule is divided into hand-fed platen, automatic platen, and letterpress cylinder equipment. The size of the press under consideration, top-rated speed, and BHR (at the specified level of productivity) are also provided.

It should be noted that letterpress printing today is generally considered a specialty operation for imprinting, numbering, die cutting, and other very specialized operations. This categorization is largely due to the technological advances in lithographic printing—both web and sheetfed—in typesetting, prepress preparations, and faster lithographic press speeds. It is not unusual, however, for most small and medium sized commercial printers to have one or more letterpress presses available for special jobs such as imprinting and numbering.

When letterpress presswork is estimated, determination of three categories of information is necessary:

Makeready and press preparation,
Press running in impressions per hour (imp/hr),
Washup time.

The assumption is made that shop conditions are normal; that equipment, paper, and ink provide no unusual production problems; and that press operators are competent workers. Printing Industries Association, Inc., of Southern California recommends classification of letterpress printed goods based on the following (starting with a form locked up for press):

Makeready and Press Preparation

Simple: Typematter or line cuts with minimum makeready; easily registered forms; news, book, bond, or coated stocks

Average: Typematter, squared or silhouetted half tones; medium register forms; book or coated book of good quality

Difficult: Typematter with vignetted halftones or process color; closely registered forms; extra-heavy ink coverage; dull-coated book or No. 1 enamel

Press Running (assume average quality and forms)

Simple: Stock 0.003 to 0.006 inch thick

Average: Stock 0.002 to 0.003 inch and 0.006 to 0.008 inch thick

Difficult: Stock lighter than 0.002 inch and heavier than 0.008 inch thick

Washup

Average: light to dark ink (use single chart figure); dark to light ink (use double chart figure)

Example 12.3

A printer has an estimate request for the imprinting of 8500 calendars for a local insurance agency. They will run 1-up on a Heidelberg High-Speed 10 × 15 Platen. Press preparation and makeready will be at the average level, press running at simple, and a washup (wp) added for each 6 hours of press time or any fraction thereof. Determine imprinting production time and cost.

Solution Use Figure 12A.1.

Step 1 Determine total press production time:

Total time for preparation and makeready:
1.00 hr

Total running time:
8500 imp ÷ 4125 imp/hr = 2.06 hr

Total washup time (each washup requires 0.15 hr):
2.06 hr ÷ 6 hr/wp = 0.34 wp (round up to 1 wp) = 0.13 hr

Total press production time:
1.00 hr + 2.06 hr + 0.13 hr = 3.19 hr

Step 2 Determine total cost of presswork:

3.19 hr × $42.64/hr = $136.02

Step 3 Summarize your findings:

Total production time: 3.19 hr

Total production cost: $136.02

12.6 Estimating Offset Lithographic Sheetfed Presswork

Even though webfed presses have begun to compete more intensely than ever with sheetfed lithographic production, sheetfed offset still represents a major force in producing the bulk of printed goods in the United States. Figures 12A.2 to 12A.7 in the appendix to this chapter are provided for estimating sheetfed lithographic times and costs. It is essential to remember that the indicated cost rates and standard production times will change from plant to plant and should be developed on an individual plant basis for accuracy.

When an estimate is being prepared, the job under consideration must be carefully analyzed and broken down in detail. This analysis ensures that the proper categories of makeready and press preparation, running, and washup are used. As will be noted, both the makeready and press preparation category and running categories are subdivided into simple, average, and difficult levels; washup is either single or double depending upon the next color on the press. It is not unusual for makeready and press preparation to be at average level while running is assessed at difficult or simple. Each job must be analyzed and categorized individually. The following list defines the parameters by which the schedule data are to be applied:

Makeready and Press Preparation

Simple: Typematter and line work in one color
Average: Typematter with half tones; simple two-color work
Difficult: Process color jobs; intricate registration work; jobs with heavy ink coverage

Press Running (add additional penalties for troublesome work)

Simple: Stock 0.003 to 0.006 inch thick; typematter or line work in one color
Average: Stock 0.002 to 0.003 inch and 0.006 to 0.008 inch thick; typematter with half tones; simple two-color work
Difficult: Stock lighter than 0.002 inch or heavier than 0.008 inch thick; solids, intricate register, process color

Washup

Average: Light to dark ink (use single chart figure)
Difficult: Dark to light ink (use double chart figure)

The following formulas should be used in conjunction with the offset presswork estimating schedules to estimate sheetfed production time and cost:

total number of press sheets required = (total number of finished pieces ÷ number-up on the press sheet) × (spoilage percentage + 100%)
makeready and press preparation production time = number of forms per job × hours per form for makeready and press preparation

hours of press time per form = total number of press sheets required ÷ number of impressions per hour per form
press-running time = hours of press time per form × number of forms per job
* number of washups per form = hours of press time per form ÷ hours of press time per form per washup
production time for washups = number of washups per form × hours of washup time per form × number of forms per job
total production time per job = makeready and preparation production time + press-running time + production time for washups
total production cost per job = total production time per job × press BHR

Example 12.4

A customer wants an estimate for 14,000 three-color letterheads to be printed in black, red, and blue. Letterheads will run 4-up on a Solna 125/80 (one-color press) on a press sheet measuring 17 1/2 × 22 1/2 inches. Finish size for the product will be 8 1/2 × 11 inches. Makeready and press preparation will be at the average level with the press running at difficult. One washup will be done for each 5 hours of press time per form and per color. Use a single wash for black and blue and a double wash for red. Add 15% spoilage to the base number of press sheets required (5% per color). The plant runs 1 shift. Determine press time and total press cost using schedule data and a BHR at 70%.

Solution Use Figure 12A.3 in the appendix to this chapter and the preceding formulas.

Step 1 Determine number of press size sheets required:

(14,000 pc ÷ 4-up) × (15% + 100%) = 3500 pss × 115% = 4025 pss

Step 2 Determine total time for makeready and press preparation for 3 fm (black, red, and blue):

3 fm × 0.5 hr/fm = 1.50 hr

Step 3 Determine total press-running time:

Hours of press time per form:
4025 pss ÷ 4000 imp/hr = 1.00 hr/fm
Press-running time:
1.00 hr/fm × 3 fm = 3.00 hr

Step 4 Determine total time for washups:

*When using the following formula, round up decimal answers to the next whole number. Use this value to locate the hours of washup time per form in the appropriate estimating schedule.

Number of washups per form:
1.00 hr/fm ÷ 5 hr/fm/wp = 0.20 wp/fm = 1 wp/fm
Time for 2 wp at single rate (black and blue):
0.40 hr/wp × 2 wp = 0.80 hr
Time for 1 wp at double rate (red):
0.40 hr/wp × 2 × 1 wp = 0.80 hr
Total washup time:
0.80 hr + 0.80 hr = 1.60 hr

Step 5 Determine total production time and cost:

Total production time:
1.50 hr + 3.00 hr + 1.60 hr = 6.10 hr
Total production cost:
6.10 hr × $60.25 = $367.53

Example 12.5

We have an order for 250,000 (3 1/2 × 5 1/2 inch) postcards to print four colors on one side with a black type backup on the other. The job will run on a 24 × 36 press size sheet, 36-up. Add 15% spoilage to the base number of press sheets. A Komori 440 Lithrone (four-color) will be used for the process run, and a Heidelberg SORS (one-color) for the backup black. Use average preparation and running for all presswork and 1 washup per each 4 hours of presswork per form set. The four-color press has an additional feeder operator to be added. The plant runs 1 shift. Determine press time and total press cost using schedule data and a BHR at 70%.

Solution Use Figures 12A.3 and 12A.5 in the appendix to this chapter and the appropriate formulas.

Step 1 Determine number of press size sheets required:

(250,000 pc ÷ 36-up) × (15% + 100%) = 6945 pss × 115% = 7987 pss

Step 2 Determine total time for makeready and press preparation on the Komori 440 Lithrone for 1 fm set of 4 colors:

1 fm set × 2.25 hr/fm set = 2.25 hr

Step 3 Determine total press-running time for the Komori 440 Lithrone:

Hours of press time per form set:
7987 pss ÷ 7800 imp/hr = 1.03 hr

Total hours of press time for 1 fm set (4 colors):
1 fm set × 1.03 hr/fm set = 1.03 hr

Step 4 Determine total washup time for the Komori 440 Lithrone:

Number of washups per form set:
1.03 hr/fm set ÷ 4 hr/fm set/wp = 0.26 wp/fm set = 1 wp/fm set
Total time for washups:
1 wp/fm set × 1 fm set × 1 hr/wp = 1 hr

Step 5 Determine total production time and cost for the Komori 440 Lithrone:

Total production time:
2.25 hr + 1.03 hr + 1 hr = 4.28 hr
Total production cost:
4.28 hr × ($201.67 + $37.42)/hr = $1023.31

Step 6 Determine total time for makeready and press preparation on the Heidelberg SORS:

1 fm × 0.70 hr/fm = 0.70 hr

Step 7 Determine total press-running time for the Heidelberg SORS:

Hours of press time per form:
7987 pss ÷ 6600 imp/hr = 1.21 hr
Total hours of press time for 1 fm:
1 fm × 1.21 hr/fm = 1.21 hr

Step 8 Determine total washup time for the Heidelberg SORS:

Number of washups per form:
1.21 hr/fm ÷ 4 hr/fm/wp = 0.30 wp/fm = 1 wp/fm
Total time for washups:
1 wp/fm × 1 fm × 0.60 hr/wp = 0.60 hr

Step 9 Determine total production time and cost for the Heidelberg SORS:

Total production time:
0.70 hr + 1.21 hr + 0.60 hr = 2.51 hr
Total production cost:
2.51 hr × $80.25/hr = $201.43

Step 10 Determine the total production time and cost for both presses:

Total production time:
4.28 hr + 2.51 hr = 6.79 hr
Total production cost:
\$1023.31 + \$201.43 = \$1224.74

Example 12.6

A local church group wants 18,000 (16 page) booklets printed to detail their church's 100 year history. Untrimmed page size is 8 3/4 × 11 3/4 inches. The press is a single-color Heidelberg SORD. The job will run in 1-16 SW if possible with black text and a spot red second color throughout for paragraph heads and duotones (to register to black half tone printers). Use average levels for makeready and press preparation and running and single washups with 1 wash per each 5 hours of press time per form or any fraction thereof. Add 12% to the base number of press sheets for spoilage (4% per color). The plant runs 1 shift; no additional operators are needed on the press. Determine press time and total press cost using schedule data and a BHR at 70%.

Solution Use Figure 12A.3 in the appendix to this chapter and the appropriate formulas. The booklet will run in 1-16 SW with 2 black-and-white and 2 spot red color forms. Each form will contain 8 pp.

	Outside	Inside	
	1	2	
	4	3	
	5	6	
Form I (black)	8	7	Form II (black)
Form III (red)	9	10	Form IV (red)
	12	11	
	13	14	
	16	15	

Step 1 Determine press size sheet and form chart data confirmation:

$$\frac{35 \times 24\ 1/4}{8\ 3/4 \times 11\ 1/4} = 4 \times 2 = 8 \text{ pp/fm}$$

8 pp/fm × 2 fm = 16 pp/sig on a 22 1/2 × 35 pss

Therefore, for production, 1 pss = 1 sig of 16 pp.

Determine total number of press sheets (signatures) required:

Step 2 18,000 pc × 112% = 20,160 pss

Step 3 Determine total time for makeready and press preparation:

0.70 hr/fm × 4 fm = 2.80 hr

Step 4 Determine press-running time:

Hours of press time per form:
20,160 pss ÷ 6600 imp/hr = 3.05 hr
Total hours of press time for 4 fm:
3.05 hr/fm × 4 fm = 12.20 hr

Step 5 Determine total washup time:

Number of washups per form:
3.05 hr/fm ÷ 5 hr/fm/wp = 0.61 wp/fm = 1 wp/fm
Total time for washups:
1 wp/fm × 4 fm × 0.60 hr/wp = 2.40 hr

Step 6 Determine total production time and cost:

Total production time:
2.80 hr +12.20 hr + 2.40 hr = 17.40 hr
Total production cost:
17.40 hr × $80.25/hr = $1396.35

Example 12.7

We want to compare the church booklet job (Example 12.6) run on a single-color Heidelberg SORD with running that same job on a new two-color Komori 240 Lithrone that we have just purchased. The job will still run in 1-16 SW on a press size sheet of 22 1/2 × 35 inches with black type and spot red throughout. Use average level for makeready and press preparation and running and single washups with 1 wash per each 5 hours of press time per form set or any fraction thereof. Add 8% to the base number of press sheets for spoilage (4% per form set). The plant runs 1 shift. Determine press time and total press cost using schedule data and a BHR at 70%. An additional feeder operator will be needed on the press.

Use Figure 12A.4 in the appendix at the end of this chapter.

Solution

Determine the configuration:

Step 1 The booklet will run in 1-16 SW on a 22 1/2 × 35 in. pss. Each form will contain 8 pp.The press will print one side of the sheet with black and red forms and then the other (backup) side with backup black and red forms.

Step 2 Determine number of press size sheets required:

18,000 pc × 108% = 19,440 pss

Step 3 Determine total time for makeready and press preparation for 2 fm sets (2 colors each):

2 fm sets × 1.20 hr/fm set = 2.40 hr

Step 4 Determine total press-running time:

Hours of press time per form set:
19,440 pss ÷ 7800 imp/hr = 2.49 hr
Total hours of press time:
2 fm sets × 2.49 hr/fm set = 4.98 hr

Step 5 Determine total washup time:

Number of washups per form set:
2.49 hr/fm set ÷ 5 hr/fm set/wp = 0.50 wp/fm set = 1 wp/fm set
Total time for washups:
1 wp/fm set × 2 fm sets × 0.90 hr/wp = 1.80 hr

Step 6 Determine total production time and cost:

Total production time:
2.40 hr + 4.98 hr + 1.80 hr = 9.18 hr
Total production cost:
9.18 hr × ($125.17 + $34.30)/hr = $1463.93

Step 7 Make a comparison of the two presses (see 6, Example 12.6):

Total production time: 17.40 hr (SORD) versus 9.18 hr (Komori) yields 8.22 hr less using Komori 240 Lithrone (two-color press).
Total production cost: $1463.93 (Komori) versus $1396.35 (SORD) yields $67.58 less using SORD (one-color press)

12.7 Presswork Estimating Tips

Because of the capital and labor intensiveness of the pressroom in most printing plants, estimating presswork may be a make-or-break proposition for even the most cautious estimator. The following pointers are offered to aid the estimator when dealing with presswork estimating:

1. Attempt, whenever possible, to fill the press sheet completely, thereby reducing expensive press time. Filling the press sheet may increase prepress costs for film duplication or step-and-repeat procedures, but can save many production dollars. When production costs are decreased, the company becomes more competitive, which stimulates greater sales.
2. Throughout the plant, and especially in the pressroom, lock production into standardized procedures, products, and processes. This production ingredient is one common to a significant number of high-profit printers. In addition, standardization allows for easy development of COP data since the BU for production comparison is the same.
3. Develop pressroom changeover charts using your own specific production data and costs.
4. Constantly monitor existing production standards and costs, especially in the capital-intensive press area. Develop a standard production manual (see Chapter 3) through accurate collection of historic production data.
5. When estimating with the applicable presswork charts and schedules, be careful to categorize each job into the appropriate section. Remember that it is possible to develop additional categories of data, such as "very simple" or "super-difficult," for types of work that will not fit into existing divisions.
6. Carefully investigate the profitability of black-and-white versus multicolor work. Gear sales toward the type of work that is most profitable at quantity and quality levels most appropriate for your company.
7. Seriously consider buyouts of press sheets, especially when presswork is not typical of your usual production. Carefully investigate cost differences comparing minimal production charges against buyouts from other printers. As production lines become more specialized, which is an important key to company profitability, cost savings on buyouts of printed press sheets may be substantial.
8. During the estimating process, check the job under consideration with pressroom production scheduling. If the job is awarded to your company, production and delivery aspects will have been given an initial review.
9. As an estimator, be able to identify common printing problems that occur in the pressroom. When a definable pattern of any single problem or bottleneck appears, modify presswork estimating schedules to compensate. This course will help to insure the balancing of estimated and actual times.
10. Develop an understanding of the duties and responsibilities of pressroom personnel. These individuals are skilled craftspersons who must deal with many interrelated printing variables during production.

Appendix to Chapter 12 Pressworl Estimating Schedules

The figures in this appendix were reproduced with permission of Printing Industries Association, Inc., of Southern California from their 1988-89 *Blue Book of Production Standards and Costs for the Printing Industry*

PRESSWORK SCHEDULE—LETTERPRESS

(Time given represents machine hours on the decimal system)

PRESS NAME	SIZE	RATED SHEETS PER HOUR	HOUR RATE At 70% Productivity	PREPARATION/MAKE-READY TIME			AVG. IMPRESSION PER RUN HR.			WASH-UP Time
				SIMPLE Time	AVERAGE Time	DIFFICULT Time	SIMPLE	AVERAGE	DIFFICULT	
LETTERPRESS										
Heidelberg, Stamp & emboss	10 x 15	5500	43.36	0.75	1.00	2.00	4125	3300	2750	N/A
Heidelberg, Stamp & emboss	13 x 18	4000	45.57	0.75	1.00	2.00	3000	2400	2000	N/A
Kluge, Stamp & emboss	14 x 22	4000	45.07	0.75	1.00	2.00	3000	2400	2000	N/A
Platen, Hand-fed*	8 x 12	1800	38.95	0.75	1.00	2.00	1350	1080	900	0.13
Platen, Hand-fed*	10 x 15	1800	38.95	0.75	1.00	2.00	1350	1080	900	0.13
Platen, Hand-fed*	12 x 18	1800	38.95	0.75	1.00	2.00	1350	1080	900	0.13
Heidelberg Platen	10 x 15	5500	42.64	0.75	1.00	2.00	4125	3300	2750	0.13
Kluge Automatic	11 x 17	4500	41.02	0.75	1.00	2.00	3375	2700	2250	0.13
Heidelberg	13 x 18	5500	44.85	0.75	1.00	2.00	4125	3300	2750	0.13
Kluge Automatic	13 x 19	5000	41.82	0.75	1.00	2.00	3750	3000	2500	0.13
Heidelberg*	15 x 20 ½	5000	40.98	0.75	1.00	2.00	3750	3000	2500	0.13
Cylinder Press*	14 x 20	5000	41.15	0.75	1.00	2.00	3750	3000	2500	0.13

(*) Reconditioned Equipment

Figure 12A.1

Letterpress presswork estimating schedule

PRESSWORK SCHEDULE—OFFSET

(Time given represents machine hours on the decimal system)

PRESS NAME	SIZE	RATED SHEETS PER HOUR	HOUR RATE At 70% Productivity	PREPARATION/MAKE-READY TIME SIMPLE Time	PREPARATION/MAKE-READY TIME AVERAGE Time	PREPARATION/MAKE-READY TIME DIFFICULT Time	AVG. IMPRESSION PER RUN HR. SIMPLE	AVG. IMPRESSION PER RUN HR. AVERAGE	AVG. IMPRESSION PER RUN HR. DIFFICULT	WASH-UP Time
SMALL PRESSES										
Multi, Chief, chute del.	10 x 15	9000	40.86	0.20	0.25	0.30	6750	5400	4500	0.25
Multi, Chief, chain del.	10 x 15	9000	41.26	0.20	0.25	0.30	6750	5400	4500	0.25
Multi, Chief, chute del.	11 x 15	9000	40.93	0.20	0.25	0.30	6750	5400	4500	0.25
Multi, Chief, chain del.	11 x 15	9000	41.31	0.20	0.25	0.30	6750	5400	4500	0.25
Multi, Chief, chute del.	11 x 17	9000	41.40	0.20	0.25	0.30	6750	5400	4500	0.25
Multi, Chief, chain del.	11 x 17	9000	41.71	0.20	0.25	0.30	6750	5400	4500	0.25
Multi, Chief, 2/c chain del.	10 x 15	9000	42.67	0.30	0.40	0.50	6750	5400	4500	0.50
Multi, Chief, 2/c chain del.	11 x 17	9000	42.97	0.30	0.40	0.50	6750	5400	4500	0.50
Multi, Chief, chute del.	14 x 18	9000	42.37	0.20	0.30	0.40	6750	5400	4500	0.25
Multi, Chief, chain del.	14 x 18	9000	42.62	0.20	0.30	0.40	6750	5400	4500	0.25
Multi, Chief, 2/c chain del.	14 x 18	9000	43.89	0.30	0.40	0.50	6750	5400	4500	0.25
Ryobi 17	11 x 17	9000	42.18	0.30	0.40	0.50	6750	5400	4500	0.50
ATF Davidson 501	11 x 15	9000	41.65	0.30	0.40	0.50	6750	5400	4500	0.40
Atf Davidson 502P	11 x 15	9000	42.07	0.40	0.50	0.60	6750	5400	4500	0.40
ATF Davidson 701	15 x 18	9000	42.27	0.30	0.40	0.50	6750	5400	4500	0.50
ATF Davidson 702P	15 x 18	9000	44.55	0.40	0.50	0.60	6750	5400	4500	0.40
ATF Davidson 901	15 x 20	9000	42.69	0.30	0.40	0.50	6750	5400	4500	0.50
Heidelberg TOM	11 x 15 ½	10000	42.24	0.30	0.40	0.50	7500	6000	5000	0.40
Heidelberg TOK	11 x 15 ½	10000	42.48	0.30	0.40	0.50	7500	6000	5000	0.50
Heidelberg GTO 52	14 x 20 ½	8000	52.02	0.30	0.40	0.50	6000	4800	4000	0.35
Hamada 500 CDA	10 x 15	9000	41.23	0.20	0.30	0.40	6750	5400	4500	0.35
Hamada 550 CDA 2/c	10 x 15	9000	42.25	0.20	0.30	0.50	6750	5400	4500	0.40
Hamada 600 CD	11 x 17	9000	41.95	0.30	0.40	0.50	6750	5400	4500	0.25
Hamada 660 CD 2/C	11 x 17	9000	42.92	0.30	0.50	0.60	6750	5400	4500	0.35
Hamada 611, 612 CD	11 x 17	12000	41.91	0.30	0.40	0.50	9000	7200	6000	0.25
Hamada 661, 662 CD 2/c	11 x 17	12000	42.94	0.40	0.50	0.60	9000	7200	6000	0.40
Hamada 700 CD	14 x 18	9000	41.80	0.30	0.25	0.30	6750	5400	4500	0.25
Hamada 770 CD 2/c	14 x 18	9000	43.73	0.30	0.40	0.50	6750	5400	4500	0.35
Hamada 800 DX	14 x 20	9000	43.72	0.30	0.40	0.50	6750	5400	4500	0.25
Hamada 800 CDX	14 x 20	8000	44.29	0.30	0.40	0.50	6000	4800	4000	0.35
Hamada 880 DX	14 x 20	9000	45.33	0.30	0.40	0.50	6750	5400	4500	0.35
Hamada 880 CDX	14 x 20	9000	45.79	0.30	0.40	0.50	6750	5400	4500	0.35
Oliver 52	14 $\frac{3}{16}$ x 20	12000	48.63	0.20	0.40	0.50	9000	7200	6000	0.35
Imperial 2200 Maxim	12 x 18	10000	41.84	0.20	0.25	0.30	7500	6000	5000	0.25
Imperial 3200 Maxim	12 x 18	10000	42.00	0.20	0.25	0.30	7500	6000	5000	0.25
Imperial 4200 Maxim	12 x 18	10000	42.19	0.20	0.25	0.30	7500	6000	5000	0.25
A.B. Dick 369	11 x 17	9000	41.89	0.30	0.40	0.50	6750	5400	4500	0.25
A.A. Dick 369 T *	11 x 17	9000	46.15	0.40	0.40	0.50	6750	5400	4500	0.40
A.B. Dick 8810	11 x 17	10000	40.67	0.20	0.25	0.30	7500	6000	5000	0.25
A.B. Dick 9840	17 ¾ x 13 ½	10000	42.19	0.20	0.25	0.30	7500	6000	5000	0.30
A.B. Dick 385	17 ½ x 22 ½	8000	43.36	0.30	0.40	0.50	6000	4800	4000	0.35
A.B. Dick 385 w/T head	17 ½ x 22 ½	8000	44.81	0.40	0.50	0.60	6000	4800	4000	0.50
Oliver 58	17 ½ x 22 ½	10000	52.36	0.30	0.40	0.50	7500	6000	5000	0.35

Figure 12A.2

Offset presswork estimating schedules: Duplicators and small presses

PRESSWORK SCHEDULE—OFFSET

(Time given represents machine hours on the decimal system)

PRESS NAME	SIZE	RATED SHEETS PER HOUR	HOUR RATE At 70% Productivity	PREPARATION/MAKE-READY TIME			AVG. IMPRESSION PER RUN HR.			WASH-UP
				SIMPLE Time	AVERAGE Time	DIFFICULT Time	SIMPLE	AVERAGE	DIFFICULT	Time
ONE-COLOR PRESSES										
Solna 125/80	18 x 25 3/16	8000	60.25	0.40	0.50	0.60	6000	4800	4000	0.40
Heidelberg KORD	18 1/8 x 25 1/4	6000	57.46	0.40	0.50	0.60	4500	3600	3000	0.40
Oliver 66	19 x 26	12000	61.90	0.40	0.50	0.60	9000	7200	6000	0.40
Royal Zenith R2 26	19 x 26	12000	60.01	0.40	0.50	0.60	9000	7200	6000	0.40
Solna 164	19 x 26	10000	62.49	0.40	0.50	0.60	7500	6000	5000	0.50
Heidelberg MO	19 x 25 1/2	11000	63.09	0.40	0.50	0.60	8250	6600	5500	0.50
Komori 126 Sprint	20 x 26	11000	65.01	0.40	0.50	0.60	8250	6600	5500	0.50
Oliver 72	20 x 28 5/16	10000	62.32	0.05	0.06	0.07	7500	6000	5000	0.05
Heidelberg SORM	20 1/2 x 29 1/8	11000	65.57	0.50	0.60	0.70	8250	6600	5500	0.50
Miehle Roland 201	20 1/2 x 28 3/8	12000	66.07	0.60	0.70	0.80	9000	7200	6000	0.50
Miller SC 74	20 1/2 x 29 1/8	13000	68.32	0.60	0.70	0.80	9750	7800	6500	0.50
Heidelberg SORD	24 1/4 x 36	11000	77.56	0.60	0.70	0.80	8250	6600	5500	0.60
Miehle Roland 36	25 x 36	10000	84.52	0.60	0.70	0.80	7500	6000	5000	0.60
Miller SC 95	25 5/8 x 37 1/2	12000	77.34	0.60	0.70	0.80	9000	7200	6000	0.60
Komori 140 Lithrone	28 x 40	13000	84.63	0.60	0.70	0.80	9750	7800	6500	0.60
Heidelberg SORS	28 3/8 x 40 1/8	11000	80.25	0.60	0.70	0.80	8250	6600	5500	0.60
Miehle Roland 40 low pile	28 3/8 x 40 3/16	10000	88.69	0.60	0.70	0.80	7500	6000	5000	0.60
Miehle Roland 40 high pile	28 3/8 x 40 3/16	10000	92.78	0.60	0.70	0.80	7500	6000	5000	0.60
Miehle Roland 40 Perfector	28 3/8 x 40 3/16	10000	123.19	0.70	0.80	1.00	7500	6000	5000	1.00

Figure 12A.3

Offset presswork estimating schedules: One-color presses

PRESSWORK SCHEDULE—OFFSET

(Time given represents machine hours on the decimal system)

INCLUDES ONLY ONE PRESSMAN

YOU MUST COMPUTE YOUR OWN COST FIGURES FOR TWO-, FOUR, FIVE- AND SIX-COLOR PRESSES. THE HOUR RATE FOR THE 2ND PRESSMAN AND FEEDER MUST BE ADDED TO THE COST CENTER HOUR RATE WHERE APPLICABLE BEFORE COMPUTING THE HOURLY COST FOR MAKE READY, IMPRESSIONS PER THOUSAND OR WASH-UP

Feeder = $34.30 per hour.
Second Pressman = $51.45 per hour.

TWO COLOR PRESSES

PRESS NAME	SIZE	RATED SHEETS PER HOUR	HOUR RATE At 70% Productivity	PREPARATION/MAKE-READY TIME SIMPLE Time	AVERAGE Time	DIFFICULT Time	AVG. IMPRESSION PER RUN HR. SIMPLE	AVERAGE	DIFFICULT	WASH-UP Time
Heidelberg GTO ZP	14 x 20 1/2	8000	70.63	0.65	0.90	1.20	6000	4800	4000	0.65
Oliver 252 RP	14 3/16 x 20 1/2	10000	77.85	0.65	0.90	1.20	7500	6000	5000	0.65
Oliver 258	17 1/2 x 22 1/2	10000	73.39	0.65	0.90	1.20	7500	6000	5000	0.65
Oliver 258 RP	17 1/2 x 22 1/2	10000	77.57	0.65	0.90	1.20	7500	6000	5000	0.65
Solna 225	18 1/8 x 25 3/16	10000	76.29	0.65	0.90	1.20	7500	6000	5000	0.65
Fuji Perfector	19 x 25	10000	85.03	0.65	0.90	1.20	7500	6000	5000	0.65
Heidelberg MOZP	19 x 25 1/4	11000	87.09	0.65	0.90	1.20	8250	6600	5500	0.65
Royal Zenith 26	19 x 26	12000	77.68	0.65	0.90	1.20	9000	7200	6000	0.65
Royal Zenith 26-2CP	19 x 26	12000	80.66	0.65	0.90	1.20	9000	7200	6000	0.65
Solna 264	19 1/16 x 26	10000	79.62	0.65	0.90	1.20	7500	6000	5000	0.65
Komori 226 Sprint	20 x 26	11000	84.36	0.65	1.00	1.30	8250	6600	5500	0.65
Komori 226 Lithrone	20 x 26	13000	91.09	0.65	1.00	1.30	9750	7800	6500	0.65
Komori 226 Sprint*	20 x 26	11000	90.44	0.65	1.00	1.30	8250	6600	5500	0.65
Akiyama 228	20 x 28	13000	92.70	0.65	1.00	1.30	9750	7800	6500	0.75
Oliver 272	20 x 28	10000	73.82	0.65	1.00	1.30	7500	6000	5000	0.75
Oliver 272 RP HP	20 3/8 x 28	10000	88.06	0.65	1.00	1.30	7500	6000	5000	0.75
Miehle Roland 28 2/c	20 1/2 x 28 3/8	10000	89.60	0.75	1.00	1.30	7500	6000	5000	0.75
Heidelberg SORMZ	20 1/2 x 29 1/8	11000	84.94	0.75	1.00	1.30	8250	6600	5500	0.75
OMSCA H226-NP 2	20 7/16 x 29 1/8	10000	89.36	0.75	1.00	1.30	7500	6000	5000	0.75
Miller TP 74	20 1/2 x 29 1/8	12000	98.57	0.75	1.00	1.30	9000	7200	6000	0.75
Akiyama 232	23 x 32 1/4	13000	111.97	0.75	1.00	1.30	9750	7800	6500	0.75
Miller TP 84*	24 x 33	10000	100.15	0.75	1.00	1.30	7500	6000	5000	0.80
Miehle Roland 36	25 x 36	10000	105.77	0.80	1.10	1.40	7500	6000	5000	0.80
Miller TP 95*	25 5/8 x 37 1/2	10000	108.85	0.80	1.10	1.40	7500	6000	5000	0.80
Heidelberg SORSZ	28 3/8 x 40 1/8	11000	97.81	0.90	1.20	1.50	8250	6600	5500	0.90
Komori 240 Lithrone	28 3/8 x 40 1/2	13000	125.17	0.90	1.20	1.50	9750	7800	6500	0.90
Akiyama 240	28 5/16 x 40 1/8	13000	125.24	0.90	1.20	1.50	9750	7800	6500	0.90
Miehle Roland 40 2/c	28 1/8 x 40 3/16	10000	111.78	0.90	1.20	1.50	7500	6000	5000	0.90
Miller TP 104*	28 3/4 x 41	11500	118.65	0.90	1.20	1.50	8625	6900	5750	0.90
Miehle Roland 800 50	39 3/8 x 50	10000	163.49	1.20	1.40	1.90	7500	6000	5000	1.00
Miehle Roland 800 55	39 3/8 x 55 1/8	10000	168.43	1.20	1.40	1.90	7500	6000	5000	1.00
Miehle Roland 800 63	43 5/16 x 63	10000	182.28	1.20	1.40	1.90	7500	6000	5000	1.00

* Denotes Perfector

Figure 12A.4

Offset presswork estimating schedules: Two-color presses

PRESSWORK SCHEDULE—OFFSET

(Time given represents machine hours on the decimal system)

PRESS NAME	SIZE	RATED SHEETS PER HOUR	HOUR RATE At 70% Productivity	PREPARATION/MAKE-READY TIME: SIMPLE Time	PREPARATION/MAKE-READY TIME: AVERAGE Time	PREPARATION/MAKE-READY TIME: DIFFICULT Time	AVG. IMPRESSION PER RUN HR.: SIMPLE	AVG. IMPRESSION PER RUN HR.: AVERAGE	AVG. IMPRESSION PER RUN HR.: DIFFICULT	WASH-UP Time
INCLUDES ONLY ONE PRESSMAN	YOU MUST COMPUTE YOUR OWN COST FIGURES FOR TWO-, FOUR, FIVE- AND SIX-COLOR PRESSES. THE HOUR RATE FOR THE 2ND PRESSMAN AND FEEDER MUST BE ADDED TO THE COST CENTER HOUR RATE WHERE APPLICABLE BEFORE COMPUTING THE HOURLY COST FOR MAKE READY, IMPRESSIONS PER THOUSAND OR WASH-UP			Feeder = $37.42 per hour. Second Pressman (to 29" press) = $51.45 per hour. Second Pressman (over 29" press) = $52.23 per hour.						
FOUR COLOR PRESSES										
Heidelberg GTOV-52	14 x 20 ½	8000	90.28	1.15	1.65	1.90	6000	4800	4000	0.80
Solna 425	18 ⅛ x 25 3/16	10000	98.56	1.40	1.90	2.10	7500	6000	5000	0.80
Heidelberg MOV	19 x 25 ½	11000	115.73	1.40	1.90	2.10	8250	6600	5500	0.80
Heidelberg MOVP	19 x 25 ½	11000	120.00	1.40	1.90	2.10	8250	6600	5500	0.80
Solna 464	19 x 26	10000	116.90	1.40	1.90	2.10	7500	6000	5000	0.80
Komori Lithrone 426	20 x 26	13000	138.07	1.50	2.00	2.20	9750	7800	6500	0.80
Komori Lithrone 426**	20 x 26	13000	152.62	1.50	2.00	2.20	9750	7800	6500	0.80
Akiyama 428	20 x 28	13000	131.90	1.50	2.00	2.20	9750	7800	6500	0.80
Heidelberg 72V	20 ½ x 28 ⅜	11000	137.78	1.50	2.00	2.20	8250	6600	5500	0.80
Miehle Roland 28	20 ½ x 28 ⅜	10000	146.41	1.50	2.00	2.20	7500	6000	5000	0.80
Miller TP 74	20 ½ x 29 ⅛	12000	143.28	1.50	2.00	2.20	9000	7200	6000	0.80
Akiyama 432	23 x 32 ¼	13000	144.41	1.60	2.10	2.30	9750	7800	6500	0.80
Miller TP 84*	24 x 33	10000	153.89	1.60	2.10	2.30	7500	6000	5000	0.80
Miehle Roland 36	25 x 36	10000	163.53	1.60	2.10	2.30	7500	6000	5000	0.80
Miller TP 95*	25 ⅝ x 37 ½	10000	160.42	1.60	2.10	2.30	7500	6000	5000	0.80
Akiyama 440	28 5/16 x 40 ⅛	13000	180.58	1.80	2.25	2.50	9750	7800	6500	1.00
Heidelberg 102V	28 ⅜ x 40 ⅛	11000	162.20	1.80	2.25	2.50	8250	6600	5500	1.00
Miehle Roland 40	28 ⅛ x 40 3/16	10000	174.75	1.80	2.25	2.50	7500	6000	5000	1.00
Komori 440 Lithrone**	28 ⅜ x 40 ½	13000	201.67	1.80	2.25	2.50	9750	7800	6500	1.00
Miller TP104	28 ⅜ x 41	10000	184.01	1.80	2.25	2.50	7500	6000	5000	1.00
Miehle Roland 50	39 ⅜ x 50	10000	254.37	2.50	2.50	2.90	7500	6000	5000	1.20
Miehle Roland 55	39 ⅜ x 55 ⅛	10000	264.70	2.50	2.50	2.90	7500	6000	5000	1.20
Miehle Roland 63	43 5/16 x 63 ¾	10000	279.03	2.50	2.50	2.90	7500	6000	5000	1.20

* Denotes Perfector
** Remote Inking Console

Figure 12A.5

Offset presswork estimating schedules: Four-color presses

PRESSWORK SCHEDULE—OFFSET

(Time given represents machine hours on the decimal system)

PRESS NAME	SIZE	RATED SHEETS PER HOUR	HOUR RATE At 70% Productivity	PREPARATION/MAKE-READY TIME: SIMPLE Time	AVERAGE Time	DIFFICULT Time	AVG. IMPRESSION PER RUN HR.: SIMPLE	AVERAGE	DIFFICULT	WASH-UP Time
INCLUDES ONLY ONE PRESSMAN	YOU MUST COMPUTE YOUR OWN COST FIGURES FOR TWO-, FOUR, FIVE- AND SIX-COLOR PRESSES. THE HOUR RATE FOR THE 2ND PRESSMAN AND FEEDER MUST BE ADDED TO THE COST CENTER HOUR RATE WHERE APPLICABLE BEFORE COMPUTING THE HOURLY COST FOR MAKE READY, IMPRESSIONS PER THOUSAND OR WASH-UP			First Feeder = $36.93 per hour. Second Feeder = $30.78 per hour. Second Pressman (to 29" press) = $50.79 per hour. Second Pressman (over 29" press) = $52.32 per hour.						
FIVE COLOR PRESSES										
Heidelberg GTOF-S2	14 x 20 ½	8000	103.14	1.30	1.80	2.00	6000	4800	4000	1.00
Heidelberg MOF	19 x 25 ½	11000	135.23	1.30	1.80	2.00	8250	6600	5500	1.00
Heidelberg MOFP	19 x 25 ½	11000	139.43	1.30	1.80	2.00	8250	6600	5500	1.00
Komori 526 Lithrone**	20 x 26	13000	176.40	1.30	1.80	2.00	9750	7800	6500	1.00
Akiyama 528	20 x 26	13000	154.65	1.30	1.80	2.00	9750	7800	6500	1.00
Miehle Roland 28	20 ½ x 28 ⅜	10000	177.75	1.30	1.80	2.00	7500	6000	5000	1.00
Miller TP 74	20 ½ x 29 ⅛	12000	170.33	1.30	1.80	2.00	9000	7200	6000	1.00
Akiyama 532	23 x 32 ¼	13000	172.43	1.30	1.80	2.00	9750	7800	6500	1.00
Miller TP 84*	24 x 33	10000	180.80	1.30	1.80	2.00	7500	6000	5000	1.00
Miller TP 95*	25 ⅝ x 37 ½	10000	192.60	1.30	1.80	2.00	7500	6000	5000	1.00
Akiyama 540	28 5/16 x 40 ⅛	13000	211.87	1.50	2.00	2.30	9750	7800	6500	1.20
Heidelberg 102F	28 ⅜ x 40 ⅛	11000	195.80	1.50	2.00	2.30	8250	6600	5500	1.20
Komori 540 Lithrone**	28 ⅜ x 40 ½	13000	238.98	1.50	2.00	2.30	9750	7800	6500	1.20
Miller TP 104 *	28 ⅜ x 41	11500	240.66	1.50	2.00	2.30	8625	6900	5750	1.20

* Denotes Perfector
** Remote Inking Console

Figure 12A.6

Offset presswork estimating schedules: Five-color presses

PRESSWORK SCHEDULE—OFFSET

(Time given represents machine hours on the decimal system)

PRESS NAME	SIZE	RATED SHEETS PER HOUR	HOUR RATE At 70% Productivity	PREPARATION/MAKE-READY TIME SIMPLE Time	PREPARATION/MAKE-READY TIME AVERAGE Time	PREPARATION/MAKE-READY TIME DIFFICULT Time	AVG. IMPRESSION PER RUN HR. SIMPLE	AVG. IMPRESSION PER RUN HR. AVERAGE	AVG. IMPRESSION PER RUN HR. DIFFICULT	WASH-UP Time
Komori 626 Lithrone**	20 x 26	13000	193.41	1.30	1.80	2.00	9750	7800	6500	1.00
Akiyama 628	20 x 28	13000	166.14	1.30	1.80	2.00	9750	7800	6500	1.00
Miehle Roland 28	20 1/2 x 28 3/8	10000	190.11	1.30	1.80	2.00	7500	6000	5000	1.00
Miller TP 74	20 1/2 x 29 1/8	12000	190.66	1.30	1.80	2.00	9000	7200	6000	1.00
Akiyama 632	23 x 32 1/4	13000	184.40	1.30	1.80	2.00	9750	7800	6500	1.00
Miller TP 84*	24 x 33	10000	206.55	1.30	1.80	2.30	7500	6000	5000	1.00
Miller TP 95*	25 5/8 x 37	10000	223.74	1.50	2.00	2.30	7500	6000	5000	1.00
Miehle Roland 40	28 1/4 x 40 3/16	10000	240.60	1.50	2.00	2.30	7500	6000	5000	1.20
Akiyama 640	28 5/16 x 40 1/8	13000	249.87	1.50	2.00	2.30	9750	7800	6500	1.20
Heidelberg 102S	28 3/8 x 40 1/8	11000	229.76	1.50	2.00	2.30	8250	6600	5500	1.20
Komori 640 Lithrone**	28 3/8 x 40 1/2	13000	276.30	1.50	2.00	2.30	9750	7800	6500	1.20
Miller TP 104*	28 3/8 x 41	11500	276.72	1.50	2.00	2.30	8625	6900	5750	1.20
Miehle Roland 50	39 3/8 x 50	10000	355.58	1.80	2.30	2.60	7500	6000	5000	1.20
Miehle Roland 55	39 3/8 x 55 1/8	10000	365.87	1.80	2.30	2.60	7500	6000	5000	1.20
Miehle Roland 63	43 5/16 x 63	10000	383.94	1.80	2.30	2.60	7500	6000	5000	1.20

INCLUDES ONLY ONE PRESSMAN

YOU MUST COMPUTE YOUR OWN COST FIGURES FOR TWO-, FOUR, FIVE- AND SIX-COLOR PRESSES. THE HOUR RATE FOR THE 2ND PRESSMAN AND FEEDER MUST BE ADDED TO THE COST CENTER HOUR RATE WHERE APPLICABLE BEFORE COMPUTING THE HOURLY COST FOR MAKE READY, IMPRESSIONS PER THOUSAND OR WASH-UP

First Feeder = $36.93 per hour.
Second Feeder = $30.78 per hour.
Second Pressman (to 29" press) = $50.79 per hour.
Second Pressman (over 29" press) = $52.32 per hour.

SIX COLOR PRESSES

* Denotes Perfector
** Remote Inking Console

Figure 12A.7

Offset presswork estimating schedules: Six-color presses

Estimating Binding and Finishing Operations

13.1 Introduction

Binding and finishing operations represent the final manufacturing step for most printed work. It is at this juncture that printed sheets are cut apart and/or trimmed, folded, gathered, jogged, and padded, or converted to the final product format. Because these operations are the last step in the manufacturing sequence, the high quality of the job can easily be destroyed if care is not taken at this point. While presswork provides the actual printed product—that is, the ink on paper—binding and finishing operations complete the product as a package ready for delivery. A beautifully reproduced job can be ruined by poor finishing.

Binding and finishing operations, as with certain prepress areas such as table stripping, sometimes require a large amount of manual labor. One reason is that binding and finishing generally involve the tailor-making of a final product which thus requires individualized, careful attention. For example, placing a coupon into a newspaper insert, attaching covers on hardbound books, or jogging, counting, or special wrapping for delivery, requires individual time and attention. Automation, generally conducive to situations oriented toward a standardized product, is not practical when many products require differing treatment during manufacturing.

In addition to the specialized product orientation of the bindery, there are a large number of small operations completed during binding and finishing procedures. Each of these operations contributes differently to the final product outcome; in addition, the required combinations of such small operations vary from job to job. For example, one job may require trimming, counting, and wrapping before shipment, and another may require cutting, trimming, folding, banding with rubber bands, and boxing prior to shipment to the customer. While standardized and specialized jobs may require much the same binding and finishing procedures, the bindery area must be prepared to complete numerous small operations using various production mixes. This situation makes semiautomatic equipment common in many binderies, while fully automated equipment is possible only in circumstances when products are specialized and processes fully integrated.

The cumulative effect of custom-tailoring the finishing of a product in the bindery, through the use of a wide array of individualized manufacturing operations, is that bindery production, similar to pressroom output, can easily become both labor-intensive and capital-intensive. Many skilled personnel, working at expensive hourly rates, are needed for efficient output of a high-quality product. Coupled with this, and in an attempt to maximize output, most printing plants standardize and automate as many binding jobs as possible.

One major change in binding and finishing is automation through robotics. Bindery equipment manufacturers have begun to standardize more routine binding and finishing processes, then build automated robotic units which can perform these operations faster and more precisely than human workers. Because of major investments in time and money to provide sophisticated robotic equipment, robotizing fits only the most high production finishing lines and higher volume manufacturing operations.

The quantity of buyouts of binding and finishing varies from plant to plant. Some companies send all work to their local trade bindery and have no bindery equipment of their own; other plants may send only jobs that require special finishing operations or use the trade bindery only when their own facilities are filled to capacity. In terms of size, the extremely large printing or publishing facility, with specialized, high-speed equipment and production, will have a tendency to do much of their own binding and finishing using in-plant facilities, allowing them complete control over the production and all other internal manufacturing operations from beginning to end. Another reason that larger plants tend to complete their own binding operations is that their quantity levels may be exceptionally high, and a trade bindery may not be able to accommodate such high volume on a continuing basis. Of course, extremely large manufacturing facilities may handle labor and capital intensiveness to their own advantage, making an in-plant bindery facility a sound financial investment. Small and medium-sized printing plants tend to buy out binding and finishing contingent upon their own product lines and available in-plant equipment.

Most printing and publishing operations, whether they are large, medium, or small in size, invest in at least two major pieces of bindery equipment: cutting machines and folders. For any printing facility doing sheetfed presswork, a cutting machine is necessary for cutting paper stock prior to presswork and for trimming out flat sheet materials before wrapping and delivery.

Buckle folders are common to many printing facilities because a large portion of printed work requires simple folding as a postpress operation. After folding, a job may go to the cutter to be trimmed or may simply be banded and boxed for shipment to the customer. Almost all other bindery equipment results in a specialized product. For example, stitching, collating, drilling, round cornering, sewing, perfect binding, and perforating equipment provide individual operations selectively used when required. These binding and finishing procedures encompass what most trade binderies would offer; of course, the trade bindery will also provide cutting and buckle folding operations if desired.

In order to improve understanding of binding and finishing estimating procedures, this chapter covers three categories of binding and finishing operations: flat sheet production, commercial booklet and pamphlet binding, and wrapping and mailing. Flat sheet production involves those finishing operations performed on a job while it is in sheet form and includes cutting, trimming, folding, jogging, counting, padding, gathering, and drilling. Commercial booklet and pamphlet binding details the binding and finishing operations that result in folded, stitched, and finished booklets and pamphlets.

Generally, this chapter covers binding and finishing operations performed in commercial printing plant binderies or trade binderies serving the commercial printing segment. Thus, many of the more sophisticated finishing operations such as edition binding and mechanical binding will not be covered. In addition, sophisticated equipment necessary to perform many of the specialized finishing and book building operations, such as equipment for casing-in, sewing, or perfect binding, will not be detailed.

It must be noted that significant variations in production output exist between companies doing binding and finishing operations. This variance may be attributed to the fact that utilization of bindery equipment changes from place to place, and differing production standards for employee performance also exist. Because of these conditions, it is recommended that schedule data, as provided in this chapter, be cautiously applied. Just as with all other production standards presented in this book for estimating and production planning, such data are intended as general information—accurate standards are best gathered and applied on an individual plant basis.

13.2 Estimating Flat Sheet Production

Cutting and Trimming

The operations of cutting and trimming are common to a large number of jobs moving through any printing facility. The term *cutting* may be defined as the separation of a large sheet of paper stock or other substrate into smaller sheets. For example, a parent (stock) sheet cut down in size into press sheets is a cutting operation. *Trimming* is a final operation in which small amounts of paper are removed from the edges of the product, insuring exact dimensions for all pieces, squaring up the paper stock, and/or providing for a neater product. Cutting may be either a prepress (before printing) or postpress (after printing) operation, while trimming is usually completed as a concluding, postpress technique.

Cutting machines are used for both cutting and trimming of paper stock as a prepress or postpress operation. This piece of equipment consists of a flat polished steel working base upon which the paper stock is slid, a cutting blade that is razor sharp, a clamp to hold the paper securely during cutting, and a back gauge that establishes the length of sheet to be cut. Paper stock is divided into hand-held lifts and slid under the cutting blade until it fits securely against the back gauge, which has been preset for the specified distance or length of cut. The hydraulic clamp is then dropped down (either manually with a foot pedal or automatically, just before the blade begins its travel), thus holding the paper securely during the slicing action of the blade. The process is repeated continuously, with changes in the back gauge corresponding to changes in sheet length. Modern cutting equipment is manufactured with electronic safety devices insuring that the sharp cutting blade does not repeat. In addition, most sophisticated cutters are now manufactured with electronic computer "spacing," allowing the operator to preset into processor memory certain desired cutting lengths. During the cutting procedure, the operator may move the back gauge to an established preset distance by engaging the computer unit. The unit saves considerable operator time in adjustment of the back gauge between cuts and insures precisely the same length since it provides exact repeatability of back gauge setting.

Figure 13.1 is provided for estimating both cutting and trimming operations.

Basically, the schedule is divided into two segments:

Lift schedule,
Standard production times—that is, cutting time per lift—for cutting machinery both with and without automatic back gauge spacing equipment.

The segments of the figure must be used in conjunction with each other.

To use Figure 13.1, the estimator begins with the lift schedule and identifies the kind of paper stock, weight (per 500 sheets of the basic size), and the caliper (thickness), thereby determining the approximate number of sheets per load (or lift). Essentially, a *lift* is a stack (pile) of paper or other material to be cut or trimmed in a height, size, or weight most easily handled by the cutter operator. Usually, a lift height of approximately 4 inches is considered a productive value. This *lift value*, or approxi-

Lift Schedule

Kind of Paper	Approx. Caliper (Thickness) in Inches	Approx. Sheets per Load
13# Bond, 50# Coated Book	0.0025	1600
16# Bond, 60# Coated Book, 50# Wove Finish Offset Book	0.0033	1300
70# Coated Book	0.0035	1200
20# Bond, 80# Coated Book, 60# Wove Finish Offset Book	0.0040	1000
24# Bond, 100# Coated Book, 70# Wove Finish Offset Book	0.0050	800
80# Wove Finish Offset Book, 60# Coated Cover	0.0060	700
100# Tagboard	0.0070	600
80# Coated Cover, 50# Antique Finish Cover, 90# Index Bristol	0.0080	500
125# Tagboard, 110# Index Bristol	0.0093	400
65# Antique Finish Cover, 150# Tagboard, 140# Index Bristol	0.0110	400
4-Ply Board	0.0180	200

Figure 13.1

Flat sheet cutting and trimming schedule (Reproduced with permission of Printing Industries Association, Inc., of Southern California from their *1988–89 Blue Book of Production Standards and Costs for the Printing Industry)*

Cutting and Trimming Time per Load (cutting without automatic spacers)

Approx. Size of Sheet before Cut or Trim Inches (sq in.)	Setup Time for Manual Spacers														
	0.03			0.07		0.10		0.15		0.20		0.25		0.05	0.07
	Number of Sheets Cut Out of Parent or Press Sheet													Trim Out Bleeds and Gutters	
	2	3	4	5-8	9-12	13-16	17-20	21-24	25-28	29-32	33-36	37-40	41-44	1-2 Sides	3-4 Sides
8½ × 11 (94)	0.03	0.05	0.07	0.09	0.11	0.13								0.04	0.07
17 × 22 (364)	0.04	0.07	0.10	0.13	0.16	0.19	0.22	0.25						0.05	0.08
22 × 29 (638)	0.08	0.11	0.14	0.17	0.20	0.23	0.26	0.29	0.32					0.06	0.09
23 × 35 (805)	0.10	0.13	0.16	0.19	0.22	0.25	0.28	0.31	0.34	0.39	0.44	0.49		0.07	0.10
25 × 38 (950)	0.10	0.15	0.20	0.25	0.30	0.35	0.40	0.45	0.50	0.55	0.60	0.65	0.70	0.08	0.11
35 × 45 (1575)	0.14	0.20	0.26	0.32	0.38	0.44	0.50	0.56	0.62	0.68	0.74	0.80	0.86	0.10	0.13
38 × 50 (1900)	0.17	0.24	0.31	0.38	0.45	0.52	0.59	0.66	0.73	0.80	0.87	0.94	1.00	0.11	0.14

Note: Schedule data are decimal hours per lift cut. Add percentages as they apply. For intermediate size sheets, use higher square inch value.

Cutting and Trimming Time per Load (cutting with automatic spacers)

Approx. Size of Sheet Before Cut or Trim in Inches (sq in.)	Setup Time for Electronic Spacers														
	0.06			0.10		0.15		0.25		0.30		0.35		0.07	0.10
	Number of Sheets Cut Out of Parent or Press Sheet													Trim Out Bleeds and Gutters	
	2	3	4	5-8	9-12	13-16	17-20	21-24	25-28	29-32	33-36	37-40	41-44	1-2 Sides	3-4 Sides
8½ × 11 (94)	0.03	0.05	0.07	0.06	0.08	0.09								0.03	0.05
17 × 22 (364)	0.04	0.07	0.10	0.09	0.11	0.13	0.15	0.18						0.04	0.06
22 × 29 (638)	0.08	0.11	0.14	0.12	0.14	0.16	0.18	0.20	0.22					0.04	0.06
23 × 35 (805)	0.10	0.13	0.16	0.13	0.15	0.18	0.20	0.22	0.24	0.27	0.31	0.34		0.05	0.07
25 × 38 (950)	0.10	0.15	0.20	0.18	0.21	0.25	0.28	0.32	0.35	0.39	0.42	0.46	0.49	0.06	0.08
35 × 45 (1575)	0.14	0.20	0.26	0.22	0.27	0.31	0.35	0.39	0.43	0.48	0.52	0.56	0.60	0.07	0.09
38 × 50 (1900)	0.17	0.24	0.31	0.27	0.32	0.36	0.41	0.46	0.51	0.56	0.61	0.66	0.70	0.08	0.10

Note: Schedule data are decimal hours per lift cut. Add percentages as they apply. For intermediate size sheets, use higher square inch value.

Penalties: Add 20% to total time for cutting perforated or scored stocks. Add 15% to total time for cutting laminated or highly polished stocks. Add 10% to total time for cutting stocks over 0.010 in. thick. Add 10% to total time for cutting stocks under 0.002 in. thick.

Figure 13.1

Continued

mate number of sheets per lift, is divided into the total number of sheets to be cut or trimmed, which yields the number of total lifts (or loads) to be cut. That is:

total number of lifts to be cut or trimmed = number of sheets to be cut or trimmed ÷ number of sheets per lift

Any fraction of a lift should be rounded to the next highest lift value.

The cutting and trimming schedule, Figure 13.1, is based on the number of cuts or trims out of the press or parent sheet and the size of the uncut or untrimmed sheet. The cutting and trimming schedule provides a cutting time per lift, which is used to determine cutting production time as follows:

cutting production time = total number of lifts to be cut × cutting time per lift

If an intermediate size of sheet is to be cut, use the next higher size. The setup time is added to the cutting time to calculate the total production time.

In instances where trimming is a separate operation from cutting, such as the trimming of a 1-up job on four sides, only that portion of the schedule should be used. When trimming operations accompany cutting, both rates can be added together since they apply to all lifts.

The addition of setup time should be made at least one time per job. Setup time includes time for reading the job ticket, making preliminary cutting calculations, moving (or programming) the back gauge, and performing any other operations required as preliminary work. Schedule data are based on a cutting crew of one person, yet it is possible that more than one operator will be utilized. In such a case, the BHR will allow for cost recovery of this additional employee. Cutting and trimming production times include a reasonable amount for jogging into the cutter back gauge. Penalties, indicated at the bottom of the schedule, should be applied as appropriate.

In addition to those just mentioned, the following formulas are used for estimating cutting and trimming production time and cost:

total number of press size sheets required = (number of finish size sheets required ÷ number-up on the press size sheet) × (percentage of spoilage + 100%)

total number of parent size sheets required = (number of press size sheets required including spoilage ÷ number-out of parent size sheet)

*total prepress cutting production time = [(number of parent size sheets required ÷ number of sheets per lift) × cutting time per lift] + cutter setup time

total cutting production time = prepress cutting production time + postpress cutting production time

total production cost = total cutting production time × cutter BHR

---

*In this formula, note that the division portion of the equations yields the number of lifts to be cut and the "cutting time per lift" portion yields the cutting production time.

Example 13.1

We want to determine cutting time and cost for 56,000 announcements, finish size sheet 6 × 8 inches, run on basis 100 tagboard. Parent size sheet from stock is 36 × 48; press size sheet is 24 × 36. The job will run 18-up on the press and cut 2-out of the parent sheet. Add 6% of base press size sheets for press spoilage. No trimming is required. The cutter is a manual-spacing, 36 inch Imperial Standard with a BHR of $51.96.

Solution Use Figure 13.1.

Step 1 Determine the number of 24 × 36 pss and 36 × 48 pars to be cut:

(56,000 fss ÷ 18-up) × (6% + 100%) = 3112 pss × 106% = 3299 pss
3299 pss ÷ 2-out = 1650 pars

Step 2 For prepress cutting, cut 2-out of a 36 × 48 pars to produce 24 × 36 pss. Determine the following:

Number of lifts:
  1650 pars ÷ 600 sht/lift = 2.75 lifts = 3 lifts
Cutting time:
  3 lifts × 0.17 hr/lift = 0.51 hr
Prepress cutting time:
  0.51 hr + 0.03 hr setup = 0.54 hr

Step 3 For postpress cutting, cut 18-out of a 24 × 36 pss to produce the total number of 6 × 8 fss pieces. Determine the following:

Number of lifts:
  3299 pss ÷ 600 sht/lift = 5.5 lifts = 6 lifts
Cutting time:
  6 lift × 0.40 hr/lift = 2.40 hr
Postpress cutting time:
  2.40 hr + 0.10 hr setup = 2.50 hr

Step 4 Determine total cutting production time and cost:

Total cutting time:
  0.54 hr + 2.50 hr = 3.04 hr
Total cutting cost:
  3.04 hr × $51.96/hr = $157.96

Example 13.2

Determine cutting time and cost for 35,000 letterheads, two colors, to run on substance 20 rag bond. Stock in inventory is 35 × 45 inches, to cut 4-out yielding

17 1/2 × 22 1/2 inch press size sheets. Add 10% spoilage to base press sheets. After cutting to approximate finish size, trim finish pieces to 8 1/2 × 11. The cutter is an automatic-spacing Wohlenberg MCSTV-2 with a BHR of $60.63.

Solution Use Figure 13.1.

Step 1 Determine number of 17 1/2 × 22 1/2 pss and 35 × 45 pars to be cut:

35,000 fss ÷ 4-up × (10% + 100%) = 8750 pss × 110% = 9625 pss
9625 pss ÷ 4-out = 2406 pars

Step 2 For prepress cutting, cut 4-out of a 35 × 45 pars to produce 17 1/2 × 22 1/2 pss. Determine the following:

Number of lifts:
2406 pars ÷ 1000 sht/lift = 2.4 lifts = 3 lifts
Cutting time:
3 lifts × 0.26 hr/lift = 0.78 hr
Prepress cutting time:
0.78 hr + 0.06 hr setup = 0.84 hr

Step 3 For postpress cutting, cut 4-out of 17 1/2 × 22 1/2 pss and then trim all cutout sheets to 8 1/2 × 11 (trim on four sides). Determine the following:

Number of lifts to be cut:
9625 sht ÷ 1000 sht/lift = 9.625 lifts = 10 lifts
Cutting time:
10 lifts × 0.14 hr/lift = 1.40 hr
Number of lifts to be trimmed:
35,000 sht ÷ 1000 sht/lift = 35 lifts
Trimming time:
35 lifts × 0.05 hr/lift = 1.75 hr
Setups:
Add 0.06 hr for cutting and 0.10 hr for trim out setup.
Postpress cutting and trimming time:
1.40 hr + 1.75 hr + 0.16 hr = 3.31 hr

Step 4 Determine total cutting production time and cost:

Total cutting time:
0.84 hr + 3.31 hr = 4.15 hr
Total cutting cost:
4.15 hr × $60.63/hr = $251.61

Folding Operations

Folding of paper stock may be completed using either buckle or knife folding equipment. *Knife folders* crease the sheet into two rotating rollers and are used with many web presses at the converting end of the press. *Buckle folders* are far more common in most printing and publishing operations. The reason is that they are flexible in the manner by which they can fold, score, perforate, and slit in varying combinations. In addition, they are consistently high output equipment items.

Buckle folders basically consist of a feeder, much like a press feeder, and folding units wherein parallel folds are put in the sheet as it bumps against folding plates. As the sheets travel down the feedboard of the buckle folding machine, closely spaced but not overlapping, they enter the first folding unit. Knurled rollers in this unit drive the sheet into the first plate, which has been adjusted for the length of fold desired. Once the sheet bumps against the plate, it is stopped momentarily; the knurled rollers then grab it and move it out of the first unit, creasing it at the fold point and placing a fold parallel to the edge of the sheet. In the same folding unit, but with a different plate, additional parallel folds may be administered to the sheet in sequence. When all parallel folds have been completed, the creased stock leaves the first folder unit and, if required, enters the second unit for additional folds. This second unit may be positioned at right angles to the first, thereby providing folds in parallel form but at right angles to the folds from the first unit. A large number of folding combinations are possible using the buckle folder, and it is important that folding considerations be carefully reviewed during the production planning and estimating phases of any job.

Buckle folding is essentially the same for both flat sheet and bookwork (booklet and pamphlet) operations; combinations of parallel and right angle folds in a planned and carefully executed sequence. It should be noted that folder production speeds may vary with respect to the type of folding sequence required, type of paper stock, and thickness and grain of paper among major factors. Figure 13.2 illustrates some of the more common folding configurations that are applicable to both flat sheet and bookwork impositions. Since equipment size and folding configuration vary considerably, these diagrams do not cover all folding possibilities.

It must be stressed that production planning should consider folding as one of the principle production factors—in fact, in some instances, all other job planning must be oriented around folding requirements. Practically every experienced estimator has had the unfortunate experience of planning and estimating an entire job, only to find during production that folding could not be properly completed. In some cases, the job might be salvaged; in others, the expensive task of redoing the entire job (perhaps at company expense) might be the only reasonable alternative. It does not take too many such episodes until the estimator or production planner carefully plans folding as a major element in job execution.

The buckle folding schedule in Figure 13.3 is divided into three segments: setup, running (parallel and right angle sections), and penalty sections. Folding machine sizes represent both small and large machines: folders up to and including 17 × 22 inches and those folders over 17 × 22 inches in capacity. Examples 13.3 and 13.4 are

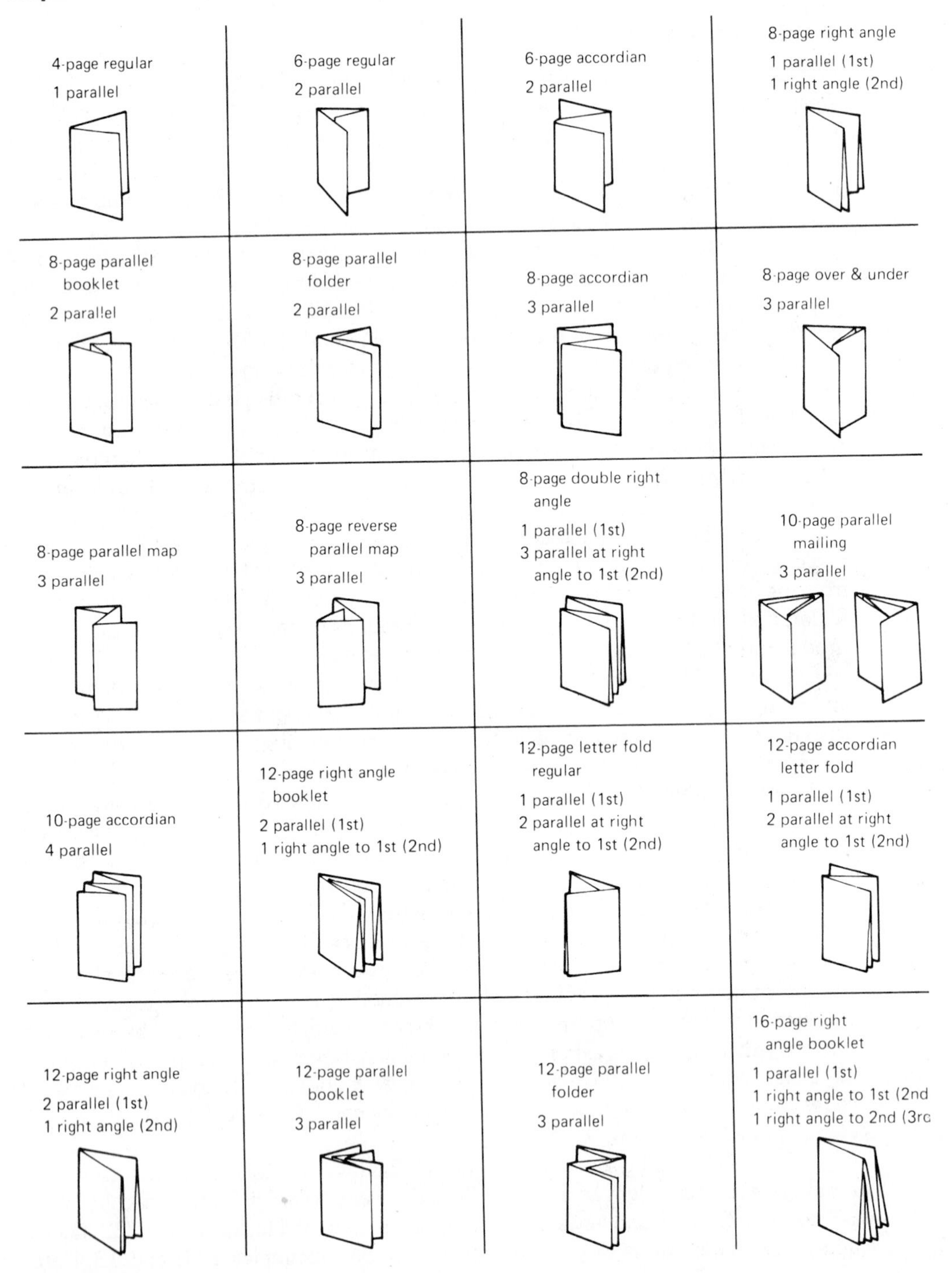

Figure 13.2

Common folding configurations

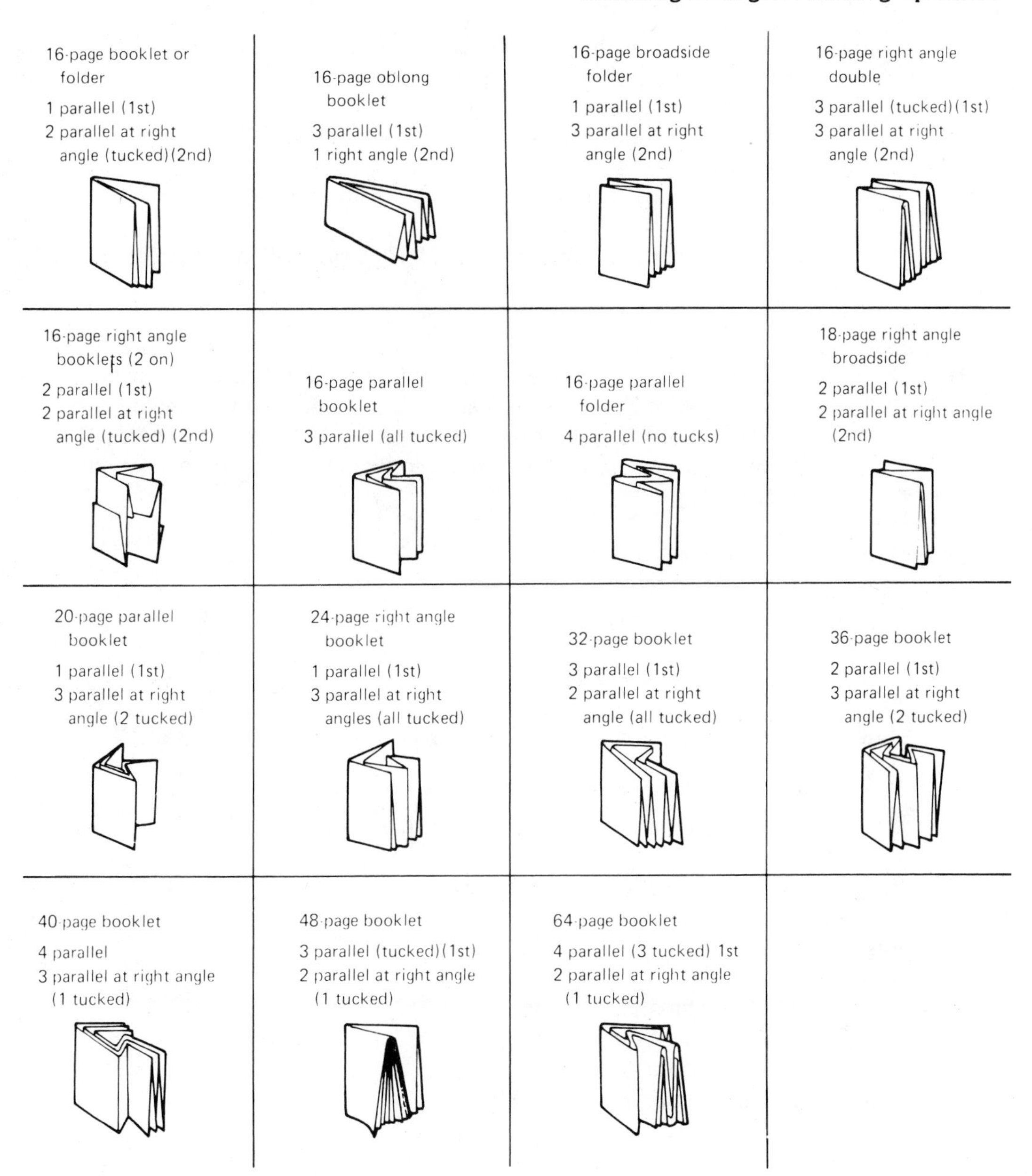

Figure 13.2

Continued

provided to show how such schedule data are used for estimating. The following formulas should be used when working the examples:

folder running time = number of pieces to be folded ÷ folder output in sheets per hour
total folding production time = folder running time + first fold setup time + additional fold setup times
total folding production cost = total folding production time × folder BHR.

Example 13.3

Using the 6 page regular fold (see Figure 13.2), we have to fold 14,000 pieces (including spoilage and extras). Unfolded size is 24 1/2 × 11 inches, to be folded to 8 1/2 × 11 inches. A Stahl C22 (22 × 35 inch maximum sheet size) with a BHR of $55.53 will be used. Stock caliper is 0.005 inch. There will be no additional folds and no perforation or scoring. Determine production time and cost for folding.

Solution Use Figure 13.3.

Step 1 Determine setup time for 2 folds:

0.40 hr + 0.10 hr = 0.50 hr

Step 2 Determine running time for folder (parallel fold section of schedule):

Folder output:
22 × 28 folder and 26 in. sheet length yield 4000 sht/hr.
Running time:
14,000 pc ÷ 4000 pc/hr = 3.50 hr

Step 3 Determine total folding production time and cost:

Total folding time:
0.50 hr + 3.50 hr = 4 hr
Total folding cost:
4 hr × $55.53/hr = $222.12

Example 13.4

Determine the time and cost to fold 21,000 (16 page) brochures, run in a 1-16 SW. (See 16 page right angle booklet in Figure 13.2.) Untrimmed page size is 6 1/4 × 9 1/2 inches, with the press sheet/folder sheet measuring 19 × 25 inches.

Our folding dummy indicates there will be 3 folds at right angles to each other. Stock is basis 70 book that calipers at 0.0055 inch. No slitting, scoring, or

Setup	Setup Time in Hours* Folders to 17 × 22 Inches	Folders to over 17 × 22 Inches
First fold	0.30	0.40
Each additional fold	0.10	0.10
Each slitter	0.20	0.20
Each perforator	0.20	0.20
Each scorer	0.20	0.20

Running Time (parallel fold section)

Sheet Length Through Machine	Sheets per Hour Machine Size to 17 × 22 Inches	Machine Size over 17 × 22 Inches
6	15,000	—
8	15,000	—
10	13,000	7,000
12	13,000	7,000
14	7,000	5,500
16	7,000	5,500
18	6,500	5,000
20	6,500	5,000
22	5,500	4,000
24	5,500	4,000
26	4,500	4,000
28	4,500	4,000
30	4,000	3,700
32	—	3,700
34	—	3,700
36	—	3,500
38	—	3,500
40	—	3,000
42	—	3,000
44	—	3,000

Running Time (right angle section)

Number of Folds	Signatures per Hour Untrimmed Page to 10 × 7 Inches	Untrimmed Page to 12 × 9 Inches
2 folds	5250	5000
3 folds	4000	3500
4 folds	3500	3000

Note: Add charges for additional personnel for banding, slitting, cartoning, and so on.

*All setup times are decimal hours.

Penalties: Add 25% to total time for stock under 0.0025 in. thick. Add 25% to total time for stock over 0.008 in. thick. Add 10% to total time for slit, perforate, or score.

Figure 13.3

Buckle folding schedule (Reproduced with permission of Printing Industries Association, Inc., of Southern California from their *1988–89 Blue Book of Production Standards and Costs for the Printing Industry)* perforating will be completed. Folding will be done on a Baumfolder 523 (23 × 36 inch maximum sheet size) with a BHR of $53.09.

Solution Use Figure 13.3.

Step 1 Determine setup time for 3 folds:

0.40 hr + 0.10 hr + 0.10 hr = 0.60 hr

Step 2 Determine running time (right angle folding section of schedule):

Folder output:
Page size 6 1/4 × 91/2 in. (under 10 × 7 in.) and 3 folds yield 4000 sig/hr.
Running time:
21,000 pc ÷ 4000 sig/hr = 5.25 hr

Step 3 Determine total folding production time and cost:

Total folding time:
0.60 hr + 5.25 hr = 5.85 hr
Total folding cost:
5.85 hr × $53.09/hr = $310.58

Hand Gathering of Flat Sheets

Hand gathering is a procedure by which forms and other printed material are assembled together. Perhaps the most common operation is the gathering of printed forms, which may be numbered and interleaved with carbon paper between sheets or may be printed on carbonless paper. The term *collating* has become synonymous with gathering procedures. Collating of printed signatures, either by hand or machine, is also common.

Figure 13.4 provides for estimating production of hand-gathered materials. Schedule basis is the number of gathered sets per hour. When used for estimating, the number of pieces to be gathered, also termed *pickups*, and the size of the sheets to be collated must be identified. Penalties, at the bottom of the schedule, apply as indicated. The schedule data are used in conjunction with the following formulas to estimate gathering production time and cost:

total gathering production time = [(number of sets to be gathered × number of pieces per set) ÷ number of gathered sets per hour] + penalty additions
total gathering production cost = total gathering production time × gathering BHR

The collating of folded signatures is detailed in the booklet and pamphlet section of this chapter (Section 13.3).

Example 13.5

We have printed 2500 copies of a four-part purchase order (white, canary, blue, and pink) on basis 50 book, caliper 0.004 inch, 8 1/2 × 11 inches. These copies must be hand gathered. Include time for carbon interleafing between the pages of each four-part set. The job has been numbered in consecutive order and must be gathered in sequence. The BHR for hand gathering is $29.80.

Solution Use Figure 13.4.

Step 1 Determine total production time to gather sets:

Time to gather:
2,500 sets ÷ 400 sets/hr = 6.25 hr

Operations	Number of Gathered Sets per Hour		
	Approx. Sheet Size 8½ × 11 Inches	Approx. Sheet Size 11 × 17 Inches	Approx. Sheet Size 17 × 22 Inches
2 pickups	1400	1200	700
3 pickups	950	820	450
4 pickups	720	620	325
5 pickups	580	500	250
6 pickups	480	420	200
7 pickups	400		
8 pickups	340		
9 pickups	300		
10 pickups	270		
11 pickups	240		
12 pickups	220		

Penalties: Add 20% to total time for stock under 0.0025 in. thick. Add 20% to total time for gathering of odd sizes of paper. Add 20% to total time for gathering of numbered pieces. Add 30% to total time for gathering of carbon interleafs (forms).

Figure 13.4

Hand gathering schedule (flat sheets) (Reproduced with permission of Printing Industries Association, Inc., of Southern California from their *1988–89 Blue Book of Production Standards and Costs for the Printing Industry*)

Step 2 Add penalties:

Penalty for numbering:
6.25 hr × 0.20 = 1.25 hr
Penalty for carbon interleaving:
6.25 × 0.30 = 1.88

Step 3 Total gathering production time:

6.25 + 1.25 + 1.88 = 9.38 hr

Step 4 Determine total production cost to gather sets:

9.38 hr × $29.80/hr = $279.52

Step 5 Summarize your findings:

Total gathering time: 9.38 hr
Total gathering cost: $279.52

Jogging

The *jogging* of paper, a procedure by which paper stock is aligned in very even piles using manual operations or a vibrating jogger, is important throughout binding and finishing operations since it assures perfect alignment of sheets as they are worked through various pieces of equipment. Jogging is also important when counting of stock is required since aligned sheets allow faster and more accurate counting operations.

The jogging schedule in Figure 13.5 is based on the production output per 1000 sheets of hand- or machine-jogged materials. The schedule is based on normal stock thickness (0.0025 to 0.008 inch) and approximate sheet size in inches. Penalties are added as they apply, especially for hard-to-handle stocks such as thin paper and gummed substrates.

For some operations performed in the bindery, such as folding and cutting, a small amount of production time has been built into the schedules for preliminary jogging operations. Use of this jogging schedule should be made only when special jogging operations must be performed or when additional jogging is anticipated.

Production time and cost for the rest of the binding and finishing operations that are covered in this chapter (jogging, counting, padding, and so on) can be determined by the following general formulas:

total binding or finishing operation production time = [(number of units to which operation applies ÷ 1000) × operation time per M units] + preparatory time as appropriate + penalty additions as necessary total binding or finishing operation production cost = total binding or finishing operation production time × binding or finishing operation BHR

Example 13.6 We must hand jog 6500 sheets of onionskin to be used as second sheets for a legal agreement we have printed. Stock measures 8 1/2 × 14 inches and calipers at 0.0020 inch. The BHR for hand jogging is $29.80.

Solution Use Figure 13.5.

Step 1 Determine total production time to hand jog:

Time to jog:
(6500 sht ÷ 1000) × 0.13 hr/M sht = 0.85 hr
Penalty for stock under 0.0025 in.:
0.85 hr × 0.25 = 0.22 hr
Time for preparation:
0.10 hr
Total jogging production time:
0.85 hr + 0.22 hr + 0.10 hr = 1.17 hr

Sheet Size in Inches	Time in Hours per (M) Sheets	
	Hand	Machine
9½ × 12½ and under	0.06	0.04
12½ × 19 and under	0.13	0.09
19 × 25 and under	0.26	0.17
25 × 38 and under	0.39	
35 × 45 and under	0.52	
38 × 50 and under	0.65	
Over 38 × 50	0.78	

Note: Preparatory time is 0.10 hr. All times are decimal hours.

Penalties: Add 25% to total time for stock under 0.0025 in. thick. Add 10% to total time for stock over 0.008 in. thick. Add 20% to total time for varnished or gummed stocks. Add 15% to total time for scored or perforated stocks.

Figure 13.5

Jogging Schedule (by hand or machine) (Reproduced with permission of Printing Industries Association, Inc., of Southern California from their *1988–89 Blue Book of Production Standards and Costs for the Printing Industry*)

Step 2 Determine total production cost to hand jog:

1.17 hr × $29.80/hr = $34.87

Step 3 Summarize your findings:

Total jogging time: 1.17 hr
Total jogging cost: $34.87

Counting

The *counting* of sheets and finished products in the bindery ensures that the quantity of finished products is accurate. Counting typically is completed either using manual techniques for small quantities or by inserting a metal gauge into a stack of jogged paper. The gauge divides the paper according to the number of sheets for which it has been preset. Gauge procedures are used for counting large quantities of paper. Regardless of the method used, the counting operations should be completed carefully yet quickly by bindery personnel.

The accompanying schedule in Figure 13.6 provides for estimating production time on the basis of hours per 1000 sheets or pieces counted. The size of paper or finish size of the product must be known, as well as the thickness of paper (0.008 inch thick or less for thin or normal paper and over 0.008 inch thick for board), whether hand or gauge procedures will be used, and if dividers (such as chipboard), markers, or flags are to be inserted during the counting operation. Schedule times should be adjusted for individual production operations and personnel.

	Time in Hours per (M) Sheets or Pieces				
	Thin and Normal (0.008 in. thick or less)		Board (over 0.008 in. thick)		Insert Dividers, Markers, or Flags
Sheet Size in Inches	Hand	Gauge	Hand	Gauge	
20 × 26 or less	0.167	0.100	0.208	0.154	0.004 ea.
Over 20 × 26	0.250	0.154	0.357	0.250	0.006 ea.

Note: Preparatory time is 0.10 hr. All times are in decimal hours.
Penalties: Add 20% for counting scored or perforated stocks.

Figure 13.6

Counting schedule (by hand or gauge)

Example 13.7

We have a job almost ready to be sent to the trade binders of 16 page signatures on a 19 × 25 press size sheet in flat sheet form. Since the job is a "no less than" order, we must be certain that we send 12,500 sheets plus 3% for folder and stitcher waste. We will use a gauge for counting. Stock is basis 60 book, 0.0045 inch. The gauge-counting BHR is $29.80.

Solution Use Figure 13.6.

Step 1 Determine the number of sheets to be counted:

12,500 sht × 103% = 12,875 sht

Step 2 Determine total production time to count sheets:

Time to count:
12,875 M sht x 0.10 hr/M sht = 1.29 hr
Time for preparation:
0.10 hr
Total production time:
1.29 hr + 0.10 hr = 1.39 hr

Step 3 Determine total production cost to count sheets:

1.39 hr × $29.80/hr = $41.42

Step 4 Summarize your findings:

Total counting time: 1.39 hr
Total counting cost: $41.42

Padding

Padding operations involve the application of a glue material for temporary binding of paper stock. Padding is generally a hand operation in most binderies.

Figure 13.7, based on hours per 1000 sheets padded, requires that the estimator know the number of sheets per pad, the number of total sheets to be padded, and the size of the sheets. Production time includes jogging of pads, counting and inserting chipboard, jogging piles of pads, and positioning for gluing, application of glue, and slicing pads apart when the adhesive is dry.

Example 13.8

We must determine the time and cost to pad 3500 four-part invoices that have been printed on precollated NCR paper, 8 1/2 × 11 inches. There will be 25 sets per pad, or 100 sheets per pad. The BHR for padding is $29.80.

Bindery	Time in Hours per M Sheets Padded		
	8½ × 11 Inches or Smaller	9 × 12, 14 × 17, 12 × 19 Inches	17 × 22, 17 × 28, 19 × 24 Inches
100 sheets to a pad	0.20	0.30	0.40
50 sheets to a pad	0.30	0.40	0.50
25 sheets to a pad	0.35	0.45	0.55
10 sheets to a pad	0.40	0.50	0.60
5 sheets to a pad	0.50	0.60	0.70

Note: Preparatory time is 0.10 hr. All times are decimal hours.

Figure 13.7

Padding schedule (Reproduced with permission of Printing Industries Association, Inc., of Southern California from their *1988–89 Blue Book of Production Standards and Costs for the Printing Industry*)

Solution Use Figure 13.7.

Step 1 Determine total production time to pad:

Number of sheets:
3500 sets × 4 sht/set = 14,000 sht
Time to pad:
14 M sht × 0.20 hr/M sht = 2.8 hr
Time for preparation:
0.10 hr
Total production time:
2.8 hr + 0.10 hr = 2.9 hr

Step 2 Determine total production cost to pad:

2.9 hr × $29.80/hr = $86.42

Step 3 Summarize your findings:

Total production time: 2.9 hr
Total production cost: $86.42

Drilling

Paper drilling equipment is used to drill holes in precise positions in the finished product. A *single-spindle drill*, also called a *single-head drill*, provides for the drilling of one hole at a time. The *three-spindle drill*, with three revolving bits, is used for high production and drills three holes in one operation. Most paper drills have semiautomatic paper-guiding mechanisms and are operated using a foot pedal for the actual drilling operation.

Figure 13.8 can be used for estimating drilling time with either single-spindle or three-spindle drilling equipment. Production output is measured in hours per 1000 sheets drilled. Production time includes handling of already jogged sheets and drilling the required number of holes. The estimator must know the number of drilling heads on the equipment, the number of holes required, and the size and kind of paper stock. Schedule data should be adjusted for individual drilling operations and equipment and skill of personnel.

Example 13.9 We must determine the time and cost to drill 150,000 notebook inserts using our three-spindle drill. Stock is basis 50 book that calipers 0.0048 inch. Stock is properly jogged and ready for drilling. The BHR for drilling is $29.80.

Solution Use figure 13.8.

Step 1 Determine total production time to drill:

Time to drill:
150 M pc × 0.04 hr/pc = 6 hr

	Time in Hours per M Sheets Drilled		
	Single-Spindle Drill		Three-Spindle Drill
Type and Caliper of Paper Stock	One Hole	Each Additional Hole	Three Holes
Lightweight paper			
13# Bond, 50# Coated Book	0.05	0.02	0.05
Stocks calipering less than 0.003 in.	0.05	0.02	0.05
Average weight paper			
16-24# Bond, 60-100# Coated Book	0.04	0.02	0.04
50-70# Wove Finish Offset Book	0.04	0.02	0.04
Stocks calipering 0.003-0.005 in.	0.04	0.02	0.04
Heavyweight paper			
Stocks calipering	0.06	0.02	0.06
0.0051-0.008 in.	0.06	0.02	0.06

Note: Preparatory time is 0.10 hr per drilling head. Time given is for sheets 11 × 17 in. or smaller. All times are decimal hours.

Penalties: Add 25% to total time for stocks measuring over 11 × 17 in.

Figure 13.8

Drilling schedule (Reproduced with permission of Printing Industries Association, Inc., of Southern California from their *1988–89 Blue Book of Production Standards and Costs for the Printing Industry*)

Time for preparation:
0.10 hr × 3 heads = 0.30 hr
Total production time:
6 hr + 0.30 hr = 6.30 hr

Step 2 Determine total production cost to drill:

6.3 hr × $29.80/hr = $187.74

Step 3 Summarize your findings:

Total drilling time: 6.30 hr
Total drilling cost: $187.74

13.3 Estimating Commercial Booklet and Pamphlet Binding Operations

This portion of the binding and finishing chapter deals with estimating for booklets and pamphlets, typically products that contain 64 or fewer pages. Some small and medium-sized printing companies complete such types of work on an infrequent basis, while others specialize in booklet and pamphlet work. However, only one or two such jobs annually is not sufficient to warrant the purchase of some of the equipment to be covered here. Typically, when volume of pamphlets or booklets is not high, finishing and binding will be completed at a trade bindery after the printer has produced the press sheets.

It should be made clear that the production of books—that is, works containing more than 64 pages—is not covered in this section. Generally, the publishing of hardbound (case-bound) and perfect-bound books is completed using sophisticated bookbinding equipment, at volumes suitable to justify such costly equipment. Case-bound books require sewing, rounding, casing-in, and other operations generally limited to specialized finishing lines in large book-publishing plants. Typically, a highspeed binding line is coordinated directly with sheetfed or web production, providing a complete manufacturing process for such book production. Equipment and facilities are specialized and sometimes almost tailor-made to the types of books produced.

This pamphlet binding and finishing section deals primarily with gathering, wire stitching, and trimming of signatured materials. Some of the operations are manually performed, while others use automated or semiautomated equipment. As with all estimating data and equipment, such standards should be carefully reviewed before direct application to an individual plant or facility.

Hand Gathering of Folded Signatures

Figure 13.9 is presented for estimating the time required to collate folded signatures into loose booklets. Schedule basis is in hours per 1000 booklets gathered. The estimator must identify the untrimmed page size of the booklet, the number of sections for signatures) to be collected, and whether the booklet is imposed in an upright or oblong format. Preparatory time and penalties should be added as they apply.

Number of Sections	Gathering Time in Hours per M Booklets					
	10 × 7 Inches and Under (untrimmed)		12 × 9 Inches and Under (untrimmed)		Over 12 × 9 Inches (untrimmed)	
	Upright	Oblong	Upright	Oblong	Upright	Oblong
2	0.90	1.00	1.00	1.20	1.20	1.30
3	1.60	1.80	1.80	2.00	2.20	2.40
4	2.40	2.70	2.60	2.90	3.10	3.50
5	3.00	3.40	3.30	3.80	4.00	4.50
6	3.70	4.20	4.10	4.60	4.90	5.50

Note: Preparatory time is 0.10 hr for 2 to 3 sections and 0.20 hr for 4 to 6 sections. All times are decimal hours.

Penalties: Add 10% to total time for signatures over 16 pages.

Figure 13.9

Hand gathering schedule (folded signatures) Reproduced with permission of Printing Industries Association, Inc., of Southern California from their *1988–89 Blue Book of Production Standards and Costs for the Printing Industry*)

Number of Sections (sig)	Stitching Time in Hours per M Booklets
1	1.10
2	1.18
3	1.25
4	1.33
5	1.43

Note: Preparatory time is 0.10 hr. All times are decimal hours.

Penalties: Add 10% to total time for each additional stitch over two. Add 5% to total time for extended cover. Add 20% to total time for booklets over 12 × 9 in. untrimmed size.

Figure 13.10

Saddle-wire stitching schedule (foot-operated machine)

It should be noted that the first dimension given with respect to the size of the booklet represents the binding edge dimension. For example, in the schedule, "10 × 7 inches and under" indicates that the edge to be bound measures 10 inches, making the booklet imposed with an upright page format. This dimensioning method is common when estimating all areas of booklet and pamphlet finishing.

Saddle-Wire Stitching

Saddle-wire stitching is common for booklets containing 64 or fewer pages, yet it can be used for books up to 1/2 inch in thickness. As noted in Chapter 6, saddle-stitched products require different imposition treatment than do books and booklets that will be side stitched during binding. Of course, such imposition decisions must be made far in advance of such stitching procedures.

Figure 13.10 is provided for estimating saddle stitching production using a foot operated stitching machine. Schedule basis is hours per 1000 booklets stitched with two wire stitches. The data apply to untrimmed sizes between 4 × 6 and 9 × 12 inches. The estimator must identify the number of sections (signatures) per booklet when applying schedule data. A cover counts as one signature. As will be noted, penalties apply for each additional stitch (over two), for extended covers, and for stitching booklets over 12 × 9 inches in untrimmed size. Add preparatory time as applicable.

Example 13.10

We want to estimate the time and cost to hand gather and saddle-wire stitch a 48 page self-cover booklet in our plant. The booklet was printed and folded with 3 (16 page) signatures in an upright format, with an untrimmed page size of 11 1/4 × 8 3/4 inches. It will require 3 stitches. The BHR for gathering is $29.80, and for saddle-wire stitching, $52.70. We must deliver 5200 copies.

Solution Use Figures 13.9 and 13.10.

Step 1 Determine total production time to gather signatures:

Time to gather:
5.2 M bkt × 1.80 hr/M bkt = 9.36 hr
Time for preparation:
0.10 hr
Total gathering production time:
9.36 hr + 0.10 hr = 9.46 hr

Step 2 Determine total production cost to gather signatures:

9.46 hr × $29.80/hr = $281.91

Step 3 Determine total production time to saddle-wire stitch signatures:

Time to stitch (3 sig):
5.2 M bkt × 1.25 hr/M bkt = 6.50 hr
Time for preparation:
0.10 hr
Production time:
6.50 hr + 0.10 hr = 6.60 hr
Penalty for third stitch:
6.60 hr × 0.10 = 0.66 hr
Total stitching production time:
6.60 hr + 0.66 hr = 7.26 hr

Step 4 Determine total production cost to saddle-wire stitch signatures:

7.26 hr × $52.70/hr = $382.60

Step 5 Determine total production time and cost to gather and stitch the booklets:

Total production time:
9.46 hr + 7.26 hr = 16.72 hr
Total production cost:
$281.91 + $382.60 = $664.50

Side-Wire Stitching

Side-wire stitching is used for books and booklets that may be up to 1 inch thick. The procedure involves the insertion of wire stitches through the side of collated signatures at the binding edge (see Chapter 6). Side-wire stitching does not provide the ease of opening offered by saddle-wire stitching procedures, but is far more practical for books and booklets printed on bulkier stocks and where a rugged, sturdy binding is desired. While saddle-stitched books and booklets have covers stitched simultaneously with the signatures, side stitching allows for different cover attachment methods. For example, a cover may be side stitched with the signatures allowing the wire stitches to show; or covers may be attached with glue after stitching, covering up the side-wire material so it does not show.

A side-wire stitching schedule, Figure 13.11, is provided for estimating stitching time using foot-operated equipment and is based on hours per 1000 books or booklets stitched with two stitches. The estimator must carefully identify the number of signatures contained in the book and the untrimmed page size of the product. Add preparatory time as it applies. Penalties apply for each stitch over the first two and when the book size is larger than the untrimmed 12 × 9 inch dimensions.

Example 13.11

We want to estimate the time and cost to gather and side-wire stitch a self-cover summary of court findings. The customer wants 2200 delivered. It is an upright publication with an untrimmed page size of 9 1/2 × 6 1/4 inches, and there are 6 (16 page) signatures to be side stitched (a 96 page book). The customer wants 3 stitches per booklet instead of the usual 2 stitches. The BHR for gathering is $29.80 and is $59.20 for the stitcher.

Solution Use Figures 13.9 and 13.11.

Step 1 Determine total production time to gather signatures:

Time to gather:
2.2 M bkt × 3.70 hr/bkt = 8.14 hr
Time for preparation:
0.20 hr
Total gathering production time:
8.14 hr + 0.20 hr = 8.34 hr

Step 2 Determine total production cost to gather signatures:

8.34 hr × $29.80/hr = $248.53

Number of Folded Signatures	Stitching Time in Hours per M Booklets (12 × 9 in. and under; 2 stitches)
3	1.70
4	1.80
5	1.90
6	1.95
7	2.00
8	2.05
9	2.10
10	2.20

Note: Preparatory time is 0.20 hr. All times are decimal hours.

Penalties: Add 10% to total time for each additional stitch over two.
Add 15% to total time if untrimmed page size is over 12 × 9 in.

Figure 13.11

Side-wire stitching schedule (foot-operated machine) (Reproduced with permission of Printing Industries Association, Inc., of Southern California from their *1988–89 Blue Book of Production Standards and Costs for the Printing Industry*)

Step 3 Determine total production time to side-wire stitch signatures:

Time to stitch:
2.2 M bkt × 1.95 hr bkt = 4.29 hr
Time for preparation:
0.20 hr
Total stitching production time:
4.29 hr × 0.20 hr = 4.49 hr
Penalty for third stitch:
4.49 hr × 0.10 = 0.45 hr
Total production time:
4.49 hr + 0.45 hr = 4.94 hr

Step 4 Determine total production cost to side-wire stitch signatures:

4.94 hr × $59.20/hr = $292.45

Step 5 Determine total production time and cost to gather and stitch the booklets:

Total production time:
8.34 hr + 4.94 hr = 13.28 hr
Total production cost:
$248.53 + $292.45 = $540.98

Semiautomatic Saddle Stitching Equipment

Christensen and Rosback are combination systems wherein both gathering and saddle stitching phases of booklet and pamphlet production are incorporated as one in-line process. This type of equipment is popular because its cost is moderate yet its production output is generally quite high.

Figure 13.12 is provided for estimating using a Christensen or Rosback system. The chart basis is hours per 1000 booklets gathered and stitched with two stitches; trimming of stitched products is completed separately. The estimator must determine the number of sections (signatures) gathered to locate the hours per M booklets for that number of sections. Add preparatory time as applicable. The cover should be counted as one signature (section).

Automatic Saddle-Wire Stitching Equipment

Figure 13.13 is provided for estimating automated saddle stitching equipment. All gathering of signatures, stitching, and trimming is completed automatically as an in-line operation. Even though equipment configurations vary, this type of machinery, especially with three-knife trimming, is found in most large trade binderies and book manufacturing plants where volume is high. Of course, such equipment is also quite expensive as a capital investment.

The schedule basis is hours per 1000 booklets gathered, stitched, and trimmed, cross-referenced by the quantity level of the job. In addition, the estimator must identify the number of sections of the machine that will be used (the number of signatures to be gathered), counting the cover as one section (signature). Then, the hours per 1000 booklets figure can be located and preparatory time can be added as applicable.

Example 13.12

We want to estimate the time and cost to automatically gather and stitch a 8 page self-cover booklet in our plant using our McCain saddle binding system. The booklet was printed and folded with 3 (16 page) signatures in an upright format, with an untrimmed page size of 11 1/4 × 8 3/4 inches. It will require 3 stitches.

Number of Sections (sig)	Stitching Time in Hours per M Booklets
1	0.2857
2	0.3571
3	0.4167
4	0.4545
5	0.5882

Note: Preparatory time is 0.30 hr for first section and 0.10 hr for each additional section. All times are decimal hours.

Figure 13.12

Christensen/Rosback saddle-wire stitching schedule (semiautomated stitching equipment)

	Stitching Time in Hours per M Booklets					
Quantity	1 Sig	2 Sig	3 Sig	4 Sig	5 Sig	6-8 Sig
5 M-24,999	0.2000	0.2222	0.2500	0.2857	0.3333	0.4000
25 M-49,999	0.1428	0.1667	0.1818	0.2000	0.2222	0.2500
50 M-100,000	0.1250	0.1333	0.1428	0.1538	0.1667	0.2000

Note: Preparatory time is 0.60 hr for first section and 0.20 hr for each additional section. All times are decimal hours.

Figure 13.13

McCain/Harris/Mueller/Macy automatic stitching schedule (large equipment for saddle stitching with three-knife trimmer)

The BHR for the unit is $72.20. We must deliver 5200 copies. Compare the answers with those of Example 13.10, which are for manually gathering and stitching this job.

Solution Use Figure 13.13.

Step 1 Determine total production time to complete job:

Gathering and stitching time:
5.2 M bkt × 0.25 hr/M bkt = 1.30 hr
Time for preparation:
0.60 (1st) + 0.20 (2nd) + 0.20 (3rd) = 1 hr
Total production time:
1.30 hr + 1 hr = 2.30 hr

Step 2 Determine total production cost to complete job:

2.30 hr × $72.20 = $166.06

Step 3 Compare your answers with those of Example 13.10:

Total production time difference: 16.72 hr versus 2.30 hr yields 14.42 hr less using fully automated McCain equipment.
Total production cost difference: $664.50 versus $166.06 yields $498.44 less using fully automated McCain equipment.

Perfect Binding for Books and Booklets

Perfect binding is a bookbinding procedure using glue and preprinted flexible covers to bind a publication. Telephone directories, catalogs, and paperback books are examples of perfect-bound publications.

The binding process is an in-line finishing operation completing collating, gluing, and trimming. Pockets, or stations, hold prefolded signatures that are typically perforated during folding. All the signatures for one finished booklet are collated automatically and then held tightly by a moving clamp system through a saw unit that cuts off from 1/16 to 1/4 inch of the spine (binding edge) of the signatures. Immediately after the sawing procedure is completed, with the booklet still tightly clamped, the signature moves across a gluing section (or wheels) that applies a glue mixture, after which a cover is wrapped to the glued area. Thus, the glue holds together all the pages of the publication and the cover to the outside of the book as well. Three-knife trimming is typically done in-line following cover attachment, and the book or booklet is completely finished and ready to be wrapped.

Perfect binding equipment is available with from 4 pockets to more than 20 pockets for signature collating and can typically handle books or booklets from 6 × 4 × 1/8 inches to those 20 × 12 × 2 inches. If a cover scoring unit is purchased, it can 2-score (two scored lines directly at the place where the edges of the cover meet the spine) or 4-score (four scored lines so that the outer scored lines hinge the cover away from the spine). Signatures must be properly imposed and folded with perforation along the binding edge to remove air easily during the binding process. Typically, signatures are from 4 to 32 pages on book stock of normal caliper. Covers, which are preprinted, are cut to the untrimmed size of the booklet and must be prescored if no scoring unit attachment is purchased with the perfect binder. Production problems and slowdowns during production generally relate to signature collating when signatures are too thin, when the type of stock is hard to feed, or when the printed stock has been vanished or is too glossy. Cover insertion and registration may be a problem when the vanish or gloss coating does not permit good control.

The schedule in Figure 13.14 is provided to estimate perfect binding production time. As with all schedules in this book, the indicated times are for example only and should be adjusted for individual plant production.

Example 13.13

Using our Muller Martini Pony 3000 perfect binding machine, we must determine the production time and cost to perfect bind 9500 chemistry books (bk). They have been run on basis 50 book in 4 (32 page) signatures with an untrimmed page size of 11 1/4 × 8 3/4inches. The cover is varnished, but will

Number of Sections (sig)	Binding Time in Hours per M Books or Booklets
1-3	0.14
4-6	0.15
7-10	0.17
11-14	0.18
15-17	0.20
17-20	0.22

Note: Preparatory time is 0.10 hr per pocket used plus 0.20 hr for cover insertion unit. All times are decimal hours.

Penalties: Add 25% to total time for signatures under 16 pages. Add 20% to total time if cover is exceptionally glossy or varnished, which may lead to insertion problems. Add 15% to total time if signatures are printed on coated stock or if signatures may represent collating problems.

Figure 13.14

Perfect binding production schedule

feed well since we have used this kind of cover and vanish before. The BHR for the Muller is $67.25.

Solution Use figure 13.14.

Step 1 Determine production time to perfect bind the job:

9.5 M bk × 0.15 hr/M bk = 1.43 hr

Step 2 Determine preparatory time:

Time for pocket (pk):
4 pk × 0.10 hr/pk = 0.40 hr
Time for covers (cv):
0.20 hr/cv
Total preparatory time:
0.40 hr + 0.20 hr = 0.60 hr

Step 3 Determine total perfect binding production time:

1.43 hr + 0.60 hr = 2.03 hr

Step 4 Determine total production cost to per the job:

2.03 hr × $67.25/hr = $136.52

Step 5 Summarize your findings:

Total perfect binding time: 2.03 hr
Total perfect binding cost: $136.52

Pamphlet and Book Trimming

Split Back Gauge Cutting Machines. Figure 13.15 is provided for estimating trimming out production of booklets and pamphlets. Schedule data apply to cutting machines with split back gauges and cutters with electronic or mechanical spacing. Production is based on a single operator. The schedule basis is the number of lifts cut per hour of production. Determine the number of booklets in a lift by dividing the number of sheets of paper in one booklet into the approximate number of sheets per load, as indicated in the schedule. For example, a 32 page booklet contains 16 sheets of paper. If substance 70 wove finish bookstock (caliper 0.005 inch) is the substrate, then 800 sheets of paper represent the lift height for production (see schedule). Dividing 800 sheets per lift by 16 sheets per booklet shows that each lift will contain about 50 booklets. At a trimming rate of 60 lifts per hour (see schedule), approximately 3000 of these booklets can be trimmed out during 1 hour of production. If the job consists of 20,000 copies, trimming time will be approximately 6.7 hours.

Three-Knife Trimmers. High-production commercial printing companies and most trade binderies typically have three-, four- and five-knife trimmers that are used to cut

Caliper and Kind of Paper	Approximate Sheets per Load	Lifts Cut per Hour
0.0025 in. 13# Bond, 50# Coated Book	1600	60
0.0035 in. 70# Coated Book	1200	60
0.0040 in. 20# Bond, 80# Coated Book, 60# Wove Finish Offset Book	1000	60
0.0050 in. 24# Bond, 100# Coated Book, 70# Wove Finish Offset Book	800	60
0.0060 in. 80# Wove Finish Offset Book, 60# Coated Cover	700	60
0.0080 in. 80# Coated Cover, 50# Antique Finish Cover, 90# Bristol	500	60
0.0110 in. 65# Antique Finish Cover, 150# Tag, 140# Bristol	400	60

Figure 13.15

Pamphlet and book trimming schedule (single-knife split back gauge or spacer cutter) (Reproduced with permission of Printing Industries Association, Inc., of Southern California from their *1988–89 Blue Book of Production Standards and Costs for the Printing Industry*)

Booklet Finish Size in Inches	Trimming Time in Hours per M Pieces
Up to and including 6 × 9	0.14
Over 6 × 9 and up to and including 8½ × 11	0.17
Over 8½ × 11	0.19

Note: Preparatory time is 0.20 hr. All times are decimal hours.

Penalties: Add 25% to total time for booklets containing 16 or fewer pages. Add 20% to total time for odd size books, booklets, and pamphlets.

Figure 13.16

Pamphlet and book trimming schedule (three-knife trimmer)

apart and trim books, booklets, and pamphlets after binding. Generally, such units are set up to operate in-line with stitching, sewing, or perfect binding units, thereby expediting production since the final product is fully finished and ready for boxing or wrapping.

Figure 13.16 is provided to estimate book, booklet, and pamphlet production using a three-knife trimmer. Speed is determined by the size of the finished book, with penalties for very thin booklets and pamphlets and for odd-sized publications. Schedule data are presented for example only and should be modified for individual plant production.

Example 13.14

We want to estimate the time and cost to three-knife trim a 48 page self-cover booklet using plant equipment. The booklet was printed and folded with 3 (16 page) signatures in an upright format with a trimmed page size of 11 × 8 1/2 inches. Add three-knife trimming production time to the automatic saddle-wire stitching time in Example 13.12 to determine the total cost for this job. The BHR for the three-knife trimmer is $52.70.

Solution Use Figure 13.16.

Step 1 Determine total production time to use a three-knife trimmer on the job:

Time to trim:
 5.2 M bkt × 0.17 hr/M bkt = 0.89 hr
Time for preparation:
 0.20 hr
Total production time:
 0.89 hr + 0.20 hr = 1.09 hr

Step 2 Determine total production cost to trim:

1.09 hr × $52.70/hr = $57.44

Step 3 Determine total production time and cost to stitch and trim the booklets (Example 13.12 and Example 13.14):

Total production time:
 2.30 hr + 1.09 hr = 3.39 hr
Total production cost:
 $166.06 + $57.44 = $223.50

Note: If the job was done in-line, production time would be 2.30 hr.

13.4 Estimating Banding, Wrapping, and Mailing Operations

The final area of binding and finishing to be estimated involves the concluding operations before shipment of the product. These operations include four separate procedures: banding and boxing (Figure 13.17), kraft and corrugated wrapping (Figure 13.18), shrink-wrapping (Figure 13.19), and mailing operations (Figure 13.20).

Type of Banding and Type of Package	Banding Time in Hours per M Packages		
	Under 25 Inches*	26–50 Inches*	Over 50 Inches*
Kraft paper band (firm, solid package wrapped with precut bands)	4.00	5.00	6.50
Kraft paper band (spongy package with precut bands)	6.50	8.00	9.25
Rubber bands: Single	2.20	2.50	2.90
Rubber bands: Double	4.00	4.50	5.00
String-tied packages (completed by hand)	5.00	5.70	6.60
String-tied packages (completed with tying machine)	2.80	3.10	3.50

Note: Preparation time is 0.10 hr. All times are decimal hours.

*Package girth plus length.

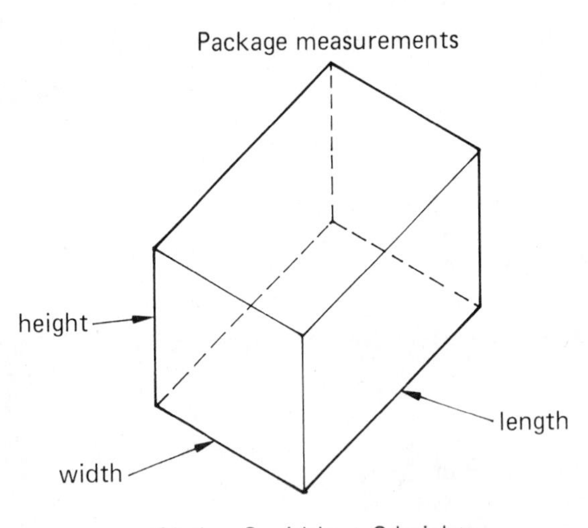

Figure 13.17

Banding and boxing schedule

Cubic Inch Volume (width × length × height)	Wrapping and Sealing Time in Hours per Package: Kraft Paper Roll and Paper Tape	Corrugated Material for Shipment; Kraft Wrap Also Included
Under 120 cubic inches (4 × 6 × 5, 3 × 5 × 6, or less)	0.020	0.033
121-200 cubic inches (4 × 6 × 6, 5½ × 8½ × 4, etc.)	0.022	0.040
201-350 inches (8½ × 11 × 3, 5½ × 8½ × 8, etc.)	0.025	0.050
350-500 cubic inches (8½ × 13 × 4, 8½ × 11 × 5, etc.)	0.029	0.067
501-600 cubic inches (11 × 14 × 3, 12 × 18 × 2, etc.)	0.033	0.100
Over 600 cubic inches (11 × 14 × 5, 19 × 25 × 3, etc.)	0.050	0.125

Note: All times are decimal hours. Schedule data assume reasonable package weight easily handled by one person.

Penalties: Add 15% if the contents of package(s) to be wrapped are spongy or unfirm.

Figure 13.18

Kraft and corrugated wrapping schedule

Type of Item	Shrink-Wrap Time in Hours per Package
Loose Sheets in Easy-to-Handle Amounts/Lifts	
Up to and including 6 × 9 in. (54 sq in.) finish size	0.003
6 × 9 in. up to and including 10 × 15 in. (150 sq in.) finish size	0.004
Individual Booklets	
Up to 10 × 12 in. (120 sq in.) finish size and 64 or fewer pages	0.003
Up to 10 × 12 in. (120 sq in.) finish size and more than 64 pages	0.004
Packages of Books or Booklets in Easy-to-Handle Amounts	
Up to and including 5½ × 8½ in. (48 sq in.) finish size	0.005
5½ × 8½ in. up to and including 10 × 12 in. (120 sq in.) finish size	0.006

Note: Preparatory time is 0.10 hr. All times are decimal hours.

Figure 13.19

Shrink-wrapping schedule (semiautomatic equipment)

Banding and Boxing and Paper Wrapping

Both banding and boxing and wrapping operations, estimated with Figures 13.17, 13.18, and 13.19, are common procedures prior to the delivery of the job to the customer. Of course, banding and boxing or wrapping should not be estimated when incorporated as a portion of an earlier bindery procedure, such as the rubber banding and boxing of pamphlets during folding. In instances when banding and boxing or wrapping represents a necessary portion of the order—for example, when the customer specifically requests shrink-wrapping of letterheads—it should be included in the estimate as a separate operation. Obviously, plant policy should dictate such packaging procedures.

Shrink-Wrapping Production

Figure 13.19 is provided for estimating shrink-wrapping production using semiautomated equipment. Typically, the sealing of the shrink-wrap film is done manually by the operator, after which the package is placed on a conveyor transport that moves it through the heating unit that causes the film to shrink and seal the package tightly. Different models and speeds of equipment provide different production output. Thus, the figures indicated in Figure 13.19 are for example only.

Example 13.15

We have been asked to shrink-wrap 3000 (120 page) books using our semiautomatic equipment. They have been perfect bound and are finished at 8 1/2 × 11 inches. The customer wants them shrink-wrapped in groups of 12 per package (pkg). The BHR for our shrink-wrap machine is $31.50. Determine time and cost to shrink-wrap this job.

Solution Use Figure 13.19.

Step 1 Determine total production time to shrink-wrap:

Number of packages:
3000 bk ÷ 12 bk/pkg = 250 pkg
Time to shrink-wrap:
250 pkg × 0.006 hr/pkg = 1.5 hr
Time for preparation:
0.10 hr
Total production time:
1.5 hr + 0.1 hr = 1.6 hr

Step 2 Determine total production cost to shrink-wrap:

1.6 hr × $31.50/hr = $50.40

Mailing Operations	Hours per M Complete Pieces	Average Production per Hour (number of pieces)
Seal envelopes by hand	1.00	1,000
Seal envelopes by machine	0.25	4,000
Affix stamps to ordinary envelopes by hand	1.30	770
Affix stamps to ordinary envelopes by machine	0.25	4,000
Attach wire clips	1.80	555
Insert 1 piece in envelope, flap tucked	2.00	500
Insert 1 piece in envelope, flap tucked and stamped by hand	3.30	300
Insert 1 piece in envelope and sealed by hand	2.70	370
Insert 1 piece in envelope, sealed and stamped by hand	4.00	250
Add for each additional piece enclosed	0.80	125 ea. piece
Insert pamphlets in clasp fastener envelopes	4.00	250
Insert catalogs in pockets	4.00	250
Insert pamphlets in string fastener envelopes	3.90	256
Insert one poster in mailing tube		
Up to 22 × 28 in.	10.50	95
Up to 25 × 38 in.	11.60	86
Up to 32 × 44 in.	12.80	78
Up to 38 × 50 in.	14.00	71
Affix addressed labels to mailing tubes	3.20	313
Address on typewriter		
1 line addresses	5.00	200
2 line addresses	5.80	175
3 line addresses	7.10	140
4 line addresses	10.00	100
Address by hand		
1 line addresses	4.10	245
2 line addresses	7.40	135
3 line addresses	8.00	125
Address by machine (Addressograph)	0.17	6,000
Address on Wing-Horton or Dick Mailer, 3 line addresses	0.50	2,000
Affix postage stamps to mailing tubes	2.30	434
Seal with gummed seals or precancelled stamps	2.80	358
Wrap 1 piece, rolled	2.70	370
Wrap 1 piece, flat	3.00	330
Sort permit mail	1.50	666

Note: All times are decimal hours.

Figure 13.20

Mailing operations schedule (Reproduced with permission of Printing Industries Association, Inc., of Southern California from their *1988–89 Blue Book of Production Standards and Costs for the Printing Industry*)

Step 3 Summaraize your findings:

Total production time: 1.6 hr
Total production cost: $50.40

Mailing Operations

Figure 13.20 provides estimating data for estimating mailing operations, of which the major production elements include inserting, addressing, and sorting mail to be sent through normal post office channels. Depending upon the setup of the binding and shipping facilities of the plant, mailing may or may not be considered a portion of the bindery area. Nevertheless, mailing operations have been included here as part of the bindery process, along with banding and wrapping, since they are concluding procedures and may utilize, or at least share, binding and shipping floor space.

Figure 13.20 is based on the number of hours per 1000 completed pieces; average production output per hour is also indicated. Select the mailing operations to be completed and then determine the number of pieces subject to such mailing procedures. Multiply the two to determine the required time for mailing.

Estimating and Pricing for the Quick Printing and Small Commercial Printing Company

14.1 Introduction

The primary focus of the book has, thus far, been toward the typical medium-size or larger commercial printer, manufacturing four-color process and more complex color work through the integration of conventional and electronic technology. However, there is no doubt today that the impact of both the higher-volume printer and the lower-volume "quick" printer, or small commercial printer, is important to understand the changing dynamics of the industry as we enter the next decade and move toward the next century. In fact, it seems fairly safe to predict that as each of these industry segments further clarifies its market niches and customers, each will grow largely by taking work away from the typical commercial printer. Thus, given no significant expansion in existing markets—in fact, a decline seems likely, based on increasing foreign competition and a shrinking U.S. population into the next decade and beyond—both the quick printer and high volume printer will compete intensely with commercial printers to increase their relative share of the market.

This chapter addresses the estimating and related costing and pricing needs of the quick printer and smaller commercial printer. It could be argued that the approach to estimating, pricing, and costing as previously detailed in this text is too complex and too demanding for a small commercial plant or quick printer. To some extent this may be true. Yet, the profitable operation of any printing company of any size—particularly the quick printer facing intense competition with other printers and philosophically using a less-than-ideal entrepreneurial profit motive—requires good business practices. This means the prudent quick printer, or small commercial printer, must have a focus on costs, prices, and profit, on methods and procedures by which cost recovery is completed, on the relationship between actual manufacturing procedures and actual costs, on estimating methods used to evaluate production costs before the job enters production, and on computers which tie all the elements of the business together, including accounting, estimating, billing, and purchasing. In fact, for any quick printer or small commercial printer wishing to computerize, a knowledge of cost estimating as presented in this text is essential.

The terms "quick printer" and "small commercial printer" will be used interchangeably throughout this chapter, even though there are subtle distinctions between them. For example, both serve largely the same customer bases and markets, work with the same sizes and kinds of equipment, generally hire and use the same caliber and type of employees, and typically begin with one small outlet. Differences between the two types of shops could be broken down essentially into two general observations: First, while some (especially small and new) quick printers are less focused as to where they "fit in" in the printing industry, shops claiming to be "small commercial printers" attempt to mirror their larger commercial printing counterparts in terms of manufacturing attitude, management philosophy, and anticipated growth. Second, there does seem to be a trend that as quick printers begin to expand their products and services and grow in sales volume and size, they begin to more philosophically view themselves as "commercial" printers, performing increasingly complex printing work, and not necessarily seeking out or focusing on producing only fast-turnaround products.

14.2 Overview of the Quick Printing Industry

Brief History

In the early 1960s, Itek Corporation, working with Eastman Kodak, developed the first "photo-direct" platemaking system. Itek manufactured the photo-direct "daylight" camera—the Itek 1015—which allowed original art to be reduced or enlarged and imaged to a silver halide plate in a one-step process; Kodak manufactured the photosensitive plate material. Multilith, AB Dick, and other small-format presses were adapted to accept this new plate technology and thus fit into this production area.

During the 1960s and into the 1970s, the quick printing industry slowly evolved. For a fairly modest investment, any start-up company could buy an Itek platemaker and a Multilith or AB Dick press and begin a printing business. The term "quick printer" was coined sometime during the early- to mid-1970s. In 1975 George Pataky founded the National Association of Quick Printers (NAQP). Based on NAQP data, in 1969 there were about 1,100 quick printers, by 1976 that number had grown to over 6,500 and by 1987 there were an estimated 24,000 quick printing firms in the U.S.

Two primary market forces have been at the root of the success of quick printing from its modest beginnings to the approximately $6 billion in estimated sales in 1988: fast-turnaround of printing orders—literally on a "while-you-wait" basis with many quick printers—and geographic convenience to customers. Both of these are customer-oriented issues and provide perhaps the primary reason for the significant growth and success of the quick printer. The quick printer realized, either by luck or wisdom or perhaps a little of both, that the key to success was to focus on products and services to meet the true needs and desires of the customer. This marketing tactic was generally ignored by larger commercial and high-volume printers until the late 1970s, perhaps inspired to some degree by the growth and dynamic structure of the quick printing industry.

It is generally acknowledged that the quick printer has not been an accepted part of the overall printing industry over the years. When quick printers began with the photo-direct Itek 1015 and small press printing in the 1960s, the general view expressed was that this type of printing would not survive since the quality of product was "low" and customers had no need for such immediate delivery. That view continued to be held by a declining number of industry observers, even as the industry began to mature in the 1970s, with improving equipment and a growing foothold in terms of customer satisfaction and profitability. In the 1980s the quick printer and smaller commercial printer came of age, serving specific markets and customers, and attracting a growing amount of printing—including four-color printing—which previously was considered only in the domain of the larger printer.

Today, a major problem in segmenting "quick printing" from "commercial printing" is defining where the quick printer title stops and the larger commercial printer title begins. Some define this by the size of press, preferring to relate the term quick printer as a "small press printer." Others continue to refer to quick printing as "rush" production, although many larger commercial printers are now performing a significant portion of their volume on a fast-turnaround basis. Still others make a quality distinction stating that quick-printed products offer "good enough" quality, as opposed to fancier and more complex graphically-produced materials. Still others define "small" by the length of run on a press in that longer runs characterize larger printers. In sum, when studying larger quick printers and commercial printers of equal size by sales volume, there appear to be few differences between the businesses. As the quick printer's company size gets smaller, its impact on the higher-volume commercial printer or high-volume web printer appears less pronounced.

Industry Structure: Independents and Franchises

There are essentially two categories of quick printing firms today—those which have been established as an independent business with no aid or support from an outside source such as a parent corporation, and franchise firms, whereby any interested person or persons is able to purchase equipment, training, and company support from a parent franchiser to open their own quick printing business using the franchiser's name.

Looking at the global perspective based on NAQP and *Quick Printing Magazine* data, in 1987 there were approximately 24,000 quick printers. Of these, approximately 6,000 quick printers—about 25 percent—were operating under franchise license. Independents, then, represent approximately 75 percent of the quick printing industry.

Figure 14.1, "Franchise Facts," is reproduced from an article in the May 1989 issue of *Quick Printing Magazine*, based on franchiser responses. It is interesting to note that six of the fifteen franchiser companies began in the 1960s, including Copies Now, Insty-Prints, Kwik-Copy, PIP, Quik Print and Sir Speedy. As of May 1989, the six 1960s franchisers represented 3,584 quick printing store locations, out of the total number of franchise-based quick printer shops of 6,035, or about 60%. Figure 14.1 also provides data related to the number of shops, the minimum start-up capital needed, total investment required and—of particular interest—average estimated sales per shop for firms over one year old.

Franchisers indicate that the advantages they offer include, and are not limited to, full start-up support, extensive, continuous and thorough training, a regional or national "image" through various advertising media, fancy store design, discounts on equipment and supplies through franchiser purchasing power, and continuous man-

FRANCHISE FACTS

Franchise	Year Franchising Began	Number of Shops Open		Franchise — Owned	Co. — Owned	Number of States	Number of Countries	New Shops in 1988	Re-sold in 1988	Closed in 1988	Minimum Start-up Capital Needed Cash* ($000)		Total investment required ($000)*		Estimated Sales per shop (average - shops over 1 year old) ($000)		
		5/88	5/89								1988	1989	1988	1989	1987	1988	1989
AlphaGraphics	1970	244	281	236	8	33	6	48	14	3	35	47	269	250	408	490	564
American Speedy	1977	469	588	588	0	41	2	140	35	29	30	30	127	133	N/A	N/A	N/A
Copies Now	1968	44	51	51	0	11	1	18	1	1	30	30	60	60	N/A	N/A	N/A
CopyMat	1985	48	57	56	1	1	1	5	22	0	120	130	230	250	556	·552	580
Franklin's	1977	85	95	95	0	14	1	10	0	0	45	50	180	184	N/A	N/A	N/A
Insty-Prints	1967	376	388	385	3	45	2	12	0	0	75	75	151	190	193	250	250
Kwik-Kopy	1967	1021	1013	1012	1	41	6	54	30	62	70	70	130	130	244	N/A	N/A
Minuteman	1975	800+	900+	ALL	0	44	2	100+	N/A	N/A	32	32	82	82	N/A	N/A	N/A
PIP	1968	1167	1224	1216	8	49	3	71	42	N/A	35	35	107	107	N/A	N/A	N/A
Printmasters	1977	80	105	105	0	3	1	26	4	1	35	40	150	150	N/A	N/A	N/A
Print Shack	1983	106	113	113	0	32	1	19	7	13	45	52	99	103	N/A	N/A	N/A
Print Three	1984	105	155	155	0	16	2	50	0	0	40	50	174	184	279	300	420
Quik Print	1963	194	200	105	95	25	1	14	1	1	73	30	100	125	N/A	N/A	N/A
Sir Speedy	1968	783	815	815	0	45	3	57	36	25	45	50	156	165	300	N/A	N/A
The Ink Well	1981	41	50	47	3	6	1	9	5	0	30	30	105	110	155	180	198

Figure 14.1

Franchise Facts (Reproduced with permission of *Quick Printing Magazine*)

agement and technical support. To provide this, the quick printer typically pays a one-time start-up cost and then an annual royalty fee of between 3% and 8% of gross sales dollars. Yet despite the fairly extensive listing of services and support provided by a typical franchiser, not all franchisees are happy. Some franchisee quick printers feel that their franchiser does not provide the promised continuous, excellent training, nor advertising support, nor substantial discounts on equipment, etc.; other franchisees feel that this support is provided, but that it is marginal at best, and does not even closely match the royalties paid to the franchiser.

The independent quick printer, who represents the majority of the industry, typically functions relative to the size of the shop. If the independent quick printer is small in size—and many quick printers have less than ten employees—they operate much like a small, entrepreneurial printing firm as described in Chapter 2 of this text. Larger quick printers—some of which claim to be "small commercial printers" or "combination printers"—number perhaps only 50 out of over 24,000 in the quick printing industry. These firms practice investment community standards, also described in Chapter 2, and have moved away from operating on a more casual business style. Based on *Quick Printing Magazine's* "Top 100" listing in the June 1989 issue, the largest quick printer was an independent with 1988 sales of $22 million, while the 50th largest had 1988 sales of $2.41 million.

Another dimension of the quick printing industry is the number of multiple shops owned and operated by a single owner. Based on NAQP's 1987 *Operating Ratio Study*, 105 shops of the 649 respondent companies (approximately 16 percent) were multiple locations, with the average multiple shop owner having 3.1 different locations. In *Quick Printing Magazine's* "Top 100" listing, the top 10 firms—a mix of independents and franchises—totaled 275 locations, with one company having 95 operating outlets, while another company in the "Top 10" listing only eight operating stores.

Changing Dynamics of the Quick Printing Industry

Most experts and quick printing insiders believe a "shake-out" in the industry is inevitable. Many observers note that as quick printing has become recognized as a profitable venture and a good way to get into a continuous-demand business, there are simply too many locations vying for the same customers, offering the same services, at approximately the same prices. This is further complicated by the fact that some small commercial printers or "combination" shops (a mixture of commercial printing and quick printing production) originally attempted to grow through competing with their larger commercial printing counterparts in the production of complex color work, which they have since abandoned to pursue quick printing as their focused market niche.

When faced with increasing competition, the typical reactions in a quick printing company include doing one or more of the following:

1. Offer customers an expanded array of services
2. Attempt to enlarge the customer base through increased (usually "outside") sales efforts
3. Cut prices

Invariably such intense competition has a ripple effect through any business community, much like a "gas war" where competing stations cut gasoline prices to increase volume. Ultimately as competition increases, the strongest survive and the weakest either go out of business or are merged or acquired by the competition.

While too many locations, too much competition, increasing services, expanding outside sales and price cutting are signs of possible industry instability, the following represent other important factors likely to affect quick printing in the next five years, and likely to contribute to the "shake-out" as well:

More customers doing their own black-and white copying—The convenience of the single-color home copier such as the Canon PC series, along with the purchase and implementation of similar stand-alone high speed copiers for businesses, have effectively removed an essential portion of this copy volume from the quick printer. Furthermore, it is likely that more and more such machines will be purchased and used in thousands of homes and businesses, further reducing this copying market segment from the quick printer.

Fear of four-color offset printing technology—Many quick printing experts—based on numerous sources, studies and discussion—predict that the next major thrust for quick printing is toward short run, fast-turnaround, multicolor offset lithographic production. While some quick printers have taken the plunge into multicolor lithographic printing through the purchase of multicolor presses, thousands of other quick printers fear entering this domain. The reasons cited for for this are varied. Some indicate that the four-color lithographic printing process is too complicated and complex. Other quick printers believe it would be hard to manage, technically difficult to control, or would require significant investment which they do not want to make given insufficient data to determine the return on such an investment. To counter this fear and yet provide the four-color services requested by some customers, some quick printers have formed close ties with local commercial printers to buy this work on a brokered basis.

In an article in the March 1989 issue of *Instant & Small Commercial Printing* (Figure 14.2) printing shops which designed themselves as "small commercial" were almost four times as likely to be doing some process color on the premises compared to quick printing shops. Furthermore, the article reported that "instant" or quick printers of all sizes of sales volumes lagged behind "small commercial" printers in terms of doing color work of either multiple color or process color. "Combination" shops, which stated they took in both quick printing and commercial work, fell largely in the middle, with an average of 17 percent completing on-site process color.

It is important to note that laser color copying has rapidly brought four-color production into the quick printing company. Images are generated using simple

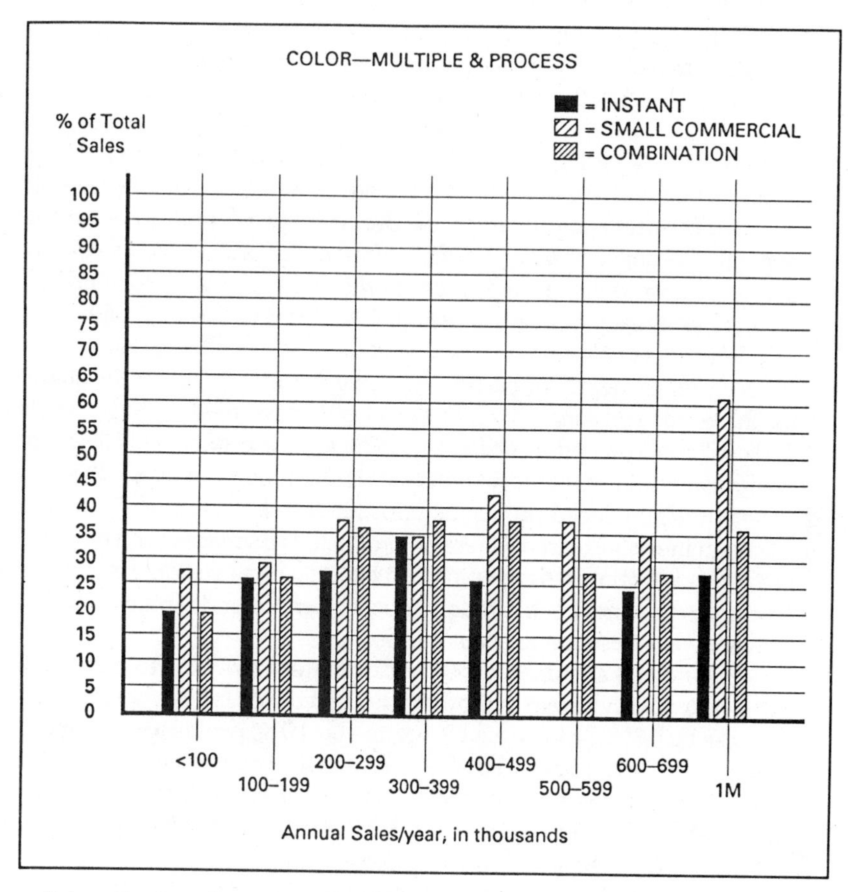

Below: Number of respondents in each category reporting in-house process color printing.

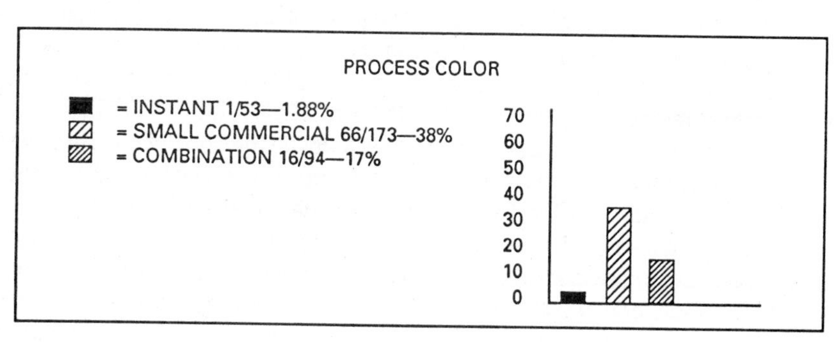

Figure 14.2

Analysis of process color done by quick printers, small commercial printers, and combination shops. (reproduced with permission of *Instant and Small Commercial Printer*)

operator controls, approximately one copy every two minutes. At this writing, costs per copy are high, but will likely drop somewhat with increased volume and competitive pressures. Whether the color copier is here to stay or simply an evolution that will slowly be tried and discarded (as happened with previous electrostatic color copiers) remains to be seen. Regardless, color copying is limited by size and the quality of color reproduction, two disadvantages not evident with four-color lithographic production.

Incomplete knowledge of the effects of cost, volume and price—In general, the quick printer, or small commercial printer, is unaware of the dynamic nature of cost, volume, and price, particularly as related to market elasticity. This subject has already been addressed in Chapter 2 in this text, with a discussion of the *1966 McKinsey Report to PIA*. As Figure 2.2 graphically demonstrates, if a typical lower-profit printer raises his prices by only 3% and customers are able to go elsewhere for the goods and services—called an elastic market—the printer would experience a 10% volume loss and a subsequent 50% loss of profits. This illustration shows that doing business in an elastic market place is highly competitive, with little pricing latitude and generally reduced profits, as opposed to the printer who has moved into a less elastic (or inelastic) market where his products and services can't be so easily duplicated. For the quick printer, developing and keeping loyal customers and encouraging new customers to come back through effective service and targeted, desired products are primary ways this can be accomplished.

Figure 2.3 shows that if a typical printer were able to reduce his material cost 10%, he would increase his profits by 66%; reduce his labor cost by 10%, profits would increase by 56%; or increase his volume by 10%, profits would jump 33%; however, if prices can be increased 10%—in an inelastic market where the customer can't run away—then profits soar to 180%. This illustration demonstrates clearly that cost modifications in labor and materials have some effect on profitability, as does volume, but that price is really the key to making significant income in the printing industry.

Not knowing when or how to computerize—The issue of computerization is of growing interest and concern to the quick printing industry in general. To some quick printers, particularly small, single location firms, a computer is a mystery tool that they can well do without. Yet as the quick printer grows in size and expands in the number of locations, the question of when and how to computerize looms large on their immediate agenda. Regardless of size, problems relative to computerization in quick printing relate to a general fear of computers by the shopowner, inability to accurately identify company needs, and a general confusion as to which system would best serve both their short- and long-term requirements. There are no quick answers to these issues. However, it is safe to say that any quick printing business, to grow, needs to develop customer and employee databases, needs accounting and other such financial information, and needs some method to track more complex jobs and work-in-process. Computers provide this, yet are still a tool largely unused by many quick printers. This subject has been addressed in Chapter 5 of this text and will be discussed specific to the quick printer later in this chapter.

Fear of change from a profitable business to one that is less profitable—Any responsible manager of a business is concerned when the profit for the business is on the decline. How should one react? What steps should be taken to reverse the situation? Should there be an immediate reaction which may have the greatest impact, or would such a knee-jerk reaction be foolish? The questions go on and on. The key—the essential element—is not to panic, not to be afraid, not to be fearful. Yet many smaller printers and quick printers—and perhaps some larger printers as well—are at a loss as to how to deal with the issue of declining profitability. Fear sets in, the panic button is pushed, mistakes are made and, in some cases, the business is lost. Survival of the "shake-out" will probably be linked to how well quick printing management responds to the challenges of reduced margins, price cutting, and lower volumes.

14.3 Overview of the Quick Printing or Small Commercial Printing Company

The following describes the general operating procedures of a typical independent quick printing or small commercial printing firm. Defining "typical" for purposes of this discussion factors in numerous sources of information and observations. Information and data will be provided from various sources including the National Association of Quick Printers (NAQP), *Quick Printing Magazine* and *Instant & Small Commercial Printer*.

General Business Environment

The overall operation of a typical quick printing establishment is largely a product of the owner's desires, opinions and past experiences. The prudent owner has learned the importance of positive customer contact with a service-oriented focus. The environment of the business provides equipment on a showplace basis, allowing customers to view work as it is produced. As much as possible, the facility is clean and neatly arranged, with ample work areas and counterspace to facilitate taking walk-in customer orders. Jobs enter the production area through direct customer contact and, increasingly, through outside sales. Employees working the counter and greeting the public should be friendly, honest and cooperative.

Products and Services Provided

The types of products and services varies among quick printing firms. Figure 14.3, reproduced from NAQP's 1987 *Member Shop Profile*, provides a listing of printing services and other services. The data is broken down into the actual 1987 percentage

Printing Services	*%NAQP Members Offering in 1987*	*% NAQP Members Offering by end of 1988*	*Net Change*
Black-and-white printing	89.2%	89.9%	+0.7%
One-color printing	87.5	88.2	+0.7
2- and 3-color printing	83.9	85.0	+1.1
Self-service copying	74.2	75.9	+1.7
High-speed copying	65.2	69.5	+4.3
Commercial printing (over 11 × 17 inches)	22.0	26.3	+4.3
Four-color printing	19.9	24.6	+4.7

Other Services	*%NAQP Members Offering in 1987*	*% NAQP Members Offering by end of 1988*	*Net Change*
Typesetting	75.3%	78.4%	+3.1%
Graphic/Design	56.2	57.8	+1.6
Photostats	49.0	51.2	+2.2
Word Processing/Typing	35.9	41.1	+5.2
Mailing Services	31.4	35.7	+4.3
Sell Supplies	25.3	27.5	+2.2
DTP/Laser Printing	23.5	46.6	+23.1
List Maintenance	17.7	22.1	+4.4
FAX/QWIP	12.3	23.7	+11.4

Type of Equipment	*%NAQP Members Having Equipment*	*% NAQP Members Increase by End 1988*
Bindery	91%	7%
Presses	87	17
Platemakers	78	6
Self-service copiers	72	19
High-speed copiers	61	11
Cameras	60	16
Typesetters	53	7
DTP/Laser Printing	23	11

Figure 14.3

Listing of printing services, other services and equipment profile of NAQP members for 1987 and projected to 1988. (Reproduced with permission of NAQP from their 1987 *Member Shop Profile*)

of that service, and then an estimated end-of-1988 percentage change anticipated by NAQP-member study respondents.

Printing Services. As will be noted, literally every type of printing service is expected to increase between 1987 and 1989 with four-color printing showing the most increase, followed with increases in both larger-sheet commercial printing and high-speed copying. It is important to note that black-and-white and 1- through 3-color printing provide the mainstay products for the NAQP membership included in this survey.

This essentially confirms that many quick printers are moving into the four-color printing market, either through manufacturing their own process color work or brokering this work to printers who have the capability and capacity to complete it.

Other Services. Included in Figure 14.3, also from NAQP's 1987 *Member Shop Profile*, is a listing of other (non-printing) services for 1987 and projected growth into 1988. Those areas which will likely experience the most significant growth are desktop publishing with laser printing and FAX/QWIP use. DTP—discussed in detail in Chapter 9, including DTP estimating procedures—represents the most dynamic change in the industry in the past ten years. Installation of DTP/laser printing equipment in the quickprinting company is normally done on a hourly rental basis for the Macintosh or IBM-AT machine, with a set cost per copy from the laser printer. In the typical quick printing firm, customers work at carrels or private workstations which, in many cases, have been comfortably designed for DTP. Popular software is provided by the quick printer with less-known programs supplied by the customer; storage of captured graphic and type images are saved to the customer's own floppy disk. Printing is tracked through the number of laser copies made, at a charge of between $0.25 and $1.00 each.

Facsimile (FAX or QWIP) equipment has become a popular method of telecommunications, as discussed in Chapter 5. While FAX procedures are not new, the convenience and written clarification from sender to receiver, as well as the speed by which such information can be transmitted, has had substantial impact on the way business communication is completed today. The growth of FAX/QWIP as a service to quick printing customers is not really surprising, and will likely continue to increase as more FAX/QWIP machines are sold and go on-line.

Comparing product and service variations among small commercial printers—Figure 14.4, from an article on print marketing, the March 1989 issue of *Instant & Small Commercial Printing*, is a comparison of one group of twelve "small commercial" shops with annual sales of $200,000 to another group of 12 "small commercial" shops with yearly sales of $500,000. The bar graphs clearly show that the shops with higher annual sales volume showed increases in almost every service category from "large copy" to "design services." Furthermore, it must be noted that while these printers classified themselves as "small commercial" and not quick printers, the services listed are largely the same services offered by many quick printers (Figure 14.3). As mentioned previously, the line separating the quick printer from the small commercial printer is difficult to define, yet there does seem to be a relationship relative to the size of shop. As quick printers grow and expand their products and services, they begin to see themselves more clearly in the printing business, and thus appear more likely to define their company as a "small commercial" venture and not a business oriented only to fast-turnaround copying and duplicating.

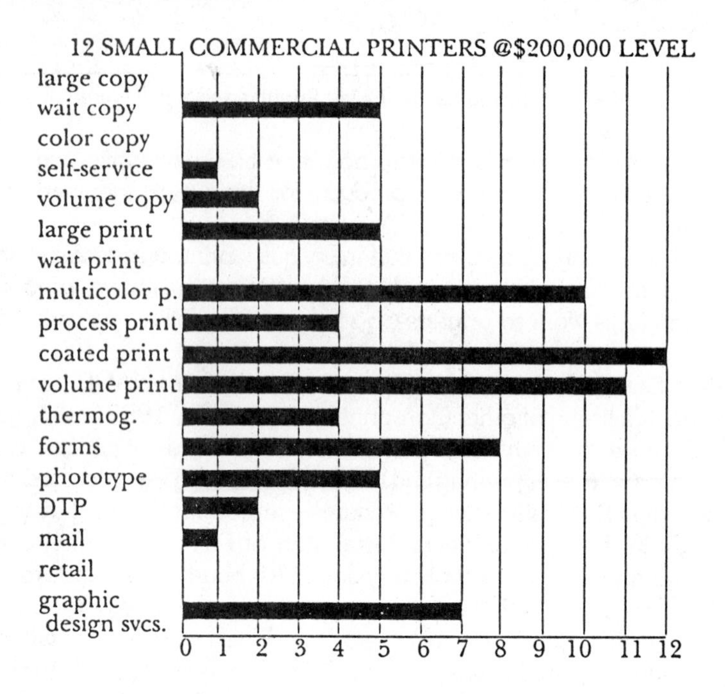

Above and below: Comparisons of respondents at identical sales volumes indicating what they provide in-house, from a choice of 18 products and services (listed at left on charts).

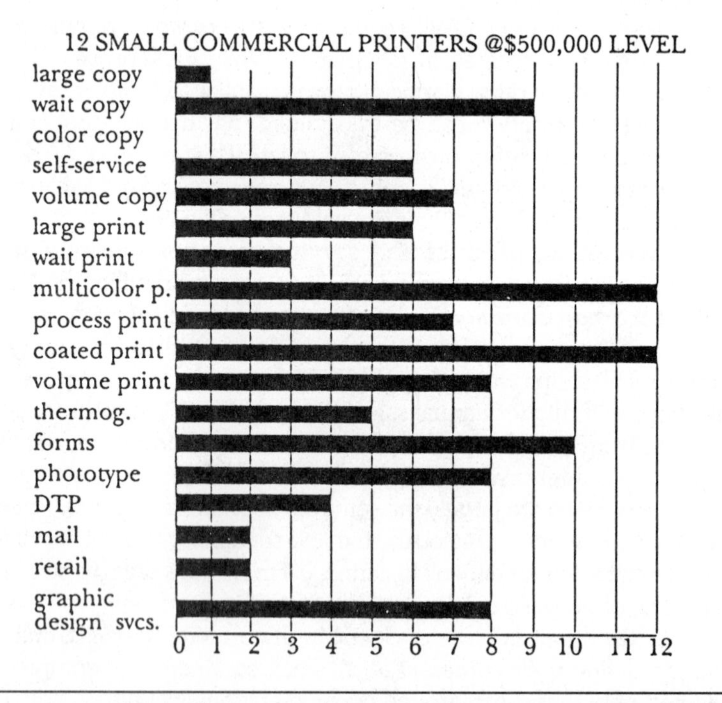

Figure 14.4

Comparison of one group of twelve small commercial shops with annual sales of $200,000 to another group of twelve small commercial shops with yearly sales of $500,000. (Reproduced with permission of *Instant and Small Commercial Printer*)

Jobbing out or brokering products and services—The process of brokering out printing is one whereby the quick printer or small commercial printer buys printed products from another printer—usually a speciality printer—and sells them to his customers as if they were produced in-house. In general, the concept is one where the quick printer thus provides his customers with needed products as a customer service, at the same time marking up the brokered work from 20 to 50 percent.

Based on a study presented in the April 1989 issue of *Instant & Small Commercial Printing*, fully 91 percent of the reader respondents indicated they jobbed out printing work. In fact, every respondent printer in the $400M to $600M annual sales category indicated they jobbed out work at least once in a while. The following represents a summarized listing from the article (arranged in descending order, beginning with the most popular item) of jobbed out products, covering shops from under $100M in annual sales to those with sales over $1 million:

1. Forms of all kinds (by far the most popular brokered product)
2. Thermography items/raised printing
3. Business cards
4. Four-color printing
5. Wedding invitations and announcements
6. Large-format printing

Other brokered items mentioned in the study (presented in no order): labels, stamps, tags, typesetting, binding, large collating, long runs (10,000 up), multicolor printing on coated stock, four-color separations, halftones and camera work, numbering, perforating, scoring, embossing, foil stamping, die cutting and bookwork, desktop publishing, laminating, heavy solid printing, volume copying, web work, presentation folders, crash numbering, graphic design services, plastic signs.

Outside sales and printing brokers—Another way quick printers and small commercial printers have found to increase volume has been to hire outside sales representatives. Operating in a manner much like their larger commercial printing counterparts, these sales representatives work outside the shop to locate customers and sell printed products and services. Chapter 4.2 provides information regarding commercial printing sales representatives including Figure 4.1 which graphically demonstrates the typical procedures for selling, estimating, quoting, producing and billing a printing order.

In larger metropolitan areas, some quick printers and smaller commercial printers work with printing brokers, who serve as middlemen by locating customers and arranging printing, then locating a printer to do the work. The broker invoices the client then pays the printer, so the customer does not deal directly with the printer. The concept of printing brokerages is discussed in Chapter 4.2 in some detail, including information as to how the typical broker operates.

Profile of the Owner

From a "one minute survey" conducted by *Quick Printing Magazine* and reported in the March 1989 issue, the following demographic profile of a "typical" quick printer is as follows: The typical quick printer is a 45-year old, Caucasian male. He is married, has two children, and has his spouse and other family members typically working with him in the business. He owns his home, attended college (and probably graduated), and worked in a "white-collar" job before entering the quick printing industry over ten years ago. He received his training on the job. Currently his shop brings in about $250,000 annually and he takes home between $30,000 and $40,000 per year.

Equipment Profile

As noted in the NAQP 1987 Member Shop Profile in Figure 14.3, bindery equipment and presses top the list of equipment in the typical NAQP-member shop, followed with platemakers, self-service copiers, high-speed copiers, cameras, typesetting systems, and DTP equipment with laser printers. Of the equipment areas expected to expand in the next few years, desktop publishing appears to offer great potential.

Employee Profile

In general the new employee of a typical quick printing firm begins as an inexperienced employee, making in the range of minimum wage to $5.00 per hour. As job skills increase, the employee's wage increases, with the typical experienced pressman making $10 to $15 per hour, not including fringe benefits. Because the typical entry-level quick printing employee begins with little experience—high school and college

	Type of Location		
	Small/Medium Community	*Suburban*	*Within City*
Word of Mouth	21 %	23%	21%
Direct Mail	19	13	14
Direct Sales Contacts	9	16	15
Yellow Page Ads	9	11	7
Over-the-Counter Sales	4	5	5
Telephone Solicitation	3	2	2
NewspaperAdvertisements	2	1	3
Radio Advertisements	1	*	*
TV Advertisements	*	*	1
Trade Journal Ads	–	1	*
Other Methods	2	2	2
Unsure/Did Not Respond	30	26	30

Figure 14.5

Listing of marketing and advertising methods used by NAQP members. (Reproduced with permission of NAQP from their 1987 *Member Shop Profile*)

programs are generally not available—the quick print shop owner is faced with continuously training employees. Generally this training is focused on either production, such as presswork, copying operations or bindery, or customer service including dealing with customers at the counter.

Advertising/Marketing Profile

In general, larger commercial printers and high volume printers have not found advertising to be particularly beneficial in increasing sales volume, largely due to the fact that such print orders are business-driven and not purchased by the public-at-large. The quick printer, however, markets his products and services directly to the public, which necessitates that he employ various methods of advertising. Figure 14.5 lists the variety of methods used by NAQP members to advertise and market their products.

Word of mouth and direct mail represent the two most effective methods to achieve sales among the NAQP respondent group, regardless of shop location. Advertisements of all types did not substantially improve sales volume. The large number of respondents who where unsure or did not respond is not surprising in that this type of data is always difficult to collect and customers, when asked, may cite more than one source or may not remember where they initially heard of the quick printing establishment.

Counter Personnel: A Key Marketing Element

There is no question that friendly, dynamic counter employees are critical to increasing volume both through repeat customers and improved sales when the customer is in the store. Counter employees take the customer's order, resolve customer complaints, answer questions, handle cash transactions and represent the company to the customer on a one-to-one basis.

In smaller quick printing firms—those with fewer than eight people—it is typical for employees to handle counter work in addition to completing production duties. Generally this is based on who is available when the customer needs assistance. In larger quick printing firms, specific individuals are usually assigned to work the counter. In either case, the relative sensitivity of the counter employee to the customer is critical to positive customer relations. While opinions regarding counter approach and customer treatment vary, there are some general rules that apply regardless of the size of the quick printing shop or how busy the shop is at any given time. Effective counter personnel should be trained to do the following:

1. Evoke a positive, friendly, cooperative attitude at all times. Smile and be courteous, polite and calm. Take interest in the customer from the first greeting. This is not only true with respect to direct customer contact, but with telephone dealings as well.

2. Deal with the customer in a non-confusing, clear manner. Find out what the customer wants. Clarify any misconceptions or areas that may appear to be confusing or lead to misunderstandings.
3. Listen carefully to customers and respond to their questions, problems, or inquiries. Attempt to "read between the lines" as necessary to really find out what the customer is saying. Be intuitive.
4. Learn how to handle customer complaints in a courteous, polite manner. Know how to get on the customer's side of the problem. Look for mutually satisfactory solutions without compromising either party.
5. Suggest ways or methods that may save the customer time or money. This is a particularly important customer service role, particularly for the quick printer. Suggest other sources for an item a customer might need. For example, when reviewing the copying work to be done, suggest ways that images might be duplicated on a master, saving the customer money.
6. Offer to provide any product or service necessary to meet the customer's graphic or imaging needs. This may lead to jobbing out printing from other sources, perhaps suggesting a designer or artist to help complete a customer's logo, or providing training on a specific DTP software problem, and so on.
7. Let the customer know you value his business. Be consistent in dealing with customers. Get to know them as people. Learn their names, likes, dislikes, and interests.

   Various seminars are available which deal in depth with training counter personnel. Individual consultants are also available to train and streamline counter procedures. The last section of this chapter provides some sources in this area.

Quick Printing Financial Profile

Analysis of the financial aspects of the quick printing business have been thoroughly studied by NAQP and presented in their 1987 *Operating Ratio Study*. The following represents some of the financial measurements used. It should be noted that the data, when broken out by kind and type of shop (single shop, multiple shop, franchise firm, and independent shop) vary somewhat from the composite profile as presented here. It should be noted that NAQP members, on the whole, are larger and more profitable than their non-member counterparts. For this 1987 report, NAQP members had average sales annual sales of $498,223 with net owner's compensation of $75,210, while non-NAQP members had annual sales of $454,044 and a net owner's compensation of $51,480. Net owner's compensation is defined as the amount of money left over after covering all business expenses, but before paying the owner a salary or providing the owner any fringe benefits.

Composite Profit & Loss Statement
All Companies

	1985		1987	
Number of firms reporting	203		649	
Sales				
Regular Printing	311,800	69.3	333,609	67.6
Copying	94,400	21.0	112,706	22.8
Brokered Sales	34,900	7.8	35,839	7.3
Other	8,600	1.9	11,439	2.3
Total Sales	449,700	100.0	493,593	100.0
Costs				
Materials	96,700	21.5	104,482	21.2
Outside Services	13,500	3.0	16,710	3.4
Outside Purchases	22,300	5.0	21,293	4.3
Total costs of sales	132,500	29.5	142,485	28.9
Payroll costs	115,900	25.7	134,747	27.3
Overhead*	135,800	30.2	143,637	29.2
Total Costs	384,200	85.4	420,869	85.4
Net owner's compensation**	65,500	14.6	72,724	14.6

*Overhead is broken down into ten separate cost components on the Ratio Study results. Each Profit & Loss comparison contains cost detail for rent, depreciation, advertising, etc.
**Net owner's compensation is defined as that money which is leftover after covering all expenses of the business, but before paying the owner a salary or giving the owner any fringe benefits.

Figure 14.6

Composite profit and loss statement for quick printing industry (Reproduced with permission of NAQP and *Quick Printing Magazine*, from NAQP's 1987 *Operating Ratio Study*).

Profit and Loss. Figure 14.6, compares the 1985 year with 1987. Analyzing the 1987 data and based on 649 firms reporting, regular printing accounted for approximately 68 percent of all sales, with copying amounting to about 23 percent; thus, more than 90 percent of all sales comes from these two categories. In terms of costs and studying the 1987 year, approximately 29 percent of the total dollar costs went to materials, outside services, and outside purchases, with 27 percent in payroll costs and 29 percent in overhead. Net owner's compensation is approximately 15 percent, which is fairly close to the national average of NAQP members.

Net Owner's Compensation. The bar graph in Figure 14.7 shows that net owner's

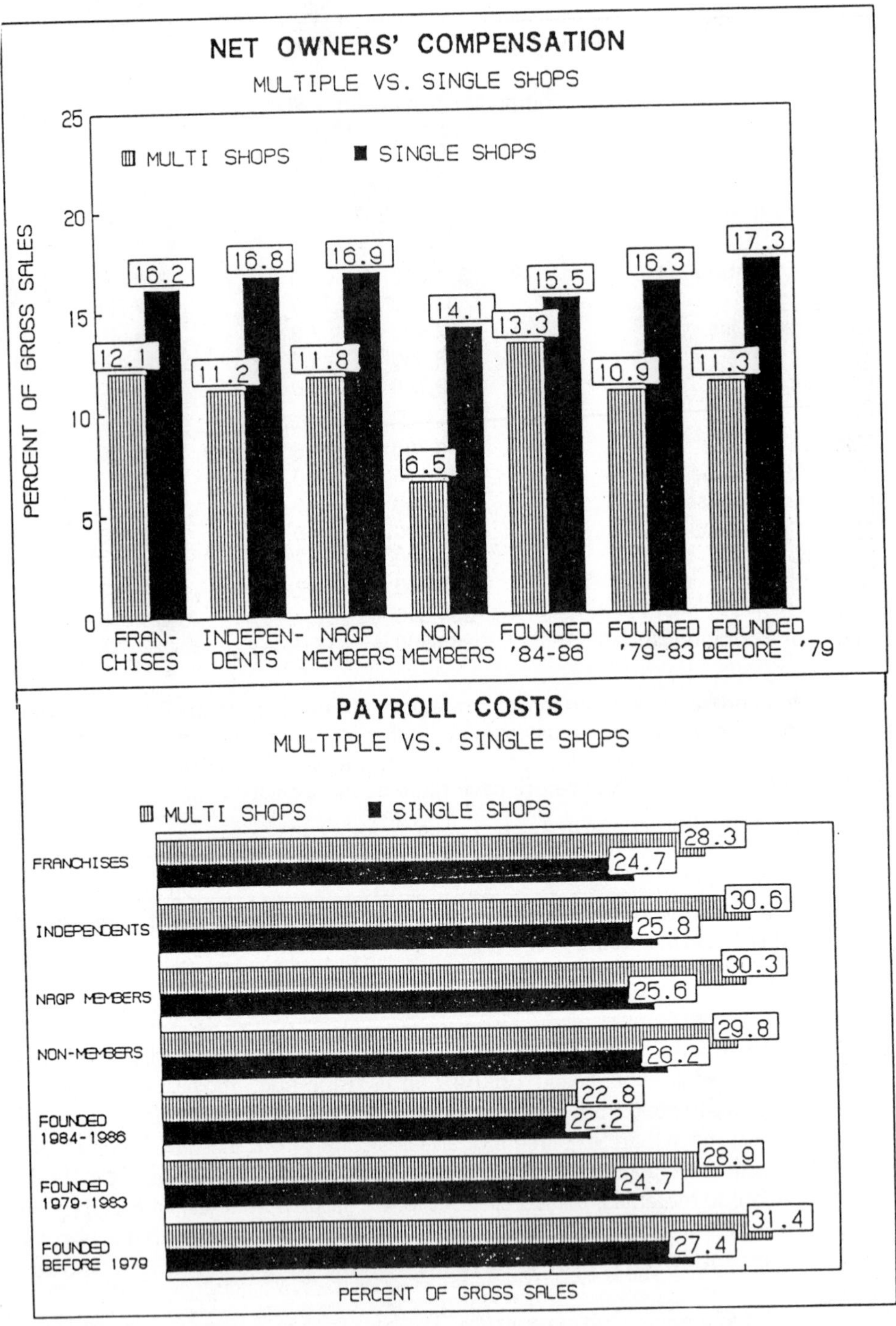

Figure 14.7

Illustrations of net owner's compensation and payroll costs for the quick printing industry (Reproduced with permission of NAQP from their 1987 *Operating Ratio Study*)

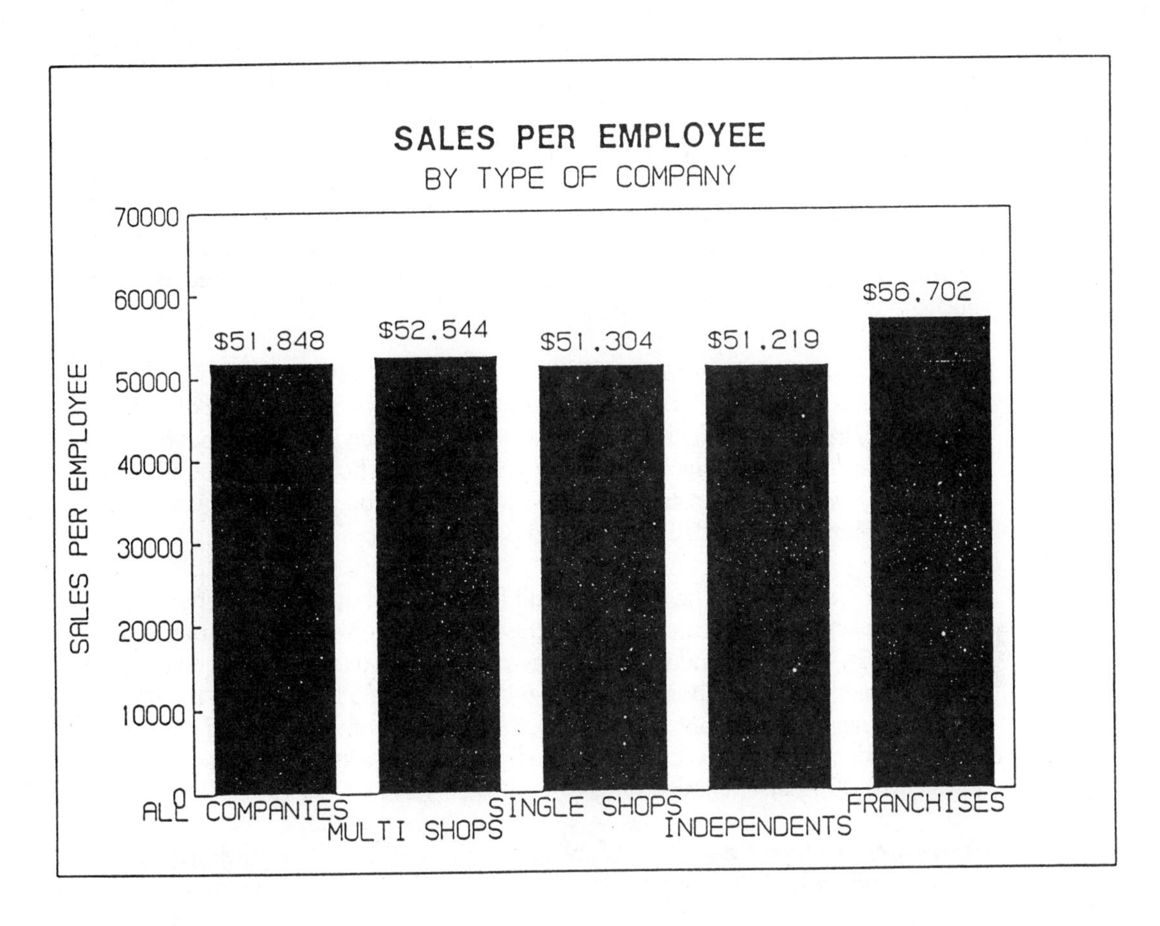

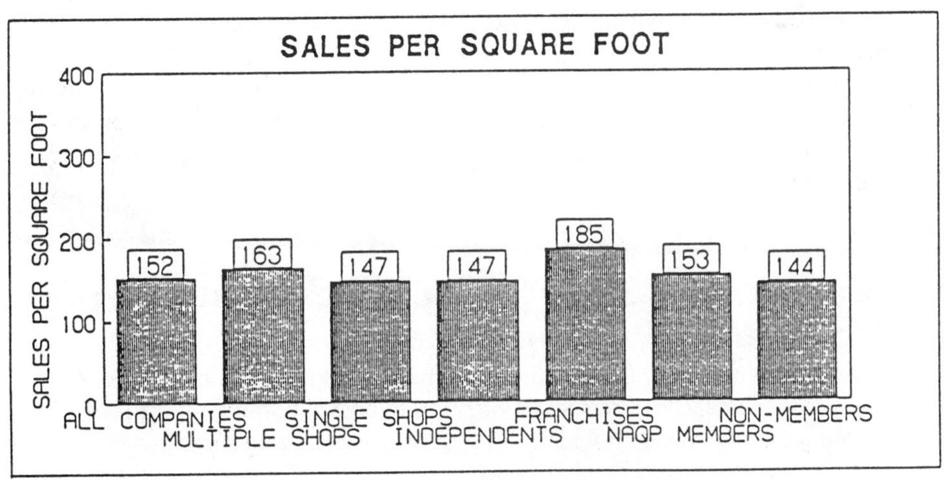

Figure 14.8

Illustrations of sales per employee and sales per square foot (Reproduced with permission of NAQP from their 1987 *Operating Ratio Study*)

compensation—based on gross sales—varies from 17.3 percent (single shops founded before 1979) to 16.9 percent (single shops, NAQP members) to 6.5 percent (multiple shops which are not NAQP members.) Clearly multiple shop net owner compensation is less in all categories, perhaps due to the fact that multiple shops require duplication of supervisory skills at each location, plus the added rent and additional other fixed costs.

Payroll Costs. The bar graph in Figure 14.7 on payroll costs shows that multiple shops founded before 1979 had the highest payroll costs at 31.4 percent of gross sales, followed by independent multiple shops which had the second highest payroll costs at 30.6 percent. The least payroll costs shown to be single shops founded between 1984 and 1986, with payroll costs of 22.2 percent. As will be noted in Figure 14.6, the average payroll costs in 1987 were 27.3 percent of gross sales.

Sales per Employee. The bar graph in Figure 14.8 shows the sales per employee (all employees), and indicates that franchises have the greatest sales per employee figure at $56,702, while independents have the smallest with $51,219. Average sales per employee for the composite group of 649 firms was $51,848. Sales per employee is considered an important measure to the general commercial and high volume printing industry as well, with the average figure for these industry segments hovering in the $80,000 per employee range.

Sales per Square Foot. This is another important figure for quickprinters, since location and accessibility are critical to market share. Figure 14.8 shows that franchises have the highest sales per square foot at $185 while non-NAQP members have the lowest sales per square foot at $144. The average sales per square foot is $152, with single shops, independents, and NAQP members all hovering close to that figure.

The intention of the preceding discussion is to provide a framework for financial analysis of a quick printing firm. Other financial measures, including profitability and return on investment data can also be used, which are more typically found with larger and high volume printers.

14.4 Management Skills for the Quick Printing or Small Commercial Printing Manager

Managing a quick printing or small commercial printing company requires skills and knowledge similar to other kinds of printing and publishing enterprises. There are seven general areas with which the typical quick printing manager must be skilled, which are summarized briefly below.

The manager must have a thorough understanding of cost and price

discipline as applied to the quick printing and commercial printing industries. As detailed in Chapter 2 of this text and briefly in this chapter, it is essential that management staff understand the dynamic relationship between cost, price, volume and company profitability. This information is valuable when considering such issues as the purchase of a new piece of equipment, hiring a new employee, instituting a cost cutting program or spoilage control program, establishing prices, setting up an estimating or costing system, or dealing with any business issue that relates to these variables.

A basic understanding of accounting, cash flow, invoicing and billing procedures and inventory is vital. Every printing manager should have a basic understanding of accounting procedures. This should include a working knowledge of basic accounting, invoicing and billing systems, the ability to monitor the flow of cash through the business and the ability to understand and monitor inventory. This is not to suggest that the manager actually complete these tasks, but more to indicate that the manager should have a thorough, basic grounding in these critical business activities.

A thorough knowledge of customer needs and ability to perceive and pursue new customers and markets is neccessary. As with the intensely competitive larger commercial and high volume printing segments, effective customer service—practicing "customer love"—may well be the major key to survival. Linked to this is the manager's ability to seek out new markets and then successfully pursue these markets in light of the risk and possible failure that might result. This is not to say that marketing should be so risky as to jeopardize the overall business, but to say that risk and innovation may well be essential tools for survival in the next decade.

It is important to have the ability to recruit, train, and handle employee issues effectively. Critical to any quick printing business is the employee base. It is essential that the quick printing manager know effective recruiting procedures, particularly how potentially excellent employees can be brought into the business at an entry-level stage. If high-quality employees are recruited, then the manager must know how to effectively train these individuals for their own personal growth and in the best interest of the company. Effective training may well be one of the major differences between survival and failure for the quick printer in the 1990s, just as with the other segments of the printing industry. Certainly, too, the manager must know how to deal with wage and salary matters, employee complaints and problems, and the many other issues related to keeping employees happy and productive.

The manager needs a good knowledge of printing technology used in the quick printing plant. On the horizon are numerous technology advances which will have major impact on the quick or small commercial printer. For quick printers wishing to enter the four-color process market, the implementation of small-scale CEPS systems might be a major consideration. Color copying advances will certainly be made, as with advances in various process and production areas, particularly related to environmental issues. The quick printing manger doesn't need to become so involved in technology that other areas are neglected, yet should have a basic knowledge of new technology, especially as it applies to his shop and his employees and customers.

The ability to handle customer complaints and deal with customer issues should be cultivated. If any one in the quick printing plant needs to know how to handle customer complaints, it is the manager of the business. Frustrated and less-experienced counter employees need to have a strong, supportive mentor in this area. Also, some customers, such as larger volume buyers, prefer to deal directly with the manager when ordering, which may be a desirable business practice. Numerous other customer/company needs—price discounts, production delays, equipment repair, make-buy decisions—are best handled by the manager.

The manager should be able to provide leadership and common-sense management. Of all the areas critical to company survival, leadership through common-sense management may be the most essential. Running any business is difficult at best. Problems continuously need resolution. Employee and customer frustration are common. Present technology is continuously being replaced by sophisticated, costly new technology. Operating any type of graphic arts business requires planning, key information analysis, a steady hand and the ability to deal with reality. Dealing with these requires leadership talent and the ability to deal in a common-sense fashion with the barrage of issues. Leadership is also mentorship, so effective leaders become good teachers and mentors to their employees.

14.5 Estimating Procedures and Counter Pricing Systems

Common Methods Used to Estimate Quick Printing

Figure 14.9 provides a comparison view of the most common estimating and pricing systems used by the quick printer or smaller commercial printer. The following provides a description of each estimating/pricing system. It is important to note that Chapter 1 provides general information on commercial printing estimating systems and Chapters 2 through 4 provide details with respect to these estimating methods and procedures as practiced in the commercial printing industry as well as information relative to implementing these systems.

Cost Estimating. The cost estimating process requires the quick printer to the formulize the estimating process into the following blocks of information:

standard production time × budgeted hour cost + material costs (+ markups) + outside services and purchases (+ markups) + profit = selling price for customer's job

	System accuracy	Easy to use/ speed of use	Complexity of basic system	Ease of modification	Computer-adaptability	Time to learn/ training time	Provides shop performance data
Cost estimating	excellent	manual: slow computer: fast	fairly high	good	excellent	may be extensive	yes (when job costing included)
Tailored pricing systems	excellent (when based on cost estimating)	moderate to fast	varies with system design	varies with system design	good	moderate to low	not extensive
Standard price books or standard price systems	good to fair	moderate to fast	fairly low	done by vendor	fair	moderate to low	not extensive
Past work basis	good	good	varies with system	good	good (but based on system design)	moderate to low	not extensive
Borrowed prices from other quick printers	poor	varies with system	varies with system	can't easily modify	varies with system	varies with system	not extensive
Ratio systems (paper ratio, cost/sheet ratio)	good to fair (depends on index)	moderate to fast	moderate	good	varies with system	varies with system	not extensive
Intuitive method	poor	fast	low	low	none	impossible to determine	none

Figure 14.9

Comparison chart for various methods for estimating and pricing quick printing

As described in Chapters 1 and 4,this system requires that the quick printer divide the plant into cost centers, or working areas of production, then establish two separate databases for these centers: standard production times which are averages of production time to complete a specific job, and budgeted hour costs which represent an hourly cost figure for the established production centers.

Material costs—paper, ink, film, plates or other such items—and outside services which are chargeable to the customer's job (such as the cost of color separations) are then added in, each with a markup to recover for the necessary servicing or inventory costs. Profit is then added, and the sum of all these yields the final selling price for the job.

The major disadvantages of cost estimating (Figure 14.9) that it is a fairly slow process when completed manually, the level of difficulty in both setting up the system and working with it on a daily basis can be high and extensive training may be required. However, this is counterbalanced by the advantages in that the process is very accurate, modification of the system is fairly easy, shop performance data is typically available when job costing is also included, and, perhaps most importantly, the system can be computerized. When computers are used, the cost estimating process is speeded up and the disadvantages of slowness and level of difficulty somewhat reduced. Chapter 5 discusses computer-assisted and computer-generated estimating systems; computer systems specifically for the quick printer or smaller commercial printer are discussed in the following section of this chapter as well.

Tailored Pricing Systems. Pricing systems which have been specifically developed to reflect the array of products and services offered by the quick printer represent an excellent way to estimate. One major pricing tool is tailored price lists, which are fairly fast, easy to modify, can be computerized (especially using a generic spreadsheet program such as Lotus 1-2-3 or Microsoft Excel) and are fairly easy to learn. All types of tailored pricing systems have a major disadvantage in that they do not provide shop performance data for scheduling production. Also, if the complexity of the system used to develop the prices is great, modification may be difficult.

There are three types of pricing presentation methods in use in the quick printing industry: (1) "circulated" price lists which are developed, typeset, printed and given to the customer so they can price their own work, (2) confidential price sheets or books that are used only by counter personnel and management to price customers' jobs inside the store or on the telephone, and (3) posted or displayed prices which can be seen by the customer when in the quick printing shop. Regardless of method of presentation, the prices should reflect all shop costs and, because of this, cost estimating provides the most suitable base and has been the most traditional and accurate method to achieve this goal. Step-by-step development of accurate, tailored prices for a quick printer is discussed in the next section of this chapter.

Standard price books or systems. Standardized pricing is a process by which a company or persons, unrelated to the direct operation of a specific quickprinting firm, develops a list of prices and a process by which these prices can be implemented into any quick printing company. As a general rule, independent quick printers purchase and use standard price books such as the *Franklin Catalog* (Franklin Estimating Systems) or *Counter Price Book* (produced and sold through Prudential Publishing Company). Chapter 1 provides a detailed description as to how both of these books are used.

Franchised quick printing outlets may either use these sources or may have franchise-wide pricing, which is a confidential set of prices developed by the franchiser and available to all franchisees in the corporation. A critical issue related to franchise-wide pricing, particularly for larger franchisers who have nationwide outlets, is the development and use of a standard nationwide pricing scheme as opposed to development and implementation of regional pricing procedures. Two advantages of standard, countrywide pricing—for example, the price of 1,000 two-color letterheads printed on sub. 20 bond—are that nationwide advertising can be more easily coordinated and customers across the country will be assured of consistency and comparability in pricing among any franchise outlet. The primary disadvantage, of course, is that the costs used to build any nationwide price—such as the cost of labor or the cost of materials—varies by region or even by specific city, thus making nationwide pricing inaccurate. Because of the highly competitive nature of quick printing, over- or under-pricing products and services via inaccurate or fixed prices, or unresponsive pricing policies, can destroy the business.

The "nationwide versus regional" franchise price issue is further complicated by the fact that, in some cases, the franchisee-owner may not be fully conversant or knowledgeable about printing when they initially purchase the franchise, and thus may initially desire a packaged pricing system that is easy to understand and fully supported by the franchiser. The franchiser is thus required to develop some type of fairly simple, uncomplicated pricing procedure, usually oriented to a national base, aimed primarily at the start-up quick print franchisee-owner. Then, as the franchisee-owner becomes more familiar with the quick printing business and begins to understand the dynamics of cost and price, he or she finds this uncomplicated (usually nationwide) pricing system does not reflect the costs of their quick printing operation.

Another disadvantage of standardized pricing systems relates to the growing complexity of printing manufacturing with many quick printing plants. For example, an increasing number of quick printers, in order to gain an advantage over their competition, are aggressively selling tailor-made printing and complex four-color products. Since these markets require individualized, tailored operations, standardized price lists don't work effectively. In sum, the less routine customers and products become, the less standardized price systems can be accurately used.

Many quick printers—independent, franchised, and small commercial printers—who have developed their own tailor-made pricing methods, use either the *Franklin Catalog* or *Counter Price Book* as a guide post or check against their own tailored prices. Opinion varies as to the accuracy and effectiveness of standard price books and systems among quick printers. It is safe to say that the more knowledgeable, product-specific, and market-oriented a quick printer becomes, the less he or she depends on standard prices and the more he or she develops his or her own tailored pricing procedures.

Past Work Basis. The process of using past work as a basis for pricing is described in Chapter 1. In sum, the procedure requires that the quick printer develops a standard group of products and a coordinated set of prices for those products, that are profitable to the company. Products are sold using "past selling price" as a basis. The past work system is most accurate when the basis for the initial prices have been developed through cost estimating, thus allowing for price changes or upgrades return to the cost estimating process.

Borrowed Prices from Other Quick Printers. The collection and use of prices charged by competing printers throughout the industry—quick printers, commercial printers, and high volume printers—is a fairly common printing industry procedure and is addressed in Chapter 1 in some detail. In sum, two major problems exist: First, does the competition know the costs upon which the prices were built, and how much profit has been added to these costs? Second, does the competition's production, labor and other manufacturing processes reflect the adopting company's procedures?

In general, as noted in Figure 14.9, the disadvantages outweigh the advantages when using competition's prices. Of course, in any given printing environment, some printing managers view price cutting as the way to most quickly and effectively increase the volume of their printing operation. It is fair to state that such price cutting—that is, beating the competition's price to increase volume—is a dangerous practice at best, and may well lead to the demise of the price cutting firm over the long term. Price negotiation and price cutting are discussed further in the next section of this chapter.

Ratio Systems. In sum, a ratio system for pricing printing is one where the price of a job is mathematically proportional to a specific material cost or other index. As described in Chapter 1, the most common index in the commercial printing industry is the cost of paper, yet the selected index could be any material or other item that provides a direct link to job price on a consistent basis. For example, in establishing the price for a high-speed copier, developing costs and prices using a "per sheet" basis as the index might be more appropriate and accurate than the cost of paper. Ratio systems, when used, can be fairly fast and may be accurate when there is little variation in production; also, they are not usually complex or hard to develop and can be quickly modified if desired.

Intuitive Method. Pricing by intuition—by guesswork—is more common than might typically be believed, particularly with smaller jobs or those that appear to require only one or two simple manufacturing operations. Certainly, guessing at the price for any printed product is an inaccurate, yet fast, pricing method, as indicated in Figure 14.9. It should be noted that one of the general rules of estimating/pricing systems is that the faster the system, the less accurate the system typically turns out to be. Nevertheless, because the "guesstimating" process is commonly accepted and practiced, it does deserve mention in a discussion on estimating and pricing systems.

14.6 Developing BHRs and Production Standards for the Quick Printer

Cost recovery procedures, regardless of size of the printing firm or markets served, follows essentially the same process: identify the working area or production center, determine every cost component, derive a method to establish the cost basis for that component, calculate and sum together all costs, then divide the total cost figure by the number of anticipated hours the equipment will be bought by various customers.

However, there are certain differences between the essential concepts used to develop BHRs and production standards for the quick printing shop and those used for larger commercial and high volume printers as presented in Chapter 3. The typical quick printer differs from its larger industry counterparts in the following areas:

NAQP cost data is generated differently compared to commercial printing industry cost data. PIA and NAPL represent a cross-section of large, medium and small printers, including some quick printers, and these associations generate cost data which are intermixes of all these firms. NAQP, the major trade association serving the quick printing industry, has evolved a somewhat different cost analysis process which breaks quick printing costs into three major categories: overhead costs, labor costs, and owner's compensation. In sum, because of the differences in data structure between the associations, and the fact that NAQP represents quick printers and smaller commercial printers almost exclusively, development of a BHR system specifically related to the quick printing industry seems practical. Of course, since the result of a BHR costing system is to identify and recover all costs, the BHR process presented in Chapter 3 could be used by quick printers and small commercial printers if they wished.

Over-the-counter sales requires a fast, accurate method of estimating and pricing. There is no doubt that most quick printers or small commercial printers do more over-the-counter sales when compared to their larger commercial or high volume printing counterparts. Because of this, the quick printer needs estimating and pricing methods that are accurate, responsive to fast-turnaround estimating/pricing situations, and adaptable to being completed in the customer's presence with minimum complexity and error.

As will be noted in Figure 14.9, the seven methods commonly used to estimate and price work in the quick printing industry begins with cost estimating, which is fairly slow but very accurate, and moves through intuition or "guesstimating" which is fairly fast but generally an inaccurate process. Ideally, most quick printers want both accuracy and speed. The most obvious solution is for the quick printer to develop and use price lists or standard pricing systems, summarized in Figure 14.6; price list development will be discussed in the next section of this chapter. A second—more encompassing—solution is for the quick printer to computerize the cost estimating process linked with a pricing program. Various software packages are available from commercial computer vendors. Quick printers can also develop their own tailor-made systems using spreadsheets or computer programming in languages available to them. Chapter 5 provides a detailed analysis of different "ways to go" for any printing company considering computerization, including a list of vendors providing such products and services to the printing industry. Later in this chapter, computerization for the quick printer will be discussed, since the computer needs for the quick printer or small commercial printer are sometimes different than their larger printer counterparts.

Many small printers and quick printers don't want or won't use complex costing and pricing systems. Complexity in any business, particularly smaller businesses where job-sharing and a large number of part-time employees is common, distracts from getting the work done efficiently and quickly. Because of this, smaller business establishments of all kinds—including quick printing—typically practice a "keep it simple" (KIS) philosophy.

Adoption of KIS practices are obvious with newer quick printing firms in three areas: estimating and pricing methods, in the way they handle production management, and in the general operation of the business. KIS practices are commonly found when the quick printing shop owner comes from a non-printing or nonbusiness background. In fact, as indicated earlier, the profile of the typical quick printing shop owner indicates that he or she has learned most of their knowledge of the quick printing business on-the-job, even though many have college training, and some graduate training or professional degrees in other fields. In sum, practical, simple, on-the-job systems are most preferred in the typical quick printing shop, as opposed to complex, involved systems and procedures more common to larger and high volume printers.

As the typical quick printer grows in size and experience—whether the shop be franchised or independent—the KIS philosophy continues to be practiced where possible. As the business expands, more complex systems and greater information is required to competently operate the shop. Yet, even as these newer systems are implemented and improved upon the older procedures, there is still a focus to keep them simple, uncluttered and easily passed-on from one person to another.

Even with a focus on simplicity, both manual and computer systems, particularly those oriented toward number-crunching duties, can be complex to the uninitiated. Thus, one of the requirements for a quick printing computer system is simplicity and ease-of-use. Of course, the relative complexity of any system—manual or computer—is a subjective determination, linked to an individual's knowledge, training, and range of experience.

QUICK PRINTING
BUDGETED HOUR
COST RATE

Title of center______________________________

SPECIFICATIONS:

Value of equipment (incl. install)	____________	Wage scale	________	per hr.
No. employees in center	____________	Paid vacation	________	days
Sq. ft. working area (incl. aisles) . . .	____________	Holidays,etc	________	days
		Weekly hours	________	

==

OVERHEAD COSTS	Calculation method	
1. Fixed overhead (rent+utilities)	$9.00/sq. ft. x no. sq. ft..	
2. Insurance on equipment	$6.00 per $1,000 equipment value.	
3. Property taxes applicable to center	$13.00 per $1,000 equipment value	
4. Depreciation or equipment rental fee	10% straight line/direct cost/other method . .	
5. Nonchargeable supplies	10% equipment value.	
6. General shop overhead	5% equipment value.	
7. Maintenance & repairs	4% equipment value.	
8. Advertising	4% equipment value.	
9. Interest on loans	3% equipment value.	
10. Miscellaneous expenses	1% equipment value.	
Total overhead costs (lines 1-10)		
LABOR COSTS		
11. Cost of employee	hourly rate x no. straight time hours.	
12. Payroll taxes and benefits	30% x cost of employee	
Total labor costs (lines 11+12)		
TOTAL OVERHEAD AND LABOR COST (lines 1-12)		
13. Owner's compensation	25% x [Total overhead and labor cost]	
TOTAL ANNUAL CENTER COST	[Total overhead and labor cost + Owner's compensation].	

Manufacturing Cost per Chargeable Hour

(Total overhead and labor cost ÷ anticipated chargeable hours)

80%:__________ annual hours . . . . . $____________ per hour

70%:__________ annual hours . . . . . $____________ per hour

60%:__________ annual hours . . . . . $____________ per hour

Budgeted Hour Cost per Chargeable Hour

(Total annual center cost ÷ anticipated chargeable hours)

80%:__________ annual hours . . . . . $____________ per hour

70%:__________ annual hours . . . . . $____________ per hour

60%:__________ annual hours . . . . . $____________ per hour

Figure 14.10

Sample budgeted hour cost sheet for quick printing

Developing Quick Printing BHRs

As noted in Chapter 3, the development of accurate budgeted hour rates is an essential process for any printer since it requires that all relevant costs of a particular center or operating area to be identified, with these costs then broken down to a chargeable hour base for use during the estimating process. Accurate budgeted hour rates are thus a result of a factual identification of all relevant costs to be borne by an identified center or operating production area, such as a small format press, high speed copier, or desktop publishing/laser printing workstation.

Figure 14.10 is a BHR worksheet tailored to the quick printer or small commercial printing shop. The procedure shown in Figure 14.10 is somewhat different than the BHR system presented in Figure 3.3 of Chapter 3, which was developed primarily for larger and high volume printers and draws upon various sources such as the PIA Ratio Studies. The Quick Printing BHR system follows the same format as the BHR procedure in Chapter 3, yet is broken into three major cost components: overhead costs, labor costs and owner's compensation. In sum, the Quick Printing BHR process has been revised to better reflect the normal cost measurements of the quick printing industry.

It must be stressed that the BHR cost recovery process is not intended as an academic exercise. It is a procedure *to accurately and completely identify every factual cost incurred by the company in order to have the center available to perform productive work on a continuing basis*. Material costs, outside purchases, and outside services are not included. Profit, including compensation to the owner for ownership risk in the form of invested dollars, is not included in the BHR but added by the owner after the cost estimate is completed.

The following points should be noted when using Figure 14.10:

1. Cost center specifications must be precisely determined. An important specification is the value of equipment, which should be accurately stated regardless of whether the equipment is owned, rented, or leased. With respect to the number of employees in the center, it is possible to have a fraction of a person operating one piece of equipment, or two or more persons working on one piece of equipment simultaneously, as a crew. Square footage of the center, wage scale, paid vacation, and holidays and weekly hours should be indicated as they apply.

2. In Figures 14.10 and 14.11, all dollar figures in the right column are on an annual basis, since chargeable hours are calculated annually. Weekly, monthly or quarterly costs and hours could be used if desired. All labor cost calculations are at straight-time rates, with no overtime. Chapter 3 addresses the effects of overtime on budgeted hour rates if such information is desired.

3. The calculation figures indicated—such as $9.00 or a specified percentage such as 10%—are for example only and will invariably change from one quick printer to another. Also, the calculation components—square footage or equipment value—may be modified or changed to better represent the true costs of the line item for the quick printer. If the total annual cost of the specific line item can be identified—such as the cost of maintenance and repairs on Line 7—that figure can be written in the column space and no calculation procedure is needed.

QUICK PRINTING
BUDGETED HOUR
COST RATE

Title of center AB Dick 3691 (11 x 17 inches)

SPECIFICATIONS:

Value of equipment (incl. install)	$ 20,470.00	Wage scale	$ 9.50	per hr.
No. employees in center	1	Paid vacation	5	days
Sq. ft. working area (incl. aisles) ...	120	Holidays,etc	9	days
		Weekly hours	40	

==

OVERHEAD COSTS	Calculation method	
1. Fixed overhead (rent+utilities)	$9.00/sq. ft. x no. sq. ft..	$ 1080.00
2. Insurance on equipment	$6.00 per $1,000 equipment value.	122.82
3. Property taxes applicable to center	$13.00 per $1,000 equipment value	266.11
4. Depreciation or equipment rental fee	10% straight line/direct cost/other method ..	2047.00
5. Nonchargeable supplies	10% equipment value.	2047.00
6. General shop overhead	5% equipment value.	1023.50
7. Maintenance & repairs	4% equipment value.	818.80
8. Advertising	4% equipment value.	818.80
9. Interest on loans	3% equipment value.	614.10
10. Miscellaneous expenses	1% equipment value.	204.70
Total overhead costs (lines 1-10)		9042.83
LABOR COSTS		
11. Cost of employee	hourly rate x no. straight time hours.	19760.00
12. Payroll taxes and benefits	30% x cost of employee	5928.00
Total labor costs (lines 11+12)		25688.00
TOTAL OVERHEAD AND LABOR COST (lines 1-12)		34730.83
13. Owner's compensation	25% x [Total overhead and labor cost]	8682.71
TOTAL ANNUAL CENTER COST	[Total overhead and labor cost + Owner's compensation].	$ 43413.54

Manufacturing Cost per Chargeable Hour

(Total overhead and labor cost ÷ anticipated chargeable hours)

80%: 1574.4 annual hours ..... $ 22.06 per hour
70%: 1377.6 annual hours ..... $ 25.21 per hour
60%: 1180.8 annual hours ..... $ 29.41 per hour

Budgeted Hour Cost per Chargeable Hour

(Total annual center cost ÷ anticipated chargeable hours)

80%: 1574.4 annual hours ..... $ 27.57 per hour
70%: 1377.6 annual hours ..... $ 31.51 per hour
60%: 1180.8 annual hours ..... $ 36.77 per hour

Figure 14.11

Completed budgeted hour cost for a small format press

4. Labor costs have two basic components: the actual cost of the employee on a "per hour" basis and any additional hidden costs such as fringe benefits, including health care costs, social security, pension, or other "perks," provided. The 30 percent used on Line 12 represents the average benefits typically received by full-time employees; if only part-time employees work in the center, and none receives major benefits, the 30 percent figure should be adjusted to reflect the reduced benefits and fringes.

5. Owner's compensation (Line 13) is included in the cost recovery process, representing the amount of money built into the system to compensate the owner for his or her direct involvement and work in the business. This does *not* compensate the owner for the risk taken but only for the work and effort the owner adds as a working member of the quick printing shop.

6. Variations of equipment use and chargeable hours should be realistically evaluated. If the quick printing firm is open 24 hours per day, then chargeable hours may need to be modified to reflect the anticipated hours of likely use. If the piece of equipment is manned with a contingent of part-time operators at different wage rates, yet remains fully used over the working time, the BHR should reflect this. Since quick printing shops vary so greatly in terms of operating hours, the best rule is to initially estimate chargeable hours, then modify this figure based on the history of the center, over time.

7. The most critical elements of BHR recovery are the "bottom line" figures of Total Overhead and Labor Cost, Total Annual Center Cost and Chargeable Hours. The accuracy of these particular figures reflects the accuracy of the overall cost recovery system.

8. It is vital to note that profit has not been added anywhere in this cost recovery system. Profit is *not* owner's compensation. In general, profit earned in a quick printing or small commercial printing shop is used to compensate the owner or company stockholders as a form of payback for the risk taken as an investment, or for business expansion or growth based on the owner's evaluation and business plan. When using cost estimating, profit represents a flexible and variable figure to be evaluated after all factual costs have been identified.

Figure 14.11 represents a completed BHR for a small offset press with a full-time pressperson.

Developing Quick Printing Production Standards

Production standards for quick printers and small commercial printers are established following largely the same procedures used by larger printing firms. Thus, there is little to add with respect to the development and application of production standards that was not detailed in Chapter 3. Figure 3.1 provides a reference with regard to the various procedures used to establish production standards.

Many quick printers establish their production standards based on a random sampling of how much time is expended in performing a certain task. In some cases this is done using quantifiable data and precise measurements, but more likely such standards evolve over a period of time, based on experience. The reason is that most equipment in the quick printing plant operates at a given production speed or specified output, and the process is so repetitively completed—day in and day out—that establishing production standards is not difficult. Also, since many quick printers have a fairly limited equipment array, the quantity of standards required is not excessive.

One major decision both quick printers and their commercial printing counterparts must make is whether they wish to implement a job costing program. Job costing is a process by which actual production times and costs are compared to estimated times and costs, for the purpose of revising production standards, streamlining production and, on the whole, operating more efficiently. Job costing data collection may be completed either manually using time cards completed during production by employees, or electronically using factory data collection (FDC). Very small firms typically have little need to job cost, since much of their work does not require estimating and the business is operated on a largely entrepreneurial profit motive. Yet as printing companies grow in size and complexity, there is increasing interest in quantitative measurements of production, including cost analysis, accurate tracking of chargeable hours, and careful evaluation of production time spent on a "by job" and "by employee" basis. The process of obtaining and implementing EFDC is further discussed in Chapter 3.3, Chapter 4.10 and Chapter 5.2.

14.7 Pricing In the Quick Printing Industry

Introduction

As indicated in the previous section referenced to Figure 14.9, there are three pricing presentation methods commonly used in the quick printing industry: (1) "circulated" price lists which are developed, typeset, printed, and given to the customer as a take-home item, (2) confidential price sheets or books that are used only by counter personnel and management to price customer's jobs inside the store or via the telephone, and (3) posted or displayed prices which can be seen by the customer and show the cost of products and services available from the shop. Some quick printers post prices for the most active products and services they provide, and use counter pricing or price lists for jobs which are beyond the scope of the posted prices.

If price lists are provided to customers as a take-home item, it is prudent for the quick printer to indicate somewhere on the price list that "prices are subject to change without notice" and date the price list with month and year when prices took effect. This will allow price changes and revisions to be made without customer recourse to the older, probably lower, prices.

The major disadvantage of counter pricing, which is perhaps the most popular of the three pricing methods, is that the customer may be unaware of the range of goods and services provided by the quick printer. To resolve this problem, some shops provide customers with a take-home flyer or handout which describes the full range of products and services available. Posted prices, which typically cover those goods and services most common to walk-in customers, are also popular and have a major advantage that they can be modified or changed at the discretion of the quick printer at any time.

Some quick printers provide "specials" on a frequent basis, and price these according to a lot-size purchase of paper or some other cost-savings. They may provide flyers or advertise this special on local radio and television. This may be an effective tool when a new location has been opened or the primary location of the quick printing shop has been moved or expanded. Some quick printers offer a "daily color choice" whereby each day of the week has an ink color running with black that can be economically added to a customer's job. Thus, on any Tuesday the second color might be reliance red and the customer can order black and reliance red letterheads produced on that day for less cost than on another day of the week.

The competition among quick printers has intensified in recent years, and in response some quick printers have added services and moderated price increases. This is much the same scenario as with the larger commercial printer doing multi-color printing and competing with other larger printers in the same vertical market. As indicated previously in this chapter, many quick printing experts predict a "shakeout" in the quick printing industry, whereby those quick printers who are unable to effectively compete will go out-of-business. Nevertheless, at this writing, the competition still remains intense and the prudent quick printer knows that expanding services beyond a reasonable range, or cutting prices simply to increase volume and reduce competition, are both unwise reactions over the long term.

Overview of Tailored Pricing Systems

There is no question that the prices charged for the products and services available through any quick printer—in fact, any printer, regardless of size or market orientation—represents the first and foremost key to making or losing money in the business. For that reason the methods used to initially establish prices, and later used to revise or change those prices, represent one of the most critical, vital areas of the business.

Let us say that the quick printer chooses to use some type of pricing/estimating system to operate his business, as opposed to intuition or standard price books. Tailored pricing, where prices are developed specifically on the basis of a given quick printing company's costs, then represents the best direction to take. As noted in Figure 14.9, in the majority of evaluation categories, price lists receive positive marks. For example, accuracy of the system is excellent when cost estimating is used as a basis; tailored pricing is generally fast and easy to use, the prices can be computerized for ease of use, and the systems are typically not difficult to learn. Complexity of the basic system and ease of modification of prices vary with the design of the system. If there is a significant disadvantage to price lists for estimating and pricing in quick printing shop, it is that no times for production are available since the basis is price and not time. Thus, the length of time to schedule production for a particular job sold using a tailored pricing system is not available.

Step-by-step Method to Establish Tailored Prices Using Cost Estimating

The following step-by-step procedure (summarized from an August 1986 article in *Quick Printing Magazine* by this author) is suggested as a framework to establish a tailored pricing system in a quick printing shop. It is presented here as a general procedure since it is difficult to address many of the details of the process.

Step 1. Identify the product(s) that will be covered by the pricing system. It is essential that the initial step in the process identifies or groups the product or products to be covered. These product lines must share similar factors: type of paper, color of paper, colors printed, quality level, equipment factors (printing, high-speed copying, color copying, etc.), level of preparatory work needed, etc.

Step 2. Set up the pricing format. The pricing format is typically a chart or listing of quantities, cross-referenced with prices (yet to be calculated) that will relate to these quantity levels. For example, the first column (on the left side of a sheet of paper) might list the quantities common to all products, beginning the minimum order quantity and numerically stepping to the highest quantity economically feasible. The remainder of the sheet would then be divided into specifically identified product categories such as 8 1/2 × 11 sub. 20 bond, 8 1/2 × 14, sub. 20 bond, or 11 × 17, sub. 60 book, and so on. If jobs will be printed on both sides, each category would be further subdivided into "printed one side" and "printed two sides."

Step 3. Use cost estimating (or a similar system) to factually determine the cost basis for the quantities and products. Accurate price lists must initially be built using accurate cost estimating, which identifies all costs that go into a specific product group (or column). The cost estimating formula is as follows:

> standard production time × budgeted hour cost + material costs
> (+ markups) + outside services and purchases (+ markups) + profit =
> selling price for customer's job

As an example, if a 10,000-quantity job took two hours on a high speed copier, and the copier's BHR was $45.00 per hour, then the production cost would be $90. Add to this the cost of paper—say $40, which includes a 20% markup as an inventory charge. Assuming there are no outside purchases or outside services, the estimated cost not the selling price—would be $130.

To save time, estimates for every third or fourth quantity level could be done, with interpolation used to fill in the quantity gaps with estimated costs. Since the basis of the system is factual cost estimating, future price list updates will revert back to this cost estimating process, ensuring complete cost recovery with each price list revision.

Step 4. Establish prices. At this point each column should have quantities cross-referenced with costs to produce that quantity. Determining the selling price of the job—that is, taking the estimated cost and adding an additional amount of profit—is perhaps the most difficult aspect of the process. While cost is "fact," price is "fiction." Many factors—some quantifiable, others less tangible and harder to evaluate clearly—enter into the profit addition for a given product in quick printing. These are discussed in depth in Chapter 2.5 under pricing determinants in the printing industry and include such factors as estimated cost, customers and markets served by the shop, quality of work, the customer's willingness and ability to pay, shop location, volume of work in the plant at any given time and so on.

Step 5. Identify and price any additional services. Additional services offered by the quick printer, when priced fairly and performed properly, have the potential to become a significant income generator. Such services might include mechanical art preparation, desktop publishing, color copying, finishing procedures such as stapling, drilling, padding, folding, gathering, trimming or round-cornering, FAX services, and self-copying. It is possible that additional services may, in some cases, be more profitable than ink-on-paper production, particularly if the service is unavailable with other quick printers or in the local community.

Step 6. Compare new prices to the shop's current prices and evaluate relative to the competition's prices. Comparison of the new prices—built on cost estimating—to the shop's current prices should be carefully completed. If the prices are about the same, the current prices could continue to be used, with the benefit that these prices have been verified by cost estimating. The next time a price revision is needed, cost estimating could be used to complete the process. If the new prices are lower than the current prices, and the shop is making an acceptable profit, then staying with the original (and higher) prices seems reasonable.

If the new prices are higher, there are three essential questions that must be asked: First, is the business currently losing money, thus requiring higher prices so that a profit—and survival of the business—is ensured? Second, assuming the business is profitable and assuming an intensely competitive situation with other quick printers, how can costs be cut to lower expenses, thus reducing these new-developed prices to more accurately match current, competitive prices? Third, if the new (higher) prices are implemented without modification, what will be the impact on the business, particularly related to the anticipated loss of volume and thus loss of profits? Cost cutting can be accomplished in many ways, too numerous to mention here. Some more obvious considerations include a reduction in the number of shop employees or a shift or reduction in compensation or fringes for shop workers, supplier discounts through quantity purchases of paper and other materials, or an automated piece of equipment that streamlines an awkward, older production operation. As indicated in Chapter 2, if a typical printer can reduce costs by 10 percent—which is a sizable reduction—profits increase by 56 percent.

Simply implementing higher prices may have a serious, negative effect on the profit of a quick printing business. As noted in Chapter 2, a major finding the the *McKinsey Report to PIA* states that in an elastic, competitive market, even a slight (3 percent) increase in prices can lead to a 10 percent drop in volume and ultimately result in a 50 percent loss of profits. If it appears the loss of volume would be substantial, one practical solution is to drop the newly-developed prices to a level approaching—but not at or below—the estimated costs, thus reducing the profit earned by the shop. This cost/price/profit dilemma is common throughout all areas of the intensely competitive printing industry. At the same time, the quick printer should begin to identify those customers and vertical markets which are most profitable, and orient the business toward these areas. This moves the quick printer away from an elastic market situation, toward an inelastic marketplace where greater price leverage is possible.

Because of the intensively competitive nature of the printing industry, including the quick printing segment, evaluating new prices in light of the competition's current prices is common. The competition will almost certainly have different costs, evidenced by their different equipment, employee compensation, location and square footage costs, and so on. Yet the competition represents a "bottom line" price that must be known and evaluated. Most successful quick printers watch the competition carefully but don't become involved in price wars or head-to-head combat. These wise quick printers focus on excellent service, good location, high quality and vertical market niches so that the competition's price is not critical to their daily operation or long-term profitability. Chapter 2 addresses the matter of pricing, profit and competition in detail.

Step 7. Implement the new prices. If the decision is made to implement the new prices, then preparing the prices in a usable form will be necessary. If a price list is to be given to customers as a take-home item, it should be designed in a non-confusing way and be typeset for good readability. Handout price lists should be dated and indicate that "prices are subject to change without notice." If the prices are to be posted or used by shop employees at the counter (and not given to customers), they should still be clearly structured and easy-to-use. Training should be provided so that every counter employee uses the pricing system the same way.

Step 8. Periodically review prices with respect to general financial health of the business and profits earned. It is important to periodically review and modify, as needed, the prices for the shop's goods and services. When cost estimating is used as the basis for price development, such modification is simplified. The price review should include an analysis of the profitability of the business, since the link between prices charged and profit made is direct. Many quick printing customers are sensitive to price changes which occur too frequently, because it appears that the quick printer doesn't know what he is doing or is inexperienced in operating the business. Given that fact, many quick printers change prices on a yearly basis regardless of rising costs during the year, which may adversely affect profit. Some quick printers prefer twice-yearly price modifications which are desirable when costs escalate or profit is marginal.

Negotiating Prices and Price Cutting

No discussion of pricing would be complete without some mention of price negotiation and price cutting. Some customers—particularly those who use price as the deciding factor in purchasing printing—understand that the quick printing industry is an extremely competitive business. They attempt to use this to their advantage by continually looking for a "better deal" through price negotiation, forcing the quick printer or small commercial printer into a price cutting role.

In general, price cutting is undesirable when the printer has a thorough knowledge of all costs, and prices have been reasonably determined and established with care. Unfortunately, because of the complexities of establishing an accurate costing program, a diversified number of shop owners who don't know their costs, and aggressive customers who push for price reductions, price cutting is fairly common in the printing industry. In terms of dealing with price cutters, one extreme is to simply refuse to negotiate and cut price on the basis that once this trend is begun with a customer it will continue. If this is the chosen method, tact and sensitivity should be used or the customer may be lost forever. The other extreme way is to negotiate and cut price when customers seek to do this. Some variations between these extremes include offering marginal or courtesy price cuts, linking price cuts with a certain volume requirement or allowing price cuts only on certain types of work or at certain times of the week.

14.8 Computer Systems for the Quick Printer

The quick printing and small commercial printing business is changing. Shop owners recognize this and, for the most part, realize that computerization is an essential means available to help address this dynamic problem. In sum, there are three primary forces acting on quick printers and small commercial printers driving this computerization interest.

1. Job complexity is increasing. No longer does the quick printer or small commercial printer handle just simple black-and-white work. Depending on the shop, production is increasingly complex: four-color printing on coated stocks, expanding product lines from envelopes to presentation folders to fancy brochures, web printing, customer-required complicated designs which may necessitate traps, reverses and other fairly complex stripping procedures, longer runs, different types of paper stocks, accurate ink matching, integration of electronic prepress methods such as desktop publishing, and so on. This complexity cannot be handled simply by customized prices or standard price lists, although some types of routine and highly standardized work can be handled this way. The quick printer and small commercial printer, to be successful, have increasingly turned to estimating, and computer systems which will aid them in completing this process.

2. Customers want increasing services including faster yet accurate prices and quotes. It has been said that a buyer's market—where the buyer controls the purchasing environment—is the long-term result in any area of intense competition in business. This seems particularly true in the printing industry. In the past ten or less years, customers (also called print buyers in the larger commercial segments of the industry) have come to expect faster turnaround on jobs, much more color, moderation of prices, faster printer response to problems, and fast quotes and prices of increasingly complex work. If one printer doesn't provide the needed services, another is happy to step in and take the customer away. To solve this problem—the need to handle complex data accurately and in a timely fashion—computers have increasingly been integrated into printing companies. In sum, it appears that customer demands will not abate in the next decade, but increase in both speed and complexity.

3. A management tool is needed to aid in running the shop. One of the most important reasons to computerize any business is to provide the owner or manager with a management tool to better aid in running the business. The quick printer or small commercial printer, faced with increasingly complex production on one hand, customer-demanded services on the other, and working in an increasingly competitive environment, needs some type of method to focus and control the business. The computer fits this need precisely. For the most part, it appears the 1989 focus of the typical quick printer or small commercial printer is toward estimating and pricing systems, with less interest in other areas such as order entry, inventory control, production management, billing, invoicing, payroll, job costing and management reports. As the quick printing industry grows in need for greater management information in the next decade, it is very likely that many additional computerized functions will be desired.

It should be noted that Chapter 5 of this text addresses in detail the process by which computers are used in the printing industry, including uses by quick printers and small commercial printers. Figure 5.1 provides an overall graphic illustration of a complete management information system for a printing company, presented in "building block" or modular form.

The purpose of this section is not to repeat the information presented in Chapter 5, but to identify those computerization aspects specific to the needs of quick printing and smaller printing shops.

Overview of Computer Systems for the Printing Industry

It is essential to begin this discussion by stating that any computer system should not make those persons working with it slaves to the system. This is true of computer systems of all sizes, and for all types and sizes of printing companies.

General Ways to Computerize. As noted in Chapter 5, there are essentially three ways to computerize a printing business: (1) buy a microcomputer and application software from a printing industry vendor (M&SW), (2) buy a turnkey system, and (3) develop a customized computer package for the printing company, which could be as simple as developing a spreadsheet program to as complex as hiring a consultant to develop a fully integrated computer package for the company. Of these choices, the quick printer or small commercial printer is most likely to begin computerization by building a spreadsheet estimating/pricing program or by purchasing application software from a printing industry computer vendor. This is the simplest and most direct computerization approach, and represents minimum out-of-pocket cost to the quick printer. With either software choice, the quick printer or small commercial printer will need a microcomputer. There are essentially two microcomputer directions now available: the IBM and IBM-compatibles typically running with an MS-DOS operating system, and the Macintosh, which utilizes a unique (and completely different) operating system written for the Macintosh when the machine was developed and introduced in 1983. Chapter 5 presents a detailed discussion for spreadsheet development and purchase of application software.

Computerized Estimating. In general there are two ways to approach computerization of printing estimating: computer-assisted estimating (CAE) and computer-generated estimating (CGE). Computer-assisted estimating requires that an estimator or other individual knowledgeable about printing production operate the computer, answering questions (or "prompts") about literally each aspect of the job being estimated. CAE systems thus break the job into detailed elements representative of the production likely to occur if the job is won by the company. CAE systems can be complex and a complicated estimate may take from ten minutes to one hour to complete. Many professionally-written CAE systems will price jobs automatically once the detailed breakdown of the job is completed. CAE systems follow the cost estimating process and formula presented earlier in this chapter and thus require budgeted hour rates and standard production times when the system is established.

Computer-generated estimating systems are essentially an automatic estimating process. CGE systems require that specific job components be identified and entered into the computer, and the computer then automatically completes a preprogrammed routine, estimating and pricing the job. CGE systems do not usually require a skilled estimator since the program's routine makes the estimating decisions. Because CGE systems are developed based on normal or routine printing production, they are typically not suitable for complex production or unique types of work.

Desired Computer Features for the Quick Printer

What does the typical quick printer want in a computer system? Based on observations and discussions with various quick printers and small commercial printers, the following list is representative of the most-identified needs (presented in no specific order):

- A fast system which can be used in the customer's presence, while taking orders at the counter
- A system that prints proposals and quotations which can be clearly understood by both experienced and less-experienced customers
- A system that is simple enough to be learned without extensive training

- A system that will handle paper costing, multiple quantities, answer "what-if" questions and handle other number-crunching items
- A system that can be used as a basis for customized pricing development
- A system that will be fairly reflective of the quick printer or small commercial printer's normal production sequence
- A system that will maintain and print customer files for analysis of jobs completed by product, process, run length, vertical market and other parameters
- System options for production analysis of individual jobs, an essential basis for job costing procedures
- System documentation that is clearly and simply written and referenced, allowing for quick problem-analysis and problem-solving
- Responsive customer support from the vendor of the program (if an application program purchased from a professional printing industry vendor)
- A modular system design, so that the system is expandable to allow for "add-on" programs as the shop grows in size and computer needs
- Cost of the system that fairly reflects the benefit to the user

Choosing the Quick Printing Computer System

Figure 5.3 in Chapter 5 provides a ten-step flowchart procedure for selecting, implementing and evaluating a computer system for a printing company. The step-by-step procedure suggested in Figure 5.3 and accompanied with the detailed text in Chapter 5.4 that describes each step, covers largely the same process needed for quick printers and small commercial printers. The following points are suggested as highlights for the quick printer or smaller commercial printer in choosing a computer system for estimating, counter pricing, quoting jobs and related production duties:

1. Determining the needs of the quick printer through a needs analysis evaluation (Step 2) is absolutely essential to making the system fit the company and the employees. This process takes time and effort, but is the only way a suitable matchup of system and company is possible. The features previously presented may serve as a basis for helping define these needs.

2. Based on observation, the typical quick printer or small commercial printer would probably focus on one of three computer solutions, presented in the order likely to be considered: (1) purchase a microcomputer and application software from a professional computer vendor, (2) purchase a microcomputer and with a conventional spreadsheet program, then write a customized spreadsheet process for estimating, pricing and desired other duties, and (3) purchase a turnkey system. Advantages and disadvantages of each computer solution are presented in Chapter 5.4. Appendix B has a list of current vendors of computer equipment serving the printing industry, some oriented to quick printing as a vertical market.

3. Shopping and selecting the system is a difficult process, made more difficult since no system will fit all the requirements and demands of the quick printer. Here it pays to be realistic and understand the old adage "you get what you pay for." Depending on the cost and financing systems available, most quick printers are likely to purchase the system outright, although renting and leasing may be options available.

4. There is no question that implementing the system and training employees thoroughly in system use is an essential part of making the system work for the quick printer. The speed with which the implementation process occurs is related to the complexity of the system, the need to have it on-line, and management's commitment to making it functional in the shop. The implementation process should be employee-focused, providing ample time for the employees to learn the system and provide full employee support during the cut-over period from the old system or the break-in period if the system is the shop's first computer.

5. One of the most difficult choices any printer faces is knowing when the current computer system has served its useful life and it is time to upgrade to a new system. This can occur for a number of reasons, including expanding production techniques, addition of sophisticated new equipment, expanding markets, need for a better information processing system or computer-related issues such as limited memory or slowness of the system. Sometimes the inadequacies of the current system are evident but the owner refuses to acknowledge these problems. Sometimes employees, who are fearful of a new system and the related changes it will make in their working situation, make the existing system appear to work well. Sometimes the decision is delayed for various other reasons: current lack of funds, waiting for a plant expansion or shop move that is yet not fully defined, or simply knowing that a new system is needed and not taking the time to investigate and find a system suitable for the expanded operation.

14.9 Literature and Sources of Information for the Quick Printer

The quick printing industry is currently served by three major sources of information and support. These include (1) the National Association of Quick Printers, (2) *Quick Printing Magazine, Instant & Small Commercial Printer,* and *Copy Magazine* (3) information services provided through books, audio cassette tapes and articles by individuals focusing on the quick printing industry.

The National Association of Quick Printers (NAQP)

NAQP was founded in 1975 by George Pataky and has about 4,500 member companies representing about 6,000 quick printing and copying shops throughout the United States, Canada and several foreign countries. In 1987 it was estimated that there were about 24,000 quick printers in the United States and Canada. Gross 1988 sales for United States quick printers is estimated to be roughly $6 billion, with NAQP members representing about $2 billion—or one third—of that sales figure.

The purpose of NAQP is to provide its members with pertinent, up-to-date information on all facets of the quick printing business. The association releases numerous reports and other information such as its 1987 Member Shop Profile, 1987 Quick Printing Industry Operating Study, 1988 Desktop Publishing Studies and 1988 Pricing Study/Final Report.

The focus of NAQP is to help their members in all areas of business operation. Since the typical NAQP member is a small business owner—1988 annual sales for franchised quick printers was $434,000 while independents had sales of $360,000—the services provided covers a wide array of information and issues. Many members offer both printing and copying services, desktop publishing and related graphic services. NAQP began their "Print Expo" in 1978, which is held annually in a major metropolitan location.

NAQP is located at 111 E. Wacker Drive, Suite 600, Chicago, IL 60601. Telephone 312/644-6610.

Magazines

Quick Printing Magazine. Established in 1977 by Robert Schweiger and currently published by Coast Publishing, 1680 SW Bayshore Blvd., Port St. Lucie, FL 34984. Telephone 407/879-6666. The 1989 circulation of QP averaged over 50,000 copies with an estimated readership of over 150,000. Provides timely articles on many quick printing topics, either written by known industry experts or other specialists. Sometimes work with NAQP to jointly sponsor and publish studies and other information pertinent to the quick printing industry. Offers numerous services to their readers such as "The Source" which is a listing of industry-wide products and services. Also sponsor the Steven Pundt Quick Printing Memorial Scholarship to financially assist students in four-year university graphic arts programs. Provide research grants of up to$1000 a year for academic research, available to graphic arts educators.

Instant & Small Commercial Printer. The magazine is published 10 times per year by Innes Publishing Company, 425 Huehl Road, Northbrook, IL 60062. Telephone 312/564-5940. Articles typically cover a broad range of issues and areas which the quick printer or small commercial printer must address. For example, case studies of successful instant printers or small commercial printers are frequent. A continuous discussion of "Jobbing Out" covers provides information on brokering work to printers and other sources of supply. Monthly features include newsbriefs and copying news and products and new products.

Copy Magazine. This magazine is located at 800 W. Huron, Chicago, IL 60622. Telephone is 312/226-5600. Copy Magazine devoted to imaging technology including copiers and laser printers and is widely distributed to many quick printers who use this technology. The magazine has timely articles on a cross-section of subjects of interest to the quick printer and small commercial printer.

Other Information Sources and Services

The individuals briefly summarized in this section, listed alphabetically by last name, provide information and services to the quick printing industry. This information is provided as reference only. Thus, reference in this text does not imply endorsement by this author of the person or service indicated, nor endorsement by the listed individual of the material contained in this chapter or in this text.

Tom Carns, PDQ Printing, 3820 S. Valley View, Las Vegas, NV 89103. Telephone: 702/876-3235. Active in the National Association of Quick Printers, including president 1986-87. Began quick printing consulting in 1989 largely patterned after the success of PDQ Printing. Conducts seminars on quick printing issues to stress his identified "marks of success" which include solid outside sales, effective advertising, good customer focus and pricing for profit.

Tom Crouser, Crouser and Associates, P.O. Box 8365, S.Charleston, WV 25303.Telephone 304/768-3000. Publishes "The Crouser Report," a twice-monthly newsletter covering a wide variety of items of current interest to the quick printer and small commercial printer. Conducts seminars and workshops on costing, pricing and various quick print management topics and an active industry consultant. Has developed audio-cassette programs titled "Understanding Costs and Pricing in the Small Press Shop" (which includes computer spreadsheet price list models), "People Management System," "Printing Primer" and "Successfully Handling Customers from the Counter." Senior contributing editor to *Quick Printing Magazine.*

Bill Friday, Prudential Publishing Company, 7089 Crystal Blvd., Diamond Springs, CA 95619. Telephone 916/622-8928. Two books are available through Prudential Publishing: *Bill Friday's Counter Price Book* and *Bill Friday's Quick Printing Encyclopedia.* The counter price book was discussed earlier in this chapter and also in Chapter 1 of this text.

Larry Hunt, Copy King, 2275 Main Street, Dunedin, FL 34698. Telephone: 813/734-5434. Author of *How Much is Your Quick Printing Business Worth?* (1982) and more recently *Keys to Successful Quick Printing* (1988). Conducts seminars and consults, generally in the area of profitability management similar to the material presented in his most recent book. Former chairman of NAQP (1983-84) and founder and technical director of NAQP's High Speed Copier User's Group. A frequent contributing writer to *Quick Printing Magazine.*

Thomas Popp, Creative Solutions, Inc., 323 S. Franklin Bldg, Suite 0-60, Chicago, IL 60606. Telephone: 312/922-1793. Contributing columnist to *Instant and Small Commercial Printer.* Active industry speaker and provides consulting services to the quick printing industry. Quick printing shop owner and member of NAQP.

John Stewart, Paragon Printers, 2110 S. Daisy Road, W. Melbourne, FL 32901.Telephone 407/727-2442. Active industry speaker and consultant. Former president of NAQP and coordinator of NAQP's 1988 Pricing Study. Contributing editor to *Quick Printing Magazine* writing a monthly column titled "The Quick Consultant."

Estimating Web Lithographic Production

15.1 Introduction

The primary difference between web and sheetfed printing is that sheetfed production requires paper stock to be passed through the press as single pieces of paper, while web printing requires paper to be imaged from a continuously unwinding roll. Web's primary popularity comes from the exceptionally high speed at which a web press can operate: A modern web can produce an output equal to that of approximately four sheetfed presses. Such high press speeds are not without limitations, however, such as excessive paper spoilage, the need for sophisticated registration controls for web equipment, and a pressroom technology different from the more commonly understood sheetfed operation. Also, the fixed cutoff of a web press—the repeat of image around the cylinder—presents a significant problem since it limits the flexibility of product size to changes in web width only. Sheetfed presses can produce almost any size of press sheet within the length and width dimensions of the maximum press size sheet and, therefore, can produce nearly any finished product size desired by the customer.

This chapter begins with a detailed discussion of the advantages and disadvantages of web production, followed by an overview of web presses, papers for web printing, and web inks. A section on basic web paper formulas, including essential formulas for estimating web paper, is presented. The final section of the chapter covers estimating web production as a whole, including paper spoilage, ink consumption, and press time and cost.

15.2 Advantages and Disadvantages of Web Printing

The discussion that follows provides an in-depth perspective on the major advantages and disadvantages of printing using the web process. The following characteristics are major advantages of web production.

1. Web presses, with stock fed from rapidly unwinding rolls and not limited by slower sheetfed registration procedures, print at very high speeds. For example, the average speed of a four-color 36 inch sheetfed perfecting press is about 7500 impressions per hour. A comparable perfecting web press, producing roughly the same 24 × 36 inch finished sheet (called a cutoff by the web printer), can realistically run at 1000 feet per minute, which translates to about 30,000 finished cutoffs per hour. Thus, compared to a sheetfed press, a typical web press can produce from three to four times the volume of printed goods in the same amount of time.

There are two categories of commercial lithographic web presses used extensively in the printing industry at this writing: full-width webs and half-width webs. Full-width presses, also known as 16 page webs, typically delver a cutoff of approximately 24 × 36 inches or sixteen page (8 1/2 × 11) signatures, imposed in a sheetwise page lay with eight pages per form. These full-width presses were popular in the early 1970s then lost ground to the more flexible and production-oriented half-width web press. Half-width webs—also known as narrow width or eight page web presses—typically accommodate a web width of approximately 26 inches, although there is no fixed definition of maximum web width for this press type. Eight-page webs were extremely popular for about a ten year period, beginning in the mid-1970s, largely due to their fast startup, quick changeover during production, and suitability for shorter runs. Most web printers found that the older full-width web press was too slow and cumbersome to stand up against the faster, more flexible narrow width press.

During the early 1980s, web press manufacturers began to modify the sixteen page web, making it a more suitable as a fast-turnaround production unit. Changes incorporated faster plate mounting techniques, improved inking systems, and new electronic peripherals which greatly speed up makeready procedures; the production output of such presses increased from a top speed of 1,500 fpm to 2,500 fpm. Thus, the full-width web became faster not only with respect to gross speed, but in the capacity to deliver twice the product because of a double maximum web width. Makeready times were significantly reduced compared to its half-width counterpart, making the full-width web the press of choice for newer web installations and for customers whose budgets are critical.

At this writing both older sixteen page webs and newer sixteen page webs, as well as a variety of half-width webs, are found throughout the industry. It is clear that older sixteen page webs can't address the same customer needs as their newer full-width counterparts, while half-width webs remain a viable, competitive production press against both older and newer sixteen page web units. It is expected that the next generation of half-width presses will improve in largely the same ways as their full-width counterparts. However, when new purchases are considered, many web printers are opting for full-width webs with the hope of being able to keep the press busy since output is so high.

All of the above addresses the most obvious advantage of web presses—both half width and full-width presses—which is high production output. There are, of course, other advantages as well.

2. Most web presses, while feeding from rolls of paper that allow for high production output, are equipped with postprinting converting attachments—equipment designed to cut the ribbon of paper into sheets at high speed, then fold these sheets, and finally collect them in sequence to make up a booklet or collection of pages for a particular job. Thus, web printing production provides a built-in bindery, where a job is not only printed, but then partially or completely finished as a single, in-line operation.

In fact, many large magazine publishers and publishers of catalogs build a complete binding line at the converting end of the press. Thus, signatures are printed, sheeted, folded, and then collected at extremely high speed, with direct in-line final binding equipment (such as perfect binding units) used to produce the final catalog, telephone book, or magazine. Since binding and finishing shares similar labor and capital intensiveness with the pressroom, having an in-line bindery linked to the press can save many hours of production time, thereby reducing manufacturing cost with high-volume usage.

It should be noted that other postprinting operations, such as gluing, slitting, perforating, numbering, and scoring, can also be done with proper web press attachments. Also, some web equipment is designed to print *roll-to-roll*—that is, the ribbon of paper is rewound after printing as a roll of paper, with no converting or finishing performed during presswork. Finishing will be completed later, perhaps during a follow-up imprinting of the roll of stock.

3. Quality of the web-printed image is good to excellent. The consideration of quality has been a much-discussed point between sheetfed and web experts, and the past 15 years have seen significant improvements in web lithographic print quality. Perhaps the two most important categories of quality measurement of printed products are image registration and color valuation. Recent years have seen many changes in web tension controls and electronic devices for precision and consistency of web registration during printing.

On-press color density measurement devices, linked to plate scanner technology for faster makeready, aid in focusing immediately on accurate ink coverage and color valuation, while also saving on makeready time and wasted paper. It is important to note that there are both heatset and nonheat-set lithographic web presses. Non-heatset webs—found largely in the newspaper and newspaper preprint industry—allow for ink drying by absorption of the ink vehicle into the substrate. Thus, the ink is somewhat wet immediately after printing, and dries slowly, over time. Heatset web presses, which produce thousands of printed products on coated and uncoated book stocks, are equipped with both chillers and dryers as post-printing operations, through which the ribbon of paper quickly moves. The printed paper passes through a chiller unit which solidifies the wet inks, then heated dryers evaporate the solvents fully, hardening the inks for immediate post-printing finishing operations. Advances in ink technology and ink distribution on press have improved both the quality and speed of drying for both heatset and non-heatset printed products. Experts generally agree that the print quality of web-produced products are equal to sheetfed images.

4. A final major advantage is that the cost of paper stock is generally from 3% to 7% less in roll form when compared to the same type of paper in sheets. Sheeting and cartoning of paper at the mill require extra time and equipment, thus increasing the cost of sheeted papers. Even though this reduction may not seem significant, since paper represents approximately 25% of the average cost of a printing order, even a slight cost reduction passed on to the customer may win the job for the web printer. Because paper is a direct material cost, the greater the quantity desired by the customer, the greater the savings.

For the competitive printing manager, the advantages of web are significant. However, there are disadvantages that must be considered.

1. Perhaps the most recognized limitation of web printing is that literally all web presses are built with an unchangeable cutoff length. Basically, *cutoff* is the measured distance around the blanket cylinder circumference and defines the maximum length of the printed image on the web sheet as it passes through the press. As the press revolves, the distance between repeated images represents the cutoff value. It is possible to plan images smaller than the cutoff, however, doing so increases white margin along the printed images that must be trimmed off and discarded, which is wasteful.

An inflexible cutoff makes it necessary for plants to have standardized product lines when they use web production, thereby making maximum use of press image area and reducing waste of paper. There is some flexibility because it is common to vary the width of the web to change product size, even though cutoff is fixed. While standardization is desirable in terms of manufacturing ease and production time savings, some printers feel too locked in under the fixed cutoff factor of a web press. Some of the narrow-width webs, such as business forms presses, do offer adjustable cutoffs for their presses, but adjustable cutoff requires interchange of printing cylinders on the press and is not considered practical at this time.

2. Paper spoilage during setup and production is generally higher with web printing than with sheetfed procedures. A number of factors contribute to this high spoilage condition. First, web breaks—the ribbon of paper moving rapidly through the printing units is torn or snapped apart—require rewebbing of the press and attendant shutdown and startup throwaways. Only a few such breaks, occurring at maximum production speed on the press, can mean the difference between profit and loss because of increased paper costs. Obviously, reduction of web breaks is a vital factor for improved web production and job profitability.

Second, since there are usually a number of printing units working in relation to one another, registering of these units to each other during press setup increases paper waste. Also, during production under full running conditions, one unit printing out-of-register can mean a large number of throwaway sheets since the integrity of the printing quality will be destroyed.

Third, web lithographic presses require intricate balancing controls between ink and water, as well as a large number of other technical setup and running considerations. During production, changes in such preestablished printing conditions can translate into an excessive number of spoiled sheets. Just as with web breaks and registration problems, lack of control of technical printing factors can adversely affect the profitability of the job through increased paper cost.

3. As already noted, ink-setting can utilize either heatset or non-heatset technology. For fancier products on book stock—which represent a major group of products purchased by advertising agencies and other customers—heatset web is the chosen process. This allows for immediate setting of inks and complete post-press finishing operations, so the entire product is manufactured in one operation. However, there are serious environmental concerns over air quality with respect to heatset web offset driers, and some states now require expensive afterburners to reduce the emissions from these presses to established levels. Adding such expensive peripheral equipment represents a major fixed expense to the printer, yet adds no significant production output to the press. Increased variable costs are also required since the afterburner requires energy to operate. Both of these increase the cost of the final product. Generally, however, the owner of the heatset web press accepts this additional cost with little complaint in an effort to cooperate in improving the environment.

4. Web lithographic printing involves many different press operations in comparison to sheetfed procedures and thus requires different skills for press operators. The problem is that the highly skilled web press operator, knowledgeable about roll preparation, technical printing factors, binding, converting, and a myriad of other technical production elements, is sometimes hard to find. Because webs operate at such high speeds and represent significant capital expense, the press operator is a vital cog to ensure efficient production with minimum paper waste and time loss. The skilled web press operator is an absolute necessity. Printing management cannot operate under profitable conditions without this skilled, yet hard-to-find, craftsperson.

15.3 Overview of Web Presses, Papers, and Inks

Types of Web Presses

There are three general classifications of web presses:

The newspaper web,
The commercial web,
The specialty web.

The *newspaper web* is commonly used for newspaper production with limited process color work and varying degrees of spot color. Newspaper webs are characterized by a, large number of *roll stands* where the rolls of paper are mounted for feeding into the printing unit. This setup implies that the majority of printing will be done as black-and-white, feeding one roll stand into one printing unit. Literally all newspaper web lithographic presses are *perfectors*, meaning that they print both sides of the sheet in one pass through the press. The usual substrate with newspaper webs is newsprint; however, both book and bond papers can be used also. Most newspaper webs contain no ink-drying units, so that all color work is completed using "wet trapping" procedures. Various equipment configurations are available as special equipment to control technical printing factors. Space considerations do not permit complete discussion of all aspects of newspaper webs and the many different equipment packages available to the printing plant.

The *commercial web*, used for most commercial types of printing production including process color, usually is equipped with fewer roll stands and more printing units. Webbing configurations (diagrams) then allow a single roll of paper to be fed from one unit into a second, and perhaps a third, each printing a different color in register with the others. Depending upon the number of printing units, commercial webs usually are very flexible in terms of webbing diagrams, so many color combinations are possible. Most commercial webs run with either coated or uncoated book stock, but newsprint or bond can also be used.

Commercial web presses—both full- and half-width varieties—usually have excellent web tensioning and print control mechanisms so that precise registration is easily completed during full press production. Heatset webs are equipped with a chiller/dryer combination, either between printing units or at the end of the last printing unit, to allow for complete setting of printed ink films before the web of paper moves into the converting/finishing process. Printing is typically done using wet-trapping techniques whereby wet ink film layers are printed one over the other, with heatset drying curing the collective wet ink layers all at one time. Since the cutoff is fixed, specialization of product and standardization during production are common. Full-width commercial webs are popular for the production of book signatures, magazines, catalogs, packaging products and many higher-volume advertising materials, while narrow-width webs are used to produce smaller signatures, a wide variety of commercial and advertising products, and generally for shorter runs. As with newspaper webs, a wide variety of specialized equipment can be purchased in conjunction with the full- or half-width web press.

Specialty webs are usually narrow-width webs used for the manufacture of specialized products such as business forms and labels. Thus, while they have built in inflexibility, they also represent a very profitable segment of the printing industry because of the increased output in producing a highly specialized, nontypical product. For example, in the production of business forms, there may be in-line printing, perforating, numbering, scoring, and carbon interleafing procedures in one press pass. Specialized webs are typically used with book and bond stocks and certain special stocks without production problems; newsprint is not usually a specialty web substrate.

Web presses may be lithographic, intaglio (gravure), letterpress, or flexographic. Typically, web lithographic presses, to which this discussion is primarily aimed, are perfecting presses, meaning that each printing unit has the capacity to print an image on both sides of the sheet in one pass. Perfectors are sometimes termed *blanket-to-blanket* presses. Intaglio, letterpress, and flexographic presses usually do not perfect; however, they share high-output capability with lithographic webs since speed of operation is comparable. Intaglio and flexographic webs are common to the packaging industry, and rotary letterpress is still used by some large daily newspapers in the world.

It is vital to note that press configurations of web lithographic units vary extensively. Consequently, the purchase of a lithographic web requires specification of the desired type of roll-changing apparatus, number of roll stands, number and configuration of printing units (they may be stacked on top of each other or, perhaps, be in line with one another), as well as a diversified selection of converting and binding equipment on the delivery end of the press. Additionally, there are special attachments, driers, web-tensioning devices, plate-dampening units, and many other special extras that must be selected. Obviously, the purchase of a lithographic web press requires printing management to determine exactly the product lines to be produced with the equipment. Web printing is a procedure by which a tailor-made product is produced for the printing customer. Web lithography is both a process and product specialization.

Web Paper

As previously stated, one of the major advantages of web printing over sheetfed is that web paper stock costs 3% to 7% less than the same paper purchased in sheeted form. If annual paper purchases for a printing company are high, this percentage of savings translates into thousands of dollars of capital available for other purposes during the year.

The three most common types of paper purchased for web lithographic printing are book (both coated and uncoated), bond, and newsprint. Coated book, used largely with the commercial web press, is popular for most glossier (fancy) publications, magazines, and annual reports, especially where process color is desired. Uncoated book is used predominantly in the production of textbooks, calendars, and for thousands of other general commercial products. Uncoated book is probably the single largest seller of all web papers. Bond stocks are generally used in conjunction with specialty web operations in business forms, but are used for many commercial products also. Newsprint has greatest application with newspaper web presses and is also used with newspaper and commercial web equipment for preprint and supplement ads and inserts and many other products requiring a less expensive substrate.

There are some general facts regarding web paper. Thin caliper stocks seem to run better and offer a large number of linear feet per roll. Heavy papers and those with thick caliper have less footage (requiring more frequent roll changes) and seem to print and convert with more difficulty. In terms of winding, all rolls should be wound as tightly and uniformly as possible at the mill, regardless of the type of paper. The cores for web paper are made of either metal or fiber (paperboard), but there are also coreless rolls. Metal cores are preferred for wide rolls and coated stocks. With either type of core roll, weight will be high. Fiber cores are preferred by many plants since they are easy to work with and do not require storage and return to the mill—a necessity with metal cores. Common inside diameters for cores are 3 and 4 inches.

Web paper can be wound at the mill either *felt* (smooth) side in or out and slit at the specific width desired. Moisture content of most web stocks resembles that of sheeted stock—that is, between 4% and 6% moisture content. Flags or roll markings will be used to indicate splices and flawed sheet formation areas; many flags mean a greater probability of web breaks and production problems. Wrapping of the rolls for shipment from the mill may be done using either moisture-proof or non-moistureproof material. When ordering paper, felt in or out, the number of acceptable flags and type of wrapping should be specified. Of course, roll dimension (width), roll weight, and type of paper should also be indicated, as should any special instructions and delivery information.

Web Inks

Web lithographic inks seem to vary from manufacturer to manufacturer. Generally, differences in print quality, production performance, and working conditions will be noticed between the more expensive inks and those in the reduced price group.

Web ink consumption is contingent upon constant and variable factors. The constant values relate to the amount of ink necessary for initially inking the system and the amount of ink lost during washups. Both of these factors, once determined for a specific web press, will change very little over time. Variable ink consumption is an entirely different matter. Two factors relate here: the mileage of the ink on the press and the absorption rate of the substrate. As with sheetfed inks, both press mileage and absorption are difficult to determine precisely since even slight changes in type of ink or substrate can result in pronounced shifts in either category.

Web inks can be purchased in 5 pound cans, in the 10 and 25 pound sizes, by the 55 gallon drum, and by the railroad tank car (especially for newsinks). Overall web ink prices are lower when compared to sheetfed ink costs. Many large web offset printers buy and store a large volume of ink so that they are assured the desired quantities when necessary.

Most web inks are estimated on the basis of the number of pounds of ink per 1000 press cutoffs. For example, a 35 inch web with a 23 inch cutoff printing a 16 page signature might be estimated for ink using a factor of "2.9 pounds of ink per 1000 cutoffs" on coated stock. This same job, printed on uncoated paper but with all other factors equal, might have a factor of "3.9 pounds per 1000 cutoffs." Such coverage data vary from plant to plant; thus, it is recommended that each company determine ink consumption based on actual plant performance. In this manner, they can fit their inks to standard paper stocks, printing conditions, and presses.

15.4 Basic Web Paper Formulas

For many estimators who understand sheet paper costing and mathematics (see Chapter 6), web estimating seems to hold an aura of mystery. This perception is unfortunate because estimating web paper (paper in roll form) is not really difficult.

The accompanying formulas have been simplified and presented in such a manner that substitution of variable data is all that is required. It is important to note that calculation of each segment, such as roll footage or the number of cutoffs per roll, should be mastered since such information coordinates with the data required when estimating the complete web problem.

Determining Roll Footage

The following formula allows for the calculation of the number of feet in a roll of paper when the roll diameter, core diameter, and stock caliper of the stock are known.

$$\text{roll footage} = \frac{0.06545(D^2 - d^2)}{T}$$

where

D = roll diameter in inches
d = core diameter in inches
T = stock caliper in inches
0.06545 = a constant that provides for conversion of inches to feet

Example 15.1

Determine the footage of a roll of substance 55 book, caliper 0.004 inch, with a roll diameter of 40 inches, and wound on a 3 inch metal core.

Solution Determine the roll footage:

$$\text{roll footage} = \frac{0.06545(40 \text{ in.}^2 - 3 \text{ in.}^2)}{0.004 \text{ in.}} = \frac{0.06545(1600 \text{ in.}^2 - 9 \text{ in.}^2)}{0.004 \text{ in.}}$$

$$= \frac{0.06545(1591 \text{ in.}^2)}{0.004 \text{ in.}} = 26{,}033 \text{ ft}$$

The roll footage formula works for all types of paper stocks in precisely the substitution procedure indicated. It should be noted that roll width plays no part in the amount of linear feet in a roll of paper and is therefore excluded from any portion of these calculations.

Determining the Number of Cutoffs per Roll

Essentially, the number of cutoffs per roll can be translated to mean the number of pieces, sheets, or signatures that the roll of paper will produce during printing. To determine this "quantity per roll" figure, the estimator must know the number of feet per roll and the cutoff size of the press to be used for printing. The following formula is used:

$$\text{number of cutoffs per roll} = \frac{\text{number of feet per roll} \times 12 \text{ inches per foot}}{\text{press cutoff}}$$

where press cutoff is given in inches.

Example 15.2 A roll of newsprint contains 26,033 linear feet of paper, to run on a web press with a cutoff of 23 1/2 inches. How many cutoffs (ct) are in the roll if the entire roll will be used?

Solution Determine the number of cutoffs in the entire roll:

$$\text{number of cutoffs} = \frac{26{,}033 \text{ ft} \times 12 \text{ in./ft}}{23.5 \text{ in.}} = 13{,}293 \text{ ct}$$

where
12 inches per foot = a constant

It is, of course, unrealistic to expect that the total roll will be used during printing. There is generally a built-in waste factor to accommodate for what is sometimes termed *core waste.*

Determining the Weight per 1000 (M) Cutoffs

The weight per 1000 cutoffs is the actual scale weight of 1000 printed and finished pieces at the delivery end of the web press. The reason such a calculation is made is twofold:

1. Ink estimating is sometimes based on the "weight/M cutoffs" figure (which will be detailed later in this chapter)
2. Such weight information can be used to calculate relative shipping weights of printed materials.

$$\text{weight per M cutoffs} = \frac{(c \times w)2b}{s}$$

where

c = cutoff in inches
w = roll width in inches
b = basis weight of paper
s = basic sheet size in square inches
2 = a constant

Example 15.3 Determine the weight per M cutoffs for a press with a cutoff of 22 3/4 inches running a roll 36 onches wide of substance 55 book. Add an additional 3% to compensate for core waste, white waste, and spoilage.

Solution Determine weight per M cutoffs:

$$\text{weight/M cutoffs} = \frac{(22\ 3/4 \text{ in.} \times 36 \text{ in.})\ 110 \text{ lb}}{25 \text{ in.} \times 38 \text{ in.}} = 94.83 \text{ lb}$$

Total weight including spoilage: 94.83 lb × 103% = 97.67 lb

To save time, some estimators prefer to calculate the weight per M cutoffs (or weight per M sheets for sheet papers) using a "constant schedule" as provided in Figure 15.1. The constant factor is calculated by doubling the basis weight of the identified shock and dividing that figure by the square inches of the basic size. For example, sub 50 book with a basic size of 25 × 38 would then calculate as sub 50 × 2 = 100 ÷ 950 sq in. = 0.1053. For use, if a cutoff of sub 50 book measured 24 × 36 inches, the weight per M cutoffs would then be 24 × 36 or 846 sq in. × 0.1053 (constant factor) which equals 90.98 or approximately 91 M.

NEWSPRINT TAG KRAFT		BONDS, WRITING & LEDGER		BOOK PAPER	
Basis Wt. 24 × 36/500	1000 Sheet Factor	Basis Wt. 17 × 22/500	1000 Sheet Factor	Basis Wt. 25 × 38/500	1000 Sheet Factor
32	.0741	8	.0428	35	.0737
34	.0787	9	.0481	40	.0842
35	.0810	11	.0588	45	.0947
40	.0926	12	.0642	50	.1053
50	.1157	13	.0695	60	.1263
60	.1389	16	.0856	70	.1474
100	.2315	20	.1070	80	.1684
125	.2894	24	.1283	100	.2105
150	.3472	28	.1497	120	.2526
175	.4051	32	.1711		
200	.4630	36	.1925		

BRISTOL		COVER PAPER		INDEX BRISTOL	
Basis Wt. 22 1/2 × 28 1/2/500	1000 Sheet Factor	Basis Wt. 20 × 26/500	1000 Sheet Factor	Basis Wt. 25 1/2 × 30 1/2/500	1000 Sheet Factor
67	.2090	50	.1923	90	.2314
80	.2489	60	.2308	110	.2829
90	.2807	65	.2500	140	.3600
100	.3119	80	.3077		
120	.3743	90	.3462		
140	.4366	100	.3846		
160	.4990	130	.5000		

Figure 15.1

Constant factors used to determine the weight per M cutoffs or weight per M sheets.

Determining the Weight of a Roll of Paper

The total weight of a roll of paper is important for shipping, warehousing, and costing purposes. Total weight of a roll of paper is dependent upon many factors: type of paper, thickness, coatings used, sizing compounds, and wrapping materials, as well as the roll dimensions themselves. The following formula is used to calculate roll weight:

$$\text{roll weight} = \frac{0.00157(D^2 - d^2) \times w \times b}{T \times s}$$

where

0.00157 = a constant

Example 15.4

Determine the weight of a roll of newsprint that measures 36 inches wide and 40 inches in diameter. It has been wound on a 3 inch core. Stock is substance 30 with a caliper of 0.0025 inch.

Solution Determine the roll weight:

$$\text{roll weight} = \frac{0.00157(40 \text{ in.}^2 - 3 \text{ in.}^2) \times 36 \text{ in.} \times 30 \text{ lb}}{0.0025 \text{ in.} \times (24 \text{ in.} \times 36 \text{ in.})} = 1249 \text{ lb}$$

Determining Slabbing Waste

Slabbing of paper rolls is a process by which the outside of the paper roll is cut away, usually with a knife or power saw, because of damage to the outer roll. The paper removed from the roll is typically discarded. Because the outside of the roll contains a significant portion of the paper on the roll, cutting off an inch or more may represent a major loss of paper. The roll will produce less cutoffs than originally estimated, perhaps requiring additional rolls and roll changes to make up for this paper loss. Thus, slabbing is generally undesirable.

The usual reason for slabbing paper from rolls is damage to the roll during transit or handling by the shipping agent or in the plant. For example, foreign matter may become embedded in the roll and require slabbing to remove it, or a gouge of the outer roll might occur due to careless forklift operation. If the damage occurs during shipping, this fact should be noted on the bill of lading when the roll is delivered to the plant. Typically, the shipper will acknowledge the imperfection and make restitution through insurance or other means once the amount for damage can be determined. Of course, any damage occurring inside the plant is not shipping-related damage and represents lost paper and lost company dollars since the slabbed paper becomes a company expense and cannot be charged to the customer's account. Not only is the slabbed paper thrown away, which represents a dollar loss to the printer, but additional paper will need to be substituted in its place, which may affect roll inventories for other jobs. The following formula is used for calculating slabbing waste:

$$\text{slabbing waste} = \frac{4(DS - S^2)}{D^2} \times ORW$$

where

D = original roll diameter

S = slab thickness in inches to be cut away

ORW = original roll weight

4 = a constant

Example 15.5

A 2 inch slab must be cut from a roll of basis 50 book due to damage to the roll by the shipper. Original roll diameter is 36 inches, and original roll weight is 1850 pounds. Cost of paper is $420 per ton. Determine poundage and cost of slabbed paper.

Solution

Step 1 Determine the weight of the slabbing waste:

$$\text{slabbing waste} = \frac{4\,[(36\text{ in.} \times 2\text{ in.}) - 2\text{ in.}^2]}{36\text{ in.}^2} \times 1850\text{ lb}$$

$$= \frac{272}{1296} \times 1850 = 388\text{ lb}$$

Note: The waste is about 21% of the total roll.

Step 2 Determine cost of paper:

388 lb × ($420.00 ÷ 2000 lb) = $81.48

15.5 Estimating Web Production

Web production estimating is divided into three major segments:

Estimating paper requirements and spoilage,
Estimating ink consumption,
Determination of press makeready and running times and costs.

Since production on the press must include paper additions, the estimate must be "built" by first determining total paper requirements. Ink quantity and press time are then calculated based on total estimated paper needs. Application of previously presented roll formulas will be made.

It must be stressed emphatically that development of all categories of web production data must be completed for each web press on an individual basis. As will be noted with the accompanying discussion regarding paper spoilage, there are a large number of equipment and production differences that exist from one web to the next and among different web plants. These variables apply not only to paper consumption, but also materially affect ink and press operations. For this reason, all schedule data in this text are presented for example only. While data may appear realistic for one web operating under a certain production-equipment discipline, they may be totally "off base" for another web operation. The only way to ensure accurate web estimating data is through individual development and continual maintenance of this information.

Three general points regarding the establishment of individual web production standards should be made. First, it is best if only one individual determines such data, working in a careful and scientific manner; this practice ensures a consistency of orientation. Of course, the individual must be fully conversant with web procedures and technology and especially aware of requirements for establishing accurate production standards and BHRs (see Chapter 3).

Second, it will probably be necessary for this individual to design web production forms. For example, ordering web paper is an important factor; the ordering procedure should be carefully detailed so that the many variables will be considered and specified accurately. In terms of estimating web production, it is important that the webbing of the press be detailed with each job or when major webbing changes are evident. This procedure requires that a webbing diagram form be completed to aid platemaking and press production. The diagram must precisely represent the printing units and on-press converting operations and should accompany the job through the plant during production. Form design for web requires careful effort to minimize costly production errors and ensure recovery of accurate production times for modification of production standards.

Third, development and maintenance of web data should be discussed frequently with the head press operator on the web for which the data are applicable. This individual is the most knowledgeable of all plant personnel regarding the press operating conditions, technical printing factors, and spoilage information. Some web estimators consult with the head press operator about complicated jobs or when estimating data do not seem to fit the job under consideration. The competent web operator is a most valuable source of information to the web estimator and should not be ignored.

Estimating Roll Paper Spoilage

Figure 15.2 has been developed to estimate the amount of extra paper necessary for spoilage during production. The basic spoilage percentage includes spoilage for core waste; wrap; normal waste for webbing, prepress setup, and makeready; usual running conditions; and on-press binding operations. Penalties indicated apply as a total percentage to the basic percentage (see Example 15.6). All data in Figure 15.2 are for example only and must be modified to reflect variations in production, press, and binding for the individual web equipment.

There are a large number of equipment and production variables affecting the database used in Figure 15.2:

1. There are differences in the types of web presses, number of roll stands, and number of printing units.
2. There are differences in technical printing attachments and special controls for improved consistency and greater production.
3. There are differences in types of products produced and special operations required.

4. There are differences in types of paper stock and other substrates. Coated stocks may blister or pick; thick papers may reduce press speeds considerably. Coordination with drying units is also important.
5. There are differences in types of roll feeding and roll change techniques. Web paper can be fed from manually loaded roll stands or from roll stands that contain automatic splicing equipment. Manual roll changes increase spoilage dramatically since the press must be stopped, while automatic pasting devices connect rolls at normal press speeds.
6. There are differences to be considered with converting and binding operations. These converting operations can be done on-press only for some jobs; other orders may require additional finishing completed separately in the bindery. Paper spoilage from converting and binding operations may be as significant as spoilage during the press run. Extra spoilage must be included when additional operations are required. For the most part, web estimators include normal on-press binding spoilage in the basic spoilage schedule.
7. There are differences in the number of stops and startups for plate changes and other identified production operations.

Number of Copies	Basic Spoilage (percentage)*
Under 10,000	18
10,001-25,000	15
25,001-50,000	13
50,001-100,000	11
100,001-250,000	9
250,001-500,000	7
Over 500,000	4

***Percentage of additional cutoffs/units includes setup of first roll and initial webbing procedures.**
Penalties:
Manual roll changes (during press run): Add 0.50% for each roll change after first (startup) roll. Automatic roll changes (during press run): Add 0.10% for each roll change after first (startup) roll. Color: Add 1.0% for color for each additional color unit.
Stock penalties:
Newsprint: Add 0% for sub 30 or 32,
Add 1.0% for sub 28 or less. Uncoated book: Add 0% for sub 45 through sub 75, Add 1.5% for sub 40 or less, Add 25% for sub 80 or more. Coated book: Add 0% for sub 45 through sub 75, Add 2% for sub 40 or less, Add 3% for sub 80 or more. Gluing: Add 1% per on-press gluing unit.
Off press finishing: Determine operations and increace paper accordingly.

Figure 15.2

Web press spoilage schedule: Full-width webs

8. There are differences in job scheduling. Spoilage can increase significantly when complete stock changeover is needed from one job to the next or when webbing conditions are completely different between jobs.
9. There are differences in job quality factors wherein special color matches, intricate register, or other quality requirements cause increased throwaways during printing.

Example 15.6

One roll stand is being fed into one perfecting printing unit. Rolls measure 35 inches wide by 40 inches in diameter and are wound on 3 inch cores. Stock is substance 50 book with a caliper of 0.0035 inch. Press cutoff is 22 1/2 inches, making the final signature size 22 1/2 × 35 inches, with pages measuring 8 3/4 × 11 1/4 inches. Each signature contains 16 pages. Estimate the amount of paper required to run 60,000 (16 page) signatures, black-and-white only, using this one-unit web press. Manual roll changes. Determine:

The number of total cutoffs (ct) that will be printed including spoilage,
The number of rolls that will be required including spoilage.

Also, if the cost of this paper is $850 per ton, what will be the total cost of paper for this job?

Solution Use the given formulas and Figure 15.2 as necessary.

Step 1 Determine basic roll paper information:

Number of linear feet per roll:
$[0.06545\ (40\ \text{in.}^2 - 3\ \text{in.}^2)] \div 0.0035\ \text{in.} = 29{,}752\ \text{ft/roll}$
Number of cutoffs per roll:
$(29{,}752 \times 12\ \text{in./ft}) \div 22.5\ \text{in.} = 15{,}867\ \text{ct}$

Step 2 Add spoilage percentage (see Figure 15.2):

Base percentage at 60,000 copies:
11%
Number of manual roll changes:
60,000 copies ÷ 15,867 ct = 3.78 rolls = 4 rolls on press −1 roll (with setup) = 3 roll changes
Percentage of roll changes:
3 rolls × 0.5% = 1.5%
Total spoilage required:
11% + 1.5% = 12.5%

Step 3 Determine total paper required using cutoff data:

Total cutoffs with spoilage:
60,000 copies × 112.5% = 67,500 ct
Number of pounds per M cutoffs using basic formula:

$$\frac{(22\frac{1}{2}\text{ in.} \times 35\text{ in.})100\text{ lb}}{(25\text{ in.} \times 38\text{ in.})} = 82.89\text{ lb}$$

Total stock poundage:
82.89 lb/M ct × 67.5 M ct = 5595 lb

Step 4 Determine total paper required using roll data:

Total rolls required:
3.78 rolls × 112.5% = 4.253 rolls
Roll weight using basic formula:

$$\frac{0.00157(40\text{ in.}^2 - 3\text{ in.}^2) \times 35\text{ in.} \times 50\text{ lb}}{0.0035\text{ in.} \times (25\text{ in.} \times 38\text{ in.})} = 1315\text{ lb}$$

Total weight of paper:
4.253 rolls × 1315 lb/roll = 5593 lb

Step 5 Determine paper cost:

5593 lb × ($850.00 ÷ 2000 lb) = $2377.03

Step 6 Summarize your findings:

Total cutoffs: 67,500 ct
Total weight: 5593 lb
Total paper cost: $2377.03

Establishing Web Ink Consumption

Just as with paper spoilage, there are many variables affecting ink consumption during web printing. Perhaps the most prevalent is the ink mileage difference evident from one web press to the next. The difference can be attributed to different press operators, different kinds of presses, variations in ink fountain setup (where the setting might be light or heavy with little visible difference in density during printing), and variations between different brands of ink.

When web ink consumption is estimated, too frequently "rules of thumb" are used instead of more clearly identified data. In addition, sometimes ink is estimated on the number of finished pieces only (or the number of ordered cutoffs), while press spoilage may have a significant bearing on increased ink requirements.

Black Ink Coverage Schedule

Percentage of Coverage of 24 × 36 Inch Form or Plate	Pounds per M Cutoffs		
	Newsprint	Uncoated (Smooth) Book and Bond	Coated Book
10	0.27	0.25	0.17
15	0.41	0.37	0.25
20	0.55	0.49	0.33
25	0.68	0.61	0.41
30	0.82	0.74	0.50
35	0.95	0.86	0.58
40	1.09	0.98	0.66
45	1.22	1.10	0.74
50	1.36	1.22	0.83
60	1.63	1.46	0.99
70	1.90	1.71	1.16
80	2.18	1.95	1.32
90	2.45	2.20	1.49
100	2.72	2.44	1.65

Note: Cutoffs measure approximately 24 × 36 in. and are printed by the web lithographic process on a full-width web.

Schedule Adjustment for Colored Inks

Color of Ink	Adjustment Factor		
	Newsprint	Uncoated (Smooth) Book and Bond	Coated Book
Process blue	1.11	1.07	1.06
Process red	1.23	1.18	1.10
Process yellow	1.16	1.13	1.09
Opaque red	1.72	1.68	1.51
Opaque blue	1.72	1.68	1.51
Opaque green	1.94	1.88	1.69
Opaque yellow	1.84	1.80	1.62

Note: Multiply the factor indicated in the schedule by the black ink coverage factor in the black ink schedule at the appropriate percentage of coverage. The result is the number of pounds of colored ink for a form or plate measuring approximately 24 × 36 in. printed by web lithography.

Figure 15.3

Web ink consumption schedule

Rules for sheetfed inks generally hold true for web inks (see Chapter 7). One difference is that web inks are purchased in larger quantities since the process is fast and volumes are usually high. Differences in cost normally translate into differences in quality: The less expensive inks characteristically perform poorly when compared to the more expensive formulations.

The accompanying web ink estimating schedule, Figure 15.3, is based on the number of pounds of ink per 1000 cutoffs per form. Cutoff sheets measure approximately 24 × 36 inches. For example, using newsink on newsprint, we have a single perfecting unit printing black only. Coverage for Form I is 35% and for Form II is 50%. Therefore, Form I will require 0.95 pound of ink per 1000 cutoffs, and Form II will consume 1.36 pounds of black ink per 1000 cutoffs printed. Because the ink type and color are the same, these figures may be added together. In this example, 1000 cutoffs measuring approximately 24 × 36 inches, printed on both sides with newsblack, will require 2.31 pounds of ink.

Adjustments of colors as related to black coverage are provided in a separate segment of Figure 15.3. When changes in cutoff dimensions are applicable, schedule data should be increased or decreased proportionately. As with other estimating schedules in this book, data must be adjusted as appropriate to fit individual operating conditions of the plant and the process used. The data can be substituted into the following formula to calculate ink poundage:

total ink poundage = (ink poundage per M cutoffs for left form + ink poundage per M cutoffs for right form) × adjustment factor, if applicable × (total number of cutoffs including spoilage ÷ 1000)

Example 15.7

A four-unit perfecting web is used in the production of process color booklets, 9 × 12 inches (untrimmed page) in 16 page signatures. Press cutoff is 24 inches; all rolls are 36 inches wide. We have an order for a process color booklet to be printed on coated book, basis 60. Ink coverage, by unit and color, is:

Unit I: Process yellow: Left form 40%, right form 50%
Unit II: Process blue: Left form 55%, right form 60%
Unit III: Process red: Left form 40%, right form 70%
Unit IV: Process black (and typematter): Left form 45%, right form 40%

Ink costs per pound are $3.25 for process blue, $3.50 for process yellow and red, and $1.65 for black. Determine ink quantity and cost of ink for a gross press run of 52,300 cutoffs (copies).

Solution Use Figure 15.3.

Step 1 Determine ink quantity and cost for process yellow:

Base pounds from black schedule:
0.66 lb/M ct + 0.83 lb/M ct = 1.49 lb/M ct

Adjustment using factor from color schedule:
1.49 lb/M ct × 1.09 = 1.62 lb/M ct
Total amount of ink:
1.62 lb/M ct × 52.3 M ct = 84.73 lb
Total cost:
84.73 lb × $3.50 lb = $296.55

Step 2 Determine ink quantity and cost for process blue:

Base pounds from black schedule:
0.91 lb/M ct + 0.99 lb/M ct = 1.9 lb/M ct

Note: You must use a chart figure halfway between 50% at 0.83 lb and 60% at 0.99 lb.

Adjustment using factor from color schedule:
1.9 lb/M ct × 1.06 = 2.01 lb/M ct
Total amount of ink:
2.01 lb/M ct × 52.3 M ct = 105.12 lb
Total cost:
105.12 lb × $3.25/lb = $341.64

Step 3 Determine ink quantity and cost for process red:

Base pounds from black schedule:
0.66 lb/M ct + 1.16 lb/M ct = 1.82 lb/M ct
Adjustment using factor from color schedule:
1.82 lb/M ct × 1.10 = 2 lb/M ct
Total amount of ink:
2 lb/M ct × 52.3 M ct = 104.6 lb
Total cost:
104.6 lb × $3.50 lb = $366.10

Step 4 Determine ink quantity and cost for black:

Base pounds from black schedule:
0.74 lb/M ct + 0.66 lb/M ct = 1.4 lb/M ct
Adjustment:
None
Total amount of ink:
1.4 lb/M ct × 52.3 M ct = 73.22 lb
Total cost:
73.22 lb × $1.65 lb = $120.81

Step 5 Summarize your findings:

Process yellow: 84.7 lb; cost, $296.55
Process blue: 105.1 lb; cost, $341.64
Process red: 104.6 lb; cost, $366.10
Black: 73.2 lb; cost, $120.81

Estimating Web Lithographic Press Production

Estimating of lithographic web press production is divided into two major segments: press preparation and press running. Some plants lump press preparation operations together to include all major preparatory elements: webbing the press, adjusting web controls, mounting rolls on roll stands, hanging plates, setting ink and dampening controls, makeready, setting converting and folding units, and all washups required during the run and afterwards. Other plants break out those preparatory operations where they find individual attention must be directed during production. For example, a plant might consider all the above items together with the exception of press washups, which would be added to the estimate on an individual basis.

Individual judgment on the part of plant management and the estimator is necessary if estimated press preparation time is to be accurate. In this book, preparatory time will be lumped, thus combining all operations into one "per printing unit" hourly figure. For example, "1.3 hours/unit" covers all preparatory time for one perfecting printing unit (two plates). There is no established time standard for press preparation since so much variation exists in equipment and production techniques throughout the industry.

Estimating press running time is simplified by Figure 15.4, which provides data for the number of cutoffs per hour related to the speed of the web press in feet per minute (fpm). Data is presented from 100 fpm to 2500 fpm, in 100 fpm increments, to cover literally all web presses in use today. However, estimating press speed is a relative matter. For example, when production is standardized and routine for the company, accurately estimating running speed and, thus, cutoffs per hour is not difficult. Yet, when work begins to depart from normal procedures—for example, differing web widths in the same press run, unusual or less-typical paper stocks, webbing changes which produce a unique product, or special in-line finishing devices which may affect press speed—estimating web running times accurately can be a difficult problem.

Most web estimators learn from early experience that it is best to check with plant production staff and pressroom personnel when deviations from normal web press running is likely. The head press operator, on whose web press the job will likely run, is an excellent source for such information, as might be the production supervisor or the web pressroom superintendent. Involved discussions and consultation allow the estimator to determine with some accuracy the expected running speeds in feet per minute, cutoffs per hour or finished pieces per hour. Feet per minute (fpm) is a common measurement of web production, although cutoffs per hour and finished pieces per hour are also used.

Once the anticipated speed has been determined, as well as the cutoff length, the estimator then consults Figure 15.4 at the applicable cutoff dimension for the press, cross-referenced by the evaluated speed. The resultant figure represents the number of cutoffs, sheets, or signatures that will be produced per hour on the press. This figure is then divided into the total number of copies to be printed (including spoilage) to determine press time required for the job. When press time and preparatory time are added and then multiplied by the BHR for the press, a total press cost is achieved. Materials such as paper, ink, and plates are added separately.

Example 15.8 is provided to demonstrate how estimating web presswork is completed. Example 15.9 is provided to show how an entire web estimating problem would be done. The following formulas may be used when working the examples:

$$\text{total web press production time} = (\text{number of units printing} \times \text{preparation time per unit}) + \frac{\text{total number of cutoffs, including spoilage}}{\text{press running speed in cutoffs per hour}}$$

$$\text{total web press production cost} = \text{total web press production time} \times \text{web press BHR}$$

Example 15.8

We have a four-unit web press feeding 2 roll stands into 2 units each, producing a two-color finished product. Preparation and makeready time per printing unit is 0.75 hour, which includes roll mounting and webbing. The press will run at 670 feet per minute (fpm). BHR cost for this web press is $152.50. Determine the time and cost for presswork for this job if the gross press run is 72,600 pieces. Press cutoff is 22 3/4 inches.

Solution Use Figure 15.4 as necessary.

Step 1 Determine press preparation time:

4 units × 0.75 hr/unit = 3 hr

Step 2 Determine press-running time at 22 3/4 in. cutoff (see Figure 15.4):

Press speed at 670 fpm:
3165 ct/hr at 100 fpm × (670 fpm ÷ 100 fpm) = 21,206 ct/hr
Press time:
72,600 ct ÷ 21,206 ct/hr = 3.43 hr

Step 3 Determine total production time and cost:

Total production time:
3 hr + 3.43 hr = 6.43 hr

Total production cost:
6.4 hr × \$152.50/hr = \$980.58

Example 15.9

A printer has a three-unit perfecting newspaper web press with 3 roll stands. Manual roll changes. Each unit can produce 8 page tabloids running from standard size rolls 34 inches (wide) × 45 inches (diameter) × 0.0025 inch caliper newsprint, substance 30. All roll stands are manually prepared. Paper is pur-

Web Press Cutoff in Inches	Number of Linear Running Feet per Minute														
	100	200	300	400	500	600	700	800	900	1000	1100	1200	1300	1400	1500
	Cutoffs per Hour														
8½	8470	16941	25412	33882	42353	50823	59294	67764	76235	84705	93176	101646	110117	118587	127058
11	6545	13090	19635	26180	32725	39270	45815	52360	58905	65450	71995	78540	85085	91630	98175
14	5143	10286	15429	20571	25714	30857	36000	41143	46286	51428	56571	61714	66857	72000	77143
17	4235	8471	12706	16941	21176	25412	29647	33882	38118	42353	46588	50823	55059	59294	63529
17 1/2	4114	8229	12343	16457	20571	24686	28800	32914	37029	41143	45257	49371	53486	57600	61714
22	3273	6545	9818	13091	16364	19636	22909	26182	29454	32727	36000	39273	42545	45818	49091
22 3/4	3165	6330	6494	12659	15924	18989	22154	25319	28483	31648	34813	37978	41143	44308	47472
23	3130	6261	9391	12522	15652	18783	21913	25043	28173	31304	34435	37565	40696	43826	46956
23 1/2	3064	6128	9191	12255	15319	18383	21447	24511	27574	30638	33702	36766	39830	42893	45957
23 9/16	3056	6111	9167	12223	15279	18334	21390	24446	27501	30557	33613	36668	39724	42780	45836
24	3000	6000	9000	12000	15000	18000	21000	24000	27000	30000	33000	36000	39000	42000	45000
25 1/2	2824	5647	8471	11294	14118	16941	19765	22588	25412	28235	31059	33882	36706	39529	42353
26	2769	5538	8308	11077	13846	16615	19385	22154	24923	27692	30462	33231	36000	38769	41538
35	2057	4114	6171	8229	10286	12344	14400	16457	18514	20571	22629	24686	26743	28800	30857
41 5/8	1730	3459	5189	6919	8649	10378	12108	13838	15567	17297	19027	20757	22486	24216	25946
44 1/2	1618	3236	4854	6472	8090	9708	11326	12944	14562	16178	17798	19416	21034	22652	24270

Note: To convert other press cutoffs per hour not in the schedule, use the following formula: $\frac{\text{No. of ft/min} \times 12 \text{ in/ft} \times 60 \text{ min/hr}}{\text{press cutoff (in.)}}$

Web Press Cutoff in Inches	Number of Linear Running Feet per Minute									
	1600	1700	1800	1900	2000	2100	2200	2300	2400	2500
8 1/2	135529	144000	152471	160941	169412	177882	186353	194824	203294	211765
11	104727	111273	117818	124364	130909	137455	144000	150545	157091	163636
14	82286	87429	92571	97714	102857	108000	113143	118286	123429	128571
17	67765	72000	76235	80471	84706	88941	93176	97412	101647	105882
17 1/2	65829	69943	74057	78171	82286	86400	90514	94629	98743	102857
22	52364	55636	58909	62182	65455	68727	72000	75273	78545	81818
22 3/4	50637	53802	56967	60132	63297	66462	69626	72791	75956	79121
23	50087	53217	56348	59478	62609	65739	68870	72000	75130	78261
23 1/2	49021	52085	55149	58213	61277	64340	67404	70468	73532	76596
23 9/16	48891	51947	55003	58058	61114	64170	67225	70281	73337	76393
24	48000	51000	54000	57000	60000	63000	66000	69000	72000	75000
25 1/2	45176	48000	50824	53647	56471	59294	62118	64941	67765	70588
26	44308	47077	49846	52615	55385	58154	60923	63692	66462	69231
35	32914	34971	37029	39086	41143	43200	45257	47314	49371	51429
41 5/8	27676	29405	31135	32865	34595	36324	38054	39784	41514	43243
44 1/2	25888	27506	29124	30742	32360	33978	35596	37213	38831	40449

Figure 15.4

Web lithographic production schedule

chased for $380 per ton. All rolls are wound on 3 inch cores. Press cutoff is 22 3/4 inches. BHR is $95. A customer wants 45,000 copies of a tabloid-sized advertising circular. It will run 8 pages, black only, with another 8 pages black with process blue as a spot color. (This means that 2 roll stands will feed 2 rolls for printing: One roll will feed into one unit and print black only; the second roll will feed into two consecutive units, the first printing black and the second printing blue on both sides of the sheet.) Ink coverage, by unit and plates, is as follows:

Unit I: Black: Left form 30%, right form 40%
Unit II: Black: Left form 25%, right form 35%
Unit III: Process blue: Left form 20%, right form 30%

Ink costs $2.25 per pound for newsblack and $2.80 per pound for process blue (use uncoated book schedule data for estimating). The press will operate at 700 feet per minute. Preparatory time is estimated to be 0.8 hour per unit. Determine:

Total number of cutoffs and rolls necessary including spoilage (for inventory purposes),
Poundage and cost of paper,
Poundage and cost of ink,
Press time and cost.

Solution Use Figures 15.2, 15.3, and 15.4 as necessary.

Step 1 Determine the configuration for running: Feed 2 rolls from 2 roll stands. First roll prints black only (Unit I), while second roll prints black and spot blue (Units II and III). There is a total of 6 plates. All units are perfect.

Step 2 Determine basic roll paper information using basic formulas:

Number of linear feet per roll:

$$\frac{0.06545\ (45\text{ in.}^2 - 3\text{ in.}^2)}{0.0025\text{ in.}} = 52{,}779\text{ ft}$$

Number of cutoffs per roll:

$$\frac{52{,}779\text{ ft} \times 12\text{in./ft}}{22\ 3/4\text{ in.}} = 27{,}839\text{ ct}$$

Step 3 Determine spoilage percentage per roll stand (see Figure 15.2):

Base percentage at 45,000 cp:
13%

Roll changes per stand:
45,000 copies ÷ 27,839 ct = 1.616 rolls = 2 rolls on stand −1 roll (setup) = 1 roll change/stand
Percentage of roll changes per stand:
1 roll × 0.5% = 0.5%
Total color spoilage:
1% (spot blue)
Total spoilage per roll stand (base percentage + roll change percentage + color percentage = total spoilage percentage):
13% + 0.05% + 1% = 14.5%

Note: Since rolls feed together, the maximum percentage above applies to all stands.

Step 4 Determine the amount of newsprint required using cutoff data:

Cutoffs per roll stand:
45,000 copies × 114.5% = 51,525 ct
Weight per M cutoffs per roll stand:

$$\frac{(22\ 3/4 \text{ in.} \times 34 \text{ in.})60 \text{ lb}}{(24 \text{ in} \times 36 \text{ in.})} = \frac{(773.5 \text{ sq in.})60 \text{ lb}}{864 \text{ sq in.}} = 53.72 \text{ lb/M ct/stand}$$

Weight per M cutoffs for both stands:
53.74 lb/M ct/stand × 2 stands = 107.44 lb/M ct
Total stock weight:
107.44 lb/M ct × 51.525 M ct = 5536 lb

Step 5 Determine amount of newsprint required using roll data:

Total rolls needed per stand:
1.616 rolls × 114.5% = 1.85 rolls
Roll weight (basic formula):

$$\frac{0.00157 \times (45 \text{ in.}^2 - 3 \text{ in.}^2) \times 34 \text{ in.} \times 30 \text{ lb}}{0.0025 \text{ in.} \times (24 \text{ in.} \times 36 \text{ in.})} = 1495 \text{ lb/roll}$$

Total rolls required (rolls per stand × stands feeding = total rolls):
1.85 rolls/stand × 2 stands = 3.7 rolls
Total paper weight:
3.7 rolls × 1495 lb/roll = 5532 lb

Note: Difference of 4 lb between two methods is not considered significant.

Step 6 Determine paper cost:

5532 lb × ($380.00 ÷ 2000 lb) = $1051.08

Step 7 Determine ink quantity and cost (see Figure 15.3):

Amount of newsblack for Unit I (30%/40%):
0.82 lb/M ct + 1.09 lb/M ct = 1.91 lb/M ct
Amount of newsblack for Unit II (25%/35%):
0.68 lb/M ct + 0.95 lb/M ct = 1.63 lb/M ct
Amount of newsblack for Units I and II:
1.91 lb/M ct + 1.63 lb/M ct = 3.54 lb/M ct
Total amount of newsblack:
3.54 lb/M ct × 51.525 M ct = 182.4 lb
Newsblack cost:
182.4 lb × $2.25/lb = $410.40
Amount of process blue for Unit III (20%/30%):
0.55 lb/M ct + 0.82 lb/M ct = 1.37 lb/M ct
Adjustment:
1.37 lb/M ct × 1.11 = 1.52 lb/M ct
Total amount of process blue:
1.52 lb/M ct × 51.525 M ct = 78.3 lb
Process blue cost:
78.3 lb × $2.80/lb = $219.24
Total ink weight:
182.4 lb + 78.3 lb = 260.7 lb
Total ink cost:
$410.40 + $219.24 = $629.64

Step 8 Determine press time and cost (see Figure 15.4):

Press preparation time:
3 units × 0.8 hr/unit = 2.40 hr
Press-running time:
51,525 ct ÷ 22,154 ct/hr = 2.33 hr
Total press time:
2.4 hr + 2.33 hr = 4.73 hr
Total press cost:
4.73 hr × $95.00/hr = $449.35

Step 9 Summarize your findings:

Total number of cutoffs: 51,525 ct/stand
Total rolls required: 3.7 rolls
Total paper weight: 5532 lb
Total paper cost: $1051.08
Total ink weight: 260 lb
Total ink cost: $629.64
Total press time: 4.73 hr
Total press cost: $449.35

Appendix A
List of Abbreviations

Measurement Terms

cm	centimeter
ctn	carton
dp	depth
ft	foot, feet
fpm	feet per minute
gsm	grams per square meter
hp	horsepower
hp-hr	horsepower hours
hr	hour
in.	inch
k	1000
K	kilobyte = 1024 bytes
kg	kilogram
kW	kilowatt
kWh	kilowatt hour
l	liter
lb	pound
m	meter
M	thousand
max	maximum
MB	megabyte = 1 million bytes
mm	millimeter
no	number
pkg	package
sq	square
sub	substance
wk	week
yr	year

Data Processing Terms

AFDC	automated factory data collection
BASIC	beginners all-purpose symbolic instructional code
bits	binary digits
CAD/CAM	computer-aided design/computer-aided manufacturing
CPU	central processing unit
COBOL	common business oriented language
CP/M	control program for microcomputers
d-base	database
DBMS	database management system
DOS	disk operating system
EXCEL	spreadsheet and database program
FAX	facsimile transmission
EFDC	electronic factory data collection
FORTRAN	formula translation
I/O	input/output hardware
IBM	International Business Machines computers
IBM-compatible	computers compatible with IBM-manufactured computer equipment
Lotus 1-2-3	spreadsheet program
Mac	Macintosh computer
modem	modulator-demodulator
MS-DOS	Microsoft Disk Operating System
Multiplan	spreadsheet program
OCR	optical character recognition
PC	personal computer
Postscript	code used to communicate with laser printing equipment

RAM	random access memory
ReadySetGo	desktop publishing program
ROM	read only memory
SBC	small business computer
SFDC	shop floor data collection
UNIX	operating language
VDT	visual display terminal
Visicalc	spreadsheet program

Printing and Estimating Terms

adj	adjustment
AIC	all-inclusive center
ARCP	accounts receivable collection period
b	basis weight of paper
BEP	breakeven point
BHR	budgeted hour cost rate
bk	book
bkt	booklet
bt	(income) before taxes
BU	basic unit
c	cutoff in inches
CAE	computer assisted estimating
ccm	color correction mask
CEPS	color electronic prepress system
CGE	computer generated estimating
char	characters
col	column
COP	changeover point
cp	copies
cps	characters per second
C1S	coated on one side
C2S	coated on two sides
CSR	customer service representative
ct	cutoff
cv	cover
CWT	weight per 100 pounds of paper
d	core diameter
D	roll diameter
$/MBU	dollar cost per 1000 basic units
dpi	dots per inch
DTP	desktop publishing
E-B	emulsion-to-base
E-E	emulsion-to-emulsion
8-page web	half-width web press
fm	front matter
fss	finish size sheet
GATF	Graphic Arts Technical Foundation
H&J	hyphenation and justification
imp	impression
imp/hr	impressions per hour
ISO-A series	metric paper dimensions for business papers
ISO-B series	metric paper dimensions for posters and large-size goods
ISO-C series	metric paper dimensions for envelopes and post cards
L300	Allied Linotype's Linotronic imagesetter
MBU	1000 basic units
MF	machine-finished
MHR	machine hour rate
ms	manuscript
msp	manuscript page
M&SW	microcomputer and software
M weight	1000 sheet weight
N-00	Enco negative-acting, number of impressions per side (i.e., N-25 = 25,000 imp per side)
NAPL	National Association of Printers and Lithographers
NAPS	Enco negative-acting proofing system
NAQP	National Association of Quick Printers
p(p)	page(s)
PAPS	Enco positive-acting proofing system
pars	parent size sheet
pc	piece
PC	production center
pi	pica
PIA	Printing Industries of America
pk	pocket
PMS	Pantone Matching System

PMT	Kodak photo-mechanical transfer
prep	preparation
pss	press size sheet
pt(s)	point(s)
QBIR	Quarterly Business Indicator Report (by NAPL)
RAO	metric paper sizes including normal trim added
repro	reproduction proof
ROI	return on investment
ROS	return on sales
s	basic sheet size in square inches
S	slab thickness in inches to be cut away
sep	separations
sht	sheet
sig	signature
16-page web	full-width web press
SP	selling price
SRAO	metric paper sizes with extra trim added
S&SC	sized and supercalendered
sss	stock sheet size
SW	sheetwise
T	stock caliper
w&t	work-and-tumble
W&T	work-and-turn impressions
w	roll width in inches
wp	washup

Appendix B

List of Estimating Formulas

Cost Estimating (Chapter 1)

(standard production time × budgeted hour cost rate) + material cost + (buyout cost + markup) = selling price of job

Company Profitability (Chapter 2)

ROI = net income ÷ gross or net assets
return on sales (ROS) = net income ÷ annual gross sales
gross asset return on investment (ROI) = net income ÷ gross assets
net asset return on investment (ROI) = net income ÷ (gross assets - depreciation)

Value Added (Chapter 2)

value added = sales - (material costs + buyout costs)

Production Standard Conversions (Chapter 3)

To convert pieces per hour to hours per M pieces:

1000 ÷ number of pieces per hour = number of hours per M pieces

To convert hours per M pieces to pieces per hour:

1000 ÷ number of hours per M pieces = number of pieces per hour

Manufacturing Cost (Chapter 3)

manufacturing cost per chargeable hour = total factory cost ÷ number of chargeable hours

BHR (Chapter 3)

BHR = total annual center cost ÷ number of chargeable hours

Cost of Electrical Power When Completing BHRs (Chapter 3)

lighting cost = [(number of annual hours worked × wattage per square foot) ÷ 1000 watts per kilowatt] × cost per kilowatt hour × number of square feet in center
arc lamp cost = (number of rated input watts of source ÷ 1000 watts per kilowatt) × number of annual hours worked × cost per kilowatt hour

Note: The number of annual hours is sometimes lowered for some types of arc lamp sources because of the intermittent nature of the source.

power cost = total horsepower of motors × number of annual hours worked × 0.0746 kilowatt per horsepower × cost per kilowatt hour

Estimating Sheetfed Paper for Flat Sheet Impositions (Chapter 6)

Flat Sheet Cutting:

number-up on a press size sheet = press size sheet ÷ finish size sheet

Note: Diagrams should be drawn.

number-out of a parent size sheet = parent size sheet ÷ press size sheet

Note: Diagrams should be drawn.

maximum number-out when stagger cutting finish sheets out of press size sheets = area of press size sheet ÷ area of finish size sheet

where area is measured in square inches.

maximum number-out when stagger cutting press size sheets out of parent size sheets = area of parent size sheet ÷ area of press size sheet

where area is measured in square inches.

Quantities of Flat Sheet Stock:

number of press size sheets needed for a certain quantity of finished sheets = number of finish size sheets ÷ number-up on the press size sheet
number of parent size sheets needed for a certain quantity of press size sheets = number of press size sheets ÷ number-out of the parent size sheet

Note: Spoilage is normally added to allow for wasted stock during production.

carton level to be purchased = number of parent size sheets ÷ number of sheets per carton

Poundage, or Weight, of Paper:

equivalent weight = (area of desired size ÷ area of basic size) × basis weight

where area is measured in square inches. *Note:* This weight is for 500 sheets weight and must be doubled to yield the proper M weight, or 1000 sheet weight.

Paper poundage using the equivalent weight:

weight = (number of parent sheets ÷ 500 sheets) × equivalent weight

Paper poundage using the M weight:

weight = (number of parent sheets ÷ 1000 sheets) × M weight

Weight in Kilograms per 1,000 Sheets:

weight = (length of sheet in millimeters × width of sheet in millimeters) × grams per square meter ÷ 1,000,000

Cost of Paper:

Note: Prices can be found in a paper price catalog.

cost of paper using CWT price = paper poundage expressed in M weight or equivalent weight × price per CWT
cost of paper using M sheet price = (number of parent sheets ÷ 1000) × price per M sheets
cost of paper using the 100 sheet price = (number of parent sheets 100 ÷ 100) × price per 100 sheets

Note: In the following two formulas, use double the equivalent weight—for example, the M weight, for sizes other than the basic size; for the basic size, use the sub weight doubled.

Changing CWT price to M sheet price:

(CWT price ÷ 100) × poundage per M sheets = M sheet price

Changing M sheet price to CWT price:

(M sheet price ÷ poundage per M sheets) × 100 = CWT price

Number of Impressions:

number of impressions = number of press size sheets × (total number of forms ÷ number of forms per press pass)

Note: For maximum impressions, spoilage should be added to the number of press size sheets.

Miscellaneous:

paper waste percentage when a portion of the parent or press sheet must be thrown away = 1 − (area of sheet used ÷ total area of sheet) × 100

where area is measured in square inches.

caliper thickness = (number of plies × 0.003 inch per ply) + 0.006 inch

Estimating Sheetfed Paper for Bookwork Impositions (Chapter 6)

Note: These formulas assume that no duplicate pages will be imposed on a press sheet, such as in the use of step-and-repeat techniques for multiple signature production.

Size of the Press Sheet, Number of Pages per Press Sheet, Number of Forms:

number of pages per form = maximum press size sheet ÷ untrimmed booklet page size

number of forms = total number of booklet pages ÷ number of pages per form

size of press sheet = untrimmed page size × number-up imposed on one side of the press size sheet

Number of Signatures:

number of sheetwise signatures for a booklet of known number of pages = total number of booklet pages ÷ (number of pages per form × 2)

number of work-and-turn signatures for a booklet of known number of pages = total number of booklet pages ÷ number of pages per form

Number of Press Size Sheets:

number of press size sheets using sheetwise imposition = (number of sheetwise signatures per booklet × 1 press size sheet per signature × number of copies) + spoilage

number of press size sheets using work-and-turn imposition = (number of work-and-turn signatures per booklet × 1/2 press size sheet per signature × number of copies) + spoilage

Number of Impressions:

number of impressions without spoilage = total number of forms × number of copies

number of impressions using a straight percentage spoilage figure = total number of forms × number of copies × percentage spoilage

Estimating Ink Quantity and Cost (Chapter 7)

number of pounds of ink = (total form area × percentage of coverage × total number of copies × anticipated percentage of ink waste) ÷ ink mileage factor

where total form area is measured in square inches, number of copies includes press spoilage, and the ink mileage factor is derived from the appropriate ink mileage schedule.

Page Size:

Note: This formula is used to determine total form area.

page size in square inches, converted from picas = (pica page width ÷ 6 picas per inch) × (pica page depth ÷ 6 picas per inch)

Alternate method:

page size in square inches, converted from picas = (pica page width × pica page depth) ÷ 36 square picas per square inch

Estimating Design, Artwork and Copy Preparation (Chapter 8)

Mechanical and Paste-Up Production Time and Cost:

total time for mechanical production = (board layout time × number of boards) + (number of pieces to be pasted × production time per piece) + (overlay attachment production time × number of boards) + (number of pieces to be pasted per overlay × production time per piece)

total production cost for mechanical art = (mechanical production time × artist BHR) + cost of materials used

Establishing Hourly Mechanical Art Page Production Time and Cost:

hourly page production time = total production time for preparing similar mechanical art ÷ number of pages of finished mechanical art

hourly page production cost = total production time for preparing similar mechanical art × employee BHR) + cost of materials used ÷ number of pages of finished mechanical art

Peelcoat Production Time and Cost:

total time for peelcoat production = (total number of linear inches to be cut ÷ 100) × production time per 100 linear inches
total cost for peelcoat production = (peelcoat production time × employee BHR) + cost of materials used

Form Ruling and Paste-Up Production:

total time for form ruling and paste-up production = (production time per form for ruling × number of forms to be produced) + (number of pieces to be pasted × production time per piece)
total production cost for form ruling and paste-up = (form ruling and paste-up production time × employee BHR) + cost of materials used

Estimating Copyfitting, Typesetting and Desktop Publishing (Chapter 9)

Number of Manuscript Characters:

number of characters per typewritten line = typewritten page width in inches × character pitch
number of lines per typewritten page = typewritten page depth in inches × spacing lines per inch
number of characters per typewritten page = number of characters per typewritten line × number of lines per typewritten page
total number of manuscript characters = number of typewritten characters per line × number of typewritten lines per page × number of manuscript pages

Derivation Formulas:

Note: The following derivation formulas are used in the development of copyfitting schedules by which copyfitting estimating is completed.

Adjusted characters per linear pica:

alphabet length in points = alphabet length in picas × percentage adjustment for point size
adjusted picas for lowercase alphabet = alphabet length in points ÷ 12 points per pica
adjusted characters per linear pica = 26 alphabet characters ÷ adjusted picas per alphabet

Characters per square pica:

characters per square pica = (12 points per pica ÷ point value of typeface including leading) × characters per linear pica

Square picas per M characters:

square picas per M characters = M characters ÷ characters per square pica

Column Depth in Picas:

number of characters per typeset line = number of characters per pica × line length in picas
number of typeset lines = total number of characters per manuscript ÷ number of characters per line
column depth in picas = number of typeset lines × body size of type ÷ 12 points per pica

Alternate formula for finding depth for typeset material in picas:

depth for all typeset material in picas = (square picas per M characters × total characters × 0.001) ÷ line length in picas

Number of Pages in Job:

Note: Use this formula when total depth of typeset manuscript in picas is given.

total depth in picas ÷ depth of page in picas = number of typeset pages per job

Suitable Size and Style of Type:

total square picas per page = pica width of page × pica depth of page
total square picas per booklet = square picas per page × number of pages per booklet
characters per square pica = total manuscript characters ÷ total square picas per booklet

Total Characters to Be Written and Number of Manuscript Pages:

total characters to be written = total square picas × characters per square pica
number of manuscript pages to be typed = total characters to be written ÷ number of characters per manuscript page

Total Ems per Page:

total ems per page = number of ems per line × number of ems per page depth
ems per line = pica page width × 12 points per pica ÷ point size of type
ems per page depth = pica page depth × 12 points per pica ÷ body size of type

Slug Machine Composition Time and Cost:

basic composition time = (total ems ÷ 1000 ems) × typeset hours per M ems

where typeset hours per M ems are found in the composition schedule.

total composition time = (basic composition time × adjustment factor) + additional machine time

where the adjustment factor and additional time values are found in the composition schedule.

composition cost = total composition time × BHR

Phototypesetting Time and Cost:

keyboard production time = number of characters ÷ 1000 × keyboard hours per M characters
keyboard production cost = keyboard production time × keyboarding BHR
phototype unit production time = keyboard production time × phototype unit multiplier
phototype unit production cost = phototype unit production time × phototype BHR
proofreading production time = number of characters ÷ 1000 × proofreading hours per M characters
proofreading production cost = proofreading production time × proofreading BHR
total production time = keyboard production time + phototype production time + proofreading production time

where the appropriate values are found in the phototypesetting composition schedule.

column depth = (square picas per M characters × M characters) ÷ line length in picas

Desktop Publishing (DTP) Production Time and Cost:

total time for DTP production = preparation/setup + typesetting & proofreading + graphics input + image manipulation + printing
total cost for DTP production = total time for DTP production × BHR cost

Estimating Photographic and Electronic Prepress Systems (Chapter 10)

Process Camera:

process camera production time = (number of pieces of art to be photographed ÷ number-up on copyboard) × production time per piece
process camera production cost = camera production time × camera BHR
material and film cost = number of pieces of material and film × number of square inches per piece × material and film cost per square inch
total cost for process camera production = camera production cost + material cost

Duplicating and Contacting Production Time and Cost:

duplicating and contacting production time = preparation and setup + contacting production time + adjustments (as required)
duplicating and contacting production cost = duplicating and contacting production time × duplicating and contacting BHR
material and film cost = number of pieces of film or paper × number of square inches per piece × film cost per square inch
total duplicating and contacting production cost = duplicating and contacting production cost + film and material cost

Digital Scanner Production Time and Cost:

digital scanner production time = image review (per image) + image mounting (per image) + scanner setup and prescan adjustment (per image) + scanning time (per scanned set) + automatic processing (per scanned set)
digital scanner production cost = digital scanner production time × digital scanner BHR
material and film cost = number of pieces of film or paper × number of square inches per piece × film cost per square inch
total digital scanner production cost = digital scanner production cost + film and material cost

Estimating Film Assembly and Platemaking and Proofing Time (Chapter 11)

Total Table Stripping Production Time and Cost:

total table stripping production time = flat preparation time (key flats + master flats + marks flat) + laying and cutting time + additions
laying and cutting time for goldenrod stripping = number of pieces of film to be laid-in and cut × production time per piece
laying and cutting time for clearbase stripping = number of separations film

images to be laid-in × production time per piece + number of windows to be cut (in common-window flat) × the production time per window
total table stripping production cost = total table stripping production time × table stripping BHR
(Material cost normally recovered through BHR cost as a department direct supply).

Gerber AutoPrep Production Time and Cost:

total Gerber AutoPrep production time = Preparation/setup time per job + (Stripping input time × the number of film flats to be produced) + Film/pen plotting time.
Gerber AutoPrep production cost = Total Gerber AutoPrep production time × Gerber AutoPrep BHR
AutoPrep film cost = number of pieces of film used × size of film in square inches × cost per square inch
total Gerber AutoPrep production cost = Gerber AutoPrep production cost + AutoPrep film cost

Platemaking and Proofing:

platemaking and proofing production time = (number of first exposure flats × first exposure production time) + (number of additional exposure flats × additional exposure production time)
platemaking and proofing production cost = platemaking and proofing production time × platemaking and proofing BHR
plate cost = number of plates × cost per plate
proof cost = number of proofs × size of proof in square inches × cost per square inch
total cost for platemaking and proofing = production cost + platemaking and proofing material cost

Estimating Sheetfed Presswork (Chapter 12)

Total Cost and Unit Cost:

total cost = fixed cost + (variable cost per unit × number of units produced)
unit cost = total cost ÷ number of units produced

Changeover (Break-Even) Points:

changeover point = (fixed costs $_1$ − fixed cost$_2$) ÷ (dollars per M basic units$_2$ − dollars per M basic units$_1$)
dollars per MBU = [(BHR ÷ impressions per hour) ÷ number-up on press sheet] × 1000 units

Production Time and Cost:

total number of press sheets required = (total number of finished pieces ÷ number-up on the press sheet) × (spoilage percentage + 100%)
preparation and makeready production time = number of forms per job × hours per form for preparation and makeready
hours of press time per form = total number of press sheets required ÷ number of impressions per hour per form
press-running time = hours of press time per form × number of forms per job
number of washups per form = hours of press time per form ÷ hours of press time per form per washup
production time for washups = number of washups per form × hours of washup time per form × number of forms per job
total production time per job = preparation and makeready time + press running time + production time for washups
total production cost per job = total production time per job × press BHR

Estimating Binding and Finishing Operations Time and Cost (Chapter 13)

Cutting:

total number of press size sheets required = (number of finish size sheets required ÷ number-up on the press size sheet) × (percentage of spoilage + 100%)
total number of parent size sheets required = number of press size sheets required, including spoilage ÷ number-out of parent size sheet
total prepress cutting production time = [(total number of parent size sheets required ÷ number of sheets per lift) × cutting time per lift] + cutter setup time
total postpress cutting production time = [(total number of press size sheets required ÷ number of sheets per lift) × cutting time per lift] + cutter setup time
total cutting production time = total prepress cutting production time + total postpress cutting production time
total production cost = total cutting production time × cutter BHR

Folding:

total folding production time = (number of pieces to be folded ÷ folder output in sheets per hour) + first fold setup time + additional fold setup times
total folding production cost = total folding production time × folder BHR

Hand Gathering:

total gathering production time = [(number of sets to be gathered × number of pieces per set) ÷ number of gathered sets per hour] + penalty additions total gathering production cost = total gathering production time × gathering BHR

Jogging:

total jogging production time = [(number of sheets to be jogged ÷ 1000) × jogging time per M sheets] + penalty additions
total jogging production cost = total jogging production time × jogging BHR

Counting:

total counting production time = [(number of sheets to be counted including spoilage ÷ 1000) × counting time per M sheets] + penalty additions
total counting production cost = total counting production time × counting BHR

Padding:

total padding production time = [(number of sets to be padded × number of sheets per set) ÷ 1000] × padding time per M sheets
total padding production cost = total padding production time × padding BHR

Drilling:

total drilling production time = [(number of pieces to be drilled ÷ 1000) × drilling time per M sheets] + preparatory time
total drilling production cost = total drilling production time × drilling BHR

Signature Gathering:

total signature gathering production time = [(number of booklets to be gathered ÷ 1000) × gathering time per M booklets] + preparatory time
total signature gathering production cost = total signature gathering production time × gathering BHR

Saddle-Wire Stitching:

total saddle-wire stitching production time = [(number of booklets to be stitched ÷ 1000) × stitching time per M booklets] + preparatory time
total saddle-wire stitching production cost = total saddle-wire stitching production time × saddle-wire stitching BHR

Signature Side-Wire Stitching:

total side-wire stitching production time = [(number of booklets to be stitched ÷ 1000) × stitching time per M booklets] + preparatory time + penalties total side-wire stitching production cost = total side-wire stitching production time × side-wire stitching BHR

Perfect Binding:

total perfect binding production time = [(number of booklets to be perfect bound ÷ 1000) × binding time per M booklets] + preparatory time
total perfect binding production cost = total perfect binding production time × perfect binding BHR

Three-Knife Trimming:

total three-knife trimming production time = [(number of booklets to be trimmed ÷ 1000) × trimming time per M booklets] + preparatory time
total three-knife trimmer production cost = total trimming production time × trimmer BHR

Shrink-Wrapping:

total shrink-wrapping production time = [(number of packages to be wrapped ÷ number of pieces per package) × shrink-wrapping time per package] + preparatory time
total shrink-wrapping production cost = total shrink-wrapping production time × shrink-wrapping BHR

Estimating Web Lithographic Production (Chapter 15)

Roll Paper:

Note: Use formulas that apply and add press spoilage as appropriate.

Roll footage:

number of linear feet = [0.06545, a constant × (outside roll diameter squared − inside roll diameter squared)] − stock caliper in inches

Number of cutoffs per roll:

number of cutoffs per roll = (number of linear feet per roll × 12 inches per foot) ÷ press cutoff in inches

Weight per M cutoffs:

weight per M cutoffs = [press cutoff in inches × roll width in inches × (basis weight of paper × 2)] ÷ basic size of paper in square inches

Weight per M cutoffs using "constant factor":

weight per M cutoffs = constant factor* × sq in of cutoff
*constant factor = basis weight × 2 ÷ sq in of basic size

Weight of a roll of paper:

roll weight = [0.0157, a constant × (outside roll diameter in square inches – inside roll diameter in square inches) × roll width in inches × basis weight of paper)] ÷ (stock caliper in inches × basic size of paper in square inches)

Slabbing waste:

slabbing waste = [4, a constant × (original roll diameter × slab thickness to be removed) – slab thickness squared ÷ original roll diameter squared] × original roll weight

Ink Poundage and Cost:

total ink poundage = (ink poundage per M cutoffs for left form + ink poundage per M cutoffs for right form) × adjustment factor, if applicable × total number of cutoffs including spoilage ÷ 1000)
total ink cost = total ink poundage × ink cost per pound

Press Production Time and Cost:

total web press production time = (number of units printing × preparation time per unit) + (total number of cutoffs including spoilage ÷ press-running speed in cutoffs per hour)
total web press production cost = total web press production time × web press BHR

Appendix C

Vendors of Printing Industry Estimating and MIS Systems

The alphabetical listing that follows has been complied as a reference to aid readers in locating vendors of software, turnkey and customized systems. There are 77 vendors represented here.

At the time this list was prepared for publication, it represented the most thorough, complete listing of vendors supplying the printing industry with MIS and information processing systems, including job costing, estimating, inventory control, scheduling, machine loading and accounting packages.

All of the information in this appendix was obtained through direct contact with the listed vendors. Every attempt has been made to ensure the accuracy of the information indicated, yet it must be understood that changes occur frequently with such vendors including address and telephone number, modification of product, expanded operating system capability, types of estimating products, number of installations and years serving the industry with such product(s).

The following codes apply to the reference data presented in this appendix:

SW = software
TK = turnkey
DOS = disk operating system
Mac = Macintosh version
FDC = factory data collection
BHR = budgeted hour rates
CGE = computer-generated estimating
CAE = computer-assisted estimating
Optional = either CGE or CAE
SF = sheetfed
W = web

Vendors of Printing Industry Estimating and MIS Systems

Vendor name and address	ABE Systems 2608 Chanticleer Avenue Santa Cruz CA 95065
Telephone	408/476-8848
Contact person	Jim Alberti
Product	SW
Operating system	DOS
Estimating products	CAE/SF&W
Number of installations as of January, 1990	200+
Years serving the printing industry	5

Vendor name and address	Accu-Graphic Systems P.O. Box 3816 Winter Springs FL 32708
Telephone	407/695-3999
Contact person	Drew Alvarez
Product	SW
Operating system	DOS
Estimating products	CAE/SF&W
Number of installations as of January, 1990	25
Years serving the printing industry	10

Vendor name and address	AccuData, Inc. 1220 Spring Creek Place Springville UT 84663
Telephone	801/377-9156
Contact person	Bert Pugh
Product	Job Costing
Operating system	DOS
Estimating products	No
Number of installations as of January, 1990	700
Years serving the printing industry	10

Vendor name and address	Advanced Computer Sys. for Printers 6746 S. Revere Pkwy, #100 Englewood CO 80112
Telephone	303/792-9779
Contact person	Vince Lawrence
Product	SW&TK
Operating system	Prop
Estimating products	CAE/SF&W
Number of installations as of January, 1990	30
Years serving the printing industry	11

Vendor name and address	ADX Timesaver 835 S.E. Hawthorne Blvd. Portland OR 97214
Telephone	503/231-8160
Contact person	Bob Lewis
Product	SW&TK
Operating system	Various
Estimating products	CAE/SF&W
Number of installations as of January, 1990	135
Years serving the printing industry	7

Vendor name and address	AHP Systems 3166 Des Plaines Avenue Des Plaines IL 60018
Telephone	708/296-6040
Contact person	Udi Arieli
Product	SW&TK
Operating system	DEC
Estimating products	CAE/SF&W
Number of installations as of January, 1990	52
Years serving the printing industry	7

Vendor name and address	AlphaImage 26945 Cabot Rd., No. 104 Laguna Hills CA 92653
Telephone	714/582-7660
Contact person	Walt Garfield
Product	SW&TK
Operating system	Prop
Estimating products	CAE/SF
Number of installations as of January, 1990	85
Years serving the printing industry	9

Vendor name and address	Asta Software Corporation 969 Monroe Avenue Rochester NY 14620
Telephone	716/473-0372
Contact person	Harry Puff
Product	TK
Operating system	RPT
Estimating products	CAE/SF&W
Number of installations as of January, 1990	5
Years serving the printing industry	1

Vendor name and address	Avanti Computer Systems 2788 Bathurst St., Suite 305 Toronto CAN M6B3A3
Telephone	416/785-0424
Contact person	Seymour Gladstone
Product	SW&TK
Operating system	DOS
Estimating products	CAE/SF&W
Number of installations as of January, 1990	185
Years serving the printing industry	5

Vendor name and address	BASCORP 116 Courtland Rd. Cherry Hill NJ 08034
Telephone	609/429-2573
Contact person	Emie Jellinek
Product	SW
Operating system	DOS
Estimating products	CAE/SF&W
Number of installations as of January, 1990	45
Years serving the printing industry	9

Vendor name and address	Behoff Systems 21 Aubrey Road Upper Montclair NJ 07043
Telephone	210/509-1846
Contact person	Brian Hoffman
Product	SW
Operating system	DOS
Estimating products	CGE/SF&W
Number of installations as of January, 1990	4
Years serving the printing industry	1

Vendor name and address	BFE Systems 625 Dallas Drive, Suite 525 Denton TX 76205
Telephone	817/383-7434
Contact person	Doug Giles
Product	SW&TK
Operating system	Various
Estimating products	CGE/W
Number of installations as of January, 1990	200
Years serving the printing industry	10

Vendor name and address	Computer Dynamics P.O. Box 490 Grimesland NC 27837
Telephone	800/535-8067
Contact person	Dale Brooks
Product	SW&TK
Operating system	Various
Estimating products	CAE/SF&W
Number of installations as of January, 1990	300
Years serving the printing industry	8

Vendor name and address	Computer Programs Unlimited (CPU) 627 American Blvd. Warner Robins GA 31093
Telephone	800/922-2278
Contact person	Lisa Robbins
Product	SW
Operating system	DOS
Estimating products	CGE/SF
Number of installations as of January, 1990	11
Years serving the printing industry	1

Vendor name and address	Computer Systems for Graphic Arts 2005 DeLaCruz Blvd, No.101 Santa Clara CA 95050
Telephone	408/980-9037
Contact person	Ray Gabler
Product	SW&TK
Operating system	Various
Estimating products	CGE/SF&W
Number of installations as of January, 1990	100
Years serving the printing industry	7

Vendor name and address	Computerized Pricing Systems 13650 Silverton Drive Broomfield CO 80020
Telephone	303/469-0557
Contact person	Mark Yelich
Product	SW
Operating system	Mac&DOS
Estimating products	CAE/SF&W
Number of installations as of January, 1990	400
Years serving the printing industry	7

Vendor name and address	Covalent Systems 47436 Fremont Blvd. Fremont CA 94538
Telephone	800/321-0405
Contact person	J.B. Wood
Product	SW&TK
Operating system	Unix
Estimating products	Optional/SF&W
Number of installations as of January, 1990	820
Years serving the printing industry	7

Vendor name and address	CRC Information Systems Mtn. View Center, 9700 N. 91st Scottsdale AZ 85258
Telephone	602/451-7474
Contact person	Hank Hebing
Product	SW&TK
Operating system	Unix
Estimating products	CAE/SF&W
Number of installations as of January, 1990	75
Years serving the printing industry	6

Vendor name and address	D.W. Smith & Associates 950 Tower Lane, Suite 1275 Foster City CA 94404
Telephone	415/349-7725
Contact person	Doug Smith
Product	Custom
Operating system	Various
Estimating products	CAE/SF&W
Number of installations as of January, 1990	9
Years serving the printing industry	4

Vendor name and address	DataCatchers 4130 S. 500 W. Salt Lake City UT 84123
Telephone	801/266-3535
Contact person	Robert Lindquist
Product	SW&TK
Operating system	Various
Estimating products	CAE/SF
Number of installations as of January, 1990	20
Years serving the printing industry	5

Vendor name and address	Djan, Inc. 7220 Trade Street, Suite 115 San Diego CA 92121
Telephone	619/695-6691
Contact person	Gordon Grant
Product	SW&TK
Operating system	PS/2
Estimating products	Optional/SF&W
Number of installations as of January, 1990	80
Years serving the printing industry	8

Vendor name and address	Dove Network, Ltd. 350 E. Michigan Plaza, Suite 12 Kalamazoo MI 49007
Telephone	800/888-3683
Contact person	Ric Knapp
Product	SW&TK
Operating system	Mac&DOS
Estimating products	CGE/SF&W
Number of installations as of January, 1990	350
Years serving the printing industry	7

Vendor name and address	EST-PAC 1830 N. 55th St, Suite C Boulder CO 80301
Telephone	800/728-7190
Contact person	Brian Anderson
Product	SW&TK
Operating system	DOS
Estimating products	CGE/SF
Number of installations as of January, 1990	12
Years serving the printing industry	5

Vendor name and address	Estimator Plus 8345 Reseda Blvd., Suite 116 Northridge CA 91324
Telephone	818/993-5833
Contact person	Marci Lewis
Product	SW&TK
Operating system	DOS
Estimating products	CAE/SF&W
Number of installations as of January, 1990	300+
Years serving the printing industry	10

Vendor name and address . . . Excalibur Systems
27281 Las Ramblas, #155
Mission Viejo CA 92691
Telephone 800/932-9320
Contact person. . . . . Ed Kirchner
Product . . . . SW
Operating system . . . . DOS
Estimating products . . . . CAE/SF
Number of installations as of January, 1990 800
Years serving the printing industry 7

Vendor name and address . . . Executive Computer Systems
1571 Fenpark
Fenton MO 63026
Telephone 314/343-5757
Contact person. . . . . Cherryl Richards
Product . . . . SW&TK
Operating system . . . . DOS
Estimating products . . . . CAE/SF
Number of installations as of January, 1990 30
Years serving the printing industry 14

Vendor name and address . . . Franklin Comp-Est
952 E. 2100 S.
Salt Lake City UT 84106
Telephone 801/486-5954
Contact person. . . . . Gary Eddington
Product . . . . SW
Operating system . . . . DOS
Estimating products . . . . Pricing Only
Number of installations as of January, 1990 680
Years serving the printing industry 7

Vendor name and address . . . Franklin Estimating Systems
952 E. 2100 S.
Salt Lake City UT 84106
Telephone 801/486-5954
Contact person. . . . . Gary Eddington
Product . . . . SW
Operating system . . . . Mac&DOS
Estimating products . . . . CAE/SF
Number of installations as of January, 1990 900
Years serving the printing industry 4

Vendor name and address . . . GAMX
497 Lighthouse Avenue
Monterey CA 93940
Telephone 408/375-9994
Contact person. . . . . Dave Christensen
Product . . . . SW&TK
Operating system . . . . DOS
Estimating products . . . . CAE/SF
Number of installations as of January, 1990 26
Years serving the printing industry 2

Vendor name and address . . . Globe-Tek
295 The West Mall, Suite 400
Etobicoke, Ontario CAN M9C4Z4
Telephone 416/622-0444
Contact person. . . . . Richard Brack
Product . . . . SW&TK
Operating system . . . . Unix
Estimating products . . . . CGE/SF&W
Number of installations as of January, 1990 16
Years serving the printing industry 2

Vendor name and address . . . Graphic Data Systems
4680 D Brownsboro Rd.
Winston-Salem NC 27106
Telephone 919/759-2211
Contact person. . . . . Mark Idol
Product . . . . SW&TK
Operating system . . . . Various
Estimating products . . . . CAE/SF&W
Number of installations as of January, 1990 90
Years serving the printing industry 12

Vendor name and address . . . Graphisoft, Inc.
242 W. 27th St, #4B
New York NY 10001
Telephone 212/727-0388
Contact person. . . . . Denis Fabre
Product . . . . SW
Operating system . . . . DOS
Estimating products . . . . CGE/SF&W
Number of installations as of January, 1990 270
Years serving the printing industry 8

Vendor name and address . . . GraphiTech
2161 Palm Beach Lakes Blvd.
W. Palm Beach FL 33409
Telephone 800/634-8324
Contact person. . . . . Scott Thatcher
Product . . . . SW&TK
Operating system . . . . Various
Estimating products . . . . Optional/SF&W
Number of installations as of January, 1990 1010
Years serving the printing industry 12

Vendor name and address . . . Great Plains Software
1701 S. W. 38th St.
Fargo ND 58103
Telephone 800/456-4417
Contact person. . . . . Joel Block
Product . . . . SW
Operating system . . . . Mac&DOS
Estimating products . . . . CAE/SF
Number of installations as of January, 1990 . . . . -
Years serving the printing industry 5

Vendor name and address . . . Green Systems
P.O. Box 15414 Biltmore Station
Asheville NC 28813
Telephone 704/274-3518
Contact person. . . . . Lynn Yarbrough
Product . . . . SW
Operating system . . . . Various
Estimating products . . . . Optional/SF&W
Number of installations as of January, 1990 125
Years serving the printing industry 18

Vendor name and address . . . Hagen Systems
6438 City W. Parkway
Eden Prairie MN 55344
Telephone 800/284-2436
Contact person. . . . . Bob Bierwagen
Product . . . . TK
Operating system . . . . Unix
Estimating products . . . . Optional/SF&W
Number of installations as of January, 1990 375
Years serving the printing industry 15

Vendor name and address . . . Hi-Tek Computer Products
308 W. Erie St., Suite 500
Chicago IL 60610
Telephone 312/787-2000
Contact person. . . . . Wayne Cohen
Product . . . . SW&TK
Operating system . . . . DOS
Estimating products . . . . Optional/SF&W
Number of installations as of January, 1990 . . . . 500+
Years serving the printing industry 8

Vendor name and address . . . Horowitz Associates
6420 Castleway Drive
Indianapolis IN 46250
Telephone 317/849-9854
Contact person. . . . . Henry Wiegand
Product . . . . SW&TK
Operating system . . . . Various
Estimating products . . . . Job Costing
Number of installations as of January, 1990 10
Years serving the printing industry 11

Vendor name and address . . . Innovative Solutions
1903 W. 169th St.
Gardena CA 90247
Telephone 213/715-6900
Contact person. . . . . Bill Crispin
Product . . . . Custom
Operating system . . . . Unix
Estimating products . . . . CAE/SF&W
Number of installations as of January, 1990 60
Years serving the printing industry 10

Vendor name and address . . . Kelley World
61501 Bremen Highway
Mishawaka IN 46544
Telephone 219/255-4926
Contact person. . . . . Karen Kelley
Product . . . . SW&TK
Operating system . . . . Various
Estimating products . . . . Optional/SF&W
Number of installations as of January, 1990 8
Years serving the printing industry 7

Vendor name and address . . . Kenex Systems
575 E. 45th St., S.
Salt Lake City UT 84107
Telephone 801/263-3276
Contact person. . . . . Dennis Johnson
Product . . . . TK
Operating system . . . . DOS
Estimating products . . . . CGE/SF&W
Number of installations as of January, 1990 160
Years serving the printing industry 6

Vendor name and address . . . Logic Associates
P.O. Box 765
White River Junction VT 05001
Telephone 802/295-5661
Contact person. . . . . Nick Orem
Product . . . . TK
Operating system . . . . Unix
Estimating products . . . . Optional/SF&W
Number of installations as of January, 1990 155
Years serving the printing industry 20

Vendor name and address . . . M-Data, Inc.
4408 N. 12th St., Suite 340
Phoenix AZ 85014
Telephone 602/234-1988
Contact person. . . . . Jim Murray
Product . . . . SW&TK
Operating system . . . . Mac
Estimating products . . . . Optional/SF&W
Number of installations as of January, 1990 250
Years serving the printing industry 4

Vendor name and address . . . Management Information Corp.
1001 Cape St. Claire Rd.
Annapolis MD 21401
Telephone 301/757-9039
Contact person. . . . . Joe Holoski
Product . . . . TK
Operating system . . . . Unix
Estimating products . . . . Optional/SF&W
Number of installations as of January, 1990 47
Years serving the printing industry 6

Vendor name and address . . . McCutcheon Graphics
88 St. Regis Crescent S.
New York, Ontario CAN M3J1Y8
Telephone 416/636-6070
Contact person. . . . . Catherine Cowley
Product . . . . SW&TK
Operating system . . . . Mac
Estimating products . . . . CGE/SF
Number of installations as of January, 1990 15
Years serving the printing industry 5

Vendor name and address . . . Micro Ink Systems
1700 W. Park Drive
Westborough MA 01581
Telephone 800/842-5667
Contact person. . . . . Tom Ticknor
Product . . . . SW&TK
Operating system . . . . DOS
Estimating products . . . . CAE/SF&W
Number of installations as of January, 1990 600
Years serving the printing industry 2

Vendor name and address . . . Microprint Developments
9011 Leslie St., Suite 301
Richmond Hill, Ontario CAN L4B3B6
Telephone 416/882-0641
Contact person. . . . . Mark Porter
Product . . . . SW&TK
Operating system . . . . Mac&DOS
Estimating products . . . . CAE/SF&W
Number of installations as of January, 1990 50
Years serving the printing industry 2

Vendor name and address . . . Modular Graphic Services
611 S. Front
Wilmington NC 28401
Telephone 919/763-2012
Contact person. . . . . Dennis Walsak
Product . . . . SW
Operating system . . . . Mac
Estimating products . . . . CGE/SF&W
Number of installations as of January, 1990 80
Years serving the printing industry 1

Vendor name and address	National Composition Assoc. 1730 N. Lynn St. Arlington VA 22209
Telephone	703/841-8165
Contact person	Clifford Weiss
Product	SW
Operating system	DOS
Estimating products	TypeEstOnly
Number of installations as of January, 1990	750+
Years serving the printing industry	6

Vendor name and address	Noguska Industries 735 N. Countyline Road Fostoria OH 44830
Telephone	418/435-0404
Contact person	Kathleen Howard
Product	SW&TK
Operating system	DOS
Estimating products	CAE/SF&W
Number of installations as of January, 1990	1000
Years serving the printing industry	19

Vendor name and address	PACE/Springfield Computer System 1721 Pearl St. Jacksonville FL 32206
Telephone	904/354-0949
Contact person	Richard Wallace
Product	SW&TK
Operating system	DOS
Estimating products	Optional/SF&W
Number of installations as of January, 1990	130
Years serving the printing industry	5

Vendor name and address	Parsec Corporation 12200 N. Pecos St., Suite 265 Denver CO 80234
Telephone	303/252-0344
Contact person	Steve Hallberg
Product	SW&TK
Operating system	Various
Estimating products	Optional/SF&W
Number of installations as of January, 1990	310
Years serving the printing industry	7

Vendor name and address	PICCA 5309 Lincoln Avenue, Suite A Skokie IL 60077
Telephone	708/674-4781
Contact person	Paul Betancourt
Product	SW&TK
Operating system	Various
Estimating products	CAE/SF&W
Number of installations as of January, 1990	30
Years serving the printing industry	10

Vendor name and address	PLANTROL Systems 71 E. Main St. Westfield NY 14787
Telephone	716/326-4900
Contact person	Ed Czajka
Product	SW&TK
Operating system	Unix
Estimating products	BusFormsOnly
Number of installations as of January, 1990	65
Years serving the printing industry	14

Vendor name and address	Pluzynski Systems 618 U.S. Highway One N. Palm Beach FL 33408
Telephone	407/845-2362
Contact person	Jim Pluzynski
Product	SW&TK
Operating system	DEC&Mac
Estimating products	CAE/SF&W
Number of installations as of January, 1990	40
Years serving the printing industry	20

Vendor name and address	Press-tige Software 501 S. First Avenue Arcadia CA 91006
Telephone	818/574-1877
Contact person	Don Jones
Product	SW
Operating system	DOS
Estimating products	CAE/SF&W
Number of installations as of January, 1990	250+
Years serving the printing industry	11

Vendor name and address	Primac Systems, Inc. 4601 Langland Rd., Suite 106 Dallas TX 75244
Telephone	214/661-9336
Contact person	Van Price
Product	TK
Operating system	PICK/Unix
Estimating products	CAE/SF&W
Number of installations as of January, 1990	65
Years serving the printing industry	7

Vendor name and address	Print Sys 1254 B High Street Auburn CA 95603
Telephone	916/823-7247
Contact person	Chris Hester
Product	SW
Operating system	Unix
Estimating products	CAE/SF&W
Number of installations as of January, 1990	50
Years serving the printing industry	9

Vendor name and address	Printer's Computer Systems P.O. Box 69 Estes Park CO 80517
Telephone	303/586-4355
Contact person	Tom Melhorn
Product	SW
Operating system	DOS
Estimating products	Optional/SF&W
Number of installations as of January, 1990	1400
Years serving the printing industry	14

Vendor name and address	Printer's Data Systems 10300 Sunset Avenue, Suite 333 Miami FL 33173
Telephone	800/447-6743
Contact person	Al Weisberg
Product	SW
Operating system	DOS
Estimating products	CGE/SF
Number of installations as of January, 1990	200
Years serving the printing industry	5

Vendor name and address . . . Printer's Management Control/Lorac
1007 First Avenue
Salt Lake City UT 84103
Telephone 801/531-8726
Contact person. . . . . Jim Webster
Product . . . . SW&TK
Operating system . . . . Various
Estimating products . . . . Optional/SF&W
Number of installations as of January, 1990 140
Years serving the printing industry 14

Vendor name and address . . . The Printer's Plan
9475 Deereco Rd., Suite 206
Timonium MD 21093
Telephone 301/561-4342
Contact person. . . . . Tulin Edev
Product . . . . SW
Operating system . . . . Mac+DOS
Estimating products . . . . CGE/SF&W
Number of installations as of January, 1990 300
Years serving the printing industry 3

Vendor name and address . . . Printer's Plus
31440 Northwestern Highway
Farmington Hills MI 48018
Telephone 800/544-1224
Contact person. . . . . Dick Zerwas
Product . . . . SW&TK
Operating system . . . . Mac+DOS
Estimating products . . . . CGE/SF
Number of installations as of January, 1990 1000
Years serving the printing industry 5

Vendor name and address . . . Printer's Shareware
5019-5021 W. Lovers Lane
Dallas TX 75209
Telephone 214/350-1902
Contact person. . . . . George Croft
Product . . . . SW
Operating system . . . . DOS
Estimating products . . . . Optional/SF&W
Number of installations as of January, 1990 4000
Years serving the printing industry 18

Vendor name and address . . . Printer's Software
3665 Bee Ridge Road
Sarasota FL 34233
Telephone 813/923-9010
Contact person. . . . . Paul Grieco
Product . . . . SW
Operating system . . . . Various
Estimating products . . . . CAE/SF&W
Number of installations as of January, 1990 . . . . 2500+
Years serving the printing industry 22

Vendor name and address . . . Printing Communication System
RFD#2, Box 4100
Farmington ME 04938
Telephone 207/778-3767
Contact person. . . . . Dave Wilkinson
Product . . . . SW&TK
Operating system . . . . Mac+DOS
Estimating products . . . . CAE/SF&W
Number of installations as of January, 1990 105
Years serving the printing industry 4

Vendor name and address . . . printLEADER Software
28 Stratford Drive
Somerset NJ 08873
Telephone 800/752-4624
Contact person. . . . . John Fleming
Product . . . . SW&TK
Operating system . . . . Mac+DOS
Estimating products . . . . CAE/SF&W
Number of installations as of January, 1990 75
Years serving the printing industry 1

Vendor name and address . . . Professional Impressions
901 S. MoPac, Bldg. 2, #500
Austin TX 78746
Telephone 512/327-3377
Contact person. . . . . Rudy Ayala
Product . . . . SW&TK
Operating system . . . . Various
Estimating products . . . . CAE/SF&W
Number of installations as of January, 1990 25
Years serving the printing industry 7

Vendor name and address . . . Profit Control/Heidelberg
73-45 Woodhaven Blvd.
Glendale NY 11585
Telephone 718/830-7900
Contact person. . . . . Niels Frederiksen
Product . . . . TK
Operating system . . . . Various
Estimating products . . . . Optional/SF&W
Number of installations as of January, 1990 650
Years serving the printing industry 13

Vendor name and address . . . Programmed Solutions
25 Third Street
Stamford CT 06905
Telephone 203/358-9955
Contact person. . . . . John Taffler
Product . . . . SW&TK
Operating system . . . . DOS
Estimating products . . . . CAE/SF&W
Number of installations as of January, 1990 . . . . 50+
Years serving the printing industry 4

Vendor name and address . . . QCX Corporation
261 Circle Ct.
Palatine IL 60067
Telephone 708/884-9292
Contact person. . . . . George Manthey
Product . . . . SW
Operating system . . . . DOS
Estimating products . . . . CGE/SF&W
Number of installations as of January, 1990 24
Years serving the printing industry 2

Vendor name and address . . . RPT Group/T.W. Rogers Consulting
29 Gervais Dr., Suite 310
Don Mills, Ontario CAN M3C1Y9
Telephone 416/391-1778
Contact person. . . . . Tom Rogers
Product . . . . SW&TK
Operating system . . . . DOS
Estimating products . . . . CAE/SF&W
Number of installations as of January, 1990 80
Years serving the printing industry 6

Vendor name and address	SDI Systems 4666 Indianola Avenue Columbus OH 43214
Telephone	614/261-7103
Contact person	William Immel
Product	SW&TK
Operating system	Various
Estimating products	Optional/SF&W
Number of installations as of January, 1990	8
Years serving the printing industry	18

Vendor name and address	Softprint 1754 Maplelawn Troy MI 48084
Telephone	313/649-9008
Contact person	Lauren Brubaker
Product	SW&TK
Operating system	DOS
Estimating products	CGE/SF
Number of installations as of January, 1990	65
Years serving the printing industry	5

Vendor name and address	State Printing Company 1210 Key Road, Box 1388 Columbia SC 29202
Telephone	803/799-9550
Contact person	Cary Czajkowski
Product	SW
Operating system	Unix
Estimating products	CAE/SF&W
Number of installations as of January, 1990	35
Years serving the printing industry	7

Vendor name and address	Trackmaster (Danyl Corp.) 1509 Glen Avenue Moorestown NJ 08057
Telephone	800/732-6868
Contact person	Randy Vanderhoof
Product	TK
Operating system	DOS
Estimating products	Job Track Only
Number of installations as of January, 1990	3
Years serving the printing industry	1

Vendor name and address	Turquoise Products 9648 E. Baker Tucson AZ 85748
Telephone	602/885-9671
Contact person	William Kastner
Product	SW
Operating system	Unix
Estimating products	CAE/SF&W
Number of installations as of January, 1990	1200
Years serving the printing industry	6

Glossary

access charge: A dollar charge made to the printer each time the computer system is accessed when using a timesharing system.
account payable: Money owed to a supplier or other vendor by the printer for materials and services.
account receivable: Money owed to a printing firm for printing goods and services.
account receivable collection period (ARCP): The amount of time, in days, from the date of mailing an invoice to the date the invoice is paid.
actual time: Clock hours required to complete all or part of the production of a printing job.
additive surface plate: A lithographic printing plate that requires the addition of a developing lacquer to improve plate life.
Adobe Illustrator: A popular graphics software program originally developed for the Macintosh, used to draw and develop illustrations on computer.
AFDC: Automated factory data collection; process by which production data is collected from shop floor employees when production is completed, then stored in a computer system for analysis and review; sometimes termed electronic factory data collection (EFDC) or shop floor data collection (SFDC).
all-inclusive center (AIC) method: A prorated costing procedure that allows administrative areas to cost data just as if they were used during printing production.
alphabet length: The measured length, in points, of the lowercase alphabet of a certain size and series of type; basis for copyfitting schedules.
American Paper Institute: Organization that tracks and monitors paper and paper-related information.
analog-digital color scanner: Older type of color scanner that scans directly to film, with fairly complex controls.
anodized plate: A lithographic plate manufactured with a passivation barrier of aluminum oxide to prevent chemical reactions with diazo coatings and provide improved press performance during printing.
applications program: Software written to perform a specific task by the computer system, such as estimating, job costing, and inventory.

aqueous plate: Plate coatings that are water-soluble and thus less harmful to the environment.
archival storage: The storing of data and information in such a manner that it will represent a permanent record.
art director: An individual working in an art studio who is in charge of a group of artists; directs artists efforts for a studio.
art work: Any image prepared for graphic reproduction.
automated factory data collection (AFDC): A process by which printing production is automatically tracked as it occurs using a shop floor computer terminal into which the employee feeds information that accumulates in storage for later analysis; also termed electronic factory data collection (EFDC) or shop floor data collection (SFDC).
automatic film processing: Equipment that provides automatic developing and fixing of silver halide film products; both older, conventional "lith" and new rapid-access processing systems are popular.
automatic plate processing: Equipment that provides automatic developing and finishing of lithographic plates.
automatic processing: Mechanical equipment used for the automated development of lithographic film and plates.
automatic proof processing: Equipment that provides automatic developing and finishing of proofing materials.
automatic roll change: Process by which rolls of paper stock are automatically changed during web printing production; devices sometimes termed "flying pasters" or "automatic pasters."
automatic spacing: Mechanical or electronic equipment used with cutting machines to precisely set the position of the back gauge.
automatic toning machine (ATM): Unit used to process Cromalin proofs, whereby powdered toners are carefully applied to the exposed photopolymer material.
back gauge: That mechanical segment of a cutting machine that moves to establish a precise sheet dimension or distance from the cutting blade.
banding: The wrapping of a package with string, rubber bands, or other material to secure the contents as a complete unit.
bar code strip: A printed series of lines used to communicate a message to a computer system through some type of optical reading device.
BASIC: A high-level computer programming language; abbreviation for Beginners All-Purpose Symbolic Instructional Code.
basic size: Established size of a paper stock upon which the basis weight is calculated; each major class of paper has one basic size.
basic unit (BU): A standard product selected as the basis for comparison of changeover points with presses.
basis weight: The scale weight of 500 sheets of the basic size of a particular class of paper; also termed *substance (sub)* weight, or *pounds.*
bastard cut: Procedure by which some small sheets cut from a larger parent sheet will

have the grain in the short direction, while other sheets cut from the same parent sheet will have grain going in the long direction; also termed *stagger cut* or *Dutch cut.*
baud: Term used to describe the speed of transmission for sending and receiving information and data by telecommunication methods.
benchmark: Selecting and using a standard item, element or routine to measure output or production when comparing or analyzing a computer or other system or process.
billing: The process of preparing and producing documents to be sent to customers indicating the amount of money owed for printing and other services.
bit: The smallest unit of computer data, represented by a 0 or 1, and the only language the CPU can understand; abbreviation for binary digit.
blanket-to-blanket: See *perfector.*
blotter paper: A paper class, basic size 19 x 24 inches, used when ink blotting is necessary (such as for checkbook backs).
blueline: A photographic proof that has blue images contained on a white background; also called a *Dylux proof.*
book: A published work containing more than 64 pages.
booklet: A published work containing 64 or fewer pages..
book paper: A class of printing paper commonly used for book manufacturing with a basic size of 25 x 38 inches; subcategories include coated, uncoated, and text groups.
bookwork: A general term describing the manufacturing procedures required for the production of books and booklets.
bond paper: A paper class, basic size 17 x 22 inches, used for many types of business communications; rag and sulfite subgroups.
break-even point (BEP): A linear graphing procedure that allows for quick comparison between outputs of similar equipment producing a similar product; also called *change-over point.*
breaking for color: Dividing artwork into specific color forms during the paste-up stage.
break-in period: The amount of time necessary to debug programs and make a computer system operational in a company.
brownline: A monochromatic photographic proof that has a brown image contained on a white background; also called a *silverprint* or *Van Dyke proof.*
buckle folder: A piece of bindery equipment with knurled rollers used to fold paper.
budgeted hour cost rate (BHR): A calculated dollar figure based on defined fixed and variable costs for a specific production area (center).
business paper: A class of printing paper commonly used for business purposes, with a basic size of 17 x 22 inches; subcategories include bond, ledger, thin, duplicator, and safety paper groups.
buyout: A service purchased by a printer from an outside source; trade services include composition, bindery, and prepress segments.
Byer's Micromodifier: A machine that produces extremely accurate spreads and chokes through the use of a carefully controlled, orbital rotating bed.
byte: A sequence of bits, generally eight bits long, that translates into one character by the computer.

C1S: A term used to describe paper stock that is coated on one side.
C2S: A term used to describe paper stock that is coated on both sides.
CAD/CAM: Computer-aided design/computer-aided manufacturing; term used in reference to computer systems for design development and manufacture of various products; sometimes CAD or CAM used to indicate design-specific or manufacturing-specific systems or operations.
CAE: Computer-assisted estimating; computer software that requires an estimator to operate the system by breaking the identified job specifications into detailed elements, and then assigning time and cost values to the component parts.
calendering: A buffing process completed during paper manufacturing that polishes the sheet surface, making it less prone to printing production difficulties.
caliper: See *paper caliper.*
calligraphy: A decorative handwriting produced manually by an artist or calligrapher.
camera-ready art: Graphic artwork that is ready for photographic reproduction.
canned software: Software that has been developed to be ready for use without further work; also called *packaged software.*
capital intensive: A term describing the condition of business where large sums of money (capital) are required to enter and work in a specific sector.
cardboard paper: A class of printing paper that is generally thick and bulky; major categories include index bristol, tagboard, blanks, printing bristol, and wedding bristol.
cash flow: The movement of working capital through a business.
casting off: A term used to describe copyfitting.
cell: A specific location on a spreadsheet program matrix.
cellulose fiber: A wood fiber used as the major ingredient in most printing papers.
centimeter: A metric measurement whereby a meter (39.37 inches) is divided into 100 equal parts thus equaling 0.3937 inches; about the thickness of a pencil.
central processing unit (CPU): The operational segment of a computer that coordinates all other computer segments.
CEPS: Color Electronic Prepress System; sophisticated electronic system used to produce complex color images using digital methods.
CGE: Computer-generated estimating; computer software that has been preprogrammed with an expected or normal production routine; requires inputting of job specifications or other job parameters, after which the computer produces a final estimate based on the preprogrammed routine.
changeover point (COP): See *break-even point.*
character: A letter or number of the alphabet; also the equivalent of one byte in computer usage.
character per square pica: A copyfitting measurement of a specified type face, incorporating the type size and leading.
character pitch: Refers to the horizontal spacing of typewritten characters as the number of characters per linear inch; a pica typewriter has 10 characters per linear inch and an elite typewriter occupies a standard 12 characters per inch.
characters per linear pica: A mathematical figure indicating the average number of characters in a certain typeface and type size that can be set in a 12 point linear measure.

character set: See *set size.*
chargeable hours: Production time spent on a customer's job that can be directly charged to that customer and for which he or she must pay.
chargeback system: A procedure wherein actual times are recorded during printing production and used as a basis to calculate the selling price of a customer's job.
chiller unit: A postprinting process in heatset web production, where the web of paper passes through a refrigeration unit following heating of the web to set the ink.
chip: An integrated circuit made of silicon etched with thousands of tiny circuits and about the size of an adult fingernail; also called a *microchip.*
choke: A film image which is slightly reduced in size; produced from a positive film original to raw film using contacting procedures or orbital techniques; also termed a "shrink" or squeeze."
clear stripping base: Transparent acetate or polyester on which imaged film negatives and positives are stripped; also termed "clearbase."
clearbase stripping: Process of assembling and taping film images to clear stripping base material.
clip-out art: Preprinted art image sold to printers, which can be cut out and used directly in a paste-up.
coated paper: Paper manufactured with a fine coat of a mineral substance to increase reflectivity or printability of the final product; a category of book papers.
COBOL: A high-level computer programming language with instructions similar to normal English; abbreviation for Common Business Oriented Language.
collating: See *gathering.*
color correction mask (ccm): A piece of film used in color separation photography to correct for deficiencies in process color inks and to balance tonal range in a set of separations.
Color Electronic Prepress System (CEPS): Process by which sophisticated electronic systems are used to produce complex color using digital methods; Crosfield, Hell, DS America and Scitex are manufacturers of such equipment.
color separation: A photographic procedure by which a color image is divided into magenta (red-blue), cyan (blue-green), yellow (red-green), and black film images using red, blue, green, and amber filters.
column depth: A copyfitting term used to denote the amount of space, as a column of type, occupied by typeset manuscript.
combination shop: A printing firm that produces both quick printed materials and commercial printed products.
commercial pin registration systems: Pin registration systems that are available to the printer for purchase from commercial sources.
commercial web: A term used to describe a type of web press used for a wide variety of commercial printing and advertising production.
common-window flat: A flat that contains only windows or cut openings in masking material and used as a knockout flat when proofing and plating process color separations stripped to clearbase.
comparative weight formula: A mathematical formula used to calculate a weight for a standard size of paper; also called *equivalent weight formula.*

component estimating: An estimating process by which a job is carefully broken into component parts or elements; used to describe one method of CEPS estimating.
composite film: A film contacting process by which a film sheet receives multiple exposures from different film images of all one color or type, thereby collecting the film images together on one film sheet; procedure termed "compositing."
composition: The typesetting of manuscript copy in a selected size and style of type.
comprehensive: A completed visualization of an image prepared by an artist; final step before preparation of artwork for graphic reproduction.
computer assisted estimating (CAE): Estimating procedures that utilize computers to determine job planning and production costs.
computer literacy: A general understanding of electronic computing, including an understanding of the general technology and application of computers to solve problems.
condensed: A term used to describe typefaces that have been designed with character elements squeezed together (and elongated) slightly.
contact screen: A piece of film containing a fine series of continuous-tone dots; used in contact with lith film for halftone image production.
continuous-tone: Any image that consists of a range of values from light to intermediate to dark with no defined breaks between values.
contract programmer: An individual who provides programming services on a professional basis, usually developing specific programs to meet defined uses.
conventional prepress: Used to describe printing production that does not utilize electronic imaging processes, systems or equipment.
conversion: The process of changing information from one form to another for computer use.
copier engine: The primary component is a copier that receives electronic signals then transmits these signals to a visible image on paper; typically a replaceable cartridge containing an electro-sensitive cylinder and powdered carbon toner.
copy: To duplicate a file or program to provide a backup; the copy is then used and the original saved. A term also used to indicate manuscript and other materials supplied to the printer by the customer, representing the essential parts of a printing job.
copy drum: The segment of a color scanner where the copy to be separated is positioned.
copyfitting: A mathematical technique used to determine the amount of space a given amount of typewritten manuscript will occupy when typeset in a certain size and style of type within a given page dimension.
copy preparation: A term generally used to describe the coordination of artwork into a form suitable for graphic reproduction.
core: A circular tube made of metal or fiberboard on which roll paper is wound.
core memory: The internal memory built into a computer system in the CPU; also termed *main memory.*
cost-benefit analysis: A procedure by which costs and benefits are weighed by management when purchasing a computer, piece of equipment, or other potentially expensive item or service.
cost estimating: The process by which a printing job is broken into detailed

manufacturing components representative of anticipated production, after which production times and costs are determined, and to which material costs and outside services are added.
cost markup: An additional dollar amount added to material, outside purchase, or labor and overhead costs for handling and servicing requirements.
Counter Pricing Book: A popular standard pricing catalog used primarily by quick and small commercial printers.
counter sales: Taking a printing order over the counter, commonly done in quick printing or small commercial printing shops.
counting: A procedure used to determine the exact quantity of a certain item.
cover paper: A class of printing paper commonly used for book covers and when thick, durable stocks are required; basic size of cover paper is 20 x 26 inches.
CP/M: The abbreviation for Control Program for Microprocessors; a disk operating system used with microcomputers for business purposes.
credit association: A firm organized to determine the credit status of companies and individuals doing business with their member firms.
credit referral service: A procedure by which credit information about a customer is given; generally a free service offered through a regional PIA office for member firms.
custom software: Software developed to fit a company's individual specifications by an in-house programmer or contract programer.
customer service representative (CSR): A person working as a liaison between customer and printing company to streamline production and ensure that the final printed product is manufactured properly and to the customer's satisfaction.
customer's ability and willingness to pay: A subjective method used by some printers to determine prices for printing goods and services.
customized systems: A computer system, pricing system or other procedure that has been developed for a specific product, company or application.
cut-in: Insertion of a film image into another piece of film cutting and removing the original segment and then taping in the replacement film image.
cutoff: The measured distance around the blanket cylinder of a web press that established the length of repeatability of the image. Also, a term used to describe the printed product as a sheet or signature in web production.
cutoff length: The measured distance around the blanket cylinder of a web press; also called *cutoff*.
cutoffs per roll: The number of sheets or signatures that can be cut from a roll of web paper; sometimes abbreviated as "cuts per roll."
cutting: The separation of larger sheets of paper into smaller sheets using a cutting machine.
cutting machine: A piece of equipment used for the cutting of larger sheets of paper into small pieces and for the trimming of finished goods to final size.
CWT: A term used in the paper industry meaning hundred weight or weight per 100 pounds of paper; common paper pricing system used to provide a price per hundred pounds.
database: A collection of related data.

database management system (DBMS): A program that manages a database.
data collection: The process by which data are gathered.
data transmission: Method by which data is sent and received, commonly referring to telecommunication or satellite media.
daylight contacting: Process of contacting film products in fairly normal lighting conditions; sometimes also termed "roomlight" or "brightlight" contacting.
debugging: Finding and fixing the errors in a computer program.
decimal hours: Division of a clock hour into hundredths or tenths.
dedicated system: A system designed and manufactured to perform a specific task or tasks, such as word processing or typesetting.
deep-etch plate: A lithographic printing plate that has the image etched into the plate surface.
delivery date: The promised time that a job will be delivered to the customer by the printer.
depreciation: A dollar amount assigned as compensation for the wearout of printing plant equipment and facilities.
desktop publishing (DTP): An integrated application of computer hardware, software, and laser printing that allows the electronic assembling and printing of type and graphic images at the same time, as a finished product.
diazo: A coal-tar by-product popular as a plate coating for presensitized plates and overlay proofs.
diffusion-transfer: A photographic material used with photographic equipment to produce paste-up art; commonly known as PMT.
digital: Numeric form of an image, color separation or type, which is electronically produced.
digital color scanner: Newer type of color scanner that scans images either for electronic storage (in digital form) or directly to film; an essential component to CEPS production.
digital data: Information or image converted to numeric format.
digitized image: An electronic file that represents an image of any kind that can be manipulated, retrieved, or massaged, in electronic form.
digitizing: Rendering typematter or a graphic image into electronic, numerical form.
digitizing platform: An integrated mix of electronic hardware and software developed and used to digitize, retrieve, manipulate, or massage electronic files.
direct access: See *random access.*
direct contacting: A film contacting procedure wherein the resultant contacted image has the same tonality as the original.
direct digital proofing: Production of a visible, hard proof directly from an electronic (digital) system such as a CEPS.
direct-input system: Typesetting or other system whereby the input is able to be immediately visible to the operator.
direct separation: The separation of process color originals without production of intermediate film images.
disk: A plastic recordlike magnetic material used to store information with a computer

system; both floppy (or flexible) and hard disks are used, depending on the computer system; also spelled as *disc.*
disk drive: A disk "player" that rotates the disk and reads and writes on it (using a read-write head) as instructed by the DOS.
diskette: Another term for a 5 1/4 inch floppy disk.
disk operating system (DOS): The program that instructs the computer's CPU in coordinating data, managing files, and transferring information to and from a disk.
disk pack: A set of hard disks piled on top of each other that is inserted into a disk drive; provides extensive magnetic storage capacity.
display type: Type over 12 points in height that is commonly used for headlines and when large typematter is required.
documentation: Operator manuals for a computer system; the preparation of documents that detail a computer system, programs usage, operating systems, and other information for user reference.
dot-for-dot registration: The placement of images in precisely required relation to other images with no variation between printed sheets.
dots per inch (dpi): The number of dots per linear inch of measure; used to describe the fineness of halftones, tinted materials and electronic display terminals; also termed "lines per inch."
double-burned: A term used to describe multiple exposures on a plate of proofing material.
double-faced proofing paper: A proofing paper that contains a photographic coating on both sides (C2S).
drilling: The operation of generating round holes in paper and other materials.
dryer unit: A postprinting process in heatset web production, where the web of paper passes through a heating unit to set the ink.
dry-tapping: The printing of a new ink film layer over a previously printed and dry film layer.
DTP: Desktop publishing; an integrated application of computer hardware, software, and laser printing, which allows the electronic assembling and printing of type and graphic images at the same time, as a finished product.
dummy: A folded sample representing a book, booklet, or image to be graphically reproduced; used for production planning and estimating purposes.
duplicator paper: A paper class, basic size 17 x 22 inches, used specifically with mimeographic reproduction units.
Dutch cut: Procedure by which some small sheets cut from a larger parent sheet will have the grain in the short direction, while other sheets cut from the same parent sheet will have grain going in the long direction; also termed *stagger cut* or *bastard cut.*
Dylux proof: A stable-based photographic proof, also called a *blueline proof* or *blueline.*
EFDC: Electronic factory data collection; process by which production data is collected from shop floor employees when production is completed, then stored in a computer

system for analysis and review; sometimes termed automatic factory data collection or shop floor data collection.

eight-bit: The word size or length of many microcomputers, which indicates they handle data in groupings of eight bits or the equivalent of one byte or one character.

eight-page web: A lithographic web press capable of producing 8-page sheetwise signatures with a cutoff size of approximately 17 x 22 inches; also termed a half-width web.

elastic market demand: A theory that states that as the price of goods increases, the volume sold decreases, and as the price of goods decreases, the volume sold increases.

electronic composition: The process of setting type whereby all input and output is completed in electronic format.

electronic factory data collection (EFDC): A process by which production data is collected from shop floor employees when production is completed, then stored in a computer system for analysis and review, sometimes termed automatic factory data collection (AFDC) or shop data collection (SFDC).

electronic front-end system: An integrated electronic prepress system, such as a CEPS unit, capable of handling a wide variety of prepress functions.

electronic imaging: The process by which images of any kind are produced electronically, in digitized form.

electronic platform: An assemblage of interrelated electronic hardware and software designed for production of digitized type and graphic images; also termed a platform.

elite typewriter: A typewriter with a spacing of 12 characters per inch.

em: A square area measuring the point size of type on all four sides.

em quad: A square block or area of a point size of type; used as a common spacing unit in composition; also called a *mutt* or *mutton quad.*

emulsion: Photosensitive silver salts or other emulsified product, typically coated on a support base.

emulsion-to-base contacting: A film contacting procedure where the emulsion of the imaged piece of film is placed in direct contact with the base of an unimaged film sheet.

emulsion-to-emulsion contacting: A film contacting procedure where the emulsion of the imaged piece of film is placed in direct contact with the emulsion of an unimaged film sheet.

enlarger: A piece of photographic equipment that enlarges images for graphic reproduction.

en quad: A spacing unit one-half the set width of the em quad with the same body size; also termed a *nut quad.*

entrepreneurial philosophy: Management philosophy whereby the company, shop, or business is operated with little regard to maximizing profit on any given job, with the hopes of making a profit over the long-term.

equivalent weight: See *comparative weight formula.*

estimate blank: A worksheet used by the estimator upon which the estimate is completed.

estimated production time: The amount of time determined during estimating for the completion of a segment of a job or the entire job in the plant.

etch-and-peel film: A photo-sensitive product which, when exposed to a line negative and processed, can have selective emulsion areas peeled away, thereby creating a knockout mask.

etched plate: A lithographic plate that has a very fine (etched) surface; major categories are deep-etched and multimetal groups.

EXCEL: A combination spreadsheet/database program written by Microsoft for the Macintosh computer system.

expanded: A term used to describe typefaces that have been designed so that character elements are spread out.

fake color: Producing a rainbow of different colors through a mix of solids, screen tints and process colors, when printed in superimposed form in a predetermined manner.

FAX: Abbreviation for facsimile transmission system, whereby any graphic image can be telecommunicated electronically from one point to another.

file: Data that can stand alone but represent a complete set of information on a topic or subject.

filler work: Printing jobs taken into the plant to smooth out peaks and valleys of the production schedule.

film contacting: The procedure by which an imaged piece of film is placed to touch an unimaged film sheet; after exposure and processing, a copy of the original is produced.

film drum: That segment of a scanner that holds the film to be imaged during the scanning process in color separation.

film dupe: Abbreviation for the final product as a result of film duplication.

film duplication: The reproduction of precisely the same film image using film contacting procedures.

filter: A colored sheet of film or glass used in color separation to absorb certain colors and allow others to pass through, thereby dividing a color image into defined color printers.

financial statement: A document representing the financial condition of a business such as a balance sheet, income statement, and ratio information; also called a *financial* by accountants and management.

finished goods inventory: A listing of all work that has been completed and is held in inventory for future delivery to a customer or customer pickup.

finish size sheet (fss): The final trimmed size of paper when a printing job is finished; also called *finish size.*

fixed costs: All cost that do *not* change with changes in production; also termed *setup costs* in changeover point calculations.

fixed overhead: All costs required to maintain the space and environment of a defined plant area or production center.
flag: A marker in a roll of web paper indicating a weak area or imperfection in the paper at that point.
flat: An assemblage of film and masking base materials produced during the image assembly (stripping) procedure.
flatbed camera: A camera that images film using a flat digital scanning process, as opposed to scanning using a cylindrical scanning drum.
flat color: A color that is specifically mixed to match a given sample swatch.
flat configuration plan: A plan developed and written for stripping a job, particularly helpful when estimating complex or difficult jobs.
flat file: A group or collection of flats making up one complete printing job.
floppy disk: A thin flexible magnetic medium disk generally popular in 5 1/4 and 8 sizes; commonly termed a *diskette.*
focal plane: That point of a camera or enlarger upon which the image is focused and where the film will be positioned for exposure; sometimes termed a *vacuum back.*
folding: A procedure by which printed paper is creased, thereby reducing it in size; a common bindery and finishing procedure.
font: A complete assortment of all characters in one size and series of type.
foreign competition: A marketplace situation where foreign countries diligently plan and strive to sell goods and services within the United States.
form: A collection of images on one plate that will be printed in the same color; also termed a *printer* or *plate.*
format: The arrangement of data for a file or report.
form chart: A planning tool showing arrangement of pages in a book or booklet with which stitching conditions may be determined.
FORTRAN: A high-level computer programming language used primarily for engineering, scientific, and mathematical applications.
foundry type: Individual type-high (0.918 in.) characters, stored in cases (drawers) and manually assembled into words.
Fourdrinier machine: A long machine with a revolving copper screen upon which a slurry is deposited, which ultimately will become paper.
franchise quick printers: A business established under a franchise agreement; common with quick printers.
franchiser: The parent company or parent corporation in a franchise agreement.
Franklin Catalog: A standard pricing catalog for products normally produced by the commercial printer.
freelance artist: Self-employed individual who produces graphic art; also called a freelancer.
full-width web: A lithographic web press capable of producing 16-page sheetwise signatures with a cutoff size of approximately 24 x 36 inches; also termed a 16-page web.

galley: A tray use to hold set type. A proof printed from the tray of type.

gathering: The collection of printed matter in a prearranged sequence; typically a binding and finishing operation; also termed *collating.*

Gerber Autoprep: An electronic CAD/CAM unit used in image assembly to enable strippers to electronically prepare film flats using a DataPrep input terminal, AutoPlot film plotter, and Pen Plotter.

glassline screen: A sandwich of two pieces of glass, each containing fine inked lines mounted at right angles to each other; used for halftone production.

gluing: The application of glue adhesives to hold products together; typically a binding and finishing or web production procedure.

goldenrod masking material: A common term used to describe yellow or amber materials used to assemble images into flats.

goldenrod stripping: The process of assembling film images into flats using goldenrod materials.

grain direction: The predominant alignment of cellulose fibers in a sheet of paper; see *paper grain.*

grammage: The metric weight of paper in grams per square meter; similar measurement to paper basis weight in the United States system.

grams per square meter: Same as grammage.

graphic artist/designer: An individual employed to plan and prepare artwork for graphic reproduction.

Graphic Arts Technical Foundation (GATF): A technically-oriented research and development operation, member-supported and serving the printing and allied industries.

GraphPro: A notebook-type publication written by the NAPL providing information and data on computer systems and vendors of such systems for the printing industry.

gross profit: The amount of dollars made by a business before administration, sales expenses, and taxes have been deducted.

groundwood paper: A paper stock that is made by grinding logs into particle form; typical groundwood product is newsprint.

halftone: The reproduction of a continuous-tone photograph or image by a pattern of fine dots of varying sizes and shapes, as related to the light or dark values of the continuous-tone image; contact or glassline screens are used for this procedure.

halftone screen: A screen made of glass (glassline) or film (contact) through which continuous-tone images are exposed and divided into dots.

half-width web: A lithographic web press capable of producing 8-page sheetwise signatures with a cutoff size of approximately 17 x 22 inches; also termed an eight-page web.

hard proof: A proof on paper or other substrate that can be mailed or physically sent to a customer, as opposed to a soft proof, which is visible on a color monitor and used to ascertain approximate color values.

hardware: A term used to describe electronic computer equipment.

heatset web: Web presses that have dryers and chillers to set ink immediately after printing, thereby allowing for finishing production to be completed in-line, immediately following the pressrun.
high contrast: A term used to describe images typically drawn in black ink on white paper; denotes extreme difference in value between two materials.
high-level language: A programming language such as FORTRAN, COBOL, BASIC, and PASCAL that generally uses English words to communicate with a computer.
high-profit printers: A term defined by PIA as those printing firms that earn 8% or more return on sales annually.
high speed copier: Black-and white copiers designed and build for extremely fast copying from original images; commonly found in quick printing and small commercial printing shops.
hotline: A telephone or telecommunications system offered by a vendor to repair or review a computer program quickly, with little lost time for the printer/user.
hot metal composition: The setting of typematter from molten lead mixture and brass matrices; includes Linotype, Intertype, Monotype, and Ludlow operations.
hundred sheet price: See *price per 100 sheets.*
hundred weight: See *CWT.*
hyphenation and justification (H&J) program: A typesetting software program that automatically provides some typesetters with the ability to hyphenate and justify characters during production.
IBM-compatibles: Microcomputer systems that have an architecture and system design identical to IBM machines, but have not been manufactured by IBM.
image assembly: The procedure by which film images are positioned in a precise order for platemaking; also called *stripping.*
imposition: The exact positions of book pages as they will fall on a press sheet, ultimately folded to produce a signature; positioning of images in a precise order on a form.
impressions per hour (imp/hr): A system used to measure the output of printing presses.
incandescent light: Light produced by a tungsten filament heated to incandescence by an electric current.
income as a percentage of sales: A mathematical comparison of dollars earned as income (profit) to total dollars of sales.
income before taxes: See *net profit.*
incompatible: A term used to describe computer software and hardware that will not work on other computer equipment and programs.
independent quick printer: A quick printer who does not have an affiliation with a franchise.
index bristol: A paper class, basic size 25 1/2 x 30 1/2 inches, used when moderately thick paper material is required.

indirect contacting: A film contacting procedure wherein the resultant contacted image has the opposite tonality of the original.
indirect separation: The separation of process color images using continuous-tone intermediates for color correction purposes.
inelastic market demand: A theory that states that there is no relationship between price and volume of goods sold; prices may rise or fall with no direct reduction or increase in volume.
in-house system: A computer system that is completely operated and supported inside the company.
initialization: The startup of a computer system; the resetting of a computer system to a starting point.
ink coverage: A percentage factor related to the size of the form to be printed and the kind of form.
ink mileage: The volume of ink required to print a certain coverage and specified number of copies.
in-line: A term used to describe production operations performed in a sequential manner.
input: The method or procedure by which data are fed into a computer system.
input-only hardware: Computer hardware that can only be used as an input medium.
input/output hardware (I/O): Computer hardware that can be used for either input or output of data.
integrated circuit: A fingernail-sized silicon chip containing thousands of transistors, diodes, capacitors, and resistors, all of which are fabricated and assembled in a single integrated process.
integrated standards: Output values reflecting the combination of manual and automated (person-machine) production.
integrated system: Any system comprised of modular or separate components that interface with each other, such as MIS computer systems for the printing industry.
interpolation: A process of mathematically determining an intermediate value between two established values.
intuitive estimating: Determining the cost or price of a printed job using guesswork or intuition.
inventory (raw material) report: A printed listing by the computer of all or part of the raw material inventory of a company.
investment community philosophy: Management philosophy whereby the company, shop, or business is operated with careful cost, manufacturing, and management controls so as to maximize profits on each job, with every expectation of making a profit over the long-term.
invoice: A written statement indicating an account to be paid for providing goods and services.
ISO: International Organization for Standards, a group created to promote the development of worldwide standards; developed metric paper standards used throughout much of the world.
Itek 1015: A photo-direct camera/platemaker developed by ITEK and Eastman Kodak in the early 1960s, which served as the genesis for the quick printing industry.

job costing: A computer software system wherein actual and estimated times and costs are compared after-the-fact to note variations in the data and to revise estimating standards.

job cost summary: A procedure completed after a job has been delivered that compares actual to estimated production times and costs; production data used to revise estimating standards.

job loading/scheduling: A process by which printing equipment is organized to complete production.

job specifications: The specific details of a job to be printed, typically indicated by the sales representative to the estimator on a job specification form; also termed *job specs.*

job summary: A technique comparing actual production times and costs to estimated times and costs; see also *job cost summary.*

job ticket: An envelope or other document that indicates all production details of a job in production and that travels with the job in the plant; sometimes called a *job docket.*

jogging: The alignment of paper stock into very even piles to facilitate production.

justify: To evenly align typematter in block form on both right and left margins during the composition process.

just-in-time management: A management concept whereby a printing company operates on an "immediate need" basis with customers or suppliers. Commonly abbreviated as JIT.

keyboard: An input-only device for a computer system.

key color: The base, or first, color pasted to illustration board when preparing mechanicals to which overlay colors are registered.

keyline art: The base or key mechanical art image to which all other images register.

keystroke: The measurement of composition output by the time required to identify and strike a key on the keyboard.

kilobyte (K): 2^{10}, or 1024, bytes of memory capacity in a computer system.

kilowatt hours (kWh): A measurement of power consumption or usage for job-costing procedures.

KIS philosophy: "Keep It Simple" management philosophy, whereby there is an emphasis on the development and implementation of less complex practices and procedures to manage a business; usually practiced by small firms and companies.

knife folder: A type of folding equipment that creases the paper and forces it into two knurled rollers that complete the fold.

knockout: Dropping out or removing an image using some type of masking material such as peelcoat, goldenrod, or film.

labor intensive: A term describing a condition of business where large numbers of employees are necessary to provide output for a company.

laser printer: Term used to refer to a generic line of computer printers that print digitized type and graphic images from a microcomputer.

LaserWriter: A computer printing device introduced by Apple Computer in 1985; first laser printer commercially available in the U.S.

leaded one point: A description of type set with white space between lines that is one point over the established body size.

leaded two points: A description of type set with white space between lines that is two points over the established body size.

leading: The amount of white space between typeset lines, typically indicated as a point value over the established body size of the type.

ledger paper: A paper class, basic size 17 x 22 inches, used for record keeping and accounting work.

letterpress: A printing process wherein type-high relief forms are inked and impressed to transfer the image directly to the paper.

letter-quality printer: An output-only device that produces characters equal in quality to those of a conventional typewriter.

lift: A hand-held stack or pile of paper to be cut, trimmed, or worked in production.

lift value: The approximate number of sheets of paper in a hand-held lift; also called *load value.*

line copy: Material prepared for reproduction containing only high contrast images, lines or dots.

line film: A category of film generally used for the photographic imaging of line work.

line work: Graphic images in line or solid block form.

Linotronic 300: A typesetting output device manufactured and sold by Allied Linotype, capable of producing high-quality photographic type and graphic images as integrated output, at 1200 dots-per-inch (dpi).

Linotype: A hot metal linecaster manufactured by Mergenthaler Corporation.

lithography: A printing process whereby image transfer is accomplished through the use of plates with a planographic (flat) surface; plates accept ink in the sensitized (image) areas and accept water, which repels ink, in the nonimage areas.

lith tape: A red tape commonly used in image assembly for the taping of film base materials together.

Lotus 1-2-3: A popular spreadsheet program for IBM, IBM-compatible and Macintosh microcomputer systems.

low-level language: A machine and assembly language that uses numbers or special command words; considered the native tongue of a computer system.

Ludlow: A hot metal linecaster that requires manual selection of characters and separate casting of the assembled characters in a solid line of type.

M sheet price: Paper price expressed as a cost per 1000 sheets of an established size; also termed *thousand sheet price.*

M weight: The scale weight, in pounds, of 1000 sheets of paper of a given size; also termed *thousand sheet weight.*

M&SW: Microcomputer and software system; one of three general ways a printing company may choose to computerize their buseness.

machine hour rates: A term used interchangeably with BHR.

machine language: A low-level language used by the CPU essentially as a pattern of "0" and "1"; a high-speed language.

machine standard: The output value for a piece of equipment that is fully automated and requires no employee operation.
Macintosh: A revolutionary microcomputer introduced by Apple Computer in 1984; represents the least expensive platform (Group V) on which electronic prepress imaging production can be completed.
Macintosh II: Group IV platform sold by Apple Computer, which shows significant promise for future lower-cost CEPS applications.
magnetic media: Computer storage media that are magnetic in structure, including floppy disks, hard disks, and magnetic tapes.
mailing operations: Companies that engage in services of addressing, inserting, and sorting printed products to be mailed.
mainframe: A complex computer system with extensive storage capacity and many peripherals.
main memory: The primary memory built into a computer system; also termed *core memory.*
management information system (MIS): A system designed to provide management with the necessary data and information to run a business.
manual roll change: Changing rolls using manual methods during a web press run; requires stopping the press, which generally increases spoilage and waste.
manual standard: The output value for an operation completed entirely manually.
manufacturing cost: The production cost of producing a printing order that does not include administrative and sales costs for the product.
manuscript page: A page of typewritten copy.
markup: See *cost markup.*
material cost: A dollar value for all materials directly required in the production of a printing order; paper, ink, film, and lithographic plates are major material costs.
mechanical: Artwork that has been fully prepared for photographic reproduction as the first step of a printing job; also termed a *paste-up.*
megabyte (MB): One million bytes (characters); used to describe the size of large storage media such as disk drives.
menu-driven: A term used to describe computer software that requires the operator to select and use data from menus displayed on the monitor or terminal.
MHR: machine hour rate; same as budgeted hour rate.
micro and software system (M&SW): One of three general ways a printing company may choose to computerize their business.
microchip: See *chip.*
microcomputer: The smallest computer system available, easily affordable to smaller businesses and home use; generally divided into IBM and IBM-compatible and Apple Computer classifications.
Micro-Graphpro: The title of a publication produced by the NAPL that provides information and data on microcomputer systems and vendors of such systems for the printing industry.

microprocessor: A chip containing the circuitry to process information and data for a microcomputer or other electronic unit.
millimeter: A metric measurement whereby a meter (39.37 inches) is divided into 1000 equal parts thus equaling 0.03937 inches; about the thickness of a paper clip wire.
minicomputer: An intermediate-sized computer system generally costing between $15,000 and $150,000.
MIS: Management Information System.
modem: The abbreviation for modulator-demodulator, an input/output device used to link systems together by telecommunications.
modular computer system: A computer system logically divided into parts, segments, or modules whereby each part can operate independently of all other segments.
moiré: A generally undesirable pattern formed when two or more dotted screen areas (tints or halftones) are incorrectly overlapped.
monitor: A visual display terminal (VDT) used for data output only; available in monochromatic and color types.
monochromatic proof: A high contrast proof typically containing a colored image on a white background; brownline and blueline proofs are in this category; also called *monotone.*
Monotype: A hot metal typesetting process in which characters are individually cast and then assembled as words and lines; requires keyboard and then follow-up caster operation.
MS-DOS: Microsoft Disk Operating System; popular with IBM and IBM-compatible microcomputer systems.
M&SW: Microcomputer and software system; one of three general ways a printing company may choose to computerize their business.
multicolor stripping: Complex image assembly procedure whereby process color and fake color techniques are integrated together; can be completed manually, electronically, or as a mix of both.
multimetal plates: Long-run lithographic plates that are layered using a base metal, copper, and stainless steel or chromium.
Multiplan: Popular spreadsheet program for both IBM and IBM-compatibles and Apple Computer systems.
multiple shops: More than one location serving the public; common with many quick printers.
NAPL: National Association of Printers and Lithographers.
NAQP: National Association of Quick Printers.
narrow-width web: A class of web presses made to run web widths in the range of 20 inches.
National Association of Printers and Lithographers (NAPL): A group of affiliated member firms doing printing manufacturing; headquartered at 7800 Palisade Avenue, Teaneck, NJ 07666.
National Association of Quick Printers (NAQP): Founded in 1975, NAQP

represents quick printers and small commercial printers throughout the U.S. and Canada.

needs criteria: A listing of criteria to be met with the purchase of a computer or other type of equipment or service.

negative: An image on film or paper where the image is reproduced as clear or white and the surrounding background is dark or black.

negative-acting: A term used to describe photographic products that reverse tonality with each photographic step.

net-owners compensation: Defined by NAQP as the amount of money left over after all expenses have been paid, but before paying the owner a salary or giving the owner any fringe benefits.

net profit: The amount of dollars remaining after all costs have been deducted, but before taxes are paid; also called *net profit before taxes* or *income before taxes (bt).*

network: The process of connecting a number of computer systems together.

newspaper web: A web press typically made and used for newspaper production.

newsprint: A paper class, basic size 24 x 36 inches, used as the printing base for newspaper and other less expensive products, see also *groundwood paper.*

nonchargeable hours: All work that cannot be directly related to a specific job or customer order nor charged to any specific account.

nondedicated: A term describing a computer system that has general-purpose applications, which may be used for word processing, tie-in to database systems, estimating, and so on.

nonheatset web: Web printing that utilizes no chiller/dryer postprinting units to set ink; term commonly applied to newspaper production where ink dries by absorption; also known as *coldset web.*

number crunching: A term used to describe the processing of numbers by a computer system when completing a largely number-oriented task.

numbering: An operation wherein consecutive numbers are printed on goods.

offset paper: Same as book paper with a basic size of 25 x 38 inches.

opaque: A liquid product painted on film with a brush to cover imperfections and small pinholes.

optical character recognition (OCR): The process by which characters are recognized using scanning devices; computer-input-only hardware.

Opti-copy: An accurate projection technique for step-and-repeating images to photographic film.

orbital (micromodifier) technique: One of two methods used to produce spreads and chokes; process required the Byer's Micromodifier, which produces extremely accurate spreads and chokes using a carefully controlled, orbital rotating bed.

order entry: The beginning data necessary to enter a job into a production system; part of a computer system in printing.

original artwork: Any graphic image designed for first-time graphic reproduction.

original equipment manufacturer (EOM): A vendor that purchases computer hardware from a primary manufacturer, combines the hardware to form a computer system, and then sells the complete system, plus software, to a printer/user.

orthochromatic material: Light-sensitive material such as film products that is not sensitive to red light.
out: The number of sheets that will be cut from a larger sheet of paper.
out-of-pocket cost: A cost incurred by the printer that must be paid immediately or within a short period of time.
output per time unit: Basis for some production standards; example would be "6000 sheets per hour."
outside sales: Describes the use of a sales representative to increase the volume of a printing company; term commonly applied to quick printing since many larger printers employ outside sales representatives on a routine basis.
outside service supplier: A vendor or firm that supplies services, such as a trade house, service bureau, or timesharing firm.
overcapacity: The concept of having too much available production time or equipment; in a purely competitive economic environment, generally leads to reduced prices.
overlay proof: Proofs that consist of two or more layers of color substrate that are exposed, processed, and mounted in register to one other.
overlay proofing material: Proofing material that has color contained on different sheets of film; each color form is exposed and processed to the colored film desired and then the sheets are positioned over one another in register.
overprint: To print over or on top of another ink color; also termed a *surprint.*
over-the-counter sales: Taking orders face-to-face from customers who present themselves at the shop or company; term commonly applies to quick printing.
packaged software: See *canned software.*
padding: The application of a glue adhesive for temporary binding of paper stock in pads.
Pagemaker: A popular desktop publishing software package developed by Aldus for both the Macintosh and IBM systems.
pallet: See *skid.*
Pantone Color Simulator: A reference book that shows production requirements allowing fake color and flat color combinations to be interrelated.
Pantone Matching System (PMS): An ink and color matching system that has become the standard for the printing industry; PMS colors are referenced by number and are formulated using exact quantity measurements so the color value is always consistent.
paper: A thin mat of cellulose or other type of fiber; major material in printing.
paper caliper: The thickness of a paper stock measured in thousandths of an inch (or points).
paper catalog: A published price book indicating all paper offered for sale by a paper merchant.
paper grain: The alignment of cellulose fibers (or other type of fibers) in a sheet of paper; when writing dimensions, "grain long" or "grain short" may be indicated with a line above or below the appropriate dimension of paper.
paper micrometer: A device used to determine accurately the thickness of paper stocks.

paper point: The equivalent of one thousandth of an inch; used to indicate paper thickness (caliper).
paper ratio system: An estimating system that uses a ratio of the cost of paper to the total job cost or price of the job.
parallel folding: A procedure by which sequential folds in paper are completed parallel to one another.
parent size sheet (pars): The size of paper to be purchased from the paper merchant; also termed *stock size sheet.*
passivation: A procedure used when manufacturing lithographic plates to reduce undesirable chemical reactions between coating materials and aluminum plate bases.
password: The assignment of codes to certain computer programs and databases; the codes provide access only to those individuals who know the key term, or password.
past work basis: An estimating/pricing system, which uses the value of past printing jobs done as a basis for quoting on current work.
paste-up: See *mechanical.*
payroll: A general term that includes all the data and information needed to complete transactions with respect to payroll calculations for company employees.
peelcoat: A red or yellow gelatin material coated on a clear carrier sheet that is manually cut and peeled to produce a graphic image.
penalties: Adjustments for special operations or less typical production procedures.
percentage markup: A method by which profit is added to a job based on a percentage of the estimated cost or some other basis; percentage markup is a top management decision.
perfector: A term used to describe a printing or press that prints an image on both sides of the sheet simultaneously; also called *blanket-to-blanket.*
perforated: A term used to describe paper that has small dashed cuts made close to one edge to facilitate tearing the paper out of a book or booklet.
peripheral: An accessory part of a computer system not considered an essential part to the system's operation.
personal computer (PC): Generally, a microcomputer designed for single-person use with primary applications to small businesses and home use.
photo-composed film: A process by which images are combined or collected on film using photographic techniques; most common reference is to the production of composite film materials.
photo-direct: A process by which a lithographic plate is made directly from original art, bypassing other prepress operations such as film production and image assembly; the Itek 1015 unit developed for the quick printing industry is an example of the photo-direct process.
photomechanical transfer (PMT): A camera-speed, diffusion-transfer product used to go directly from camera copy to paste-up or film positive.
photopolymer coating: A light-sensitive coating used principally for presensitized lithographic plates.
photostat: An image produced using a mix of photographic and electrostatic methods; sometimes termed a *stat.*

phototypesetting: A procedure by which type is set photographically; requires keyboarding or punching of a paper input tape on a typewriter unit, with sequential typesetting in a photographic unit containing a light source, a font on film or magnetic medium, and photographic paper or film.
PIA: Printing Industries of America
"PIA Production PAR": A notebook containing a wide assortment of production times for most major pieces of printing equipment; produced by PIA; smaller condensed volume called "Simplified PAR," or "SimPar."
PIA Ratio Studies: An annual report containing financial information and comparisons with respect to production and finances; published by PIA.
pica: A common unit of measurement in copy preparation and typesetting; one pica equals 12 points, and there are approximately 6 picas in 1 inch.
pica typewriter: A typewriter with spacing of 10 characters per inch.
pin registration: The use of metal or plastic pins to align flats, film, or prepared copy during prepress operations.
platemaker: A vacuum frame with a light source used principally to expose lithographic plates to film images and flats.
platemaking: The procedure of preparing, exposing, and developing lithographic plates.
platemaking and proofing configuration plan: A plan developed and written for plating and proofing a job, particularly helpful when estimating complex or difficult jobs.
plate scanners: An electronic device used to read lithographic plate printing densities prior to plate mounting and press makeready; the densities are then input into the press computer console, which automatically adjusts ink coverage relative to the density values.
plate screen: A piece of film used to divide normally solid areas into a design or dot pattern; same as *screen tint.*
platform: An assemblage of interrelated electronic hardware and software designed for production of digitized type and graphic images; also termed an *electronic platform.*
ply: A term used to indicate thickness of cardboard stocks; to convert ply thickness to caliper, multiply the ply value by 3 and then add 6 to that result.
PMS: Pantone Matching System
PMT: A camera-speed diffusion-transfer product used to go directly from camera copy to paste-up or film positive.
point (paper): The equivalent of one thousandth of an inch; used as a measurement of paper thickness.
point (printer's point): The basis for typesetting, where one point equals 1/72 inch; U.S. Bureau of Standards indicates the equivalent of 1 point is 0.013837 inch.
polarity: A terms used to describe whether a film image is positive or negative; also called *tonality.*
portable: A term used to describe computer programs designed to run on different computers without modification.

positive: An image on film or paper wherein black copy is reproduced on a light or clear background.

positive-acting: A term used to describe photographic products that maintain the same tonality (or polarity) from image to image.

Postscript code: The most common language for driving laser printers and typesetting units such as the Allied L300.

pound: See *basis weight.*

prepress operation: Any printing manufacturing procedure that occurs before press-work; general categories include copy preparation, photography, image assembly, and platemaking.

presensitized plates: A lithographic surface plate that is coated during manufacture and shipped to the printer in coated form.

press-on type: Letters that are transferred from a carrier sheet to artwork by pressing or rubbing; also called *rub-on* or *transfer type.*

press preparation and makeready: That segment of press operation that sets up the press for the particular job to be run; includes such factors as setting grippers and guides, adjusting ink distribution, and setting feeder and delivery mechanisms.

press proof: A proof that has been manufacturing using conventional printing production including color separation, stripping, plating, and press operations.

press running: The continuous operation of a printing press by which acceptable press sheets are printed.

press size sheet (pss): The size of paper stock that will receive the image during press-running operations.

press spoilage: An additional amount of paper included at the beginning of the production of a job to compensate for throwaways and unusable sheets generated during presswork and finishing operations.

presswork: The mechanical reproduction of copy using a printing press.

price collusion: The illegal establishment of prices for a product, good, or service within a geographic region; price fixing.

price cutting: Reducing or lowering the price on a given printing job; typically found when competition for customers is intense, although considered an unwise practice if costs and prices have been carefully predetermined.

price list estimating: The development and use of job prices for product lines standard to a particular company.

price per 100 sheets: A paper catalog price expressed as a dollar amount per each 100 sheets; also called *hundred sheet price.*

price per 1000 sheets: A paper catalog price expressed as a dollar amount per each 1000 sheets; also called *price per M sheets.*

price proposal: A written offer from the printer to the print buyer or customer proposing a tentative price and job specifications that describe the work to be done.

price quotation: A final, written offer from the printer to the print buyer or customer, locking up price, quantity desired, and all other vital job particulars.

pricing: The establishment by the top management of a printing plant of the selling price of a printed good, product, or service.

print buyers: A representative of a business, advertising agency, or corporation who purchases printing on a consistent basis.
printer (computer): A device connected to a CPU that transforms electronic data into hard copy form.
printer (printing): A terms used interchangeably with *form* to describe a graphic image produced in a specific color.
printing broker: A person who sells printing goods and services to customers and print buyers, then arranges for printing manufacturing through printers willing to produce the job.
Printing Brokerage Association: An association of brokers who sell and service customers desiring to purchase printed goods.
printing consultants: A person retained by the customer or print buyer to ensure consistent print quality and coordinate printing projects at different stages of manufacturing.
printing cost estimating: A mathematical procedure used to determine the cost of a printing job prior to production using BHRs and standard production times and adding all additional material costs.
printing estimator: An individual who completes cost estimates in a printing company.
Printing Industries of America (PIA): A group of affiliated member firms engaged in printing manufacturing; headquartered in Arlington, Virginia.
Printing Trade Customs: A compilation of operating and business practices held as legal based on court precedent; typically printed on the back of a quotation or proposal.
procedural standard: A recognized technique or procedure used to complete a given job or task; also termed a *technical standard.*
process camera: A piece of equipment used to reduce or enlarge artwork as a major step in prepress operations of printing manufacturing.
process color: A specifically formulated pigment used to print process color separations; magenta (red-blue), cyan (blue-green), yellow (red-green) and black as transparent colors; also called process red, process blue, process yellow, and process black.
processor: A device used to process data and information using an electronic microchip.
process specialization: Concentrating on a market based on a specific process available from a printing company.
process standardization: Focus on similar production techniques that allow for streamlined manufacturing procedures.
production center: A defined work area where a certain segment of printing production is completed.
production center method (PC): A prorated costing procedure that does not include administrative areas, thereby spreading job costs over only those areas that have a defined printing production function.
production coordinator: An individual who works with sales representatives, estimators, and production personnel to ensure smooth movement of jobs through plant production.

production planner: An individual who determines the production flow for a specific printing job; also called a *job planner.*

production scheduling: The assignment of jobs into the sequence of production in a printing plant.

production standard: An hourly value representing the average output of a particular operating area producing under specified conditions.

production standards manual: A book or notebook containing all production standard times and procedures for a printing facility.

product specialization: Concentrating on a market based on a specific product manufactured by a printing company.

product standardization: Focus on similar production techniques that allow for streamlined production of a printed product.

profit: The excess dollars remaining after all costs have been accounted for in a particular job.

profit markup: The addition of an amount of dollars over the cost of a job.

proofing: A visual check of a job in production; may be completed in monochromatic or multicolor form depending upon the job.

proof to satisfaction: Process that requires the printer or prepress trade house to provide whatever number of proofs will be necessary to obtain customer approval.

proposal: An offer make to a customer for production of printing goods and services that specifically states all job requirements, specifications, and prices for such goods; may be accepted, rejected, or modified by the customer.

prorated costing: A costing process by which fixed and variable costs are identified, divided up using some index reflective of the company, and then distributed back over the company's operating areas.

purchase order control: A computer report providing management with information and data regarding purchases made by company employees for company materials and supplies.

purchasing: The arrangement for buying of materials and services for a job in production; may be assigned to the estimator as a related job duty.

Quarterly Business Indicator report (QBIR): A four-times-per-year publication from NAPL which provides timely financial data and information on the printing and allied industries.

quick printer: A printing company serving customers with fast-turnaround printing and related graphic services.

quotation: A legal, binding agreement between customer and printer for specified printing goods and services; typically a written form.

rag content paper: A paper manufactured with a specified percentage of cotton fiber.

random access memory (RAM): Erasable memory used by the computer on a temporary basis; also called *read and write* and *scratch pad memory.*

random access storage: The input or retrieval of data from a storage medium such as a disk drive where the time for access is independent of the data most recently input or retrieved; also called *direct access.*

rapid-access processing: A film processing system whereby exposed film is developed and fixed using predetermined chemical controls, allowing for fast development and consistent final film images.
read only memory (ROM): Permanent memory on chips wherein information can be retrieved but not stored; memory is not lost when the power to the computer system is turned off.
Ready-Set-Go: A popular desktop publishing software package developed by Letraset for both the Macintosh and IBM systems.
real-time plus materials estimating: Used to estimate CEPS, this process requires the estimator to carefully review job components and sometimes provide the customer with a "ballpark" estimate, then track actual time and materials as the job is completed; similar to the chargeback system.
recycled paper: A paper made from previously printed stock.
reel-to-reel magnetic tape: A magnetic medium using tape that is spooled on reels.
registration mark: A mark or symbol used throughout printing production to align (register) images; typically appears as two perpendicular lines in crossed form centered in a circle or oval.
regular cutting: The procedure by which all small sheets cut from a larger parent sheet will have the same grain direction.
repeat work: Printing jobs that are reordered by a customer with little or no change.
report generator: Software that allows report data to be formatted in various ways to be customized to the database available.
reproduction proof (repro): A proof ot typeset material, made on nonglare paper, to be used as paste-up copy.
request for estimate form: A form used by printing sales representatives to clarify and detail job specifications for a printing order.
return on investment (ROI): The ratio of income (profit) earned as compared to to assets (capital) required to earn that income; can be based on either gross or net assets.
return on sales (ROS): See *income as a percentage of sales.*
right angle folding: The procedure by which each fold is made at right angles to the preceding fold.
right-reading: A term used to describe images that can be read or deciphered from left to right; typically implies that the negative or positive reads correctly through the base side.
Robinson-Patman Act: Actually a segment of the Clayton Act, this federal law prohibits the seller of a product, good, or service from offering goods or services of like grade and quality to two or more buyers at different prices.
robotics: The process of using automated mechanical-electrical devices to perform numerous sequential production operations.
roll: Paper stock that remains as one continuous ribbon; also called a *web* of paper.
roll footage: The number of linear feet in a roll of paper.
roll-to-roll: A web production technique where rolls of paper are printed and then rewound in roll form for shipment and future conversion to sheets.

rub-on-type: See *press-on type.*
Rubylith: A popular peelcoat product that is red in color, made by the Ulano Company.
running charge: A dollar charge for computer time used for processing jobs; used with timesharing systems.
rush work: Printed work that is time-sensitive to the customer, thus requiring a fast-turnaround by the printer.
saddle stitching: The placement of wire or string along the binding spine (saddle) of gathered signatures, holding them together as a book or booklet.
safety paper: A paper class, basic size 17 x 22 inches, used when no change of written image is desired, such as for many types of documents requiring financial exchange (checkbooks).
sales per employee: A dollar figure derived by taking the gross sales dollars of the company and dividing by the number of employees required to generate that sales volume; considered an important measurement tool although the average figure varies by industry segment.
sales per square foot: A dollar figure derived by taking the gross sales dollars of the company and dividing by the number of productive square feet required to generate that sales volume.
sales representative: A person who meets with customers, determines their printing needs, and then arranges for production of products to meet these needs; also termed *salesperson* or *sales rep.*
scanner: A piece of photographic equipment with rotating copy, film drums, and a light source that translates copy to film as a light is drawn slowly across the image; popular for color separation production.
scanning wand: A hand-held device that is rubbed across a bar code to input data into a computer system.
scored: A term used to describe paper stock that has a crease to provide a crisp line for folding.
screen tint: A piece of film used to divide normally solid areas into a design or pattern; two major categories are dot and special effect patterns.
scribing: The manual removal of film emulsion using a scribing tool or knife.
self-service copying: Providing walk-in customers with stand-alone copy machines on which they can complete their own copying requrements.
selling price (SP): The amount of money a customer pays for particular printing goods and services; includes all costs incurred by the printer and a profit margin.
separation master flat: The first set of separation images assembled to which all other separations will be registered.
service bureau: A company offering computer, typesetting, or electronic imaging services.
set size: The linear (horizontal) distance a typeset character occupies, which varies from character with most typefaces; also called *character set* or *set width.*
set solid: A description of typematter set with no leading between lines.
set width: See *set size.*

SFDC (shop floor data collection): A process by which production data is collected from shop floor employees when production is completed, then stored in a computer system for analysis and review; sometimes termed automated factory data collection (AFDC) or electronic factory data collection (EFDC).
sheetfed: Printing or feeding sheets of paper; used to describe types of presses that are designed around printing paper in sheet form.
sheeting: The conversion of rolls of paper into sheet form with the use of sheeting equipment (called sheeters).
sheetwise (SW) impostion: A press sheet that has one-half of the pages of a signature printed on one side and the other half on the opposite side in direct relation; also called *work-and-back* imposition.
shop floor data collection (SFDC): A process by which production data is collected from shop floor employees when production is completed, then stored in a computer system for analysis and review; sometimes termed automatic factory data collection (AFDC) or electronic factory data collection (EFDC).
side stiching: The placement of wire or string through the side of gathered signatures, holding them together as a book or booklet.
signature: A collection of printed pages folded in a prearranged sequence to make all or part of a book or booklet.
silverprint: See *brownline.*
single-faced proofing paper: A paper coated with a photographic material on one side(C1S).
single-sheet proof: A proof upon which all proofing layers or colors are adhered or laminated to a base carrier sheet; DuPont's Cromalin and 3M's Matchprint are popular proofs in this category.
single shops: A quick printing company that operates from one location, as opposed to quick printing films that operate multiple locations.
sixteen-bit: The word size or length for many microcomputers and minicomputers, which indicates they handle data in groupings of 16 bits.
sixteen-page web: A lithographic web press capable of producing 16-page web sheetwise signatures with a cutoff size of approximately 24 x 36 inches; also termed a *full-width.*
sized and supercalendered (S&SC) paper: A paper stock that has been sized with necessary products and then highly calendered to produce a paper that has excellent surface requirements for printing.
sizing: Chemical products added to paper during manufacture that physically make the paper more suitable for printing production or ultimate product usage.
skid: A wooden or metal base with runners upon which paper is stacked in large quantities and then wrapped for shipment; a pallet is similar to a skid, but is about half the size.
slabbing waste: Unusable paper stock cut (or "slabbed") away from the outside of a roll of paper, required because of damage to the roll.
slit: To cut paper apart during folding or a special operation.
slug: A solid line of lead material containing relief characters (or blank, to be used for

leading), producted from a Linotype, Intertype, or Ludlow linecaster.
slug machine composition: A typesetting procedure utilizing slug linecasters such as Linotype, Intertype, or Ludlow.
slurry: A mixture of cellulose fiber, additives, and water that is flowed to the Fourdrinier machine in the papermaking process.
small business computer (SBC): A microcomputer system intended for use by a small business.
small commercial printer: A printing company that has limited full-time employees and generally pursues local customers on a walk-in or neighborhood basis; may offer quick printing and copying in addition to other printing services.
small press printer: A printing company which produces work on press sheets measuring no more than 11 x 17 inches in dimension.
soft-proof: An image on a color monitor or screen to ascertain approximate color values of an image, as opposed to a hard proof on paper or other substrate which can be mailed or physically sent to a customer.
software: A term generally used to describe the program by which a computer operates.
software consultant: A person hired to write programs and provide other software advice to a user.
specialization: The concentration of effort into a certain market area; in printing, specialization is oriented toward product, process, or both general categories.
speciality web: A web press, typically narrow-width, built and used for production of a specialized product.
specified form: A form used for ordering and detailing papers; a specified information sequence indicating all important ordering data.
splicing equipment: Web equipment that provides for automatic change of rolls during a pressrun; sometimes called *flying pasters* or *automatic pasters.*
split invoice: A technique whereby the customer pays a deposit for out-of-pocket costs before a job begins production; after the job is delivered, the customer is invoiced for the remaining costs.
spoilage: Usuable, throw-a-way, or spoiled paper from the pressrun portion of the production of the job.
spread: A film image that is slightly larger in size; produced from a negative film original to raw film using film contacting procedures; also termed a *fatty* or *swell.*
spreadsheet program: Packaged software for microcomputers wherein a large matrix of rows and columns of numerical data can be built to answer "what if" questions as the data are changed.
square pica: A square block measuring 12 points (one pica) on all four sides; used as a copyfitting measure.
staff management: Those individuals working in management who perform duties in a plant that are advisory in nature and not directly related to production.
stagger cutting: A procedure by which some small sheets cut from a large parent sheet will have grain in the short direction while other sheets cut from the same parent sheet will have grain going in the long direction; also termed *dutch* or *bastard cutting.*

stand-alone console: A computer system that is entirely self-contained.
standard catalog: A publication that provides standard job prices for printed goods; most common is the *Franklin Catalog* (Porte Publishing Company).
standard data: Information such as production standard times and BHRs that remain constant over a period of time.
standardization: The establishement of a defined production procedure that does not vary from day to day; also refers to products that have established and unchanging requirements.
standard production data: See *production standard.*
standard production time: An hourly value representing the average output of a particular operating area producing printing under specified conditions.
standard size paper: A paper in sizes or dimensions other than the basic size that is offered for sale by paper merchants.
statistical process control management: A management concept developed to improve the quality of the printed product through the establishement of a set of conditions, which are maintained throughout the production of the printed item by employees working on the product. Commonly abbreviated as SPC.
step-and-repeat: A procedure by which a single film image or a group of the same image is exposed to a lithographic plate or sheet of film in a defined manner to provide multiple images on the plate or film sheet.
stock size sheet (sss): See *parent size sheet.*
straight cutting: Paper stock cuts where the resultant cut sheets all have the grain going in the same direction, as opposed to stagger cutting where some of the cut sheets have grain direction one way, and other cut sheets having grain direction the opposite way.
straight matter: Material typeset in justified, block form without any typographic variation.
strike-on composition: Typematter produced by the direct impression of characters on paper, such as with a typewriter.
stripper: An individual who mounts film images into flats to complete image assembly in a printing plant.
stripping: See *image assembly.*
stripping tab: A small piece of film, with a prepunched hole at one end; pairs of these are attached to the edges of masking sheets to provide a pin registration procedure.
studio artist: An individual affiliated with an art studio and involved in the preparation of materials for graphic reproduction.
substance weight: See *basis weight.*
subtractive surface plate: A presensitized lithographic plate that requires removal of nonimage (unexposed) areas during processing.
sulfite papers: A large group of paper products made using chemical breakup of cellulose fibers with subsequent cooking, beating, and refining procedures.
supercalendered: A term used to describe paper stock that has been calendered extensively by manufacturing procedures that produce a paper surface that is very smooth and has excellent printability.

surface plates: A major category of lithographic plates that carry their image on the surface; divided into presensitized and wipe-on subgroups.
surprint: Overprint of one image over another.
SW: Abbreviation for sheetwise imposition.
systems analyst: An individual who studies the activities, methods, and procedures of an organization or system to determine actions to be taken and to recommend how such tasks can be accomplished.
systems house: A collection of individuals who are in business to gather data and write programs for a client's specific needs.
table stripping: The process of assembling images manually using a light table, masking materials, and other tools, as opposed to electronic stripping using CEPS or CAD/CAM systems such as the Gerber Autoprep.
tagboard: A paper class, basic size 24 x 36 inches, used generally for the making of tags and other products where inexpensive yet strong papers are needed.
tailored prices: Dollar figures representing selling prices, developed around the specific production and market dynamics of a printing company.
telecommunications: Communications via telephone or other signal system.
terminal: A connected keyboard and monitor used as an input/output device with some computer systems; also called a *visual display terminal (VDT).*
text-only files: Electronically-captured, digitized characters that can be stored in a manner so they are transportable to other digitized, electronic systems.
thermography: The immediate application of a special powder to printed (wet) images, which are then exposed to heat causing the powder to swell, thus producing raised images.
text paper: A paper class, basic size 25 x 38 inches, used for the production of books and booklets; subgroup of book paper; very colorful stocks typically offered for sale with matching cover papers.
text type: A type measuring up to 12 points in body size used commonly for reading material as straight composition.
thousand sheet price: See *M sheet price.*
thousand sheet weight: See *M sheet weight.*
thumbnail: A rough sketch produced by an artist during the initial phase of artwork development for a graphic image.
time and motion study: A procedure by which employees are studied during production, with resultant recommendations for improved efficiency and establishment of production time standards.
time per output quantity: Conversion of a production standard into a calculated time value related to output; an example would be "0.1667 hours per thousand sheets."
timesharing: The use of a portion of computer time, with the rest available to other individuals and firms.
tonality: See *polarity.*
total cost: With changeover points, the resultant cost when fixed and vaiable costs are summed together.

trade customs: See *Printing Trade Customs.*
trade house: A firm or company that produced a segment of a printing order; common trade services are composition, prepresss, bindery, and special operations.
transfer type: See *press-on type.*
trap: An area where color is to be printed, which is surrounded by another printed area; normally requires a spread or choke to provide for a slight overlap or undercutting of the two printed areas.
trimming: The removal of edges or segments of paper stock to bring the final printed product to the desired size.
turnkey vendor: A company providing computer systems that have been designed, programmed, checked, and then sold as a package to a customer; the client "turns the key" to begin to operate the system.
type-high: A term used to describe the height of relief characters for leterrpress printing measuring 0.918 inch as standard.
typeset material: Copy produced with some kind of composition equipment.
ultraviolet: A specific range of light wavelengths that is not visible and to which many photographic films and lithographic plates are sensitive.
uncoated paper: A paper stock, in the book class, that does not receive an additional coating of clay.
unit cost: The cost for an individual item or product.
UNIX: Computer operating system popular with some printing industry computer vendors, originally developed by Bell Laboratories.
up: A term used to describe the number of finish size sheets that can be positioned (imposed) on a press size sheet.
upgrade: To switch to a larger, more powerful computer system.
user-friendly: A general term to describe programs and operating systems that are comfortable to work with.
utilization: A percentage of time that a production center is used for chargeable work, as compared to total time available.
vacuum back: A vacuum board used to hold film securely during exposure in a process camera or enlarger; essentially the same as the focal plane.
vacuum frame: A unit for film contacting that utilizes a vacuum drawdown to obtain intimate contact between pieces of film; sometimes termed a *contact frame.*
value added: A concept used by printing management to measure the dollar amount added by the manufacturing process using a "sales minus materials and buyouts" formula.
variable costs: Costs that *do* change with charges in ouput.
variable data: Information that changes with each estimate, such as a customer's name.
vendor: A company selling a product.
Visicalc: One of the first spreadsheet programs developed for microcomputer use.
visual display terminal: See *terminal.*
W&B: Abbreviation for work-and-back, or sheetwise, impostion technique.

W&T: Abbreviation for work-and-turn imposition technique.
washup: The procedure used to remove ink from a press.
waste: A percentage comparison of paper that must be thrown away to the total surface area of the sheet.
watermark: A design or image pressed into the wet paper sheet by a dandy roll during manufacture.
webbing diagram: A picture that is drawn by a press operator or production planner to show exactly how a web will be run through the press.
web break: The accidental tearing apart of web paper during production.
web printing: The production of printed goods from rolls of paper that are passed through the press in one continous piece.
weight per 1000 cutoffs. The scale weight of 1000 cutoff sheets or signatures; used for estimating ink consumption and determination of shipping weight factors.
weight per M (1000) sheets: The actual scale weight of 1,000 sheets of paper of a given size and type; can be be calculated from the substance weight using a comparative weight formula.
wet-trapping: Printing wet colors of ink over other wet colors of ink; common with both multicolor sheetfed and web production.
window: An area in a masking sheet that has been cut away and through which light will pass.
wipe-on plate: A kind of surface lighographic plate that is coated ("wiped-on" with coating) as the first step in platemaking.
word: A group of bits, characters, or bytes considered to be an entity and capable of being stored in a computer system in one location.
word processing: The act of inputting, massaging, and outputting manuscript, typically using a word processor or word processing software in a computer system.
work-and back impostion: See *sheetwise imposition.*
work-and-turn imposition: A procedure wherein all pages for a signature are positioned in one form that is then printed on both sides of the sheet; after printing, the sheet is cut apart, producing two signatures exactly alike.
wrapping: The enclosing of a completed product in a package or box for shipment.
wrong-reading: A term used to describe images that read from right to left; typically implies that the film emulsion faces upward toward the viewer.

INDEX

T

V

W